An Introduction
to Probability
and
Stochastic Processes

PRENTICE-HALL INFORMATION AND SYSTEM SCIENCES SERIES

Thomas Kailath, *editor*

Berger	*Rate Distortion Theory: A Mathematical Basis for Data Compression*
Di Franco and Rubin	*Radar Detection*
Downing	*Modulation Systems and Noise*
Dubes	*The Theory of Applied Probability*
Franks	*Signal Theory*
Golomb et al.	*Digital Communications with Space Applications*
Lindsey	*Synchronization Systems in Communication and Control*
Lindsey and Simon	*Telecommunications Systems Engineering*
Melsa and Sage	*An Introduction to Probability and Stochastic Processes*
Patrick	*Fundamentals of Pattern Recognition*
Raemer	*Statistical Communication Theory and Applications*
van der Ziel	*Noise: Sources, Characterization, Measurement*

An Introduction to Probability and Stochastic Processes

JAMES L. MELSA and ANDREW P. SAGE

Information and Control Sciences Center
Institute of Technology
Southern Methodist University

PRENTICE-HALL, Inc. Englewood Cliffs, New Jersey

10 9 8 7 6 5 4 3 2 1

ISBN: 0–13–034850–3

Library of Congress Catalog Card Number 72–3132
Printed in the United States of America

PRENTICE-HALL INTERNATIONAL, INC., *London*
PRENTICE-HALL AUSTRALIA, PTY. LTD., *Sydney*
PRENTICE-HALL OF CANADA, LTD., *Toronto*
PRENTICE-HALL OF INDIA PRIVATE LIMITED, *New Delhi*
PRENTICE-HALL OF JAPAN, INC., *Tokyo*

To Susan, Elisabeth, Peter, Jon, and Jennifer
JLM

To Theresa, Karen, and Philip
APS

Contents

Preface

Probability theory and stochastic processes are playing an ever more popular role in all fields of engineering. This fact is due, in part, to the growing realization that many natural phenomena can be accurately described only by probabilistic means. In addition, increased interest in the study of complex systems has led to the use of probabilistic concepts as a method of description. The purpose of this book is to introduce the student to the concepts of probability and stochastic processes that are needed for the areas of system analysis, control and communication. Because emphasis has been placed on preparing the reader to apply the concepts discussed, certain interesting areas have been omitted if no direct relation to the intended areas of application is known.

In writing a book, one may strive either to impress his colleagues or to educate students. In preparing this manuscript, a dedicated effort has been directed toward the latter goal. To assist the student, several simple exercises with answers have been placed at the end of each technical section to help in the mastery of that section's material. Additional problems, which may involve the use of material from several sections, appear at the end of each chapter. It is hoped that these problems will broaden the student's understanding of the subject through imaginative use of the material. Solutions for these problems are presented in a solutions manual.

Numerous examples have been included in the text to illustrate concepts. In addition, several examples detail applications of the theory developed in the book to meaningful practical problems. While these examples are not essential to the understanding of concepts, they will hopefully provide some motivation for the reader. While every effort has been made to present cencepts in simple terms, it has been the aim of the authors not to compromise on technical accuracy.

This book is intended to be used for a one-semester course in probability and stochastic processes offered to seniors or first year graduate students. The only prerequisite is some experience with system analysis including state-variable and Laplace-transform concepts. For the student who needs a review of these concepts, two brief appendices are included. The material in this book has been taught for the past several years over closed circuit TV at the Information and Control Sciences Center of the Southern Methodist University Institute of Technology.

The writing of a book is, at best, a difficult and trying experience. However, the facilities and atmosphere provided by the farsighted and inspiring leadership of Thomas L. Martin, Jr., Dean of the Institute of Technology at Southern Methodist University, were of significant benefit, and the authors are deeply grateful.

The authors express their appreciation to their numerous colleagues and students who provided many helpful suggestions for improving the preliminary versions of this manuscript. The authors also wish to acknowledge the assistance of the reviewers, especially Dr. Thomas Kailath and Dr. Daniel Alspach. Finally, the authors offer their sympathy to the several typists who helped prepare the manuscript; any errors which remain are almost certainly random phenomena.

James L. Melsa
Andrew P. Sage

1. Introduction

1.1. Introduction to probability and stochastic processes

In any system or in natural situations, conditions inherently arise that can only be described by statistical models. In fact, the signals and commands that a system must process must have at least some components which are unknown at the outset or there would be no real transmittal of information or any "new" purpose to be served by the system.

Fortunately, some of the simplest and most intuitive laws of nature are those which state conditions under which events of interest to us will or will not occur. If an event always occurs under a given set of conditions, the event is called a certain or deterministic event. An event that will sometimes occur and sometimes not occur under a given set of conditions is called a *random event*. The randomness of an event does not in any sense demonstrate the lack of any relation between the set of conditions and the event.

As a simple example, let us consider that the majority of automobiles from a specific manufacturer will run more than 25 miles but less than 50,000 miles without repair. We could thus state, perhaps, that the trouble-free life of an automobile (from the specified manufacturer) is between 25 and 50,000 miles. This statement would not seem to be as meaningful as the statement that in 95 per cent of cases an automobile would give 2,500 miles of trouble-free life. A still more meaningful evaluation of the repair desirability of the automobiles would be to show, for any number of miles M driven, the percentage of the total number of automobiles $n(M)$ that run for M or more miles. This is often accomplished in graphical form, as indicated in Fig. 1.1-1. As we might imagine, this curve might be determined in practice by observing the number of miles a large number of cars are driven without

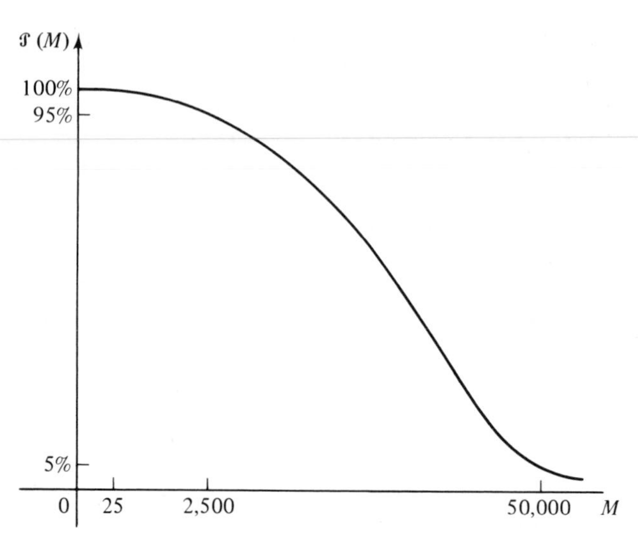

FIG. 1.1-1. The probability of an automobile running M miles without repair.

repair. The curve obtained in this fashion would be of value if we could conclude that it represents a probability law governing not only the given sample of automobiles tested, but that a new experiment repeated under the same conditions with a different sample of automobiles would lead to the same probability law.

The probability law to which we refer is given by a function $\mathcal{P}(M)$, which is the probability that a single automobile will run at least M hours without repair.

We must use statistical descriptions of situations when deterministic or causal descriptions are not valid. Various considerations may lead to these situations. A statistical model is most satisfactory when events of a similar nature occur in rather large numbers. There are several physical ways in which this may happen:

1. The same event may be repeated over and over again (as in coin tossing).
2. An event with time-varying properties may be sampled at a number of different times.
3. Different events of the same kind may be observed simultaneoulsy.

In our work to follow, we shall obtain statistical models for physical phenomena using each of these ways.

Many probability laws appear quite obvious physically. We must exercise caution, however, in over relying on physical insight in problems with random phenomena. Considerable care must accompany probabilistic model making or numerous difficulties will ensue.

To illustrate this point, let us consider an example in which the probability that one will be killed on an airline flight is 10^{-5}. We desire the probability of being killed in 10^5 airline flights. Here it turns out that we have what we shall later call a binomial probability distribution with the number of flights, n, equal to 10^5, and p, the probability of being killed on a specific flight, equal to 10^{-5}. One error to avoid is that of determining $np = 1$ and then stating that the probability of being killed on 10^5 flights is 1. Certainly the probability of being killed on any specified number of flights must be less than 1. The number np does have significance. We shall see later that it is, for this example, the expected value, and we shall see that it is the average number of fatal flights per 10^5 flights. This must be interpreted properly. What we are saying is that if we had many many times 10^5 flights, we would have one fatal crash per 10^5 flights. Thus in 10^7 flights we would expect $10^7 \times 10^{-5} = 10^2$ fatal crashes. It does not, in any sense, state that there will be exactly one fatal flight for every 10^5 flights.

A second error to avoid is computing the probability of getting killed exactly once. Obviously, a human cannot be killed more than once, but computing the probability of getting killed only once leads to the wrong answer, whereas determining the probability of getting killed one or more times leads to the correct answer. Clearly, this is so, since what we desire to determine is 1 minus the probability of not getting killed on 10^5 flights. This is the probability of getting killed on 10^5 flights.

The foregoing statements may appear somewhat paradoxical, although the explanation is quite simple. To compute the correct probability of being killed *exactly* once, we must compute the probability of getting killed on the first flight, plus the probability of surviving the first flight and getting killed on the second, plus the probability of not getting killed on the first two flights and getting killed on the third flight, and so forth. If we let p be the probability of getting killed on any one flight, then $q = 1 - p$ is the probability of not being killed; so we have

$$\begin{aligned}
\mathcal{P}(M) &= p + qp + q^2p + q^3p + \cdots + q^{M-1}p \\
&= p(1 + q + q^2 + q^3 + \cdots + q^{M-1}) \\
&= p\sum_{i=0}^{M-1} q^i = p\frac{1 - q^M}{1 - q} \\
&= 1 - q^M
\end{aligned}$$

Thus we see that by calculating the lengthy sum of probability expressions just elaborated, we simply obtain an expression equivalent to 1 minus the probability of not being killed on 10^5 flights. For this particular example, we obtain $\mathcal{P}(10^5) = 0.63$, and see that there is only a slightly better-than-even chance of being killed in 10^5 flights.

The first half of our presentation will discuss probability theory and random variables with some of the applications of this theory to the information and control sciences. Of perhaps somewhat greater concern will be our study of random or stochastic processes in the last half of this text. Most of the random effects in systems are modeled by stochastic processes, and much of the advanced study in systems is devoted to a study of various stochastic processes in systems.[1]

Our work with stochastic processes will be principally concerned with physical processes that can be represented by means of ordinary differential equations. We shall let the state of a system under consideration be defined at any instant of time t by N parameters x_i, $i = 1, 2, \ldots, N$, which we shall denote for convenience by the vector $\mathbf{x}$:

$$\mathbf{x} = \begin{bmatrix} x_1 \\ x_2 \\ \cdot \\ \cdot \\ \cdot \\ x_N \end{bmatrix} \tag{1.1-1}$$

The rate of change of the state of the system with respect to time will be expressed by the derivative relation

$$\dot{x}_i = \frac{dx_i}{dt} \qquad i = 1, 2, \ldots, N$$

which may be written in vector form as

$$\dot{\mathbf{x}} = \frac{d\mathbf{x}}{dt} = \begin{bmatrix} \dfrac{dx_1}{dt} \\ \dfrac{dx_2}{dt} \\ \cdot \\ \cdot \\ \cdot \\ \dfrac{dx_N}{dt} \end{bmatrix} \tag{1.1-2}$$

These rates will be assumed to be functions of the instantaneous values of the states, the time t, known inputs $u_i(t)$, $i = 1, 2, \ldots, M$, and random

[1] For example, see A. P. Sage, *Optimum Systems Control*, Prentice-Hall, Englewood Cliffs, N.J., 1968; A. P. Sage and J. L. Melsa, *Estimation Theory with Applications to Communication and Control*, McGraw-Hill, New York, 1971; A. P. Sage and J. L. Melsa, *System Identification*, Academic Press, New York, 1971; or any of the many available books on random process applications.

inputs $w_i(t)$, $i = 1, 2, \ldots, J$, such that we have a system of differential equations

$$\dot{x}_i = f_i[x_1(t), x_2(t), \ldots, x_N(t), u_1(t), u_2(t), \ldots, u_M(t), w_1(t), w_2(t), \ldots, w_J(t)]$$

with initial conditions that may, in fact, be random:

$$x_i(t_0) = x_{i0} \qquad i = 1, 2, \ldots, N$$

These can be expressed conveniently in the form of a vector differential equation and initial condition vector

$$\dot{\mathbf{x}} = \mathbf{f}[\mathbf{x}(t), \mathbf{u}(t), \mathbf{w}(t)] \qquad \mathbf{x}(t_0) = \mathbf{x}_0 \tag{1.1-3}$$

The vast majority of the laws of nature can be expressed in the form of the foregoing differential equation. Galileo Galilei was among the first to consider models for natural physical phenomena in terms of differential equations, although it remained for Sir Isaac Newton to give Galileo's results specific interpretation in terms of differential equations as we know them today. Gauss recognized the essentially stochastic nature of problems such as orbit determination.

In mechanics and other fields it is customary to express laws of motion as sets of differential equations of the second order. We shall find it much more convenient to express phenomena as sets of first-order differential equations. Let us consider the fall of a heavy body in the atmosphere of the earth. We shall consider that the body is only a short distance above the surface of the earth so that the resistance of the atmosphere depends on the velocity of the body only and not upon the height. Two parameters, x_1 the distance of the body from the surface of the earth, and the velocity of the body x_2, define the two states (parameters) of the motion of the body. Galileo was interested in discovering certain features concerning falling bodies, swinging pendulums, and other natural phenomena. The following quotation from Galileo's writing describes beautifully his thoughts on the subject of freely accelerated motion.

> And first of all it seems desirable to find and explain a definition best fitting natural phenomena. For anyone may invent an arbitrary type of motion and discuss its properties; thus, for instance, some have imagined helices and conchords as described by certain motions which are not met with in nature, and have very commendably established the properties which these curves possess in virtue of their definitions; but we have decided to consider the phenomena of bodies falling with an acceleration such as actually occurs in nature and to make this definition of accelerated motions. And this, at last, after repeated efforts we trust we have succeeded in doing. In this belief we are confirmed mainly by the consideration that experimental results are seen

> to agree with and exactly correspond with those properties which have been, one after another, demonstrated by us. Finally, in the investigation of naturally accelerated motion we are led, by hand as it were, in following the habit and customs of nature herself, in all her various other processes, to employ only those means which are most common, simple, and easy.

Galileo effectively postulated that the change of position of the body with respect to the earth's surface $[x_1(t)]$ and the velocity of the body with respect to the earth $[x_2(t)]$ can be represented by the differential equations

$$\dot{x}_1 = -x_2(t) \tag{1.1-4}$$

$$\dot{x}_2 = g - h[x_2(t)] \tag{1.1-5}$$

where g is the accleration due to gravity and $h(x_2)$ is the effect of atmospheric resistance for the given body.

If the function $h(x_2)$ is monotone increasing for increasing x_2 and if the body falls indefinitely, the velocity x_2 approaches a limiting value V, which is obtained from

$$g = h(V) \tag{1.1-6}$$

The velocity of the falling body thus increases up to the time when the acceleration force of gravity is balanced by air resistance. This stationary velocity V is quite dependent upon the characteristics of the particular body. A man in a parachute might attain a velocity V of 15 feet (ft) per second fairly quickly, whereas a bowling ball would attain a much higher stationary velocity after a much longer period of time.

For a very heavy object of small dimension close to the surface of the earth, the resistance term $h(x_2)$ may be much smaller than the acceleration due to gravity g and thus may be neglected. There may perhaps be a noise-corrupted position estimate, possibly obtained from a radar. We shall thus write the system equations as

$$\dot{x}_1 = -x_2(t) \tag{1.1-7}$$

$$\dot{x}_2 = g \tag{1.1-8}$$

and the observation of $x_1(t)$ is corrupted by a noise $v(t)$ so that

$$z(t) = x_1(t) + v(t) \tag{1.1-9}$$

The problem that we shall pose is to estimate the position $x_1(t)$ and the velocity $x_2(t)$ from the noise-corrupted observation $z(t)$.

The system differential equations can be solved easily to obtain

$$x_2(t) = x_2(t_0) + (t - t_0)g \tag{1.1-10}$$

$$x_1(t) = x_1(t_0) - x_2(t_0)(t - t_0) - \tfrac{1}{2}(t - t_0)^2 g \tag{1.1-11}$$

where t_0 is the initial time of interest. We shall assume that the initial conditions $x_1(t_0)$ and $x_2(t_0)$ are unknown and random. We shall estimate them from a knowledge of the observation $z(t)$.

Denote the estimate of $\mathbf{x}(t)$ (the vector composed of x_1 and x_2) based on the observation $z(\tau)$, for $t_0 \leq \tau \leq t$, as $\hat{\mathbf{x}}(t)$. The *residual* or *innovation* is defined as

$$\tilde{z}(t) \triangleq z(t) - \hat{x}_1(t) \tag{1.1-12}$$

The least-squares method of estimation theory is concerned with determining the most probable value of $\mathbf{x}(t)$, that is, $\hat{\mathbf{x}}(t)$. This most probable value is defined as the value that minimizes the integral of the square of the residual. Thus we choose $\hat{\mathbf{x}}(t)$ such that

$$J = \frac{1}{2}\int_{t_0}^{t} [z(\tau) - \hat{x}_1(\tau)]^2 \, d\tau \tag{1.1-13}$$

is minimized. Gauss, in 1795, first recognized that this least-squares method was an extremely potent one for the resolution of many practical problems. Gauss did not simply hypothesize this method of least squares. It turns out that if one assumes the distribution of the measurement noise $v(t)$ is gaussian and independent from sample to sample, then the problem of determining the best initial conditions to maximize the joint probability density function of the observations $z(\tau)$ for $t_0 \leq \tau \leq t$ is equivalent to minimizing the least-squares curve-fit cost function, Eq. (1.1-13).

If we substitute the solutions of Eqs. (1.1-10) and (1.1-11) in Eq. (1.1-13), we obtain

$$J = \frac{1}{2}\int_{t_0}^{t} [z(\tau) - \hat{x}_1(t_0) + \hat{x}_2(t_0)(\tau - t_0) + \tfrac{1}{2}(\tau - t_0)^2 g]^2 \, d\tau \tag{1.1-14}$$

We minimize Eq. (1.1-14) with respect to $\hat{x}_1(t_0)$ and $\hat{x}_2(t_0)$ by setting

$$\frac{\partial J}{\partial \hat{x}_1(t_0)} = \frac{\partial J}{\partial \hat{x}_2(t_0)} = 0 \tag{1.1-15}$$

and obtain

$$\frac{\partial J}{\partial \hat{x}_1(t_0)} = -\int_{t_0}^{t} [z(\tau) - \hat{x}_1(t_0) + \hat{x}_2(t_0)(\tau - t_0) + \tfrac{1}{2}(\tau - t_0)^2 g] \, d\tau = 0 \tag{1.1-16}$$

$$\frac{\partial J}{\partial \hat{x}_2(t_0)} = \int_{t_0}^{t} (\tau - t_0)[z(\tau) - \hat{x}_1(t_0) + \hat{x}_2(t_0)(\tau - t_0) + \tfrac{1}{2}(\tau - t_0)^2 g]\, d\tau = 0 \tag{1.1-17}$$

Simultaneous solution of these two equations easily leads to a smoothing solution for the estimate of $\mathbf{x}$ at time t_0 with observations $z(\tau)$ over the time interval $t_0 \leq \tau \leq t$. This solution is

$$\hat{x}_1(t_0 \mid t) = \frac{2}{(t - t_0)^2} \int_{t_0}^{t} (2t - 3\tau + t_0) z(\tau)\, d\tau - \tfrac{1}{12}(t - t_0)^2 g \tag{1.1-18}$$

$$\hat{x}_2(t_0 \mid t) = \frac{6}{(t - t_0)^2} \int_{t_0}^{t} (t - 2\tau + t_0) z(\tau)\, d\tau - \tfrac{1}{2}(t - t_0) g \tag{1.1-19}$$

These estimates are the optimum least-squares error estimates for the initial conditions of our very simple trajectory problem. The estimate of the actual position and velocity at time t based upon information up to time t would generally be more desirable than the initial-condition estimates. These may be obtained by incorporating the estimates of Eqs. (1.1-18) and (1.1-19) in Eqs. (1.1-10) and (1.1-11), which results in

$$\hat{x}_1(t \mid t) = \hat{x}_1(t) = \frac{2}{(t - t_0)^2} \int_{t_0}^{t} (3\tau - t - 2t_0) z(\tau)\, d\tau - \tfrac{1}{12}(t - t_0)^2 g \tag{1.1-20}$$

$$\hat{x}_2(t \mid t) = \hat{x}_2(t) = \frac{6}{(t - t_0)^3} \int_{t_0}^{t} (t - 2\tau + t_0) z(\tau)\, d\tau + \tfrac{1}{2}(t - t_0) g \tag{1.1-21}$$

Unfortunately, these solutions are a bit more difficult to implement than we might initially believe. The time-varying multiplications associated with the integrations in the foregoing equations are not particularly easy to implement. The problem posed here is basically a very simple one. For a more complex estimation problem, it would not be possible to obtain the estimation equations in anywhere near as simple a form as we have done here.

A possible remedy to our realization dilemma might be to differentiate Eqs. (1.1-20) and (1.1-21) a sufficient number of times and hope to obtain a differential equation whose solution yields the desired estimate. It turns out that this is a useful approach, and we do obtain a set of differential equations more readily implemented than Eqs. (1.1-20) and (1.1-21). This method of solution, phrased in more general terms, is precisely the solution presented by R.E. Kalman and R.S. Bucy to the problem of obtaining linear sequential estimation algorithms. These algorithms represent one of the most useful results obtained from the theories of stochastic processes and estimation. It will be the purpose of our studies here to develop the necessary background in probalility theory and stochastic processes such that we may fully appreciate this accomplishment, as well as many other marvelous accomplishments in real-world engineering, a world that can only be viewed with assistance from a sound foundation in random phenomena.

1.2. *The sequel*

In the chapters to follow, a coverage of those topics in the theory of probability and stochastic processes most important to further efforts in the systems area are presented. We develop not only various theoretical concepts but present numerous solved problems using the developed techniques, and indicate application areas for some of the topics discussed. Our specific chapter-by-chapter goals are as follows.

CHAPTER TWO—PROBABILITY THEORY

Some of the most fundamental, and certainly most interesting, laws of science are those which tell us the conditions under which an event of interest will or will not occur. The study of these laws is known as the study of probability. In Chapter 2 we shall define probability theory as the study of mathematical models of random phenomena. The use of the word *model* is deliberate here and, as we shall see throughout our efforts in this text, models are at the heart of the entire study of probability, random variables, and stochastic processes. For example, an intuitive concept of probability would allow us to define the probability that an event will occur as the ratio of the number of outcomes in which the event is observed to the number of all possible outcomes of the experiment. Certainly we believe that the probability of obtaining a head on the toss of a true coin is 0.5. We obtain this by noting that there are two outcomes of the toss of a coin—a head and a tail. One of these, occurrence of a head, is the event we are seeking. Thus we obtain 0.5 for the probability of obtaining a head on the toss of a coin.

We introduce some concepts from set theory that will be of value to us in forming appropriate mathematical models for random phenomena. Then we present various definitions useful in a study of probability and look at various avenues by which the study of probability may be approached, including the classical and relative-frequency definitions of probability. Several sections are devoted to a reasonably detailed exposition of probability theory with numerous illustrative examples to assist in our study. These include discussions of discrete probability models and the important subject of conditional probability and combined experiments.

Chapter Three—Random Variables

Unfortunately, the basic concepts of probability theory as presented in Chapter 2, although quite general, are not easy to manipulate. The basic goal of the third chapter is to translate these general concepts into a more easily handled mathematical model: the random variable and its associated probability functions. Once this is done, a vast array of analysis tools may be brought to bear on probabilistic problems. To accomplish this, we first provide precise definitions for random variables. Then we discuss probability distribution and density functions and provide several examples of these functions. The important subjects of joint, marginal, and conditional distribution and density functions will play a prominent role in our efforts in this chapter, as will the subject of independence of random variables.

Chapter Four—Functions of a Random Variable

In this chapter we extend the concepts of a random variable by examining in some depth several different types of operations on random variables. These operations greatly extend the usefulness of random-variable concepts and are invaluable for applications efforts in information and control sciences. First we consider several methods that resolve the problem of determining density and distribution functions for a random variable defined as a function of another random variable. Finally, we consider sequences of random variables and associated convergence questions for these sequences.

Chapter Five—Stochastic Processes

This chapter is concerned with some of the basic concepts in the theory of stochastic processes. As we shall see, stochastic processes are generalizations of random variables; so almost all the work of the three previous chapters can be applied directly to stochastic processes. Following the definition of a stochastic process, we shall discuss the minor modifications needed in the work of the previous chapters to consider stochastic processes. After introducing the concepts of covariance, autocorrelation, and cross-correlation functions, we shall discuss errors in actual measurements of these functions. Orthogonal and spectral representations of random processes will be developed, and the concepts of stationarity and ergodicity will be introduced. The extremely important concept of white noise will be developed.

Chapter Six—Linear System Response to Stochastic Processes

Many problems of interest in the information and control sciences may be posed as problems involving the calculation of the output response of a linear system subject to a stochastic process input. In this chapter we first give consideration to the state-space representation of the response of linear continuous systems to random inputs. Then we shall consider the frequency-domain approach, which is most useful for single input–single output systems. Finally, discrete systems will be considered, and the equivalence and differences between continuous-time and discrete-time representation of random processes will be explored. From an applications viewpoint, the material of this chapter is most useful, and the results form the basis for much further work in many applications areas for stochastic processes.

Chapter Seven—Gaussian Processes

In previous chapters we introduced the concept of a gaussian random variable. Since the gaussian process is so vital in numerous areas for theoretical and application efforts, it seems appropriate to give more explicit consideration to it. We begin the study with a discussion of the central limit theorem, which guarantees that many processes will be gaussian, and then discuss vector (multidimensional) gaussian processes. The most important topic of conditional gaussian distributions and densities will be developed. This will lead to the *orthogonal projection lemma*, which is vital for a proper understanding of linear estimation theory.

Chapter Eight—Markov Processes and Stochastic Differential Equations

When a stochastic process is Markov, many problems are considerably simplified. This Markov assumption states that a knowledge of the present separates the past from the future and therefore lends particular significance to conditional density functions, to which we shall give special emphasis in the first part of this chapter. Some of the process models from Chapter 6 will be reexamined, and we shall show that most of the models of that chapter are, in fact, Markov models. The Wiener process, a particularly important Markov process, will be developed.

In this chapter random processes in nonlinear systems will also be studied. We shall now find that ordinary rules of differential and integral calculus do not, unfortunately, always apply to stochastic processes. For instance, the integration of a product of two random variables is not defined by the rules of ordinary calculus. We shall develop the Ito stochastic calculus, which will allow us to resolve many fundamental problems concerning random process in nonlinear systems. The Fokker–Planck partial differential equation, whose solution characterizes the evolution of the probability function of the state vector of a broad class of nonlinear systems with a white gaussian noise input, is also developed.

Finally, a discussion of mean and variance propagation for random processes in nonlinear systems will be presented.

Appendix A—Transform Methods in Linear Systems

Although we have assumed that the reader has an undergraduate background in transform methods, it is highly likely that this background may differ in breadth and depth from student to student. This appendix concerning transform methods in linear systems is presented to summarize the concepts from this important subject needed for a study of Chapters 5 and 6.

Appendix B—System Representation by State Variables

This appendix summarizes the various vector, matrix, and state-variable operations and notation encountered in this text. Since this subject is likely to be less familiar to readers than the subject of transform methods, the material is considerably more self-contained.

2. *Probability Theory*

2.1. *Introduction*

Some of the most fundamental, and certainly most interesting, laws of science are those which tell us the conditions under which an event of interest will or will not occur. The study of these laws is known as the study of probability. We shall define *probability theory* as the *study of mathematical models of random phenomena.* The use of the word *model* is deliberate here and, as we shall see throughout our efforts in this text, models are at the heart of the entire study of probability, random variables, and stochastic processes. The choice of a model for random phenomena is rather critical. For example, an intuitive concept of probability would allow us to define the probability of an event occurring as the ratio of the number of outcomes of an experiment in which the event is observed to the number of all possible outcomes of the experiment. Certainly we believe that the probability of obtaining a head on the toss of a true coin is 0.5. We obtain this by noting that there are two outcomes of the toss of a coin—a head and a tail. One of these, occurrence of a head, is the event we are seeking. Thus we obtain 0.5 for the probability of obtaining a head on the toss of a coin. In a similar way, we might reason that the probability of obtaining a spade in drawing a card at random from a standard deck is 0.25. This would be obtained by assuming that there are four outcomes of the draw of a card: spade, diamond, heart, or club. One of the four, the spade, is the event sought, and thus a probability of 0.25 is obtained. An alternative reasoning could have led us to the conclusion that there are two events: obtaining a spade, and not obtaining a spade. One of these two events is favorable, and we obtain a probability of 0.5 for obtaining a spade from a random draw of a card from a 52-card deck. In the latter analysis we have, somehow, chosen the wrong mathematical model for computation of the desired probability.

In the next section we shall introduce some concepts from set theory that will be of value to us in forming appropriate mathematical models for random phenomena. Then we shall present various definitions useful in a study of probability and look at various avenues by which the study of probability may be approached. Several sections will then be devoted to a reasonably detailed exposition of probability theory with numerous illustrative examples to assist in the study.

2.2. *Spaces and sets*

To be reasonably precise in the descriptions of mathematical models for random phenomena, it is necessary to introduce some defining notation such that we may formulate postulates concerning random phenomena or experiments. We shall refer to each performance of a random experiment as a *trial* or *sample.* As a result of each trial, we obtain or observe a single, well-defined outcome ξ_i.

Definition 2.2-1. ***Sample Space:*** The set of all possible outcomes is referred to as the sample space $\mathcal{S}$ or the *sure event.*

In other words, the outcomes are elements of the set $\mathcal{S}$. In addition to the outcomes and the sample space, we wish to deal with *events* that are defined in the following manner:

Definition 2.2-2. ***Events:*** An event $\mathcal{E}$ is a subset[1] of $\mathcal{S}$; that is, $\mathcal{E} = \{\xi : \xi \text{ satisfies some property}\}$.

It should be noted that outcomes are *not* sets but are *elements* of $\mathcal{S}$, whereas events are sets whose elements are outcomes. When we say that an an event $\mathcal{E}$ is an outcome ξ, we really mean that $\mathcal{E}$ is the set consisting of the single element ξ; that is, $\mathcal{E} = \{\xi\}$. When we say that an event $\mathcal{E}$ occurs, we mean that the outcome ξ of our random experiment is an element of $\mathcal{E}$. Events may be *simple* or indecomposable, or they may be *compound* or decomposable. An event is said to be simple if it contains at most one outcome.

We shall present techniques for construction of sample spaces for random phenomena in later sections. However, a few remarks here will serve to give better understanding to the definitions.

Example ***2.2-1.*** If our random experiment is the single toss of a coin, there are two possible outcomes: head H or tail T. Hence the sample

[1] However, it should be noted that all subsets of $\mathcal{S}$ are not taken as events. The reason why certain subsets may not be events will be examined in Section 2.3.

space is $\mathcal{S} = \{H, T\}$. We might define events $\mathcal{E}_1 = \{H\}$ and $\mathcal{E}_2 = \{T\}$. These are clearly simple events.

Example *2.2-2.* If the random experiment consists of drawing a single card from a standard deck, the outcome may be defined in several ways. We might define the outcome of the experiment as the suit of the card, that is, spade S, diamond D, heart H, or club C, without regard for the denomination of the card. In this case, the sample space has four elements and is given by $\mathcal{S}_1 = \{S, D, H, C\}$. Conversely, we might define the outcome of the experiment as the denomination of the card without regard for its suit, in which case $\mathcal{S}_2$ becomes $\{1, 2, 3, \cdots, 11, 12, 13\}$. We could also define the outcome as the suit *and* the denomination; for example, the ace of spades 1S or the ten of diamonds 10D. In this case, the sample space has 52 elements and is given by $\mathcal{S}_3 = \{1S, \ldots, 13S, 1D, \ldots, 13D, 1H, \ldots, 13H, 1C, \ldots, 13C\}$. We note that we can obtain different sample spaces for a given experiment, depending on how the outcomes of the experiment are defined. When we deal with a specific random experiment, we must be certain that the outcomes are clearly and explicitly defined; once this is done, the sample space is uniquely defined.

Let us suppose that we define the outcome as the suit and the denomination so that $\mathcal{S}_3$ is the appropriate sample space. Then we might define an event $\mathcal{E}_1$ that a spade was drawn. The event $\mathcal{E}_1$ then contains the thirteen outcomes 1S, 2S, . . . , 13S, so that $\mathcal{E}_1 = \{1S, 2S, \ldots, 13S\}$. We might define an event $\mathcal{E}_2$ that an ace was drawn. This event contains the four outcomes 1S, 1D, 1H, 1C, so that $\mathcal{E}_2 = \{1S, 1D, 1H, 1C\}$. As still another event, we might define $\mathcal{E}_3$ as the event that the three of spades, the five of clubs, or the ten of hearts was drawn. Hence $\mathcal{E}_3 = \{3S, 5C, 10H\}$. When we say that $\mathcal{E}_3$ occurred, we mean that either the three of spades, the five of clubs, or the ten of hearts was drawn. It must be emphasized that we do *not* mean that *all* the outcomes occurred. Each trial has *only* one outcome, and it is impossible that more than one outcome will occur. All the above events are compound, since they contain more than one outcome.

Example *2.2-3.* In the toss of a pair of dice, a sample space is composed of the 36 outcomes of the toss, such that we have

$$
\begin{aligned}
\mathcal{S} = \{&(1, 1), (1, 2), (1, 3), (1, 4), (1, 5), (1, 6),\\
&(2, 1), (2, 2), (2, 3), (2, 4), (2, 5), (2, 6),\\
&(3, 1), (3, 2), (3, 3), (3, 4), (3, 5), (3, 6),\\
&(4, 1), (4, 2), (4, 3), (4, 4), (4, 5), (4, 6),\\
&(5, 1), (5, 2), (5, 3), (5, 4), (5, 5), (5, 6),\\
&(6, 1), (6, 2), (6, 3), (6, 4), (6, 5), (6, 6)\}
\end{aligned}
$$

It is tedious to write down these 36 outcomes; so we might write instead $\mathcal{S} = \{(x, y): x \text{ and } y \text{ are integers 1 to 6}\}$.

Example 2.2-4. It is not at all necessary that the number of events in a set be finite. If our experiment consists of drawing a number from the unit interval, the sample space may be defined as

$$\mathcal{S} = \{x: 0 \leq x \leq 1\}$$

which contains an uncountably[1] infinite number of elements. In this case, we might define an event $\mathcal{E}_1$ as $\mathcal{E}_1 = \{x: 0.5 \leq x \leq 0.75\}$, which also has an uncountably infinite number of elements. On the other hand, we might define $\mathcal{E}_2$ as $\mathcal{E}_2 = \{x: x = 0, \frac{1}{2}, 1\}$, which contains three elements or outcomes.

Definition 2.2-3. ***Null Set:*** The notation $\mathcal{E} = \varnothing$ will denote the set that contains no outcomes. This is often written as $\varnothing = \{\xi: \xi \neq \xi\}$; since there is no outcome that does not equal itself, $\varnothing$ contains no elements. The null set $\varnothing$ is also referred to as the impossible event, since there is no outcome contained in $\varnothing$.

Definition 2.2-4. ***Complement:*** The complement $\mathcal{A}'$ of a set $\mathcal{A}$ is the set consisting of all elements of $\mathcal{S}$ not in $\mathcal{A}$; that is, $\mathcal{A}' = \{\xi: \xi \notin \mathcal{A}\}$.

Note that $\mathcal{S}' = \varnothing$; that is, the complement of the sample space is the null set.

In many problems of interest we want to consider events that are combinations of other events. Of interest to us will be conditions upon which both $\mathcal{E}_1$ and $\mathcal{E}_2$ occur, and conditions under which either $\mathcal{E}_1$ or $\mathcal{E}_2$ or both occur. We shall denote the event that both $\mathcal{E}_1$ and $\mathcal{E}_2$ occur by the symbol $\mathcal{E}_1\mathcal{E}_2$. Thus $\mathcal{E}_3 = \mathcal{E}_1\mathcal{E}_2$ is the event that $\mathcal{E}_1$ *and* $\mathcal{E}_2$ occur. The event that $\mathcal{E}_1$ or $\mathcal{E}_2$ or both occur will be denoted by the symbol $\mathcal{E}_1 \cup \mathcal{E}_2$. Thus $\mathcal{E}_3 = \mathcal{E}_1 \cup \mathcal{E}_2$ is the event that $\mathcal{E}_1$ or $\mathcal{E}_2$ or both occur. The set $\mathcal{E}_1\mathcal{E}_2$ is often spoken of as the *intersection*, whereas the set $\mathcal{E}_1 \cup \mathcal{E}_2$ is the *union* of $\mathcal{E}_1$ and $\mathcal{E}_2$.

Definition 2.2-5. ***Intersection:*** The intersection of two sets $\mathcal{A}$ and $\mathcal{B}$ is the set $\mathcal{C} = \mathcal{A}\mathcal{B}$ consisting of all elements that are in both $\mathcal{A}$ and $\mathcal{B}$; that is, $\mathcal{C} = \{\xi: \xi \in \mathcal{A} \text{ and } \xi \in \mathcal{B}\}$. The intersection is sometimes referred to as the product of two sets and is also written $\mathcal{A} \cap \mathcal{B}$.

Definition 2.2-6. ***Union:*** The union of two sets $\mathcal{A}$ and $\mathcal{B}$ is the set $\mathcal{C} = \mathcal{A} \cup \mathcal{B}$ consisting of all elements that are in $\mathcal{A}$ or $\mathcal{B}$ or both; that is, $\mathcal{C} = \{\xi: \xi \in \mathcal{A} \text{ and/or } \xi \in \mathcal{B}\}$. The union is sometimes referred to as the sum of two sets.

[1] A set is denumerable if its elements can be put in one-to-one correspondence with the positive integers. A set is countable if it is either finite or denumerable.

The concepts of mutually exclusive sets and set inclusion are also of importance. From these ideas follow the concept of equality of sets.

Definition 2.2-7. ***Mutually Exclusive:*** Two sets $\mathcal{A}$ and $\mathcal{B}$ are said to be mutually exclusive (or disjoint) if they have no elements in common; this is, $\mathcal{A}\mathcal{B} = \varnothing$.

Definition 2.2-8. ***Set Inclusion:*** The set $\mathcal{A}$ is said to be included or contained in $\mathcal{B}$, or $\mathcal{A}$ is a subset of $\mathcal{B}$, written as $\mathcal{A} \subset \mathcal{B}$[1], if every element of $\mathcal{A}$ is also an element of $\mathcal{B}$.

Definition 2.2-9. ***Equality:*** We say that two sets $\mathcal{A}$ and $\mathcal{B}$ are equal if every element of $\mathcal{A}$ is also an element of $\mathcal{B}$, and every element of $\mathcal{B}$ is an element of $\mathcal{A}$; that is, $\mathcal{A} = \mathcal{B}$ implies that $\mathcal{A} \subset \mathcal{B}$ and $\mathcal{B} \subset \mathcal{A}$.

It is possible to make several additional definitions regarding sets and events. All these, which will be of use in the sequel, may be obtained from the definitions presented here. Some are presented in examples, which we shall now present to increase confidence in and understanding of the definitions just given.

Example 2.2-5. The set of all points in a two-dimensional rectangular coordinate system for which x and y are nonnegative and the points are on or below the line $y = 1 - x$ is represented by $\mathcal{E}_1$. This set is illustrated in Fig. 2.2-1 and is described by the notation $\mathcal{E}_1 = \{(x, y): x \geq 0, y \geq 0, y \leq 1 - x\}$.

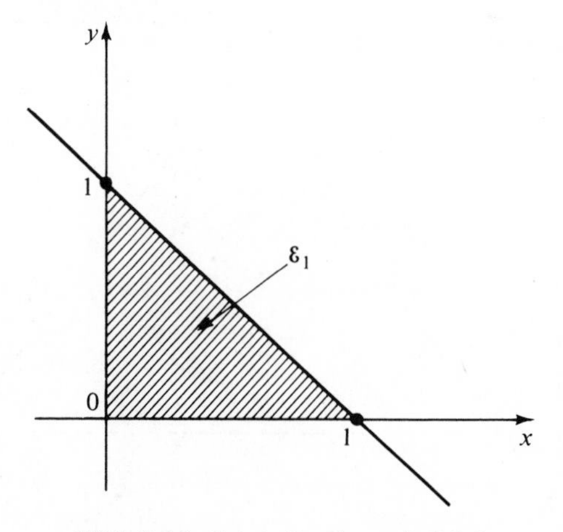

FIG. 2.2-1. Set $\mathcal{E}_1$ for Example 2.2-5.

[1] Care should be taken to distinguish between $\subset$ (set inclusion) and $\in$ (element inclusion).

Example 2.2-6. Let us suppose that a sample space $\mathcal{S}$ is the set $\mathcal{S} = \{1, 2, 3, 4, 5\}$. Events $\mathcal{E}_1 = \{1, 3\}$ and $\mathcal{E}_2 = \{3, 5\}$ are both contained in $\mathcal{S}$. The intersection of events $\mathcal{E}_1$ and $\mathcal{E}_2$ is $\mathcal{E}_3 = \mathcal{E}_1\mathcal{E}_2 = \{3\}$. The union of these events is $\mathcal{E}_4 = \mathcal{E}_1 \cup \mathcal{E}_2 = \{1, 3, 5\}$. The intersection and union of the complementary events are $\mathcal{E}_5 = \mathcal{E}'_1\mathcal{E}'_2 = \{2, 4\}$ and $\mathcal{E}_6 = \mathcal{E}'_1 \cup \mathcal{E}'_2 = \{1, 2, 4, 5\}$.

It is true for this example (and it is true in general) that $\mathcal{E}'_6 = \mathcal{E}_3$ since we may easily show that $\mathcal{E}_1\mathcal{E}_2 = [\mathcal{E}'_1 \cup \mathcal{E}'_2]'$.

Example 2.2-7. Two events $\mathcal{E}_a$ and $\mathcal{E}_b$ (or sets) are disjoint if their intersection is the null set $\mathcal{E}_a\mathcal{E}_b = \varnothing$.

In Example 2.2-6 the events $\mathcal{E}_3$ and $\mathcal{E}_5$ are disjoint as are events $\mathcal{E}_3$ and $\mathcal{E}_6$. However, $\mathcal{E}_5$ and $\mathcal{E}_6$ are *not* disjoint.

Example 2.2-8. Many interesting problems involving the use of sets may be resolved by use of Venn diagrams. These are also of great utility in the solution of problems in the area of switching circuits and other application areas based on the laws of logical algebra. Let us consider events such that $\mathcal{E}_1 \subset \mathcal{E}_2 \subset \mathcal{E}_3 \subset \mathcal{S}$.

These events may be represented by the Venn diagram of Fig. 2.2-2. This particular example is especially simple since one set is included in another. Still, several useful nontrivial sets may be generated from these sets. For example, the set $\mathcal{E}'_2\mathcal{E}_3$ is the points inside the triangle of Fig. 2.2-2 but outside the circle. The event $\mathcal{E}'_2\mathcal{E}_3\mathcal{E}_1$ is then the null set.

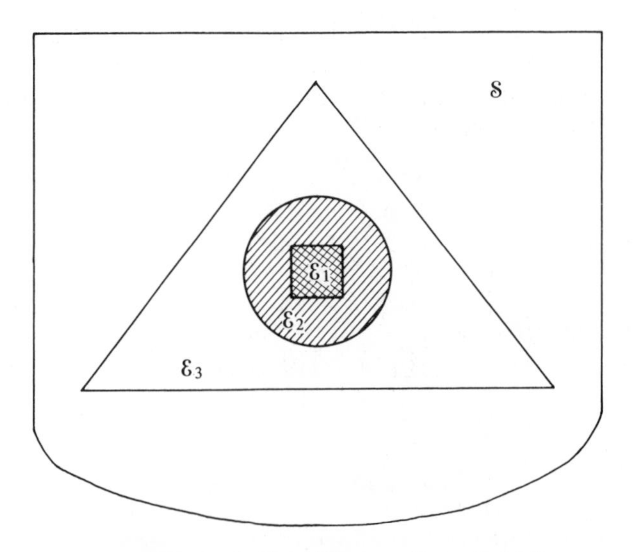

FIG. 2.2-2. Venn Diagram, Example 2.2-8.

Example ***2.2-9.*** The concept of a difference set is sometimes useful. If events $\mathcal{E}_1$ and $\mathcal{E}_2$ are as indicated in Fig. 2.2-3, it is desirable to define the difference set $\mathcal{E}_1 - \mathcal{E}_2$ as the shaded area in the figure. This notation can be avoided however by use of the equivalent expression $\mathcal{E}_1 - \mathcal{E}_2 = \mathcal{E}_1\mathcal{E}_2'$ as the reader may easily show.

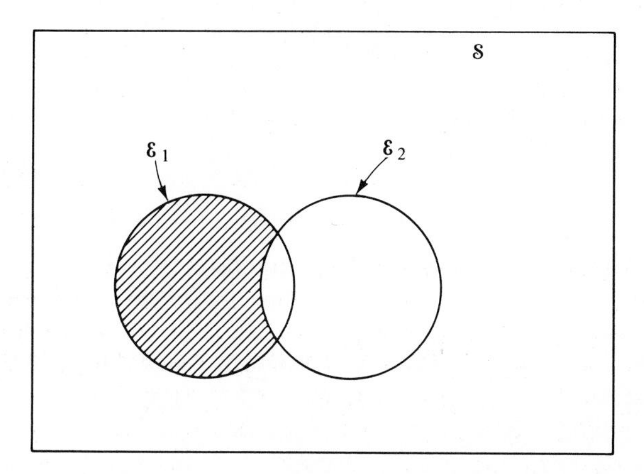

FIG. 2.2-3. Venn Diagram, Example 2.2-9.

It is also convenient to note that, in terms of the difference set, $\mathcal{S} - \mathcal{E}_1 = \mathcal{S}\mathcal{E}_1' = \mathcal{E}_1'$. Thus an equivalent expression for $\mathcal{E}_1 - \mathcal{E}_2$ is $\mathcal{E}_1 - \mathcal{E}_2 = \mathcal{E}_1\mathcal{E}_2' = \mathcal{E}_1[\mathcal{S} - \mathcal{E}_2]$.

Since we can avoid using the difference set, we shall not use it in our discussions. With this brief discussion of spaces and sets as a background, let us turn our attention to definitions of probability and probability measures in which we shall see the need for the concepts presented here.

EXERCISES 2.2

2.2-1. Plot the following sets:

(a) $\mathcal{E}_1 \cup \mathcal{E}_2$.
(b) $\mathcal{E}_1\mathcal{E}_2$.
(c) $\mathcal{E}_1\mathcal{E}_3'$.
(d) $\mathcal{E}_3\mathcal{E}_2'\mathcal{E}_1$.

if

$$\mathcal{E}_1 = \{(x, y): 0 < x < 1, 0 < y < x\}$$
$$\mathcal{E}_2 = \{(x, y): x^2 + y^2 \leq 1\}$$
$$\mathcal{E}_3 = \{(x, y): |x| \leq 0.5, |y| \leq 0.5\}$$

2.2-2. How many possible outcomes are there if a pair of dice is tossed twice?

Answer: $(36)^2$.

2.2-3. List all the possible outcomes if an experiment consists of tossing a coin and throwing a single die.

2.2-4. Show that if $\mathcal{A} \subset \mathcal{B}$, then $\mathcal{B}' \subset \mathcal{A}'$.

2.2-5. Verify the following relations:
(a) $(\mathcal{A} \cup \mathcal{B})' = \mathcal{A}'\mathcal{B}'$.
(b) $(\mathcal{A}\mathcal{B})' = \mathcal{A}' \cup \mathcal{B}'$.

2.3. Probability definition and measure

At the beginning of the last section we noted that although every event was a subset of $\mathcal{S}$, every subset of $\mathcal{S}$ was not necessarily an event. We wish to examine the nature of this restriction in more detail now. The reason for making this restriction will be discussed after we have defined the probability measure.

If the set $\mathcal{A}$ is an event, we shall require that the complement of $\mathcal{A}$, that is, $\mathcal{A}'$, is also an event. In addition, if $\mathcal{A}$ and $\mathcal{B}$ are events, require that $\mathcal{A} \cup \mathcal{B}$ is also an event. A nonempty class (collection) of sets that satisfies the above properties is referred to as a *field* $\mathfrak{F}$. If $\mathcal{A}_i \in \mathfrak{F}$ implies that $\bigcup_{i=1}^{\infty} \mathcal{A}_i \in \mathfrak{F}$, then $\mathfrak{F}$ is said to be a *sigma field.* Hence we have the following definition.

Definition 2.3-1. ***Sigma Field:*** A sigma field $\mathfrak{F}$ is a nonempty class of sets such that
1. If $\mathcal{A} \in \mathfrak{F}$, then $\mathcal{A}' \in \mathfrak{F}$.
2. If $\mathcal{A} \in \mathfrak{F}$ and $\mathcal{B} \in \mathfrak{F}$, then $\mathcal{A} \cup \mathcal{B} \in \mathfrak{F}$.
3. If $\mathcal{A}_i \in \mathfrak{F}$, then $\bigcup_{i=1}^{\infty} \mathcal{A}_i \in \mathfrak{F}$.

It is easy to show if $\mathcal{A} \in \mathfrak{F}$ and $\mathcal{B} \in \mathfrak{F}$, then $\mathcal{A}\mathcal{B} \in \mathfrak{F}$. Hence one often says that a sigma field is a class of sets that is closed under the operations of countable intersection, union, and complement. By *closed* we mean that the result of any of these operations is still a member of the set.

Note that every field must contain $\varnothing$ and $\mathcal{S}$. This fact is easily established. Since the field is nonempty, it must contain at least one set $\mathcal{A}$; but $\mathcal{A}'$ is also in $\mathfrak{F}$, so $\mathfrak{F}$ contains two sets. However, $\mathcal{A} \cup \mathcal{A}' = \mathcal{S}$, so $\mathcal{S} \in \mathfrak{F}$ and $\mathcal{S}' = \varnothing$, which must also be in $\mathfrak{F}$. Hence we see that every field must contain $\mathcal{S}$ and $\varnothing$, and this is the minimum number of elements that $\mathfrak{F}$ may have.

Example ***2.3-1.*** Let us suppose that $\mathcal{S}$ has four elements so that $\mathcal{S} = \{\xi_1, \xi_2, \xi_3, \xi_4\}$. The class of sets

$$\varnothing, \mathcal{S}, \{\xi_1, \xi_3\}, \{\xi_2, \xi_3, \xi_4\}$$

is not a field since $\{\xi_1, \xi_3\}\{\xi_2, \xi_3, \xi_4\} = \{\xi_3\}$, which does not belong to the original collection. Suppose that we add this set so that the collection becomes

$$\varnothing, \mathcal{S}, \{\xi_1, \xi_3\}, \{\xi_2, \xi_3, \xi_4\}, \{\xi_3\}$$

This is still not a field since $\{\xi_1, \xi_3\}' = \{\xi_2, \xi_4\}$, which is not in the collection.

Suppose that we consider the collection

$$\varnothing, \mathcal{S}, \{\xi_1, \xi_2\}, \{\xi_3, \xi_4\}$$

It is quite easy to show that this collection of sets is a field.

From the example it should be obvious that any arbitrary collection of sets will not, in general, be a sigma field. It should be clear that the class of all subsets of $\mathcal{S}$ is a sigma field. However, there are cases in which we do not wish to consider all subsets; this is particularly true in the case of sample spaces that contain an uncountably infinite number of elements. We note that given any class of sets, we can always form a field from this class, if necessary, by including all subsets of $\mathcal{S}$. However, it is possible to show that there exists a smallest sigma field containing any given class of sets.

We shall assume that events are a certain class of (possibly all) subsets of $\mathcal{S}$ which form a sigma field $\mathcal{F}$. Such classes of sets are also referred to as measurable. The reason why we wish the events to form a measurable class will become clear after we have defined a probability space.

Definition 2.3-2. ***Probability Space:*** The probability space associated with a random experiment consists of three items $[\mathcal{S}, \mathcal{F}, \mathcal{P}]$:

1. The sample space $\mathcal{S}$ containing all possible outcomes ξ.
2. A sigma field $\mathcal{F}$ of events $\mathcal{E}$ that are subsets of $\mathcal{S}$.
3. A probability measure $\mathcal{P}$ defined on $\mathcal{F}$ and such that the following three axioms are satisfied:
   I. $\mathcal{P}\{\mathcal{E}\} \geq 0$ for all $\mathcal{E} \in \mathcal{F}$.
   II. $\mathcal{P}\{\mathcal{S}\} = 1$.
   III. $\mathcal{P}\{\mathcal{E}_1 \cup \mathcal{E}_2\} = \mathcal{P}\{\mathcal{E}_1\} + \mathcal{P}\{\mathcal{E}_2\}$ if $\mathcal{E}_1\mathcal{E}_2 = \varnothing$.

The third axiom indicates that the probability of the union of two mutually exclusive events is the sum of the probability of each of the two events. It will be desirable to strengthen Axiom III to include denumerable unions so that we have

$$\text{IIIa.}\quad \mathcal{P}\left\{\bigcup_{i=1}^{\infty} \mathcal{E}_i\right\} = \sum_{i=1}^{\infty} \mathcal{P}\{\mathcal{E}_i\} \qquad \text{if } \mathcal{E}_i\mathcal{E}_j = \varnothing \text{ for } i \neq j$$

It must be emphasized that the probability measure is a *set* function; that is, it is defined for events. It is *not* defined for outcomes. This distinction is an important one especially in the case when the number of outcomes is uncountably infinite.

The reason why the events must belong to a field can now be established. If $\mathcal{E}_1$ and $\mathcal{E}_2$ were two disjoint events but $\mathcal{E}_1 \cup \mathcal{E}_2$ was *not* an event, then $\mathcal{P}\{\mathcal{E}_1 \cup \mathcal{E}_2\} = \mathcal{P}\{\mathcal{E}_1\} + \mathcal{P}\{\mathcal{E}_2\}$ would be meaningless, since $\mathcal{E}_1 \cup \mathcal{E}_2$ would not have a probability measure, since it is not an event. In addition, consider the event $\mathcal{E}_1$ and its complement $\mathcal{E}_1'$. These events are disjoint so that $\mathcal{P}\{\mathcal{E}_1 \cup \mathcal{E}_1'\} = \mathcal{P}\{\mathcal{E}_1\} + \mathcal{P}\{\mathcal{E}_1'\}$; however, $\mathcal{E}_1 \cup \mathcal{E}_1' = \mathcal{S}$ and $\mathcal{P}\{\mathcal{S}\} = 1$, so we have $\mathcal{P}\{\mathcal{E}_1'\} = 1 - \mathcal{P}\{\mathcal{E}_1\}$. Therefore, $\mathcal{E}_1'$ must be an event, since $\mathcal{P}$ is defined for it. By the use of Axiom IIIa, we could establish, in a similar manner, the need for the events to form a sigma field.

It may surely appear that we have posed an unduly abstract definition of probability as numerous "better" intuitive definitions of probability must immediately come to mind. We shall examine some of these ancillary probability definitions in Section 2.4 after we have developed some additional properties of our axiomatic definition. We shall do this by means of several rather important examples.

Example *2.3-2.* From the three axioms of probability, we may show that *the probability of the impossible event is zero.* This may be accomplished by showing that the certain event $\mathcal{S}$ and the null or impossible event $\varnothing$ are mutually exclusive and, in fact, that

$$\mathcal{E}\varnothing = \varnothing$$

Also, the union of the certain event $\mathcal{S}$ and the null event $\varnothing$ is the certain event $\mathcal{S}$. In fact,

$$\mathcal{E} \cup \varnothing = \mathcal{E}$$

Since $\mathcal{S}$ and $\varnothing$ are mutually exclusive, Axiom III may be applied such that

$$\mathcal{P}\{\mathcal{S} \cup \varnothing\} = \mathcal{P}\{\mathcal{S}\} = \mathcal{P}\{\mathcal{S}\} + \mathcal{P}\{\varnothing\}$$

By simple substraction we have that

$$\mathcal{P}\{\varnothing\} = 0$$

Example ***2.3-3.*** Axiom III may also be used to establish the relation for the probability of the complement of an event. We can easily establish the fact that

$$\mathcal{E}_1 = \mathcal{E}_1\mathcal{E}_2 \cup \mathcal{E}_1\mathcal{E}_2'$$

Now the events $\mathcal{E}_1\mathcal{E}_2$ and $\mathcal{E}_1\mathcal{E}_2'$ are mutually exclusive, since

$$(\mathcal{E}_1\mathcal{E}_2)(\mathcal{E}_1\mathcal{E}_2') = \varnothing$$

and thus Axiom III may be applied to yield

$$\mathcal{P}\{\mathcal{E}_1\} = \mathcal{P}\{\mathcal{E}_1\mathcal{E}_2 \cup \mathcal{E}_1\mathcal{E}_2'\} = \mathcal{P}\{\mathcal{E}_1\mathcal{E}_2\} + \mathcal{P}\{\mathcal{E}_1\mathcal{E}_2'\}$$

which is an important relation in its own right. Now if $\mathcal{E}_1$ is the certain event $\mathcal{S}$, then $\mathcal{P}\{\mathcal{S}\} = 1$, $\mathcal{S}\mathcal{E}_2 = \mathcal{E}_2$, $\mathcal{S}\mathcal{E}_2' = \mathcal{E}_2'$, and the foregoing equation becomes

$$1 = \mathcal{P}\{\mathcal{E}_2\} + \mathcal{P}\{\mathcal{E}_2'\}$$

Thus

$$\mathcal{P}\{\mathcal{E}_2'\} = 1 - \mathcal{P}\{\mathcal{E}_2\}$$

This important relation states that the probability of the event "not $\mathcal{E}_2$" is 1 minus the probability of the event $\mathcal{E}_2$. Thus if the probability of getting two heads on the toss of two coins is 0.25, the probability of not getting two heads on the toss of two coins is 0.75.

Example ***2.3-4.*** It may appear that our axioms are not a complete set since, at first glance, they do not appear to tell us how to compute the probability of the union of two events $\mathcal{E}_1 \cup \mathcal{E}_2$ unless $\mathcal{E}_1$ and $\mathcal{E}_2$ are mutually exclusive. This is not so, since it turns out that the event $\mathcal{E}_1 \cup \mathcal{E}_2$ may be written as the union of two mutually exclusive events. It is easy to verify, using Venn diagrams, that

$$\mathcal{E}_1 \cup \mathcal{E}_2 = \mathcal{E}_1 \cup (\mathcal{E}_1'\mathcal{E}_2) = \mathcal{E}_2 \cup (\mathcal{E}_1\mathcal{E}_2')$$

Now the events $\mathcal{E}_2$ and $\mathcal{E}_1\mathcal{E}_2'$ are mutually exclusive; so we may apply Axiom III to the foregoing and obtain

$$\mathcal{P}\{\mathcal{E}_1 \cup \mathcal{E}_2\} = \mathcal{P}\{\mathcal{E}_2 \cup (\mathcal{E}_1\mathcal{E}_2')\} = \mathcal{P}\{\mathcal{E}_2\} + \mathcal{P}\{\mathcal{E}_1\mathcal{E}_2'\}$$

The ancillary result of the previous example

$$\mathcal{P}\{\mathcal{E}_1\mathcal{E}_2'\} = \mathcal{P}\{\mathcal{E}_1\} - \mathcal{P}\{\mathcal{E}_1\mathcal{E}_2\}$$

may be combined with the previous expression such that we obtain

$$\mathcal{P}\{\mathcal{E}_1 \cup \mathcal{E}_2\} = \mathcal{P}\{\mathcal{E}_1\} + \mathcal{P}\{\mathcal{E}_2\} - \mathcal{P}\{\mathcal{E}_1\mathcal{E}_2\}$$

as the desired relation for the probability of the union of two non-mutually exclusive events. This relation means, for example, that if the probability of obtaining an ace on a single draw from a 52-card deck is $\frac{1}{13}$, the probability of obtaining a spade in a single draw from a 52-card deck is $\frac{1}{4}$, and the probability of obtaining an ace of spades on the single draw is $\frac{1}{52}$, then the probability that a single card drawn from a 52-card deck will be an ace or a spade (or both) is $\frac{1}{4} + \frac{1}{13} - \frac{1}{52} = \frac{4}{13}$.

Example ***2.3-5.*** Axiom I states that the probability of an event is nonnegative and Axiom II states that the probability of the certain event is 1. It is natural to inquire concerning an upper bound on the probability of an event. Let $\mathcal{E}_1$ and $\mathcal{E}_2$ be any two events on a probability space $\mathcal{S}$ such that

$$\mathcal{E}_2 \subset \mathcal{E}_1 \subset \mathcal{S}$$

Since $\mathcal{E}_2 \subset \mathcal{E}_1$, it follows that

$$\mathcal{E}_1\mathcal{E}_2 = \mathcal{E}_2$$

Also from Axiom I

$$\mathcal{P}\{\mathcal{E}_1\mathcal{E}_2'\} \geq 0$$

We again consider the ancillary result of Example 2.3-3:

$$\mathcal{P}\{\mathcal{E}_1\} = \mathcal{P}\{\mathcal{E}_1\mathcal{E}_2\} + \mathcal{P}\{\mathcal{E}_1\mathcal{E}_2'\}$$

which becomes

$$\mathcal{P}\{\mathcal{E}_1\} - \mathcal{P}\{\mathcal{E}_2\} \geq 0$$

from the foregoing two relations. This shows that

$$\mathcal{P}\{\mathcal{E}_2\} \leq \mathcal{P}\{\mathcal{E}_1\} \qquad \text{if } \mathcal{E}_2 \subset \mathcal{E}_1$$

In the particular case in which $\mathcal{E}_1 = \mathcal{S}$, this relation becomes

$$0 \leq \mathcal{P}\{\mathcal{E}_2\} \leq 1 \qquad \text{for any event } \mathcal{E}_2$$

Example ***2.3-6.*** It is of interest to consider the union of three or more events, $\mathcal{E}_1 \cup \mathcal{E}_2 \cup \mathcal{E}_3$. From the Venn diagram of Fig. 2.3-1 it is

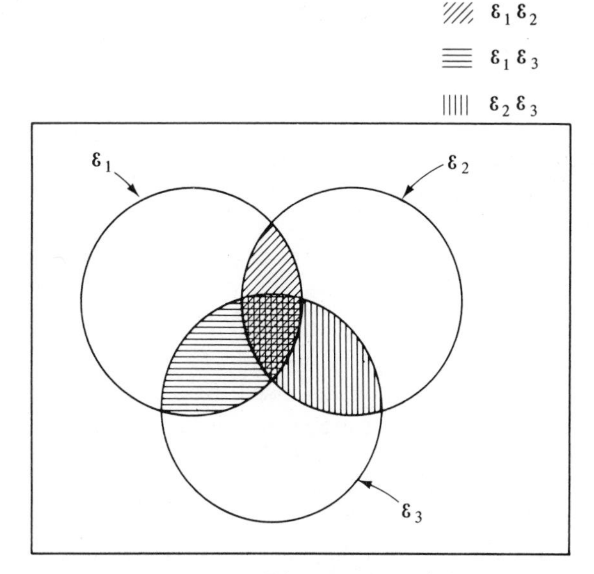

FIG 2.3-1. Venn Diagram, Example 2.3-6.

clear that the events $\mathcal{E}_1\mathcal{E}_2$, $\mathcal{E}_1\mathcal{E}_3$, and $\mathcal{E}_2\mathcal{E}_3$ have been added in twice and should therefore be subtracted out once. Also, $\mathcal{E}_1\mathcal{E}_2\mathcal{E}_3$ has been added in three times and subtracted out three times. It must be added in again. Thus we have

$$\begin{aligned}\mathcal{P}\{\mathcal{E}_1 \cup \mathcal{E}_2 \cup \mathcal{E}_3\} = \mathcal{P}\{\mathcal{E}_1\} + \mathcal{P}\{\mathcal{E}_2\} + \mathcal{P}\{\mathcal{E}_3\} - \mathcal{P}\{\mathcal{E}_1\mathcal{E}_2\} \\ - \mathcal{P}\{\mathcal{E}_1\mathcal{E}_3\} - \mathcal{P}\{\mathcal{E}_2\mathcal{E}_3\} + \mathcal{P}\{\mathcal{E}_1\mathcal{E}_2\mathcal{E}_3\}\end{aligned}$$

We can also establish this result directly by the use of Example 2.3-4 since

$$\begin{aligned}\mathcal{P}\{\mathcal{E}_1 \cup \mathcal{E}_2 \cup \mathcal{E}_3\} &= \mathcal{P}\{\mathcal{E}_1 \cup (\mathcal{E}_2 \cup \mathcal{E}_3)\} \\ &= \mathcal{P}\{\mathcal{E}_1\} + \mathcal{P}\{\mathcal{E}_2 \cup \mathcal{E}_3\} - \mathcal{P}\{\mathcal{E}_1(\mathcal{E}_2 \cup \mathcal{E}_3)\} \\ &= \mathcal{P}\{\mathcal{E}_1\} + \mathcal{P}\{\mathcal{E}_2\} + \mathcal{P}\{\mathcal{E}_3\} - \mathcal{P}\{\mathcal{E}_2\mathcal{E}_3\} \\ &\quad - \mathcal{P}\{\mathcal{E}_1\mathcal{E}_2\} - \mathcal{P}\{\mathcal{E}_1\mathcal{E}_3\} + \mathcal{P}\{\mathcal{E}_1\mathcal{E}_2\mathcal{E}_3\}\end{aligned}$$

For the general case of N events, the logical extension of our result yields

$$\begin{aligned}\mathcal{P}\{\mathcal{E}_1 \cup \mathcal{E}_2 \cup \cdots \cup \mathcal{E}_N\} = \sum_{i=1}^{N} \mathcal{P}\{\mathcal{E}_i\} - \sum_{i=1}^{N-1} \sum_{j=i+1}^{N} \mathcal{P}\{\mathcal{E}_i\mathcal{E}_j\} \\ + \sum_{i=1}^{N-2} \sum_{j=i+1}^{N-1} \sum_{k=j+1}^{N} \mathcal{P}\{\mathcal{E}_i\mathcal{E}_j\mathcal{E}_k\} + \cdots\end{aligned}$$

In the important case in which the events are mutually exclusive, that is, $\mathcal{E}_i\mathcal{E}_j = \varnothing$ for all $i \neq j$, this result becomes

$$\mathcal{P}\{\mathcal{E}_1 \cup \mathcal{E}_2 \cup \cdots \cup \mathcal{E}_N\} = \sum_{i=1}^{N} \mathcal{P}\{\mathcal{E}_i\}$$

Example *2.3-7.* As our final example of this section, let us consider N mutually exclusive events $\mathcal{E}_i$, $i = 1, 2, \ldots, N$, such that $\bigcup_{i=1}^{N} \mathcal{E}_i = \mathcal{S}$. Since the N events span the sample space $\mathcal{S}$ and the events are mutually exclusive, the result of Example 2.3-5 yields

$$\sum_{i=1}^{N} \mathcal{P}\{\mathcal{E}_i\} = 1$$

In this section we have discussed the axiomatic approach to probability. Several examples were presented that demonstrated important results obtained from the three probability axioms. For convenience, the results of this section are summarized in Table 2.3-1.

TABLE 2.3-1. Axioms and Important Probability Relations

Axiom I	$\mathcal{P}\{\mathcal{E}_i\} \geq 0$	(2.3-1)
Axiom II	$\mathcal{P}\{\mathcal{S}\} = 1$	(2.3-2)
Axiom III	$\mathcal{P}\{\mathcal{E}_1 \cup \mathcal{E}_2\} = \mathcal{P}\{\mathcal{E}_1\} + \mathcal{P}\{\mathcal{E}_2\}$ if $\mathcal{E}_1\mathcal{E}_2 = \varnothing$	(2.3-3)
Axiom IIIa	$\mathcal{P}\left\{\bigcup_{i=1}^{\infty} \mathcal{E}_i\right\} = \sum_{i=1}^{\infty} \mathcal{P}\{\mathcal{E}_i\}$ if $\mathcal{E}_i\mathcal{E}_j = \varnothing, i \neq j$	(2.3-4)
Probability of null event is zero	$\mathcal{P}\{\varnothing\} = 0$	(2.3-5)
Probability of intersection law	$\mathcal{P}\{\mathcal{E}_1\mathcal{E}_2\} = \mathcal{P}\{\mathcal{E}_1\} - \mathcal{P}\{\mathcal{E}_1\mathcal{E}_2'\}$	(2.3-6)
Probability of complement law	$\mathcal{P}\{\mathcal{E}_2'\} = 1 - \mathcal{P}\{\mathcal{E}_2\}$	(2.3-7)
Probability of nonmutually exclusive events	$\mathcal{P}\{\mathcal{E}_1 \cup \mathcal{E}_2\} = \mathcal{P}\{\mathcal{E}_1\} + \mathcal{P}\{\mathcal{E}_2\} - \mathcal{P}\{\mathcal{E}_1\mathcal{E}_2\}$	(2.3-8)
Bounds on probability	$\mathcal{P}\{\mathcal{E}_1\} \leq \mathcal{P}\{\mathcal{E}_2\}$, $\mathcal{E}_1 \subset \mathcal{E}_2$ $0 \leq \mathcal{P}\{\mathcal{E}\} \leq 1$	(2.3-9) (2.3-10)
Sum of probability law	$\sum_{i=1}^{N} \mathcal{P}\{\mathcal{E}_i\} = 1$	(2.3-11)
	$\mathcal{S}$ is composed of N mutually exclusive events $\mathcal{E}_i$, $i = 1, 2, \ldots, N$	

We have remarked that other definitions of probability, other than the axiomatic definition, could be used. Since these other definitions have considerable intuitive appeal, and since we can use these other definitions successfully for a variety of problems if we are aware of their limitations, we now present other approaches to the definition of probability.

EXERCISES 2.3

2.3-1. You and a friend each toss a fair coin. You win if your outcome matches your friend's outcome.
(a) List the possible outcomes.
(b) What is the probability that you win?

Answer: (b) 0.5.

2.3-2. An event $\mathcal{A}$ can happen only if one of two mutually exclusive events $\mathcal{B}_1$ or $\mathcal{B}_2$ does; that is, $\mathcal{A} \subset \mathcal{B}_1 \cup \mathcal{B}_2$.
(a) Show that $\mathcal{A} = \mathcal{A}\mathcal{B}_1 \cup \mathcal{A}\mathcal{B}_2$.
(b) Express $\mathcal{P}\{\mathcal{A}\}$ in terms of $\mathcal{P}\{\mathcal{A}\mathcal{B}_1\}$ and $\mathcal{P}\{\mathcal{A}\mathcal{B}_2\}$.

2.3-3. Show that if $\mathcal{A}\mathcal{B} = \varnothing$, then $\mathcal{P}\{\mathcal{A}\} \leq \mathcal{P}\{\mathcal{B}'\}$.

2.3-4. Find the probability that in three tosses of a fair coin you will obtain two heads in a row.

Answer: 0.25.

2.4. *Classical and relative frequency definition of probability*

All of us have intuitive notions concerning probability. Most of these notions are simple and often useful in discussions of models for probabilistic models. Unfortunately, a precise definition of probability is difficult to formulate using these intuitive concepts. In some cases, definitions so formulated are inherently circular. In this section we shall discuss two definitions of probability based upon intuitive concepts—the classical or equally likely definition and the relative-frequency definition. Since these concepts are exceptionally useful in many engineering problems, and since these two definitions strengthen our understanding of the axiomatic definition, we present them here and attempt to relate them to the axiomatic definition of probability. The classical and relative-frequency definitions are not, however, the only possible intuitive definitions of probability. Certainly, personal or subjective probability in which we simply use an "our best guess" for the probability of an event is another intuitive sort of probability.

Definition 2.4-1. ***(Classical) Probability:*** If there are $N_{\mathcal{S}}$ possible mutually exclusive and equally likely outcomes of a random experiment, and if N_i of these correspond to event $\mathcal{E}_i$, then the probability of event $\mathcal{E}_i$, called $\mathcal{P}\{\mathcal{E}_i\}$, is

$$\mathcal{P}\{\mathcal{E}_i\} = \frac{N_i}{N_{\mathcal{S}}} \tag{2.4-1}$$

This definition of probability, credited to Laplace in 1812, may well have originated to resolve certain gambling problems. We can easily determine the probability of obtaining a head on the toss of a coin or the probability of obtaining a 2 on the toss of a die. A bit more care must be exercised, however, in determining the probability of obtaining a head and a tail on the toss of two coins or a 7 on the toss of a pair of dice. With care, though, these problems can be resolved using this classical definition of probability, as the following examples show. The classical or equally likely definition of probability is but a special case of the axiomatic definition in which each outcome is equally likely.

Example ***2.4-1.*** Let us consider the probability of obtaining a head and a tail on the toss of two coins. There are three events here: two heads, two tails, and a tail and a head. The probability of occurrence of these events is "obviously" not equally likely however. There are four outcomes for the toss of two coins and two of these are the desired tail and head. Thus the probability of obtaining a head and a tail is $\frac{2}{4} = 0.5$.

An alternative method of solving this simple problem is instructive. We may define four events

$$\mathcal{E}_{\mathrm{HH}} = \{(\mathrm{H}, \mathrm{H})\} \qquad \mathcal{E}_{\mathrm{TT}} = \{(\mathrm{T}, \mathrm{T})\}$$

$$\mathcal{E}_{\mathrm{TH}} = \{(\mathrm{T}, \mathrm{H})\} \qquad \mathcal{E}_{\mathrm{HT}} = \{(\mathrm{H}, \mathrm{T})\}$$

that are mutually exclusive and "obviously" equally likely. Thus the probability of each event is $\frac{1}{4}$. The desired probability is just the union of events $\mathcal{E}_{\mathrm{TH}}$ and $\mathcal{E}_{\mathrm{HT}}$, and this is

$$\mathcal{P}\{\mathcal{E}_{\mathrm{TH}} \cup \mathcal{E}_{\mathrm{HT}}\} = \mathcal{P}\{\mathcal{E}_{\mathrm{TH}}\} + \mathcal{P}\{\mathcal{E}_{\mathrm{HT}}\}$$

which is just $\frac{1}{2}$, since the events are mutually exclusive.

Example ***2.4-2.*** Let us now obtain the probabilities for various events associated with the toss of two dice. Possible events, the ways in which each can occur, and the total number of ways are listed on the next page:

Sum of Reading of Two Dice	Ways in Which the Event Can Occur	Number of Ways	Probability of Occurrence
2	1-1	1	$\frac{1}{36}$
3	2-1, 1-2	2	$\frac{1}{18}$
4	2-2, 3-1, 1-3	3	$\frac{1}{12}$
5	3-2, 2-3, 4-1, 1-4	4	$\frac{1}{9}$
6	3-3, 4-2, 2-4, 5-1, 1-5	5	$\frac{5}{36}$
7	4-3, 3-4, 5-2, 2-5, 6-1, 1-6	6	$\frac{1}{6}$
8	4-4, 5-3, 3-5, 6-2, 2-6	5	$\frac{5}{36}$
9	5-4, 4-5, 6-3, 3-6	4	$\frac{1}{9}$
10	5-5, 6-4, 4-6	3	$\frac{1}{12}$
11	6-5, 5-6	2	$\frac{1}{18}$
12	6-6	1	$\frac{1}{36}$

There are 36 possible outcomes, each of which is equally likely. Thus the probability of occurrence of the various events is as listed. Seven is the most likely number to occur on the toss of a pair of dice.

Example *2.4-3.* There are 2000 different numbers in a hat. We desire to determine the probability that the first digit of a number selected at random will be the digit 1. At first glance, it might appear that the first integer, assuming that zero is excluded, would be one of an equally likely set of 9 digits in the sample space $\mathcal{S} = \{(1, 2, 3, 4, 5, 6, 7, 8, 9)\}$. Thus the probability of the digit 1 would be $\frac{1}{9}$.

However, if we assume that the numbers in the hat are consecutively numbered from 1 to 2000, we see that the following table is descriptive of the numbers whose first digit is 1.

Range of Numbers	Numbers in This Range
1	1
10–19	10
100–199	100
1000–1999	1000

Thus the sample space $\mathcal{S}$ has 2000 elements and the number of elements with first digit 1 is 1111. The desired probability is then

$$\mathcal{P}\{\mathcal{E}\} = \frac{1111}{2000} = 0.5555$$

which is quite different from the $\frac{1}{9}$ computed earlier. The reason is that the probability models are quite different. No amount of reasoning can resolve our dilemma; we must either obtain a better description of how the 2000 numbers in the hat were obtained or resort to experimentation to determine the desired probability. This leads us to the subject of the relative-frequency definition of probability.

Definition 2.4-2. ***(Relative Frequency) Probability:*** An experiment is performed n times under identical conditions. If the event $\mathcal{E}_i$ occurs n_i times, the probability of event $\mathcal{E}_i$ is

$$\mathcal{P}\{\mathcal{E}_i\} = \lim_{n \to \infty} \frac{n_i}{n} \tag{2.4-2}$$

It is important to note that, although Eqs. (2.4-1) and (2.4-2) appear quite similar in form, they are really quite different, since $N_{\mathcal{S}}$ and N_i are quite different from n and n_i. $N_{\mathcal{S}}$ and N_i are a priori numbers, numbers based on intuitive reasoning and not on experiment, whereas n and n_i are based on experiment. Note that the experiments must be performed under identical conditions. This is particularly troublesome in many application areas, such as the biological and societal sciences.

This definition is obviously quite valuable in experimental work and in applications. However, since we can never perform an infinite number of experiments, it is not possible to *measure* probability by this definition. An *estimate* of probability can be obtained, and this is often quite valuable.

2.5. *Sampling and combinatorial analysis*

In this section we shall consider random sampling and combinatorial analysis. Consider, for example, an urn containing M balls from which N balls are drawn at random. We say that this is a *random sample* of size N. If, in addition, we note the order in which the elements were drawn, we obtain an *ordered sample* of size N. If after each ball is drawn and noted, it is replaced before the next sample is obtained, the sampling is said to be with *replacement*. If, on the other hand, the ball is not replaced, the sampling is *without replacement*.

Example ***2.5-1.*** Let us consider an urn of three balls numbered 1 to 3 and take a sample of size 2. If the sampling is done with replacement the possible outcomes are

(1, 1)	(2, 1)	(3, 1)
(1, 2)	(2, 1)	(3, 2)
(1, 3)	(2, 3)	(3, 3)

If the sampling is done without replacement, the possible outcomes are

$$\begin{matrix} (1, 2) & (2, 1) & (3, 1) \\ (1, 3) & (2, 3) & (3, 2) \end{matrix}$$

There are thus nine outcomes if the sample of two balls from a three-ball urn is done with replacement, and six outcomes if the sampling is done without replacement.

For low-dimensioned problems, such as Example 2.5-1, counting is an acceptable method of determining the number of possible outcomes. For more complex problems, a more systematic method is desirable. This leads us to a study of combinatiorial analysis.

Suppose that we wish to obtain a sample of size N. The first element of the sample is drawn from a set that has N_1 elements; there are clearly N_1 possible ways in which this first may be selected. Now if we chose the second element of our sample from a set with N_2 elements, for each of the N_1 possible first elements there are now N_1 possible second elements. Hence, the total number of ways in which the first two elements may be drawn in N_1N_2. If we continue in a similar manner to obtain all N samples, the total number of possible outcomes is

$$K = N_1 N_2 N_3 \cdots N_N \tag{2.5-1}$$

Example ***2.5-2.*** There are 6 balls in urn number 1 (numbered 1 through 6) and 2 balls in urn number 2 (numbered 7 and 8). From Eq. 2.5-1) we have

$$K = 6 \cdot 2 = 12$$

ways in which we can obtain a sample of size 2 as described above. These ways are (1, 7), (1, 8), (2, 7), (2, 8), (3, 7), (3, 8), (4, 7), (4, 8), (5, 7), (5, 8), (6, 7), and (6, 8).

This same principle may be used to determine the number of ways in which we can choose an ordered sample of size N from a total of M distinguishable objects if the sampling is done without replacement. There are M choices for the first object, $M - 1$ for the second, $M - 2$ for the third, and so forth until we have $M - (N - 1)$ choice for the last Nth object. The number of ways in which we may choose N objects from M distinguishable objects is therefore

$$(M)_N = M(M - 1) \cdots (M - N + 1) = \frac{M!}{(M - N)!} \tag{2.5-2}$$

where $M!$, M factorial, is defined as

$$M! = M(M-1)(M-2)\cdots(1) \tag{2.5-3}$$

with $0! = 1$ by convention.

If the samples are drawn with replacement, there are M possible distinguishable objects that may be chosen on each draw, and the number of ways in which the N objects may be drawn is

$$(M)^N = M^N \tag{2.5-4}$$

Let us consider the number of possible outcomes associated with an unordered sample of size N drawn from a set of M elements without replacement. If we choose an ordered sample of size N without replacement from the set of M distinguishable objects, by Eq. (2.5-2) there are $(M)_N$ possible outcomes. The elements of each ordered sample of size N may be arranged $N!$ ways, since the first element may be any of the N elements, the second any of the remaining $N-1$ elements, and so forth. Because we are not interested in the ordering of the elements in the sample, there are only $(M)_N$ divided by $N!$ distinguishable outcomes, which we shall designate as $\binom{M}{N}$ or

$$\frac{\text{ordered outcomes}}{\text{ways of ordering}} = \frac{M!}{N!\,(M-N)!} = \binom{M}{N} \tag{2.5-5}$$

The quantity $\binom{M}{N}$ is called the *binomial coefficient* because of its appearance in the binomial theorem

$$(x+y)^N = \sum_{i=0}^{N} \binom{N}{i} x^{N-i} y^i \tag{2.5-6}$$

Example *2.5-3*. Let us reconsider Example 2.5-1. From Eq. (2.5-4) we have, for the number of ways in which the two balls may be drawn with replacement from the urn containing three balls,

$$(M)^N = 3^2 = 9$$

The number of ways in which the two balls may be chosen without replacement is determined from Eq. (2.5-2) as

$$(M)_N = \frac{3!}{1!} = 6$$

The number of different unordered samples of size 2 drawn without replacement is

$$\binom{3}{2} = \frac{3!}{2!\,1!} = 3$$

The possible outcomes with unordered samples are (1, 3), (2, 1), and (2, 3).

Example 2.5-4. The number of different 5-card poker hands is desired. We can easily determine this form Eq. (2.5-5) as

$$\binom{52}{5} = \frac{52!}{5!\,47!} = \frac{48\cdot49\cdot50\cdot51\cdot52}{2\cdot3\cdot4\cdot5} = 2{,}598{,}960$$

since ordering is not important and sampling is without replacement.

Of interest also is the number of ways in which three 5-card poker hands may be dealt. The number of different hands that the first player may receive is $\binom{52}{5}$. There are 47 cards left after 5 are dealt the first player. Therefore, the second player may receive any of $\binom{47}{5}$ different hands and the third player $\binom{42}{5}$ different hands. There are therefore

$$\binom{52}{5}\binom{47}{5}\binom{42}{5} = \frac{52!}{5!\,5!\,37!} = 8.924505967\cdot10^{16}$$

different hands.

Example 2.5-5. Let us suppose that we have an urn containing three white balls and one black ball. We choose two balls from the urn and wish to determine the probability that
(a) both balls are white = event $\mathcal{E}_1$.
(b) one ball is black and the other white = event $\mathcal{E}_2$.
(c) both balls are black = event $\mathcal{E}_3$.
It is convenient to make the balls distinguishable by numbering the white balls 1, 2, and 3, and numbering the black ball 4. We shall consider the problem with and without replacement.

The problem with replacement is considered first. We assume that all outcomes in $\mathcal{S}$ are equally likely. The number of ways in which we may select 2 white balls from the 3 white balls with replacement is given by Eq. (2.5-4) as 3^2. The total number of possible outcomes is the number of ways in which a sample of size 2 may be selected from a set of 4 objects with replacement, or 4^2. Therefore, the probability of $\mathcal{E}_1$ is given by

$$\mathcal{P}\{\mathcal{E}_1\} = \frac{3^2}{4^2} = \frac{9}{16}$$

In a similar manner, the probability of $\mathcal{E}_3$ is given by

$$\mathcal{P}\{\mathcal{E}_3\} = \frac{1^2}{4^2} = \frac{1}{16}$$

Since the events $\mathcal{E}_1$, $\mathcal{E}_2$, and $\mathcal{E}_3$ are mutually exclusive and cover $\mathcal{S}$, we know from the law of total probability that

$$\mathcal{P}\{\mathcal{E}_2\} = 1 - \mathcal{P}\{\mathcal{E}_1\} - \mathcal{P}\{\mathcal{E}_3\} = \tfrac{6}{16}$$

If we repeat this problem, assuming that the samples are not replaced, we obtain

$$\mathcal{P}\{\mathcal{E}_1\} = \frac{3\cdot 2}{4\cdot 3} = \frac{1}{2},$$

$$\mathcal{P}\{\mathcal{E}_3\} = \frac{1\cdot 0}{4\cdot 3} = 0$$

$$\mathcal{P}\{\mathcal{E}_2\} = 1 - \mathcal{P}\{\mathcal{E}_1\} = \tfrac{1}{2}$$

Example ***2.5-6.*** As our next example, we consider the celebrated problem of the probability of repetition for samples drawn with replacement. We consider an urn with M balls numbered 1 through M, each of which is equally likely to be drawn. N balls are chosen with replacement. The number of samples of size N without repetition is $(M)_N$ since there are M possibilities for the first number chosen, $(M - 1)$ for the second, since it must not duplicate the first, and so on. The total number of possible samples of size N (with replacement) is M^N; so the probability of no repetition in a sample of size N is

$$\mathcal{P}\{\text{no repetitions}\} = \frac{(M)_N}{M^N} = \frac{M!}{M^N(M-N)!}$$

EXERCISES 2.5

2.5-1. A box contains 5 red and 5 black balls. Compute and plot the probability that a sample of size $N = 1, 2, \ldots, 10$ taken without replacement will contain exactly 2 red balls.

2.5-2. A five-card sample is taken from a 52-card deck. What is the probability that the sample contains exactly three aces?

2.5-3. There are three red and six black balls in an urn. Calculate the probability that a sample of size a taken without replacement will contain (a) exactly one red ball, (b) at least one red ball, and (c) two red balls.

2.5-4. Repeat Problem 2.5-3, but assume that the sampling is done with replacement.

2.6. Conditional probability

In previous sections we considered the probability of discrete events associated primarily with a single experiment. In many applications, however, an experiment is composed of subexperiments. For example, we have been concerned with problems such as the one in which we desire the probability of choosing a black ball from an urn containing M_B black balls and M_W white balls. An obvious extension is to consider that the first ball chosen is black and to inquire concerning the probability that the second ball will also be black. This involves the notion of conditional probability, the subject of this section. Our discussions are perhaps best motivated by considering first the classical or equally likely definitions of probability. In terms of these definitions we have for the probabilities of events $\mathcal{E}_1$, $\mathcal{E}_2$, and $\mathcal{E}_1\mathcal{E}_2$, respectively,

$$\mathcal{P}\{\mathcal{E}_1\} = \frac{N_1}{N_\mathcal{S}} \tag{2.6-1}$$

$$\mathcal{P}\{\mathcal{E}_2\} = \frac{N_2}{N_\mathcal{S}} \tag{2.6-2}$$

$$\mathcal{P}\{\mathcal{E}_1\mathcal{E}_2\} = \frac{N_{12}}{N_\mathcal{S}} \tag{2.6-3}$$

where N_{12} is the number of outcomes favorable to both $\mathcal{E}_1$ and $\mathcal{E}_2$. We shall define the symbol $\mathcal{P}\{\mathcal{E}_1 \mid \mathcal{E}_2\}$ as representing the probability that event $\mathcal{E}_1$ occurs, conditioned upon the knowledge that $\mathcal{E}_2$ has occurred. This is a *conditional* probability since we are considering only the outcomes favorable to $\mathcal{E}_2$ and not all possible outcomes. The conditional probability of $\mathcal{E}_1$ given $\mathcal{E}_2$ is the number of times that both $\mathcal{E}_1$ and $\mathcal{E}_2$ occur divided by the number of times the event $\mathcal{E}_2$ occurs (assuming that $\mathcal{E}_2$ occurs at least once), so that

$$\mathcal{P}\{\mathcal{E}_1 \mid \mathcal{E}_2\} = \frac{N_{12}}{N_2} \tag{2.6-4}$$

We divide both numerator and denominator of the right-hand side of this expression by $N_\mathcal{S}$ and then use Eqs. (2.6-2) and (2.6-3) to obtain

$$\mathcal{P}\{\mathcal{E}_1 \mid \mathcal{E}_2\} = \frac{N_{12}/N_\mathcal{S}}{N_2/N_\mathcal{S}} = \frac{\mathcal{P}\{\mathcal{E}_1\mathcal{E}_2\}}{\mathcal{P}\{\mathcal{E}_2\}} \tag{2.6-5}$$

This expression is known as the *conditional probability law* and may also be written

$$\mathcal{P}\{\mathcal{E}_2 \mid \mathcal{E}_1\} = \frac{\mathcal{P}\{\mathcal{E}_1\mathcal{E}_2\}}{\mathcal{P}\{\mathcal{E}_1\}} \tag{2.6-6}$$

By combining Eqs. (2.6-5) and (2.6-6) such as to eliminate $\mathcal{P}\{\mathcal{E}_1\mathcal{E}_2\}$, we obtain a most useful expression known as *Bayes' rule:*

$$\mathcal{P}\{\mathcal{E}_1 \mid \mathcal{E}_2\} = \frac{\mathcal{P}\{\mathcal{E}_2 \mid \mathcal{E}_1\}\mathcal{P}\{\mathcal{E}_1\}}{\mathcal{P}\{\mathcal{E}_2\}} \tag{2.6-7}$$

It is common to speak of $\mathcal{P}\{\mathcal{E}_1\mathcal{E}_2\}$ as a joint probability. The expression $\mathcal{P}\{\mathcal{E}_1 \mid \mathcal{E}_2\}$ is spoken of as a *conditioned probability*. In other words, $\mathcal{P}\{\mathcal{E}_1 \mid \mathcal{E}_2\}$ represents the probability of event $\mathcal{E}_1$ occurring with knowledge of the information that $\mathcal{E}_2$ has occurred. Bayes rule relates the two conditional probabilities $\mathcal{P}\{\mathcal{E}_1 \mid \mathcal{E}_2\}$ and $\mathcal{P}\{\mathcal{E}_2 \mid \mathcal{E}_1\}$ in terms of the marginal probabilities $\mathcal{P}\{\mathcal{E}_1\}$ and $\mathcal{P}\{\mathcal{E}_2\}$. We may also use the conditional density to express the joint density as

$$\begin{aligned}\mathcal{P}\{\mathcal{E}_1\mathcal{E}_2\} &= \mathcal{P}\{\mathcal{E}_1 \mid \mathcal{E}_2\}\mathcal{P}\{\mathcal{E}_2\} \\ &= \mathcal{P}\{\mathcal{E}_2 \mid \mathcal{E}_1\}\mathcal{P}\{\mathcal{E}_1\}\end{aligned} \tag{2.6-8}$$

If the probability of $\mathcal{E}_2$ conditioned on $\mathcal{E}_1$ is not dependent upon $\mathcal{E}_1$, we say that the *events* $\mathcal{E}_1$ and $\mathcal{E}_2$ are *independent* and have

$$\mathcal{P}\{\mathcal{E}_1 \mid \mathcal{E}_2\} = \mathcal{P}\{\mathcal{E}_1\} \tag{2.6-9}$$

for independent events. This also requires, from Eqs. (2.6-6) and (2.6-7), that

$$\mathcal{P}\{\mathcal{E}_2 \mid \mathcal{E}_1\} = \mathcal{P}\{\mathcal{E}_2\} \tag{2.6-10}$$

$$\mathcal{P}\{\mathcal{E}_1\mathcal{E}_2\} = \mathcal{P}\{\mathcal{E}_1\}\mathcal{P}\{\mathcal{E}_2\} \tag{2.6-11}$$

We should note very carefully that independence of events $\mathcal{E}_1$ and $\mathcal{E}_2$ *does not* imply that the events are mutually exclusive. In fact, just the opposite is true. If the events are mutually exclusive, occurrence of $\mathcal{E}_2$ means that $\mathcal{E}_1$ cannot occur; thus events $\mathcal{E}_1$ and $\mathcal{E}_2$ cannot be independent and mutually exclusive unless one or both have zero probability.

Three or more events are said to be independent if and only if the probability that any specified two or more events will occur is the product of the probability of the events specified. Hence, three events $\mathcal{E}_1$, $\mathcal{E}_2$, and $\mathcal{E}_3$ are independent if

$$\begin{aligned}\mathcal{P}\{\mathcal{E}_1\mathcal{E}_2\} &= \mathcal{P}\{\mathcal{E}_1\}\mathcal{P}\{\mathcal{E}_2\} \\ \mathcal{P}\{\mathcal{E}_2\mathcal{E}_3\} &= \mathcal{P}\{\mathcal{E}_2\}\mathcal{P}\{\mathcal{E}_3\} \\ \mathcal{P}\{\mathcal{E}_1\mathcal{E}_3\} &= \mathcal{P}\{\mathcal{E}_1\}\mathcal{P}\{\mathcal{E}_3\}\end{aligned}$$

and

$$\mathcal{P}\{\mathcal{E}_1\mathcal{E}_2\mathcal{E}_3\} = \mathcal{P}\{\mathcal{E}_1\}\mathcal{P}\{\mathcal{E}_2\}\mathcal{P}\{\mathcal{E}_3\}$$

If only the first three relations are true, we say the events are *pairwise independent*. We note that three events may be pairwise independent but not independent.

We might well question whether or not conditional probabilities as we have obtained them are "really" probabilities. We did obtain them from the classical definition and have previously made the assertion that valid probabilities must satisfy the axiomatic definition. Thus we need to see if our three fundamental probability axioms are satisfied.

The first axiom

$$\mathcal{P}\{\mathcal{E}_1 \mid \mathcal{E}_2\} \geq 0 \tag{2.6-12}$$

is obviously satisfied since

$$\mathcal{P}\{\mathcal{E}_1 \mid \mathcal{E}_2\} = \frac{\mathcal{P}\{\mathcal{E}_1\mathcal{E}_2\}}{\mathcal{P}\{\mathcal{E}_2\}}$$

and the ratio of two positive numbers must surely be positive. The second axiom

$$\mathcal{P}\{\mathcal{S} \mid \mathcal{E}_2\} = 1 \tag{2.6-13}$$

follows by use of the conditional probability law

$$\mathcal{P}\{\mathcal{S} \mid \mathcal{E}_2\} = \frac{\mathcal{P}\{\mathcal{S}\mathcal{E}_2\}}{\mathcal{P}\{\mathcal{E}_2\}} = 1$$

since the intersection of the events $\mathcal{S}$ and $\mathcal{E}_2$ is just the event $\mathcal{E}_2$.

The third axiom for conditional probabilities,

$$\mathcal{P}\{\mathcal{E}_1 \cup \mathcal{E}_3 \mid \mathcal{E}_2\} = \mathcal{P}\{\mathcal{E}_1 \mid \mathcal{E}_2\} + \mathcal{P}\{\mathcal{E}_3 \mid \mathcal{E}_2\} \qquad \text{if } \mathcal{E}_1\mathcal{E}_3 = \varnothing$$

may also be shown to be valid by use of the conditional probability law, which yields

$$\mathcal{P}\{\mathcal{E}_1 \cup \mathcal{E}_3 \mid \mathcal{E}_2\} = \frac{\mathcal{P}\{[\mathcal{E}_1 \cup \mathcal{E}_3]\mathcal{E}_2\}}{\mathcal{P}\{\mathcal{E}_2\}} = \frac{\mathcal{P}\{\mathcal{E}_1\mathcal{E}_2 \cup \mathcal{E}_3\mathcal{E}_2\}}{\mathcal{P}\{\mathcal{E}_2\}}$$

Since $\mathcal{E}_1\mathcal{E}_2\mathcal{E}_3\mathcal{E}_2 = \varnothing$, we may use the third axiom for unconditional probabilities such that this expression simplifies to the desired result:

$$\begin{aligned}\mathcal{P}\{\mathcal{E}_1 \cup \mathcal{E}_3 \mid \mathcal{E}_2\} &= \frac{\mathcal{P}\{\mathcal{E}_1\mathcal{E}_2\} + \mathcal{P}\{\mathcal{E}_3\mathcal{E}_2\}}{\mathcal{P}\{\mathcal{E}_2\}} \\ &= \mathcal{P}\{\mathcal{E}_1 \mid \mathcal{E}_2\} + \mathcal{P}\{\mathcal{E}_3 \mid \mathcal{E}_2\}\end{aligned}$$

These results may be extended to the case of three or more events.

The results of this section are most important for our work to follow. We shall now illustrate the results of this section by several examples.

Example ***2.6-1.*** Consider the probabilities associated with obtaining various denominations in drawing cards from a 52-card deck. We may define 13 events as

$$\begin{array}{lll} \mathcal{A}_1 = \text{ace} & \mathcal{A}_5 = \text{five} & \mathcal{A}_9 = \text{nine} \\ \mathcal{A}_2 = \text{two} & \mathcal{A}_6 = \text{six} & \mathcal{A}_{10} = \text{ten} \\ \mathcal{A}_3 = \text{three} & \mathcal{A}_7 = \text{seven} & \mathcal{A}_{11} = \text{jack} \\ \mathcal{A}_4 = \text{four} & \mathcal{A}_8 = \text{eight} & \mathcal{A}_{12} = \text{queen} \\ & & \mathcal{A}_{13} = \text{king} \end{array}$$

All the $\mathcal{A}$'s are equally likely, and so we have

$$\mathcal{P}\{\mathcal{A}_i\} = \tfrac{1}{13} \qquad i = 1, 2, \ldots, 13$$

Also we have

$$\sum_{i=1}^{13} \mathcal{P}\{\mathcal{A}_i\} = 1$$

Alternatively, we could speak of the events of obtaining one of the four suits on a draw. These events will be defined by

$$\begin{array}{ll} \mathcal{B}_1 = \text{spade} & \mathcal{B}_3 = \text{diamond} \\ \mathcal{B}_2 = \text{heart} & \mathcal{B}_4 = \text{club} \end{array}$$

These events are also likely equal, so that

$$\mathcal{P}\{\mathcal{B}_i\} = \tfrac{1}{4} \qquad i = 1, 2, 3, 4$$

We may note the event $\mathcal{A}_3$ and $\mathcal{B}_2$ is the event of obtaining a three of hearts.

The joint probability is

$$\mathcal{P}\{\mathcal{A}_i\mathcal{B}_j\} = \tfrac{1}{52} \qquad i = 1, 2, \ldots, 13; j = 1, 2, 3, 4$$

and since $\mathcal{P}\{\mathcal{A}_i\mathcal{B}_j\} = \mathcal{P}\{\mathcal{A}_i\}\mathcal{P}\{\mathcal{B}_i\}$, the events $\mathcal{A}_i$ and $\mathcal{B}_j$ are independent.

To determine the probability of obtaining an ace or a heart for this problem, we observe that it is possible to obtain an ace and a heart. We then note that

$$\mathcal{P}\{\mathcal{A}_1\} = \tfrac{1}{13} \qquad \mathcal{P}\{\mathcal{B}_2\} = \tfrac{1}{4} \qquad \mathcal{P}\{\mathcal{A}_1\mathcal{B}_2\} = \tfrac{1}{52}$$

Eq. (2.3-8) then yields

$$\mathcal{P}\{\mathcal{A}_1 \cup \mathcal{B}_2\} = \tfrac{1}{13} + \tfrac{1}{4} - \tfrac{1}{52} = \tfrac{4}{13}$$

as the desired probability.

Example *2.6-2.* We shall illustrate the use of Bayes' theorem with a very simple example. Failure of an electronic circuit may occur because of overheating, wrong circuit components, or being dropped. We note that overheating, wrong components, and being dropped may occur without failure. The probability of failure due to more than one cause is assumed to be zero, or, in other words, the causes of failure are assumed to be mutually exclusive. We consider the following events:

$$\begin{aligned}
\mathcal{A} &= \text{event circuit fails} \\
\mathcal{B}_1 &= \text{event of overheating} \\
\mathcal{B}_2 &= \text{event of wrong circuit component} \\
\mathcal{B}_3 &= \text{event of dropping}
\end{aligned}$$

and that

$$\begin{aligned}
\mathcal{P}\{\mathcal{A} \mid \mathcal{B}_1\} &= 0.1 \qquad & \mathcal{P}\{\mathcal{B}_1\} &= 0.05 \\
\mathcal{P}\{\mathcal{A} \mid \mathcal{B}_2\} &= 0.5 \qquad & \mathcal{P}\{\mathcal{B}_2\} &= 0.001 \\
\mathcal{P}\{\mathcal{A} \mid \mathcal{B}_3\} &= 0.5 \qquad & \mathcal{P}\{\mathcal{B}_3\} &= 0.1
\end{aligned}$$

We can compute the probability of failure from any cause from Eqs. (2.6-6) and (2.6-7) as

$$\mathcal{P}\{\mathcal{A}\} = \sum_{j=1}^{3} \mathcal{P}\{\mathcal{A}\mathcal{B}_j\} = \sum_{j=1}^{3} \mathcal{P}\{\mathcal{A} \mid \mathcal{B}_j\}\mathcal{P}\{\mathcal{B}_j\} = 0.0555$$

We determine the probability of failure from specific causes from Bayes' rule, Eq. (2.6-7), which yields

$$\mathcal{P}\{\mathcal{B}_1 \mid \mathcal{A}\} = \frac{\mathcal{P}\{\mathcal{A} \mid \mathcal{B}_1\}\mathcal{P}\{\mathcal{B}_1\}}{\mathcal{P}\{\mathcal{A}\}} = \frac{0.005}{0.0555} \cong 0.09$$

$$\mathcal{P}\{\mathcal{B}_2 \mid \mathcal{A}\} = \frac{\mathcal{P}\{\mathcal{A} \mid \mathcal{B}_2\}\mathcal{P}\{\mathcal{B}_2\}}{\mathcal{P}\{\mathcal{A}\}} = \frac{0.0005}{0.0555} \cong 0.009$$

$$\mathcal{P}\{\mathcal{B}_3 \mid \mathcal{A}\} = \frac{\mathcal{P}\{\mathcal{A} \mid \mathcal{B}_3\}\mathcal{P}\{\mathcal{B}_3\}}{\mathcal{P}\{\mathcal{A}\}} = \frac{0.05}{0.0555} \cong 0.9$$

This result indicates that nine times out of ten when the electronic circuit fails, the circuit has been dropped.

Example 2.6-3. To illustrate the concept and methods of conditional probabilities, let us compute the probability that on a draw of 5 cards from an ordinary deck, exactly 2 cards will be aces. Let $\mathcal{A}_i$ be the event that we obtain an ace on the ith draw. The probability of obtaining an ordering of ace, ace, non-ace, non-ace, non-ace is

$$\begin{aligned}\mathcal{P}\{\mathcal{A}_1\mathcal{A}_2\mathcal{A}'_3\mathcal{A}'_4\mathcal{A}'_5\} &= \mathcal{P}\{\mathcal{A}_1\}\mathcal{P}\{\mathcal{A}_2 \mid \mathcal{A}_1\} \\ &\quad \cdot \mathcal{P}\{\mathcal{A}'_3 \mid \mathcal{A}_1\mathcal{A}_2\}\mathcal{P}\{\mathcal{A}'_4 \mid \mathcal{A}_1\mathcal{A}_2\mathcal{A}'_3\}\mathcal{P}\{\mathcal{A}'_5 \mid \mathcal{A}_1\mathcal{A}_2\mathcal{A}'_3\mathcal{A}'_4\}\end{aligned}$$

which is just the probability that the first card drawn is an ace times the probability that the second card is an ace given that the first card was an ace, and so forth. Thie relation is a direct extension of Eq. (2.6-8). We have for this specific problem

$$\mathcal{P}\{\mathcal{A}_1\} = \tfrac{4}{52} \qquad \mathcal{P}\{\mathcal{A}_2 \mid \mathcal{A}_1\} = \tfrac{3}{51}$$

$$\mathcal{P}\{\mathcal{A}'_3 \mid \mathcal{A}_1\mathcal{A}_2\} = \tfrac{48}{50} \qquad \mathcal{P}\{\mathcal{A}'_4 \mid \mathcal{A}_1\mathcal{A}_2\mathcal{A}'_3\} = \tfrac{47}{49}$$

$$\mathcal{P}\{\mathcal{A}'_5 \mid \mathcal{A}_1\mathcal{A}_2\mathcal{A}'_3\mathcal{A}'_4\} = \tfrac{46}{48}$$

So

$$\mathcal{P}\{\mathcal{A}_1\mathcal{A}_2\mathcal{A}'_3\mathcal{A}'_4\mathcal{A}'_5\} = \tfrac{4}{52}\cdot\tfrac{3}{51}\cdot\tfrac{48}{50}\cdot\tfrac{47}{49}\cdot\tfrac{46}{48} \approx 0.004$$

Now this is the probability of a specific ordering of two aces and three other cards. There are, from Eq. (2.5-5),

$$\binom{M}{N} = \binom{5}{2} = 10$$

ways of arranging the 5 cards. Thus the probability of obtaining 2 aces and 3 non-aces is

$$10\mathcal{P}\{\mathcal{A}_1\mathcal{A}_2\mathcal{A}'_3\mathcal{A}'_4\mathcal{A}'_5\} \cong 0.04$$

There are several ways of arriving at this same result. A most useful alternative way is to note that there are $\binom{52}{5}$ nonordered ways of obtaining 5 cards from a 52-card deck. There are $\binom{4}{2}$ nonordered ways of obtaining 2 aces from the 4 aces in the deck, and there are $\binom{48}{3}$ nonordered ways of obtaining the other 3 cards. The desired probability of obtaining 2 aces and 3 non-aces is therefore

$$\frac{\binom{4}{2}\binom{48}{3}}{\binom{52}{5}} \cong 0.04$$

If we had sampled with replacement, that is, replaced the card just obtained before taking the next card, the events would be independent and we would have

$$\mathcal{P}\{\mathcal{A}_1\mathcal{A}_2\mathcal{A}_3'\mathcal{A}_4'\mathcal{A}_5'\} = \mathcal{P}\{\mathcal{A}_1\}\mathcal{P}\{\mathcal{A}_2\}\mathcal{P}\{\mathcal{A}_3'\}\mathcal{P}\{\mathcal{A}_4'\}\mathcal{P}\{\mathcal{A}_5'\}$$

$$= \left(\frac{4}{52}\right)^2\left(\frac{48}{52}\right)^3 \approx 0.0047$$

The required probability is again 10 times this value, or 0.047. An alternative way of solving the problem is to note that there are 52^5 ways of obtaining 5 cards from the deck, 4^2 ways of obtaining the aces, 48^3 ways of obtaining the non-aces, and 10 ways of arranging these 5 cards. Thus the desired probability is

$$\frac{10(4)^2(48)^3}{(52)^5} \approx 0.047$$

In this section we have considered probabilistic systems in which there were a finite number of events and have discussed conditional probabilities and independence requirements.

EXERCISES 2.6

2.6-1. There are two urns (urn α and β). In urn α there are red balls only and in urn β there are an equal number of red (R) and white (W) balls. A coin toss is used to select an urn and a ball drawn from the urn is red. What is the probability that the urn contained red balls only?

Answer: $\frac{2}{3}$.

2.6-2. If a head results from the toss of a coin, a ball is drawn from urn 1. If a tail results from the toss of a coin, a ball is drawn from urn 2. Urn 1 contains five white and three black balls. Urn 2 contains two white and six black balls. What is the probability that the ball drawn will be white?

Answer: $\frac{7}{16}$.

2.6-3. A sample of size 2 is obtained as follows. A ball is selected at random from an urn containing three white and five black balls. That ball is returned to the urn together with a ball of the same color. Find the probability that the sample of size 2 will contain

(a) no white balls.

(b) exactly one white ball.

(c) two white balls.

Answers: (a) $\frac{15}{36}$, (b) $\frac{15}{36}$, (c) $\frac{1}{6}$.

2.6-4. Given $\mathcal{P}\{\mathcal{A}\} = \frac{1}{2}$, $\mathcal{P}\{\mathcal{B}\} = \frac{1}{3}$, and $\mathcal{P}\{\mathcal{A}\mathcal{B}\} = \frac{1}{4}$, find $\mathcal{P}\{\mathcal{A} \cup \mathcal{B}\}$, $\mathcal{P}\{\mathcal{A} \mid \mathcal{B}\}$, $\mathcal{P}\{\mathcal{B} \mid \mathcal{A}\}$, and $\mathcal{P}\{\mathcal{A} \cup \mathcal{B} \mid \mathcal{B}\}$.

Answers: $\mathcal{P}\{\mathcal{A} \cup \mathcal{B}\} = \frac{7}{12}$, $\mathcal{P}\{\mathcal{A} \mid \mathcal{B}\} = \frac{3}{4}$, $\mathcal{P}\{\mathcal{B} \mid \mathcal{A}\} = \frac{1}{2}$, $\mathcal{P}\{\mathcal{A} \cup \mathcal{B} \mid \mathcal{B}\} = 1$.

2.7. Conclusions

We have examined in this chapter a very old subject, that of the theory of probability. As might be guessed, the subject had its origins in games of chance and many of our examples were taken from this area. Although we have not cited many applications to engineering problems of interest, the material in this chapter is prerequisite to the later application-oriented material. There are numerous references for the subjects of this chapter; only textbooks will be cited here. Those by Feller and Parzen are especially illuminating for their treatment of basic probability. The book by Drake is particularly easy to read and is recommended for this reason.

2.8. References

(1) Drake, A. W., *Fundamentals of Applied Probability Theory*, McGraw-Hill, New York, 1967.

(2) Dubes, R. C., *The Theory of Applied Probability*, Prentice-Hall, Englewood Cliffs, N.J., 1968.

(3) Eisen, M., *Introduction to Mathematical Probability Theory*, McGraw-Hill, New York, 1969.

(4) Feller, William, *An Introduction to Probability Theory and Its Applications*, Wiley, New York, Vol. I, 1957, Vol. II, 1966.

(5) Mood, A. M., and F. A. Graybill, *Introduction to the Theory of Statistics*, 2nd ed., McGraw-Hill, New York, 1963.

(6) Papoulis, Athanasios, *Probability, Random Variables and Stochastic Processes*, McGraw-Hill, New York, 1964.

(7) Parzen, Emanuel, *Modern Probability Theory and Its Applications*, Wiley, New York, 1960.

(8) Thomasian, A. J., *The Structure of Probability Theory with Applications*, McGraw-Hill, New York, 1969.

(9) Tucker, H. G., *A Graduate Course in Probability*, Academic Press, New York, 1967.

(10) USPENSKY, J. V., *Introduction to Mathematical Probability*, McGraw-Hill, New York, 1937.

2.9. *Problems*

2.9-1. Sketch the sets defined by

(a) $\mathcal{E}_1 = \{x: |x| < 1\}$
(b) $\mathcal{E}_2 = \{x: |x| \leq 1\}$
(c) $\mathcal{E}_3 = \mathcal{E}_1'\mathcal{E}_2$
(d) $\mathcal{E}_4 = \{x: \frac{1}{2} \leq x \leq 2\}$
(e) $\mathcal{E}_5 = \mathcal{E}_1\mathcal{E}_4$
(f) $\mathcal{E}_6 = \mathcal{E}_1'\mathcal{E}_4$
(g) $\mathcal{E}_7 = \mathcal{E}_1 \cup \mathcal{E}_4$
(h) $\mathcal{E}_8 = \mathcal{E}_1' \cup \mathcal{E}_4$

2.9-2. Two sets are defined as

$$\mathcal{E}_1 = \{(x, y): x > y, \quad y > 0\}$$
$$\mathcal{E}_2 = \{(x, y): x + y \leq 2\}$$

Show the sets defined by

(a) $\mathcal{E}_3 = \mathcal{E}_1\mathcal{E}_2$
(b) $\mathcal{E}_4 = \mathcal{E}_1 \cup \mathcal{E}_2$
(c) $\mathcal{E}_5 = \mathcal{E}_1 \cup \mathcal{E}_2'$

2.9-3. Find simplified expressions for

(a) $[\mathcal{E}_1 \cup \mathcal{E}_2][\mathcal{E}_1 \cup \mathcal{E}_2']$
(b) $[\mathcal{E}_1 \cup \mathcal{E}_2][\mathcal{E}_2 \cup \mathcal{E}_3]$
(c) $\mathcal{E}_1\mathcal{E}_3'\mathcal{E}_4' + \mathcal{E}_1\mathcal{E}_3'\mathcal{E}_4 + \mathcal{E}_1\mathcal{E}_2\mathcal{E}_3$

2.9-4. Show which of the following relations are correct and which are incorrect.

(a) $[\mathcal{E}_1 \cup \mathcal{E}_2]' = \mathcal{E}_1'\mathcal{E}_2'$
(b) $[\mathcal{E}_1 \cup \mathcal{E}_2] - \mathcal{E}_2 = \mathcal{E}_1\mathcal{E}_2'$
(c) $\mathcal{E}_1\mathcal{E}_2\mathcal{E}_3 = \mathcal{E}_1\mathcal{E}_2[\mathcal{E}_2 \cup \mathcal{E}_3]$

2.9-5. For a finite number of events, $\mathcal{E}_1, \mathcal{E}_2, \ldots, \mathcal{E}_N$, show that

$$\mathcal{P}\{\mathcal{E}_1 \cup \mathcal{E}_2 \cup \cdots \mathcal{E}_N\} \leq \sum_{i=1}^{N} \mathcal{P}(\mathcal{E}_i)$$

2.9-6. Show that if $\mathcal{E}_1$ and $\mathcal{E}_2$ are events, then

$$\mathcal{P}\{\mathcal{E}_1\mathcal{E}_2' \cup \mathcal{E}_2\mathcal{E}_1'\} = \mathcal{P}\{\mathcal{E}_1\} + \mathcal{P}\{\mathcal{E}_2\} - 2\mathcal{P}\{\mathcal{E}_1\mathcal{E}_2\}$$

2.9-7. Show that for $\mathcal{E}_1$ and $\mathcal{E}_2$

$$\mathcal{P}\{\mathcal{E}_1\mathcal{E}_2\} \leq \mathcal{P}\{\mathcal{E}_2\} \leq \mathcal{P}\{\mathcal{E}_1 \cup \mathcal{E}_2\} \leq \mathcal{P}\{\mathcal{E}_1\} + \mathcal{P}\{\mathcal{E}_2\}$$

2.9-8. Prove that

$$\begin{aligned}\mathcal{P}\{\mathcal{E}_1 \cup \mathcal{E}_2 \cup \mathcal{E}_3\} = {} & \mathcal{P}\{\mathcal{E}_1\} + \mathcal{P}\{\mathcal{E}_2\} + \mathcal{P}\{\mathcal{E}_3\} \\ & - \mathcal{P}\{\mathcal{E}_1\mathcal{E}_2\} - \mathcal{P}\{\mathcal{E}_1\mathcal{E}_3\} - \mathcal{P}\{\mathcal{E}_2\mathcal{E}_3\} \\ & + \mathcal{P}\{\mathcal{E}_1\mathcal{E}_2\mathcal{E}_3\}\end{aligned}$$

using the method of Example 2.3-4.

2.9-9. Show that

(a) $\binom{M}{N-1} + \binom{M}{N} = \binom{M+1}{N}$

(b) $\binom{M}{N} = \binom{M}{M-N}$

2.9-10. How many four-letter words are there in the English language in which the first, third, and fourth letters are consonants and the second letter is a vowel?

2.9-11. Every one of 100 objects is tested and found defective or nondefective. How many outcomes are there?

2.9-12. Three balls are drawn from an urn containing 5 while balls and 3 black balls. Find the probability that
(a) all balls are white.
(b) all balls have the same color.
(c) at least 1 ball will be white.
Consider sampling with and without replacement.

2.9-13. Two numbers are chosen at random from the telephone book. What is the probability that the last digit of each will be different?

2.9-14. A family has two children. Each child is as likely to be a boy as it is a girl. What is the conditional probability that both children are girls given that (a) the oldest child is a girl and (b) at least one child is a girl.

2.9-15. A sample of size 5 is drawn with replacement from an urn containing 8 white balls and 5 black balls. Find the conditional probability that the ball drawn on the fourth draw was black, given that the sample contains 3 black balls.

2.9-16. Repeat Problem 2.9-15 for the case in which the sampling is without replacement.

2.9-17. Show that $\mathcal{P}\{\mathcal{E}_1\mathcal{E}_2 \mid \mathcal{E}_2\} = \mathcal{P}\{\mathcal{E}_1 \mid \mathcal{E}_2\}$.

2.9-18. Show that $\mathcal{P}\{\mathcal{E}_1 \mid \mathcal{E}_2\mathcal{E}_3\} = \mathcal{P}\{\mathcal{E}_1\mathcal{E}_2\mathcal{E}_3\}/\mathcal{P}\{\mathcal{E}_2\mathcal{E}_3\}$.

2.9-19. What are the independence requirements for four events $\mathcal{E}_1$, $\mathcal{E}_2$, $\mathcal{E}_3$, and $\mathcal{E}_4$?

2.9-20. Show that

$$\mathcal{P}\{\mathcal{E}_1\mathcal{E}_2\mathcal{E}_3\mathcal{E}_4\} = \mathcal{P}\{\mathcal{E}_1 \mid \mathcal{E}_2\mathcal{E}_3\mathcal{E}_4\}\mathcal{P}\{\mathcal{E}_2 \mid \mathcal{E}_3\mathcal{E}_4\}\mathcal{P}\{\mathcal{E}_3 \mid \mathcal{E}_4\}\mathcal{P}\{\mathcal{E}_4\}$$

2.9-21. Two dice are tossed. What is the probability of
(a) obtaining a sum of 7?
(b) obtaining no 1?
(c) obtaining one 1?
(d) obtaining at least one 1?

2.9-22. $\mathcal{A}$ is the set of positive even integers.
$\mathcal{B}$ is the set of positive odd integers.
$\mathcal{C}$ is the set of positive integers divisible by 3.
Give a description of the following sets:
(a) $\mathcal{A} \cup \mathcal{B}$

(b) $\mathcal{A}\mathcal{B}$
(c) $\mathcal{A}\mathcal{C}$
(d) $\mathcal{A} \cup \mathcal{B}\mathcal{C}$

2.9-23. A message is coded in a two-symbol (binary) code. The probability of transmission of the symbol α is 0.48, and the probability of transmission of β is 0.52. In the communication channel the symbol α is distorted into β with probability 0.1. The β's are distorted into α with probability 0.2. What is the probability that a received α has not been distorted? What is the probability that a received β has not been distorted?

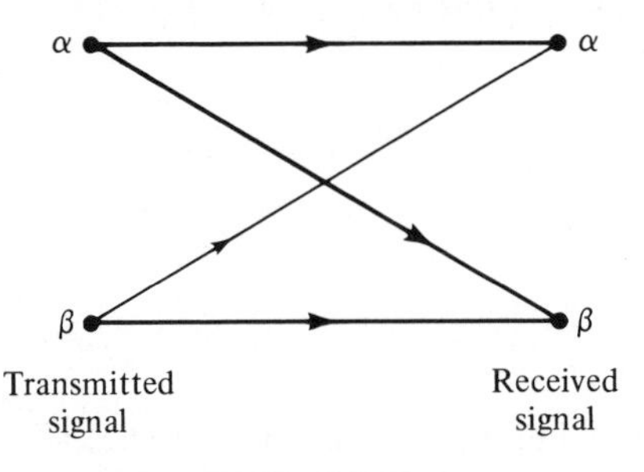

PROB. 2.9-23

2.9-24. Two balls are drawn without replacement from an urn containing 6 balls of which 4 are black and 2 are white. What is the probability that
(a) both balls are white?
(b) both balls have the same color?
(c) at least 1 ball is white?

2.9-25. A man tosses two fair coins. What is the conditional probability of tossing two heads, given that he has tossed at least one head.

2.9-26. It is known that 10 per cent of the girls on campus are blondes. You are standing in the student union watching the girls go by.
(a) What is the probability that out of the first five girls that pass, at least one will be a blonde?
(b) What is the probability that out of the first five girls that pass, exactly one will be a blonde?
(c) What is the probability that exactly three nonblondes pass before you see the first blonde?
(d) What is the probability that the third girl who passes is not a blonde?

3. *Random Variables*

3.1. *Introduction*

In the previous chapter we introduced the basic concepts of probability theory. Unfortunately, the theory in the form presented in Chapter 2, although quite general, is not easy to apply. The basic goal of this chapter is to translate these general concepts into a more easily handled mathematical model: the random variable and its associated probability functions. Once this is done, a vast array of analysis tools may be brought to bear on probabilistic problems.

3.2. *Definition of random variable*

In simple terms, a random variable is a function x that assigns to each outcome ξ of an experiment a real number $x(\xi)$. There are some mild restrictions that the function must satisfy to be allowable; however, before elaborating on these restrictions, let us consider some simple examples to illustrate the concept of a random variable.

> **Example** ***3.2-1.*** Suppose that the experiment is a single roll of a die and that the outcome is the face which is on the top. Here the sample space consists of six elements, $\xi_1, \xi_2, \ldots, \xi_6$, where ξ_i implies that face i was the outcome. In this case, there is a simple, obvious way in which we might define a random variable for this experiment: we let the value of the random variable be equal to the value of the face

that is on top so that

$$x(\xi_i) = i \qquad i = 1, 2, \ldots, 6$$

Now we have established a rule that assigns to each outcome, which is *not* a number, a real number.

Example 3.2-2. Now let us consider the simple coin-toss experiment with $\mathcal{S} = \{T, H\}$. Let us assign to the outcome tail (T) the number 0 and to head (H) the value 1. In this case, our random variable takes on the two values 0, 1. An equally valid approach would be to define a random variable that had the value 0 for H and 1 for T.

Now suppose that we toss two coins so that the sample space has four elements $\{T, T\}$, $\{T, H\}$, $\{H, T\}$, and $\{H, H\}$. The simplest way to represent the random variable in this case might be through a table of the form

Outcome	Value of Random Variable
T, T	0
T, H	1
H, T	3
H, H	−2

The point to be emphasized here is that the random variable is the table of relations and not the numbers 0, 1, 3, and −2.

Example 3.2-3. To illustrate that more than one random variable may be defined for a given experiment, let us examine the die experiment of Example 3.2-1 again. Now let us suppose that we define a random variable by the rule

$$x(\xi_i) = \begin{cases} 1 & \text{for } i = 1, 3, 5 \\ -1 & \text{for } i = 2, 4, 6 \end{cases}$$

Here we note that several different outcomes may be assigned the same real number, but that each outcome is assigned only one number.

With these simple examples as background we are ready to return to the problem of developing a more specific definition of a random variable. As a first step, we shall briefly review some concepts and definitions associated with a function.[1]

[1] We shall need these concepts again when we discuss functions of random variables in Chapter 4.

Functions. In its most general terms, a *function* is a mapping or rule of correspondence between the elements of one set and the elements of another set. More specifically, a function consists of three objects: two non-empty sets $\mathfrak{X}$ and $\mathfrak{Y}$ (which may or may not be equal) and a rule that assigns to each element $x \in \mathfrak{X}$ a single unique element $y \in \mathfrak{Y}$. This statement is expressed concisely by the symbolism $f: \mathfrak{X} \to \mathfrak{Y}$. The $y \in \mathfrak{Y}$, which corresponds to a given $x \in \mathfrak{X}$, is usually written as $y = f(x)$ and is called the *value* of f at x. The reader is cautioned to distinguish between the function (rule of correspondence) f and the value of f for a given x. This distinction, which is often not emphasized, is important for proper understanding of the concept of a random variable.

The set $\mathfrak{X}$ is called the *domain* of the function, and $\mathfrak{Y}$ is often referred to as the *codomain*. The set $\{f(x): x \in \mathfrak{X}\}$ is called the *range* of the function and is contained in $\mathfrak{Y}$. We note that the range space (set) is not necessarily equal to the codomain space; in fact, it is possible that all elements in $\mathfrak{X}$ map into a single element in $\mathfrak{Y}$. In other words, we allow a "many-to-one" mapping, but do, however, require that each element in $\mathfrak{X}$ map to only one element in $\mathfrak{Y}$ (often referred to as *single-valued* functions). We emphasize the fact that the function may be a "many-to-one" mapping by saying that f maps $\mathfrak{X}$ *into* $\mathfrak{Y}$. If we wish to restrict consideration to the case in which the range of f equals $\mathfrak{Y}$, we say f maps $\mathfrak{X}$ *onto* $\mathfrak{Y}$; that is, for every $y \in \mathfrak{Y}$ there exists one or more $x \in \mathfrak{X}$ such that $f(x) = y$.

Even if f is "onto," it may still be possible that many elements in $\mathfrak{X}$ map to the same element in $\mathfrak{Y}$, for example, when $\mathfrak{X}$ has more elements than $\mathfrak{Y}$. When the function is such that different elements in $\mathfrak{X}$ always map to different elements in $\mathfrak{Y}$, we say that the function is *one to one*. Note that a function does not need to be onto to be one to one. If $\mathfrak{Y}$ has more elements than $\mathfrak{X}$, no function that maps $\mathfrak{X}$ into $\mathfrak{Y}$ can be onto. When f is both one to one and onto, we say that it is *invertible*, meaning that an *inverse mapping* f^{-1} exists which maps $\mathfrak{Y}$ onto $\mathfrak{X}$ in the sense that for every $y \in \mathfrak{Y}$, there exists a *unique* element $x \in \mathfrak{X}$ such that $f(x) = y$; we define this x as $f^{-1}(y)$.

Random variables, therefore, are a special class of functions that have the sample space as their domain, the real line $\mathfrak{R}$ as codomain, and a subset of the real line as the range. Note that, in general, random variables are neither onto nor one to one. In Example 3.2-1 the mapping is one to one since each of the six outcomes map to distinct values, but the mapping is not onto since the range is the set $\{1, 2, \ldots, 6\}$, which is obviously not the real line. In Example 3.2-3 the random variable is neither one to one nor onto.

Notation. We shall use a symbol, x or y for example, without an argument to indicate a random variable, that is, the mapping or function. The symbol $x(\xi)$, on the other hand, will be used to represent the value of the random variable for a specific outcome ξ. One primary objective of this chapter is to associate the probabilistic structure developed in the preceding chapter with probabilistic measures defined in terms of random variables. To do this conveniently, we have need of some shorthand notation. We shall use the notation $\{x \leq \alpha\}$ to represent the set of outcomes ξ such that $x(\xi) \leq \alpha$. In other words,

$$\{x \leq \alpha\} = \{\xi \in \mathcal{S}: x(\xi) \leq \alpha\} \tag{3.2-1}$$

In terms of the random variable of Example 3.2-1, we would have $\{x \leq 2\} = \{\xi_1, \xi_2\}$, whereas $\{x \leq 0\} = \varnothing$, since there are no events for which $x(\xi) \leq 0$.

In a similar way, we define the set $\{\alpha \leq x \leq \beta\}$ by

$$\{\alpha \leq x \leq \beta\} = \{\xi \in \mathcal{S}: \alpha \leq x(\xi) \leq \beta\}$$

In general, if $\mathcal{A}$ is any subset of the real line $\mathcal{R}$, then

$$\{x \in \mathcal{A}\} = \{\xi \in \mathcal{S}: x(\xi) \in \mathcal{A}\} \tag{3.2-2}$$

It must be emphasized that $\{x \in \mathcal{A}\}$ is a set of *outcomes* from the sample space and not a set of *numbers* from $\mathcal{R}$. This is a very important distinction!

To establish a probabilistic foundation for random variables, it is necessary to relate statements concerning the random variable to *events* from the sigma field $\mathfrak{F}$ associated with the probability space rather than just outcomes. It is necessary that the statements be related to events because, as we recall from the preceding chapter, probability is only defined for events. To be more specific, we must be certain that the subset of the sample space defined by $\{x \in \mathcal{A}\}$ for any $\mathcal{A}$ of interest is an event from $\mathfrak{F}$. For simplicity, we shall restrict attention to those cases in which $\mathcal{A}$ is any set from $\mathcal{R}$ that can be written as a countable union or intersection of half-open intervals of the form $(\alpha, \beta]$; that is, $\alpha < x \leq \beta$. We shall refer to any set of this nature as *admissible*.[1] Included in this class of sets are the sets $x \leq \alpha$, $\alpha < x \leq \beta$, $x = \alpha$, and all countable unions and intersections of these sets.

Since by definition any admissible set may be written as a countable sum or product of sets of the form $x \leq \alpha$, it is only necessary that the random variable be defined such that the sets $\{x \leq \alpha\}$ are events for all α. Because the set of events form a sigma field, $\{x \in \mathcal{A}\}$ is an event for all admissible sets $\mathcal{A}$ if the set $\{x \leq \alpha\}$ is an event for all α; this is a restriction that our definition of the random variable function must reflect.

[1] This restriction is not essential but does considerably simplify the definition of a random variable. In terms of practical application, this restriction is of no consequence since all realistic applications would involve events which could be written as admissible sets.

Lest the preceding discussion appear to be nothing more than mathematical abstraction, let us consider a simple example to illustrate the difficulty. Consider the random variable of Example 3.2-1 defined for the die-rolling experiment; however, we wish to assume that the sigma field of events contains only the four events $\varnothing$, $\mathcal{S}$, $\{\xi_i : i \text{ even}\}$, and $\{\xi_i : i \text{ odd}\}$. If we examine the subset of $\mathcal{S}$ defined by $\{x \leq 2\} = \{\xi_1, \xi_2\}$, we see that this subset is *not* one of our four events, and hence there is no way to relate the probability measure to this subset. Now let us suppose that for this probability space, the random variable is defined as in Example 3.2-3. In this case, we see that the subsets $\{x \leq \alpha\}$ are always events, and hence a probability measure may be related to this random variable.

The basic difficulty which this discussion points out is that the probability measure is defined for events, whereas the random variable is defined in terms of outcomes. Only if each outcome is also an event[1] is there any assurance that an arbitrarily defined random variable will lead to admissible sets that are always events.

There is one other restriction that we must place on our definition of a random variable. The subsets $\{x = +\infty\}$ and $\{x = -\infty\}$ are events and, in general, are not necessarily equal to the null space $\varnothing$. We shall however require that the probability of these events be zero. This restriction will considerably simplify our later developments, since it will eliminate certain troublesome properties at $\pm\infty$ and the associated special consideration of these points.

We are now prepared to make a rigorous definition of a random variable.

Definition 3.2-1. ***Random Variable:*** A (real-valued) random variable associated with a probability space $[\mathcal{S}, \mathcal{F}, \mathcal{P}]$ is any function x that maps the sample space $\mathcal{S}$ into the real line $\mathcal{R}$, that is, assigns a real number $x(\xi)$ to each outcome $\xi \in \mathcal{S}$, such that (1) the set $\{x \leq \alpha\}$ is an event in $\mathcal{F}$ for every $\alpha \in \mathcal{R}$, and (2) $\mathcal{P}\{x = +\infty\} = \mathcal{P}\{x = -\infty\} = 0$.

As we have noted earlier, it is quite possible to define more than one random variable for a given experiment. In general, we might define a random vector by assigning a vector $\mathbf{x}(\xi)$ from the n-dimensional Euclidean space $\mathcal{R}^n$ to each outcome $\xi \in \mathcal{S}$. In other words, an n-dimensional random vector $\mathbf{x}$ maps $\mathcal{S}$ into $\mathcal{R}^n$. Here a boldfaced symbol, such as $\mathbf{x}$, is used to represent a vector; subscripted symbols such as $x_1, x_2, \ldots$ will be used to represent the elements of $\mathbf{x}$.

[1] This is almost always the case in practice, and hence the discussion is somewhat academic. There are treatments of probability in which it is automatically assumed that each outcome is an event; this approach is restrictive and leads to cumbersome efforts when there is an uncountable number of outcomes.

We shall use the notation $\{\mathbf{x} \in \mathcal{A}\}$, where $\mathcal{A} \subseteq \mathcal{R}^n$, to indicate the set of outcomes such that $\mathbf{x}(\xi) \in \mathcal{A}$; that is,

$$\{\mathbf{x} \in \mathcal{A}\} = \{\xi : \mathbf{x}(\xi) \in \mathcal{A}\} \tag{3.2-3}$$

As before, we wish to define $\mathbf{x}$ such that we may find a probability measure for it. This may be done if $\{\mathbf{x} \leq \boldsymbol{\alpha}\}$ is an event from $\mathcal{F}$ for every $\boldsymbol{\alpha} \in \mathcal{R}^n$. Here we have employed the notation $\mathbf{x} \leq \boldsymbol{\alpha}$ to imply that the set of inequalities $x_1 \leq \alpha_1, x_2 \leq \alpha_2, \ldots, x_n \leq \alpha_n$, must be simultaneously satisfied. Once again we require that the probability of the events[1] $\{\mathbf{x} = -\infty\}$ and $\{\mathbf{x} = +\infty\}$ be zero. Our formal definition of a random vector[2] therefore takes the following form:

> **Definition 3.2-2.** ***Random Vector:*** A real-valued random vector associated with $[\mathcal{S}, \mathcal{F}, \mathcal{P}]$ is any vector function $\mathbf{x}$ which maps $\mathcal{S}$ into $\mathcal{R}^n$ such that (1) $\{\mathbf{x} \leq \boldsymbol{\alpha}\} \in \mathcal{F}$ for all $\boldsymbol{\alpha} \in \mathcal{R}^n$, and (2) $\mathcal{P}\{\mathbf{x} = -\infty\} = \mathcal{P}\{\mathbf{x} = +\infty\} = 0$.

When we, on occasion, discuss several random variables or vectors, we shall assume that they are defined on the same probability space $[\mathcal{S}, \mathcal{F}, \mathcal{P}]$ unless noted otherwise.

The next step in our investigation of random variables is to develop a probability measure for random variables. This is the subject of the next two sections.

EXERCISES 3.2

3.2-1. Suppose that sample space $\mathcal{S} = \{\xi_1, \xi_2, \xi_3, \xi_4\}$. Find the remaining elements in the field that contains $\{\xi_1\}$ and $\{\xi_3, \xi_4\}$.
Answers: $\mathcal{S}$, $\varnothing$, $\{\xi_1, \xi_3, \xi_4\}$, $\{\xi_2, \xi_3, \xi_4\}$, $\{\xi_1, \xi_2\}$, and $\{\xi_2\}$.

3.2-2. A coin is tossed three times. Suppose that we are only interested in the event that there were exactly two heads. Define an appropriate random variable.

3.2-3. We have a sample space that consists of $\mathcal{S} = \{\xi_1, \xi_2, \xi_3, \xi_4\}$ and a function x that maps $\mathcal{S}$ into $\mathcal{R}$ by the rule $x(\xi_1) = x(\xi_2) = 0$ and $x(\xi_3) = x(\xi_4) = 1$. For which of the following fields is x a random variable?
(a) $\varnothing$, $\mathcal{S}$, $\{\xi_1, \xi_2\}$, $\{\xi_3, \xi_4\}$.
(b) the set of all subsets of $\mathcal{S}$.

[1] By $\mathbf{x} = \pm\infty$, we mean $x_i = \infty$ for some i.

[2] We shall use the term "random variables" to loosely refer to both random variables (scalars) and random vectors, except when the distinction is important. We will however continue to use boldface notation for random vectors.

(c) $\varnothing$, $\mathcal{S}$, $\{\xi_1, \xi_3\}$, $\{\xi_2, \xi_4\}$.
(d) $\varnothing$, $\mathcal{S}$, $\{\xi_1, \xi_4\}$, $\{\xi_2, \xi_3\}$.

Answer: (a), (b)

3.2-4. Show that the only random variables which may be defined for the field $\mathcal{F} = \{\mathcal{S}, \varnothing\}$ are constants.

3.2-5. Show that the only random variables which may be defined for the field $\mathcal{F} = \{\mathcal{S}, \varnothing, \mathcal{A}, \mathcal{A}'\}$ are of the form $x(\xi) = c_1$ for $\xi \in \mathcal{A}$ and $x(\xi) = c_2$ for $\xi \in \mathcal{A}'$, where c_1 and c_2 are real constants.

3.3. Distribution function

As noted in the preceding section, any admissible set, that is, any set which is a countable union or intersection of half-open intervals $(a, b]$, can be formed from a countable union or intersection of semi-infinite interval $x \leq \alpha$. Hence sets of the form $\{x \leq \alpha\}$ play a fundamental role, and we begin by defining a probability measure for sets of this form. By the definition of a random variable, the set $\{x \leq \alpha\}$ is an event for all $\alpha \in \mathcal{R}$, and hence $\mathcal{P}\{x \leq \alpha\}$ is a well-defined number for all $\alpha \in \mathcal{R}$.

Definition 3.3-1. ***Distribution Function:*** The distribution function F_x of a random variable x is defined by

$$F_x(\alpha) = \mathcal{P}\{x \leq \alpha\} = \mathcal{P}\{\xi \in \mathcal{S}: x(\xi) \leq \alpha\} \tag{3.3-1}$$

for all $\alpha \in \mathcal{R}$.

In simple terms, $F_x(\alpha)$ is the probability that the random variable x will take a value $x(\xi)$ which is less than or equal to α. Often we shall just say $F_x(\alpha)$ is the probability that $x \leq \alpha$.

Note that F_x is a mapping from $\mathcal{R}$ into $\mathcal{R}$,[1] in which the rule of correspondence is determined by reference to events from $\mathcal{F}$ and their associate probability measure $\mathcal{P}$. In some cases, it will be notationally convenient to write F_x as $F(x)$; this notation must be interpreted as the distribution function of the random variable x and not as the probability that $x \leq x$, which is meaningless. Before considering some examples, let us examine a few of the properties of distribution functions.

Properties of Distribution Functions. We wish to show that distribution functions have four basic properties:

[1] Later it will be shown that the range of F_x is the closed unit interval [0, 1].

1. $F_x(+\infty) = 1$.
2. $F_x(-\infty) = 0$.
3. F_x is a nondecreasing function; that is, $F_x(\alpha_1) \leq F_x(\alpha_2)$ for $\alpha_1 \leq \alpha_2$.
4. F_x is continuous from the right in the sense that[1] $\lim_{\alpha \downarrow \alpha_0} F_x(\alpha) = F_x(\alpha_0^+) = F_x(\alpha_0)$.

Each of these properties is easily established by the use of Definition 3.3-1 and the basic properties of probability spaces, as shown below.

It is easy to see that the first and second properties must be true. By definition, $F_x(+\infty) = \mathcal{P}\{x \leq +\infty\}$, and for every outcome $x(\xi) \leq +\infty$, we see that $F_x(+\infty) = \mathcal{P}\{\mathcal{S}\} = 1$. On the other hand, $F_x(-\infty) = \mathcal{P}\{x \leq -\infty\} = \mathcal{P}\{x = -\infty\}$, since there is no outcome for $x(\xi) < -\infty$. From the definition of a random variable we know that $\mathcal{P}\{x = -\infty\} = 0$, so $F_x(-\infty) = 0$.

To establish the third property, we use Definition 3.3-1 to write

$$F_x(\alpha_2) = \mathcal{P}\{x \leq \alpha_2\} = \mathcal{P}\{\{x \leq \alpha_1\} \cup \{\alpha_1 < x \leq \alpha_2\}\}$$

Since the events $\{x \leq \alpha_1\}$ and $\{\alpha_1 < x \leq \alpha_2\}$ are disjoint, the use of Axiom III from the probability-space definition yields

$$F_x(\alpha_2) = \mathcal{P}\{x \leq \alpha_1\} + \mathcal{P}\{\alpha_1 < x \leq \alpha_2\}$$

but $\mathcal{P}\{x \leq \alpha_1\} = F_x(\alpha_1)$ by definition, so we have

$$F_x(\alpha_2) - F_x(\alpha_1) = \mathcal{P}\{\alpha_1 < x \leq \alpha_2\} \tag{3.3-2}$$

This result is quite useful in itself, and we shall return to it later. The application of Axiom I (nonnegativity of the probability measure) to Eq. (3.3-2) gives the desired rssult:

$$F_x(\alpha_2) - F_x(\alpha_1) \geq 0 \qquad \text{for } \alpha_2 \geq \alpha_1 \tag{3.3-3}$$

The combination of Properties 1 through 3 indicates that the distribution function starts at zero and increases monotonically to 1. Thus the range of the distribution function is the closed unit interval [0, 1].

The proof of the fact that the distribution function is continuous from the right (Property 4) follows almost directly from the definition and Property 3. Because the distribution function is nondecreasing on $[-\infty, \infty]$, it is known from advanced calculus that the unilateral limits

$$F_x(\alpha_0^+) \triangleq \lim_{\alpha \downarrow \alpha_0} F_x(\alpha) = \lim_{\alpha \downarrow \alpha_0} \mathcal{P}\{x \leq \alpha\} \tag{3.3-4}$$

[1] The notation $\lim_{\alpha \downarrow \alpha_0} F_x(\alpha)$ is used to indicate that α approaches α_0 from above; that is, $\alpha - \alpha_0 > 0$.

and

$$F_x(\alpha_0^-) \triangleq \lim_{\alpha \uparrow \alpha_0} F_x(\alpha) = \lim_{\alpha \downarrow \alpha_0} \mathcal{P}\{x \leq \alpha\} \tag{3.3-5}$$

exist for all α_0 in $(-\infty, +\infty)$. If $F_x(\alpha_0^+) = F_x(\alpha_0^-)$, then F_x is continuous at α_0 and hence is obviously continuous from the right. If, on the other hand, $F_x(\alpha_0^+) \neq F_x(\alpha_0^-)$, there exists a jump[1] at α_0, and F_x is not continuous at α_0. However, we wish to establish that F_x is still continuous from the right at α_0; that is, $F_x(\alpha_0) = F_x(\alpha_0^+)$.

Let us consider two points α_1 and α_2 such that $\alpha_1 < \alpha_0 < \alpha_2$; then using Eq. (3.3-2) we may write

$$\mathcal{P}\{\alpha_1 < x \leq \alpha_2\} = F_x(\alpha_2) - F_x(\alpha_1)$$

Now we shall let α_1 approach α_0 from below and α_2 approach α_0 from above so that we have

$$\lim_{\substack{\alpha_1 \uparrow \alpha_0 \\ \alpha_2 \downarrow \alpha_0}} \mathcal{P}\{\alpha_1 < x \leq \alpha_2\} = \lim_{\alpha_2 \downarrow \alpha_0} F_x(\alpha_2) - \lim_{\alpha_1 \uparrow \alpha_0} F_x(\alpha_1)$$

The limit on the left side of this equation becomes just $\mathcal{P}\{x = \alpha_0\}$, whereas the limits on the right side are just the upper and lower limits of F_x at α_0, so we have

$$\mathcal{P}\{x = \alpha_0\} = F_x(\alpha_0^+) - F_x(\alpha_0^-) \tag{3.3-6}$$

To complete our proof, let us write $F_x(\alpha_0)$ as

$$F_x(\alpha_0) \triangleq \mathcal{P}\{x \leq \alpha_0\} = \mathcal{P}\{x \leq \alpha\} + \mathcal{P}\{\alpha < x \leq \alpha_0\} \tag{3.3-7}$$

If we take the limit as α approaches α_0 from below, Eq. (3.3-7) becomes simply

$$F_x(\alpha_0) = F_x(\alpha_0^-) + \mathcal{P}\{x = \alpha_0\} = F_x(\alpha_0^+) \tag{3.3-8}$$

and F_x is thus continuous from the right at α_0, which is the desired result.

There are several other properties that we could establish for distribution functions. For example, it can be shown that the set of discontinuities of F_x must be countable. We shall not prove this statement nor state any of the other properties since they are beyond the scope of interest here. The interested reader is directed to the literature for a more extensive treatment.

[1] We say that a function f has a *jump* at x if and only if $f(x^+)$ and $f(x^-)$ exist but are unequal. Because of the above discussion, we see that the only type of discontinuity which a distribution function may have is a jump.

Before continuing the development of distribution functions, let us pause briefly to examine some simple examples that illustrate these concepts.

Example *3.3-1.* We consider first the die-rolling experiment of Example 3.2-1. There the random variable x was defined by

$$x(\xi_i) = i \qquad i = 1, 2, \ldots, 6$$

Now let us assume that each outcome is an event and that $\mathcal{P}\{\xi_i\} = \frac{1}{6}$ for $i = 1, 2, \ldots, 6$.

The distribution function $F_x(\alpha)$ is zero for $\alpha < 1$, since $\{x < 1\} = \varnothing$, $F_x(\alpha) = \mathcal{P}\{x \leq \alpha\} = \mathcal{P}\{\varnothing\}$, for $\alpha < 1$. If α is greater than or equal to 1 but less than 2, we have

$$F_x(\alpha) = \mathcal{P}\{x \leq \alpha\} = \mathcal{P}\{\xi_1\} = \tfrac{1}{6} \qquad \text{for } 1 \leq \alpha < 2$$

When $2 \leq \alpha < 3$, $\{x \leq \alpha\} = \{\xi_1, \xi_2\}$, and since these events are disjoint, we have

$$F_x(\alpha) = \mathcal{P}\{\xi_1, \xi_2\} = \mathcal{P}\{\xi_1\} + \mathcal{P}\{\xi_2\} = \tfrac{1}{3} \qquad \text{for } 2 \leq \alpha < 3$$

We can construct the rest of the distribution function in a similar manner with the result that

$$F_x(\alpha) = \begin{cases} 0 & \text{for } \alpha < 1 \\ \frac{1}{6} & \text{for } 1 \leq \alpha < 2 \\ \frac{1}{3} & \text{for } 2 \leq \alpha < 3 \\ \frac{1}{2} & \text{for } 3 \leq \alpha < 4 \\ \frac{2}{3} & \text{for } 4 \leq \alpha < 5 \\ \frac{5}{6} & \text{for } 5 \leq \alpha < 6 \\ 1 & \text{for } 6 \leq \alpha \end{cases}$$

which can also be written (where $\mathcal{s}$ indicates a unit step function) as

$$F_x(\alpha) = \tfrac{1}{6} \sum_{i=1}^{6} \mathcal{s}(\alpha - i)$$

The graph of this function is shown in Fig. 3.3-1. Note that this function satisfies all four properties just discussed.

Example *3.3-2.* Let us suppose that we own a candy factory which manufactures candy sticks 4 inches (in.) in length. However, one of the machines is performing erratically so that it produces sticks which vary in length from 3–5 in. Each value in this range is equally likely

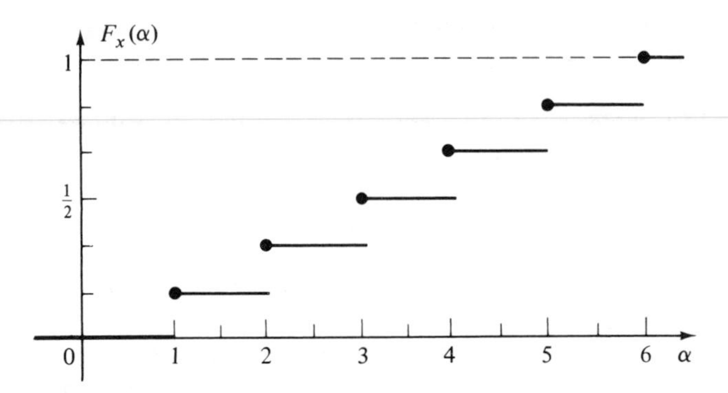

FIG. 3.3-1. Graph of distribution function for Example 3.3-1.

to occur. Our experiment is to measure one stick and the outcome is the length of the stick l. Because each outcome is equally likely, we find that the probability of the event $\{\alpha_1 < l \leq \alpha_2\}$ is $(\alpha_2 - \alpha_1)/2$ if $3 \leq \alpha_1 \leq \alpha_2 \leq 5$.

We define a random variable x as being equal to the length l so that $x(l) = l$. Now we wish to find the distribution function F_x. If $\alpha < 3$, then

$$F_x(\alpha) = \mathcal{P}\{x \leq \alpha\} = \mathcal{P}\{\varnothing\} = 0 \qquad \text{for } \alpha < 3$$

since there is no outcome for which $x < 3$. When $3 \leq \alpha \leq 5$, we have

$$F_x(\alpha) = \mathcal{P}\{x < 3\} + \mathcal{P}\{x = 3\} + \mathcal{P}\{3 < x \leq \alpha\} \qquad 3 \leq \alpha \leq 5$$

However, $\mathcal{P}\{x < 3\} = 0$ from the above argument, and $\mathcal{P}\{x = 3\} = (3 - 3)/2 = 0$ from the definition of the experiment. Hence $F_x(\alpha)$ becomes

$$F_x(\alpha) = \mathcal{P}\{3 < x \leq \alpha\} = \frac{\alpha - 3}{2} \qquad 3 \leq \alpha \leq 5$$

By a similar argument we find that

$$F_x(\alpha) = 1 \qquad \text{for } \alpha > 5$$

so the complete distribution function is

$$F_x(\alpha) = \begin{cases} 0 & \alpha < 3 \\ \dfrac{\alpha - 3}{2} & 3 \leq \alpha \leq 5 \\ 1 & 5 < \alpha \end{cases}$$

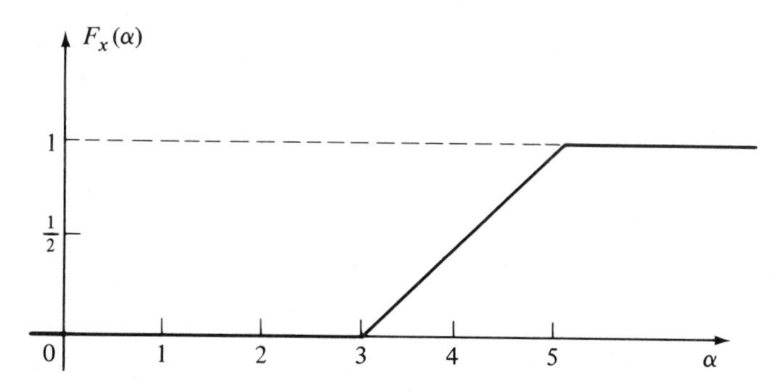

FIG. 3.3-2. Graph of distribution function for Example 3.3-2.

The graph of this function is shown in Fig. 3.3-2.

Example *3.3-3.* Let us consider the experiment defined in the preceding example again. Now, however, let us define a random variable such that

$$x(l) = \begin{cases} 0 & \text{for } 3 \leq l \leq 4 \\ l & \text{for } 4 < l \leq 5 \end{cases}$$

Now $F_x(\alpha) = 0$ for $\alpha < 0$, but becomes $F_x(\alpha) = \frac{1}{2}$, if $0 \leq \alpha \leq 4$, since

$$F_x(\alpha) \triangleq \mathcal{P}\{x \leq \alpha\} \triangleq \mathcal{P}\{l : x(l) \leq \alpha\} = \tfrac{1}{2} \qquad 0 \leq \alpha \leq 4$$

and $x(l) \leq \alpha$, for $0 \leq \alpha \leq 4$, for any l such that $3 \leq l \leq 4$; so

$$F_x(\alpha) = \mathcal{P}\{3 \leq l \leq 4\} = \tfrac{1}{2} \qquad 0 \leq l \leq 4$$

When $4 < \alpha \leq 5$, we have

$$\begin{aligned} F_x(\alpha) &= \mathcal{P}\{3 \leq l \leq 4\} + \mathcal{P}\{4 < l \leq \alpha\} \\ &= \frac{1}{2} + \frac{\alpha - 4}{2} = \frac{\alpha - 3}{2} \qquad 4 < \alpha \leq 5 \end{aligned}$$

and finally $F_x(\alpha) = 1$ when $5 < \alpha$. The final complete distribution function is therefore

$$F_x(\alpha) = \begin{cases} 0 & \text{for } \alpha < 0 \\ \frac{1}{2} & \text{for } 0 \leq \alpha \leq 4 \\ (\alpha - 3)/2 & \text{for } 4 < \alpha \leq 5 \\ 1 & \text{for } 5 < \alpha \end{cases}$$

with the graph given by Fig. 3.3-3.

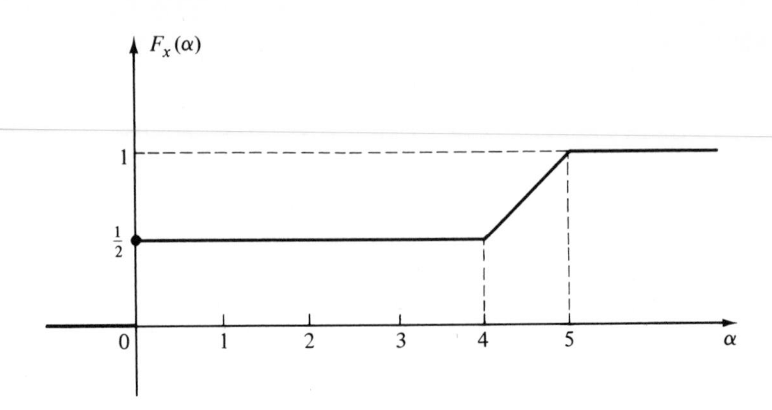

FIG. 3.3-3. Graph of distribution function for Example 3.3-3.

Once a random variable has been defined and its associated probability distribution function determined, there is no further need to have reference to the probability space $(\mathcal{S}, \mathfrak{F}, \mathcal{P})$. We have generated from the probability space model an abstract, numerically oriented model that contains an equivalent amount of information. The advantage of abstract model formulation is that almost all uses of probabilistic concepts are more easily developed in this framework, since all the elegant mathematical background of real analysis and integral and differential calculus is now available for simplifying the computational work. Another advantage is that all physical experiments which admit the same random-variable definition and distribution function can now be handled together. Because of this fact, we shall begin most of the rest of the discussion in this text with a description of the random behavior in terms of a distribution; it will no longer be necessary to return to the fundamental statement of the physical experiment. Lest the reader be lulled into a false sense of security by the seeming simplicity of this approach, let us emphasize that real problems require that we start with a definition of an appropriate probability space. Random problems do not in practice come with distribution functions attached.

Although the distribution function is very convenient for determining the probability of events, such as $\{x \leq \alpha\}$ and $\{\alpha_1 < x \leq \alpha_2\}$, it is in general not convenient for determining $\mathcal{P}\{x \in \mathcal{A}\}$, where $\mathcal{A}$ is any admissible set. To compute $\mathcal{P}\{x \in \mathcal{A}\}$, it is necessary to represent $\mathcal{A}$ in terms of sets of the form $\{x \leq \alpha\}$, which is not always easy. In Section 3.4 we shall consider the definition of another "probability function" that will considerably simplify such problems—the probability density function.

EXERCISES 3.3

3.3-1. A fair coin is tossed until a head is obtained. Let N be the number of the toss on which the first head appears. Find the probability distribution function for N. Plot the function.

Answer:

$$F_N(n) = \sum_{i=1}^{n} (\tfrac{1}{2})^i \mathscr{u}(n - i)$$

3.3-2. Given the probability distribution function for x as

$$F_x(\alpha) = \begin{cases} 1 - e^{-2\alpha} & \text{for } \alpha \geq 0 \\ 0 & \text{for } \alpha < 0 \end{cases}$$

find the probability that $2 < x \leq 3$.

Answer: $e^{-4} - e^{-6}$

3.3-3. Given the probability distribution function for x as

$$F_x(\alpha) = \tfrac{1}{4}(\alpha + 1)\mathscr{u}(\alpha) - \tfrac{1}{4}(\alpha - 1)\mathscr{u}(\alpha - 1)$$

find (a) $\mathcal{P}\{x = 0\}$, (b) $\mathcal{P}\{x = \frac{1}{2}\}$, and (c) $\mathcal{P}\{x = 1\}$.

Answers: (a) $\frac{1}{4}$, (b) 0, (c) 0.

3.4. Density functions

As noted at the end of Section 3.3, we now have two valid, equivalent mathematical models for a random experiment—probability space and the distribution function associated with a random variable. In this section we wish to define the density function as still another method of describing a random experiment. We shall find that the density function is a valuable tool in analyzing probabilistic situations. The distribution function serves, in part, as a conceptual tool to help us in the transition from the probability space model to the density-function description of a random experiment. In addition, the distribution function occupies a position of mathematical rigor and generality that can only be obtained in the case of density functions by the use of some rather advanced mathematical concepts. For our purposes, it will be possible to avoid these problems of mathematical rigor by carefully sidestepping at the right time. It is only when one wishes to study abstract probabilistic behavior that these pathological cases begin to pose a problem; we shall avoid consideration of such problems since they have little to do with the problems of engineering practice.

Let us begin the study of the probability density function by giving a definition.[1]

Definition 3.4-1. ***Density Function:*** The probability density function for a random variable x having a distribution function F_x is any function p_x defined on $\mathcal{R}$ and such that

$$F_x(\alpha) = \int_{-\infty}^{\alpha} p_x(\gamma)\, d\gamma \tag{3.4-1}$$

for all $\alpha \in \mathcal{R}$ and $p_x(\alpha) \geq 0$ for all $\alpha \in \mathcal{R}$.

Just as in the case of the distribution function, we use the subscript on p_x to indicate the random variable associated with the density function. We may also omit the subscript if there is no chance of confusion; and we may employ, as before, the notation $p(x)$ to represent p_x on occasion.

It may appear that the density function is just the derivative of the distribution function, and to a certain extent this is true. In another sense, p_x is not the derivative of F_x, and it is for that reason that we chose to make the definition in this indirect form. To more easily carry out our study of the probability density function, it is desirable to divide random variables into three disjoint classes: (1) continuous, (2) discrete, and (3) mixed, which are defined in the following manner:

Definition 3.4-2. ***Continuous Random Variable:*** A random variable x is said to be continuous if its probability distribution function F_x is continuous at every $\alpha \in \mathcal{R}$ and its derivative exists except at a countable number of points.[2]

Definition 3.4-3. ***Discrete Random Variable:*** A random variable x is said to be discrete if the derivative of its probability distribution function F_x exists and equals zero except at a countable number of jump points of F_x.

Stated in another way, a random variable is a discrete random variable if its distribution function can be written as a sum of step functions such as

$$F_x(\alpha) = \sum_{i=1}^{n} a_i \mathscr{s}(\alpha - \alpha_i)$$

[1] We shall first assume scalar random variables and defer consideration of vector random variables until Section 3.6.

[2] We shall assume that all distribution functions have only a countable number of points for which the derivative does not exist. Although it is quite possible to construct distribution functions that do not have derivatives at an uncountable number of points, such problems have little practical significance and we shall not discuss them.

where n may be either finite or infinite. Here α_i, $i = 1, 2, \ldots$, are the jump points of $F_x(\alpha)$, so that $a_i = F_x(\alpha_i+) - F_x(\alpha_i-)$. Since $F_x(+\infty) = 1$, we see that

$$\sum_{i=1}^{n} a_i = 1$$

> **Definition 3.4-4.** ***Mixed Random Variable:*** A random variable x is said to be mixed if it is neither continuous nor discrete.

Typical forms for the three types of random variables are shown graphically in Fig. 3.4-1. In terms of the three examples of Section 3.3, the first example

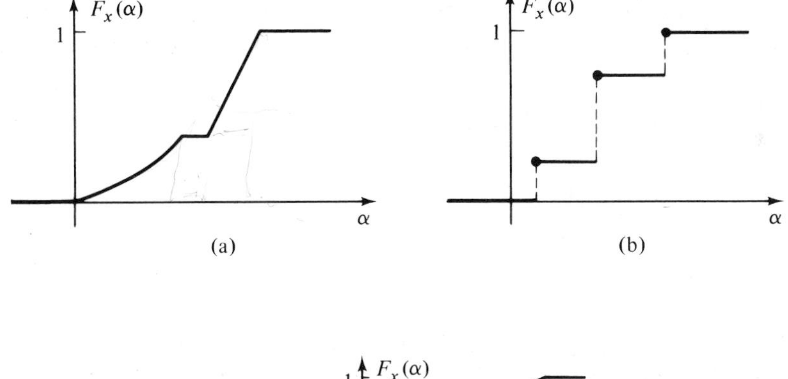

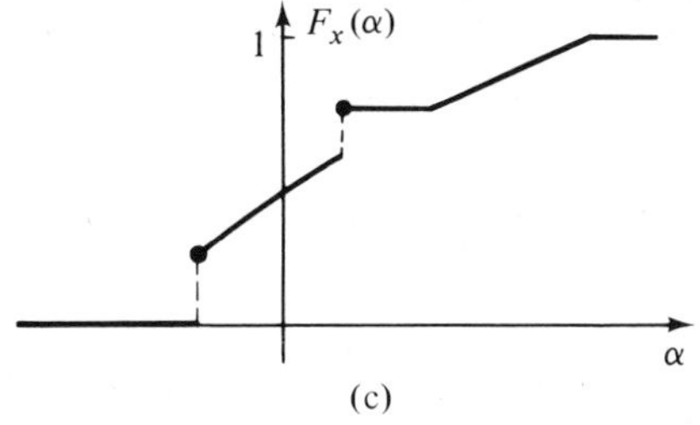

FIG. 3.4-1. Typical forms of distribution functions: (a) continuous; (b) discrete; and (c) mixed.

(Example 3.3-1) was a case of a discrete random variable, the second was a continuous random variable, and Example 3.3-3 was a mixed random variable. To continue the study of density functions, it is desirable to consider each of the three classes of random variables separately.

Continuous Random Variables. In the case of continuous random variables, the derivative of the distribution function F_x exists except at a countable number of points. Hence we may define the density function

p_x by

$$p_x(\alpha) = \frac{dF_x(\alpha)}{d\alpha} \tag{3.4-2}$$

at all points for which the derivative exists. At the countable points at which the derivative does not exist we may assign any (positive) value to p_x; the integral expression of Eq. (3.4-1) will be satisfied since the countable points have zero integral. Thus we may assume that $p_x(\alpha)$ is defined for all $\alpha \in \mathcal{R}$. We note that p_x is normally not uniquely defined by Eq. (3.4-1); however, the isolated points at which it is arbitrary [where $dF_x(\alpha)/d\alpha$ does not exist] are of no importance. Because of the monotonicity of F_x, $p_x(\alpha) \geq 0$ for all $\alpha \in \mathcal{R}$.

The use of the density function allows us to express the probability of the event $\{x \in \mathcal{A}\}$, where $\mathcal{A}$ is any admissible set,[1] quite conveniently as

$$\mathcal{P}\{x \in \mathcal{A}\} = \int_{\mathcal{A}} p_x(\alpha)\, d\alpha \tag{3.4-3}$$

In particular, if $\mathcal{A} = \mathcal{R}$, we have

$$\mathcal{P}\{x \in \mathcal{R}\} = \mathcal{P}\{\mathcal{S}\} = 1 = \int_{\mathcal{R}} p_x(\alpha)\, d\alpha = \int_{-\infty}^{\infty} p_x(\alpha)\, d\alpha \tag{3.4-4}$$

This is a property that must be true of every density function.

To develop another interesting property, let us consider the probability of the event $\{\alpha < x \leq \alpha + d\alpha\}$. By the use of Eq. (3.4-3), we have

$$\mathcal{P}\{\alpha \leq x \leq \alpha + d\alpha\} = \int_{\alpha}^{\alpha + d\alpha} p_x(\beta)\, d\beta \tag{3.4-5}$$

The use of the mean-value theorem for integrals[2] (assuming that p_x is continuous in $[\alpha, \alpha + d\alpha]$) indicates that there exists a β', $\alpha \leq \beta' \leq \alpha + d\alpha$, such that

$$\mathcal{P}\{\alpha \leq x \leq \alpha + d\alpha\} = p_x(\beta')\, d\alpha \tag{3.4-6}$$

[1] We must actually restrict $\mathcal{A}$ to be a finite union or intersection of half-open intervals in order to use the ordinary Riemann integral in Eq. (3.4-3). To consider the general case when $\mathcal{A}$ is admissible, it would be necessary to employ the more general Lebesgue–Stieltijes integral in Eq. (3.4-3).

[2] There are several mean-value theorems for integrals. The most common one is used here. If $f(x)$ is continuous on $[a, b]$, there exists a value X of x in (a, b) such that

$$\int_a^b f(x)\, dx = f(X)(b - a)$$

If we take the limit as $d\alpha$ approaches zero, we have

$$\lim_{d\alpha \to 0} \mathcal{P}\{\alpha \leq x \leq \alpha + d\alpha\} = \mathcal{P}\{x = \alpha\} = 0 \tag{3.4-7}$$

for any $\alpha \in \mathcal{R}$. In other words, for a continuous random variable, the probability that x will equal any single value α is zero! Because of this fact we see that $\mathcal{P}\{\alpha \leq x \leq \beta\} = \mathcal{P}\{\alpha < x \leq \beta\} = \mathcal{P}\{\alpha \leq x < \beta\} = \mathcal{P}\{\alpha < x < \beta\}$, since the addition or deletion of any single point from the interval cannot change the probability. It should be noted that this property is *not* true in general, but is only valid for continuous random variables.

Equation (3.4-7) also points up a fact that tends to be confusing at first. It is important to note that $p_x(\alpha)$ *does not* equal the probability of the event $\{x = \alpha\}$; that is, $p_x(\alpha) \neq \mathcal{P}\{x = \alpha\}$. From Eq. (3.4-7) we see that $\mathcal{P}\{x = \alpha\} = 0$ for any $\alpha \in \mathcal{R}$ (for a continuous random variable). However, $p_x(\alpha)$ is in general not equal to zero.

In addition, it is quite possible for $p_x(\alpha) > 1$, once again indicating that $p_x(\alpha)$ is *not* the probability that $x = \alpha$ since that probability cannot be greater than unity for any type of event. We can, however, relate $p_x(\alpha)$ to a probability since, for small $d\alpha$, Eq. (3.4-6) yields

$$p_x(\alpha)\, d\alpha \simeq \mathcal{P}\{\alpha \leq x \leq \alpha + d\alpha\} \tag{3.4-8}$$

The symbol $\simeq$ is used to indicate a first-order approximation that is only valid as $d\alpha$ becomes infinitesimal. Equation (3.4-8) is easily derived from Eq. (3.4-6) by an application of the Taylor series. It should be noted that Eq. (3.4-8) is only valid if α is a point at which the derivative of F_x exists; otherwise, $p_x(\alpha)$ can be arbitrarily defined giving a nonunique value for $\mathcal{P}\{\alpha \leq x \leq \alpha + d\alpha\}$.

Example *3.4-1*. To illustrate the determination of a density function for a continuous random variable, let us examine the random variable of Example 3.3-2 again. There we found that the distribution function F_x was given by

$$F_x(\alpha) = \begin{cases} 0 & \alpha < 3 \\ \dfrac{\alpha - 3}{2} & 3 \leq \alpha \leq 5 \\ 1 & 5 < \alpha \end{cases}$$

Here we see that the derivative of F_x exists, except for $\alpha = 3$ and 5, and is given by

$$\frac{dF_x(\alpha)}{d\alpha} = \begin{cases} 0 & \alpha < 3 \\ \frac{1}{2} & 3 < \alpha < 5 \\ 0 & 5 < \alpha \end{cases}$$

At $\alpha = 3$ and 5 we may select any value for p_x. If we let $p_x(3) = \frac{1}{2}$ and $p_x(5) = 0$, we have the density function

$$p_x(\alpha) = \tfrac{1}{2}[\mathscr{s}(\alpha - 3) - \mathscr{s}(\alpha - 5)]$$

which is represented graphically in Fig. 3.4-2.

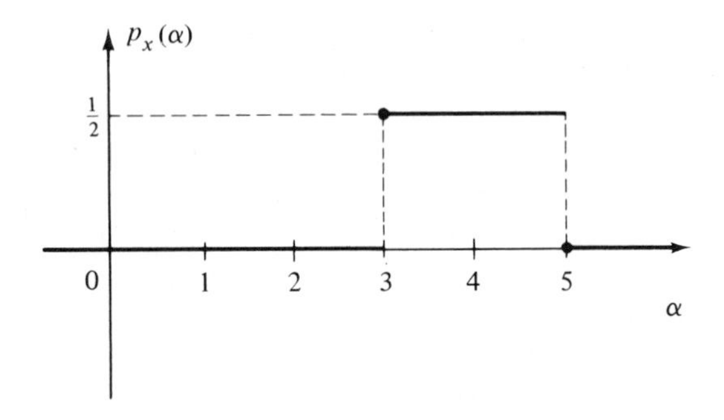

FIG. 3.4-2. Density function for Example 3.4-1.

Let us suppose now that we wish to find the probability that $I - \epsilon \leq x \leq I + \epsilon$, for $I = 1, 2, \ldots$, and $\epsilon > 0$. By the use of Eq. (3.4-3), we have

$$\mathcal{P}\{I - \epsilon \leq x \leq I + \epsilon, I = 1, 2, \ldots\} = \sum_{I=1}^{\infty} \int_{I-\epsilon}^{I+\epsilon} p_x(\alpha)\, d\alpha$$

However, $p_x(\alpha) = 0$ except for $3 \leq \alpha < 5$, so we have

$$\mathcal{P}\{I - \epsilon \leq x \leq I + \epsilon, I = 1, 2, \ldots\} = \int_{3}^{3+\epsilon} \tfrac{1}{2} d\alpha + \int_{4-\epsilon}^{4+\epsilon} \tfrac{1}{2} d\alpha$$
$$+ \int_{5-\epsilon}^{5} \tfrac{1}{2} dx = 2\epsilon$$

Note the ease with which we were able to handle this problem by the use of the density function.

Discrete Random Variables. Because the derivative of the distribution function for a discrete random variable is zero except at the jump points of F_x where the derivative is undefined, it would appear that there is no ordinary function p_x such that Eq. (3.4-1) is satisfied for a discrete random variable. However, because of the usefulness of the density function in the study of random phenomena, it is desirable to extend the class of functions in some way so that we may use density functions for discrete random variables.

One possible approach is to make use of the Lebesgue–Stieltjes integral and write

$$F_x(\alpha) = \int_{-\infty}^{\alpha} dF_x$$

This approach, although mathematically quite elegant, gives us little insight into the manipulation of the density function. Hence we reject this approach as not appropriate for our study.

Our approach to the problem of forming a density function for a discrete random variable will employ the Dirac delta function. Although this approach lacks some mathematical rigor, it provides a very convenient and simple method of handling density functions for discrete random variables. It will be assumed that the reader is familiar with the use of the Dirac delta function; however, it may be appropriate to recall briefly the nature of this very useful and unusual "function."

The delta function $\delta_D(x)$ is not really a function at all, at least not in the ordinary sense, but is more properly referred to as a functional, that is, a function of functions, or a generalized function. The Dirac delta function is defined by the properties that $\delta_D(x) = 0$ for $x \neq 0$ and

$$\int_{-\infty}^{\infty} f(x)\delta_D(x - a)\, dx = f(a) \tag{3.4-9}$$

for any f which is continuous at $x = a$. We see that the use of the Dirac delta function allows us to generate a jump in the integral at the point of occurrence of the delta function. This suggests the use of the delta function to represent a density function for a discrete random variable.

Let us suppose that we have a discrete random variable x with a distribution function F_x which has jump points at α_i, $i = 1, 2, \ldots$. Then the density function for x may be written as

$$p_x(\alpha) = \sum_i [F_x(\alpha_i^+) - F_x(\alpha_i^-)]\delta_D(\alpha - \alpha_i) \tag{3.4-10}$$

such that Eq. (3.4-1) is satisfied.

When using density functions containing an impulse to calculate probabilities by the use of Eq. (3.4-3), it is very important to handle the end points of the intervals of interest carefully, particularly if they are jump points of the distribution function. This problem may be most easily resolved by the use of limiting from above or below to include or exclude a point as needed. For example, the $\mathcal{P}\{\alpha_1 \leq x < \beta_1\}$ is written as

$$\mathcal{P}\{\alpha_1 \leq x < \beta_1\} = \lim_{\alpha \uparrow \alpha_1} \lim_{\beta \uparrow \beta_1} \int_{\alpha}^{\beta} p_x(\gamma)\, d\gamma \tag{3.4-11}$$

Here we note that the point α_1 is always included in the interval, as desired, and β_1 is excluded. If, on the other hand, we wish to compute $\mathcal{P}\{\alpha_1 < x \leq \beta_1\}$, we would write

$$\mathcal{P}\{\alpha_1 < x \leq \beta_1\} = \lim_{\alpha \downarrow \alpha_1} \lim_{\beta \downarrow \beta_1} \int_{\alpha}^{\beta} p_x(\gamma)\, d\gamma \tag{3.4-12}$$

Here we exclude α_1 and include β_1. Of course, if the function p_x is well behaved at α_1 and β_1, there is no difficulty, since the two integrals in Eqs. (3.4-11) and (3.4-12) are in fact equal.

Example 3.4-2. Let us illustrate the use of the delta function to represent the density function for a discrete random variable by considering a simple example. Consider a random experiment with four equally likely outcomes ξ_1, ξ_2, ξ_3, and ξ_4; in addition, each outcome is an event, so we have $\mathcal{P}\{\xi_1\} = \mathcal{P}\{\xi_2\} = \mathcal{P}\{\xi_3\} = \mathcal{P}\{\xi_4\} = \frac{1}{4}$. Now we define a random variable x by

$$x(\xi) = \begin{cases} 1 & \text{for } \xi = \xi_1 \\ 2 & \text{for } \xi = \xi_2 \\ 3 & \text{for } \xi = \xi_3 \\ 4 & \text{for } \xi = \xi_4 \end{cases}$$

It is quite easy to show that the distribution for this random variable is given by

$$F_x(\alpha) = \tfrac{1}{4}[\mathcal{s}(\alpha - 1) + \mathcal{s}(\alpha - 2) + \mathcal{s}(\alpha - 3) + \mathcal{s}(\alpha - 4)]$$

so that its graph takes the form shown in Fig. 3.4-3a.

There are four jump points and $F_x(\alpha_i^+) - F_x(\alpha_i^-) = \frac{1}{4}$ for each, so the density function is simply

$$p_x(\alpha) = \tfrac{1}{4}[\delta_D(\alpha - 1) + \delta_D(\alpha - 2) + \delta_D(\alpha - 3) + \delta_D(\alpha - 4)]$$

and is represented graphically by Fig. 3.4-3b.

Let us suppose that we wish to determine the probability of the event $\{1 < x \leq 3\}$ by the use of the density function. The application of Eq. (3.4-3) yields

$$\begin{aligned}\mathcal{P}\{1 < x \leq 3\} = \lim_{\alpha \downarrow 1} \lim_{\beta \downarrow 3} \int_{\alpha}^{\beta} \tfrac{1}{4}[&\delta_D(\gamma - 1) + \delta_D(\gamma - 2) \\ &+ \delta_D(\gamma - 3) + \delta_D(\gamma - 4)]\, d\gamma\end{aligned}$$

Only the delta functions at $\gamma = 2$ and 3 are covered by the interval of

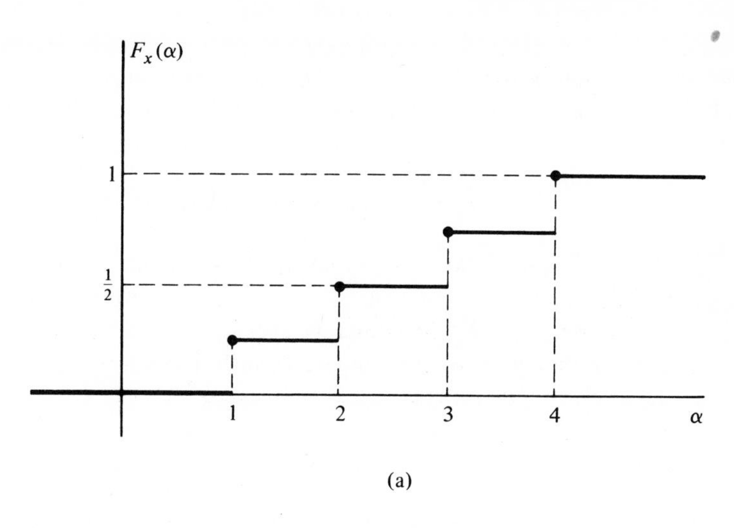

(a)

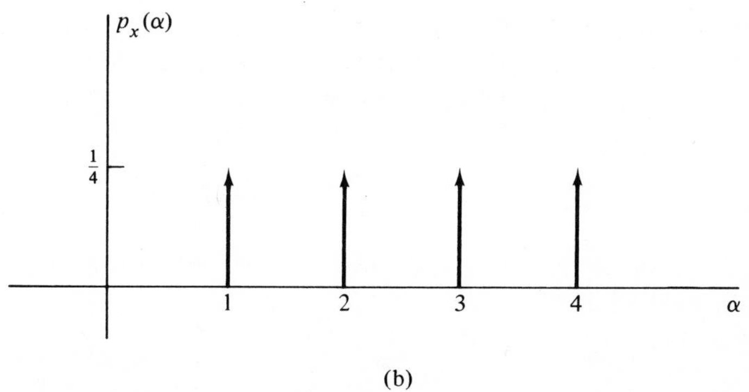

(b)

FIG. 3.4-3. Distribution and density function for Example 3.4-2: (a) distribution function, (b) density function.

interest, since the point $\gamma = 1$ is excluded, so the desired probability becomes

$$\mathcal{P}\{1 < x \leq 3\} = \lim_{\alpha \downarrow 1} \lim_{\beta \downarrow 3} \int_{\alpha}^{\beta} \tfrac{1}{4}[\delta_D(\gamma - 2) + \delta_D(\gamma - 3)]\, d\gamma = \tfrac{1}{2}$$

Mixed Random Variables. The determination of the density function for a mixed random variable is quite easy. We need only recognize that the distribution function for a mixed random variable can be written as a linear combination of a discrete and a continuous distribution function. In other words, if F_x is the distribution function for a mixed random variable, then F_x can always be written as

$$F_x(\alpha) = a_1 F_x^d(\alpha) + a_2 F_x^c(\alpha) \tag{3.4-13}$$

with $a_1 + a_2 = 1$. Since we have already treated the problems of determining the density function for discrete and continuous random variables, we have only to combine the methods developed previously for the individual cases to find the density function for the mixed random variable as

$$p_x(\alpha) = a_1 p_x^d(\alpha) + a_2 p_x^c(\alpha) \tag{3.4-14}$$

The decomposition of Eq. (3.4-14) is normally quite easy to accomplish. We start by forming a piecewise constant function f_x^d that contains all the discontinuities of F_x. The difference between F_x and f_x^d is a continuous function f_x^c, so that we have decomposed F_x into the sum of a piecewise constant and a continuous function as

$$F_x(\alpha) = f_x^d(\alpha) + f_x^c(\alpha) \tag{3.4-15}$$

The functions f_x^d and f_x^c are not distributions, although they are monotone increasing and have $f_x^d(-\infty) = f_x^c(-\infty) = 0$, since $f_x^d(+\infty)$ and $f_x^c(+\infty) \neq 1$. We now divide $f_x^d(\alpha)$ by $a_1 = f_x^d(+\infty)$ and divide $f_x^c(\alpha)$ by $a_2 = f_x^c(+\infty)$, so that Eq. (3.4-15) becomes

$$F_x(\alpha) = a_1\left[\frac{1}{a_1} f_x^d(\alpha)\right] + a_2\left[\frac{1}{a_2} f_x^c(\alpha)\right] \tag{3.4-16}$$

The functions f_x^d/a_1 and f_x^c/a_2 are now distribution functions, and the required decomposition has been achieved. Note that the requirement that $a_1 + a_2 = 1$ is satisfied, since $a_1 + a_2 = f_x^d(+\infty) + f_x^c(+\infty) = F_x(+\infty) = 1$.

Example *3.4-3*. As an example, let us determine the density function for the mixed random variable of Example 3.3-3. There we found that the distribution function was given by

$$F_x(\alpha) = \begin{cases} 0 & \text{for } \alpha < 0 \\ \frac{1}{2} & \text{for } 0 \leq \alpha \leq 4 \\ \dfrac{\alpha - 3}{2} & \text{for } 4 < \alpha \leq 5 \\ 1 & \text{for } 5 < \alpha \end{cases}$$

We see that there is one discontinuity in this function at $x = 0$, so the function f_x^d becomes

$$f_x^d(\alpha) = \begin{cases} 0 & \text{for } \alpha < 0 \\ \frac{1}{2} & \text{for } 0 \leq \alpha \end{cases}$$

whereas $f_x^c = F_x - f_x^d$ is given by

$$f_x^c(\alpha) = \begin{cases} 0 & \text{for} \quad \alpha \leq 4 \\ \dfrac{\alpha - 4}{2} & \text{for } 4 < \alpha \leq 5 \\ \frac{1}{2} & \text{for} \quad 5 < \alpha \end{cases}$$

Here we find that $a_1 = a_2 = \frac{1}{2}$, and the component discrete and continuous distributions are

$$F_x^d(\alpha) = \begin{cases} 0 & \text{for } \alpha < 0 \\ 1 & \text{for } 0 \leq \alpha \end{cases}$$

and

$$F_x^c(\alpha) = \begin{cases} 0 & \text{for} \quad \alpha < 4 \\ \alpha - 4 & \text{for } 4 < \alpha \leq 5 \\ 1 & \text{for} \quad 5 < \alpha \end{cases}$$

Applying the methods previously developed for determining the density function for discrete and continuous random variables, we find that

$$p_x^d(\alpha) = \delta_D(\alpha)$$

and

$$p_x^c(\alpha) = \mathscr{s}(\alpha - 4) - \mathscr{s}(\alpha - 5)$$

So the final density function for the mixed random variable is given by Eq. (3.3-14) as

$$p_x(\alpha) = \tfrac{1}{2}\delta_D(\alpha) + \tfrac{1}{2}[\mathscr{s}(\alpha - 4) - \mathscr{s}(\alpha - 5)]$$

In the last three sections we have examined the important concepts of random variables and distribution and density functions. These concepts will greatly facilitate the remainder of the study of random phenomena. Before continuing the study, we pause briefly to consider some of the more important distributions found in nature.

EXERCISES 3.4

3.4-1. Show that each function given below is a valid probability density function:

(a) $p_x(\alpha) = \begin{cases} 0.5 & -1 < \alpha < 1 \\ 0 & \text{otherwise} \end{cases}$

(b) $p_x(\alpha) = \begin{cases} 0.25 + 0.5\delta_D(\alpha) & -1 < \alpha < 1 \\ 0 & \text{otherwise} \end{cases}$

(c) $p_x(\alpha) = \begin{cases} 1/2\sqrt{\alpha} & 0 < \alpha < 1 \\ 0 & \text{otherwise} \end{cases}$

(d) $p_x(\alpha) = \begin{cases} 2\alpha & 0 < \alpha < 1 \\ 0 & \text{otherwise} \end{cases}$

(e) $p_x(\alpha) = \begin{cases} 1/[\pi(1 - \alpha^2)^{1/2}] & -1 < \alpha < 1 \\ 0 & \text{otherwise} \end{cases}$

(f) $p_x(\alpha) = \sum_{i=2}^{6} \frac{\binom{8}{i}\binom{4}{6-i}}{\binom{12}{6}} \delta_D(\alpha - i)$

3.4-2. The number of transistors (in pounds) that a certain factory is able to produce in a day is a numerical-valued random phenomenon with a probability density function

$$p_x(\alpha) = \begin{cases} A\alpha & 0 \le \alpha < 5 \\ A(10 - \alpha) & 5 \le \alpha < 10 \\ 0 & \text{otherwise} \end{cases}$$

(a) Find the value of A that makes $p_x(\alpha)$ a valid probability density function.
(b) What is the probability that the number of pounds to be produced tomorrow is between 2.5 and 7.5 pounds?

Answers: (a) 0.04, (b) 0.75.

3.4-3. Find the probability density function associated with each of the following distribution functions:

(a) $F_x(\alpha) = \frac{1}{4}\mathscr{u}(\alpha) + \frac{1}{2}\mathscr{u}(\alpha - 1) + \frac{1}{4}\mathscr{u}(\alpha - 2)$.
(b) $F_x(\alpha) = (1 - e^{-2\alpha})\mathscr{u}(\alpha)$.
(c) $F_x(\alpha) = \frac{1}{2}(\alpha + 1)\mathscr{u}(\alpha) - \frac{1}{2}(\alpha - 1)\mathscr{u}(\alpha - 1)$.

Answers

(a) $p_x(\alpha) = \frac{1}{4}[\delta_D(\alpha) + 2\delta_D(\alpha - 1) + \delta_D(\alpha - 2)]$.
(b) $p_x(\alpha) = 2e^{-2\alpha}\mathscr{u}(\alpha)$.
(c) $p_x(\alpha) = \frac{1}{2}\delta_D(\alpha) + \frac{1}{2}[\mathscr{u}(\alpha) - \mathscr{u}(\alpha - 1)]$.

3.5. Example of distribution and density functions

There is a large number of probability distributions in common use. Some of these distributions describe basic physical phenomena; others describe the distribution of various combinations of other random variables and are used primarily for statistical studies. In this section we shall consider only a few of the more widely used distributions. We shall begin with distributions for discrete random variables.

Discrete Uniform Distribution. Without referring to it by name, we used the discrete uniform distribution several times in the last chapter. Coin-tossing and dice-rolling experiments can both be described by the discrete uniform distribution. We say that a discrete random variable x has a uniform distribution if

$$p_x(\alpha) = \frac{1}{n} \sum_{i=1}^{n} \delta_D(\alpha - \alpha_i) \tag{3.5-1}$$

In other words, x can only take n values $\alpha_1, \alpha_2, \ldots, \alpha_n$, each with a probability $1/n$. We previously displayed the graph of the probability density and distribution function for the dice-rolling experiment in Section 3.3 as Fig. 3.3-1, which the reader should refer to for a specific example of the discrete uniform distribution. The distribution function for the general case is easily obtained by integrating Eq. (3.5-1):

$$F_x(\alpha) = \frac{1}{n} \sum_{i=1}^{n} \mathscr{s}(\alpha - \alpha_i) \tag{3.5-2}$$

A special case of the discrete uniform distribution occurs when $n = 1$. Then $p_x(\alpha)$ is

$$p_x(\alpha) = \delta_D(\alpha - \alpha_0)$$

and $x = \alpha_0$ with probability 1. This special distribution is sometimes referred to as the *causal distribution* and can be used to represent a constant in a probabilistic manner. It is also the limiting case of a number of other distributions.

Binomial Distribution. Consider a set of n *independent* experiments each of which has two possible outcomes ξ_0 and ξ_1 with $\mathcal{P}\{\xi_1\} = p$. A physical example might be n tosses of a coin with ξ_1 = head and ξ_0 = tail and with the probability of a head on each toss being p, where $0 \leq p \leq 1$. Experiments of the above type, where the outcomes are independents from one experiment to another and the probability of ξ_1 (and consequently ξ_0) remains constant over all trials, are known as *Bernoulli trials*. Let us define a random variable on the ith experiment such that $x_i(\xi_1) = 1$ and $x_i\{\xi_0\} = 0$, where $i = 1, 2, \ldots, n$. In other words, we have n identical random variables, each with a density function

$$p_{x_i}(\alpha) = p\delta_D(\alpha - 1) + (1 - p)\delta_D(\alpha)$$

Now we define another random variable y as the sum of the x_i's or

$$y = \sum_{i=1}^{n} x_i$$

Note that y is defined on the combined sample space of the n experiments, whereas each x_i is defined on the sample space of a single experiment.

We see that y may take any value from 0 to n. It is desired to obtain the probability density function for y. Let us suppose that $y = m$, $0 \leq m \leq n$; this value could be obtained if $x_i = 1$, for $i = 1, 2, \ldots, m$, and $x_i = 0$, for $i = m + 1, m + 2, \ldots, n$. The probability of this specific set of outcomes is given by

$$\mathcal{P}\{x_i = 1, i = 1, 2, \ldots, m, \text{ and } x_i = 0, i = m + 1, m + 2, \ldots, n\}$$
$$= [\mathcal{P}\{x_i = 1\}]^m [\mathcal{P}\{x_i = 0\}]^{n-m} = p^m(1-p)^{n-m}$$

Of course, this is only one possible set of outcomes for which $y = m$; there are many others. In general, the numbers of possible sets of outcomes for which $y = m$ is given by the number of combinations of n things taken m at a time, which is expressed as[1]

$$\binom{n}{m} \triangleq \frac{n!}{(n-m)!m!}$$

where $j! = \prod_{i=1}^{j} i$ and $0! = 1$ (by definition). Hence we see that the probability that $y = m$ is given by

$$\mathcal{P}\{y = m\} = \binom{n}{m} p^m (1-p)^{n-m}$$

From this result we can easily determine that the density function of y is

$$p_y(\alpha) = \sum_{i=0}^{n} \binom{n}{i} p^i (1-p)^{n-i} \delta_D(\alpha - i) \tag{3.5-3}$$

A random variable with this density function is said to have a *binomial distribution* with parameters n and p. Because the distribution has two parameters, we can get a whole family of graphs for $p_y(\alpha)$, depending on the values of n and p chosen. The density $p_y(\alpha)$ is plotted for a few different values of n and p in Fig. 3.5-1.

The binomial distribution has been tabulated in several sources[2]

[1] See Section 2.5

[2] See, for example, *Tables of the Binomial Probability Distribution*, National Bureau of Standards, Washington, D.C., Applied Mathematics Series 6, 1950. $p = 0.01(0.01)0.50$, $n = 2(1)49$. Here the notation $p = 0.01(0.01)0.50$ means p ranges from 0.01 to 0.50 in steps of 0.01.

Tables of the Cumulative Binomial Probability Distribution, Annals of the Computation Laboratory, XXXV, Harvard University Press, Cambridge, Mass., 1955. $p = 0.01(0.01)0.50$, $n = 1(1)50(2)100(10)500(50)1000$.

Tables of the Cumulative Binomial Probabilities, Ordnance Corps Pamphlet ORDP-1, 1952, U.S. Department of Commerce, Office of Technical Services, Washington, D.C. $p = 0.01(0.01)0.50$, $n = (1)150$.

50–100 *Binomial Tables*, H. G. Romig, Wiley, New York, 1953. $p = 0.01(0.01)0.50$, $n = 50(5)100$.

for various values of n and p. If one does not have access to these tables and wishes to compute $p_y(\alpha)$ for large values of n, Stirling's approximation

$$N! \simeq \sqrt{2\pi}\, N^{N+1/2} e^{-N} \tag{3.5-4}$$

may be used to simplify the computation. Other approximations to the binomial distribution for large n will be discussed later.

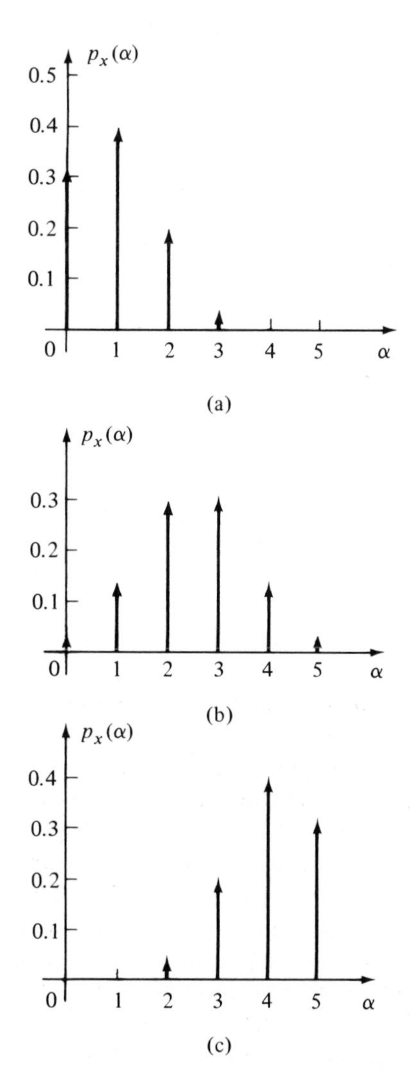

FIG. 3.5-1. Example of density functions for a binomial distribution: (a) $n = 5, p = 0.2$, (b) $n = 5, p = 0.5$, (c) $n = 5, p = 0.8$, (d) $n = 10$, $p = 0.2$, (e) $n = 10, p = 0.5$, (f) $n = 10, p = 0.8$.

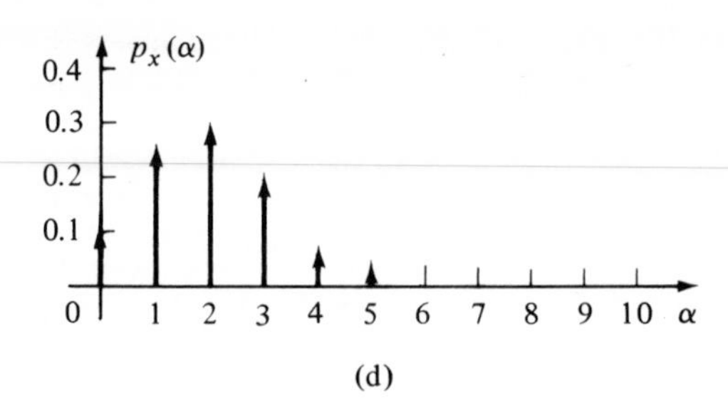

(d)

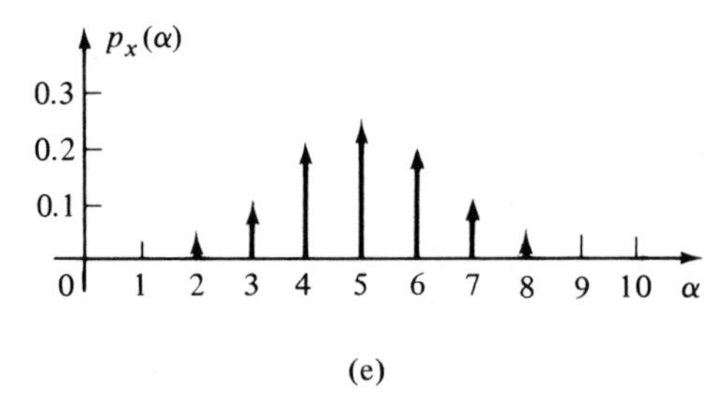

(e)

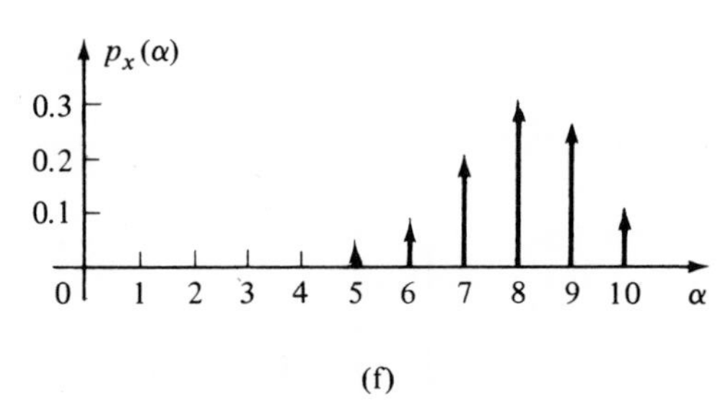

(f)

FIG 3.5-1. (cont.)

Poisson Distribution.[1] The Poisson distribution is used to model a wide variety of physical processes from shot noise in vacuum tubes to the occurrence of calls in a communication system. Let us suppose that we observe a random phenomenon over a time interval $[0, t]$ and designate as "event points" the times $t_0, t_1, \ldots, \in [0, t]$ at which some previously specified "event" occurs. Since the phenomenon that we are observing is random, the number of event points in the time interval $[0, t]$ is random; let us define a random variable x as the number of event points in $[0, t]$. Note that x is a discrete random variable which can assume the values $0, 1, 2, \ldots$.

We wish to determine the probability distribution of x with the length of the time interval, t, as a parameter. It is convenient to use the notation $\mathcal{P}\{x = k: t_1, t_2\}$ to represent the probability that $x = k$ for the observation

[1] Additional development and discussion of the Poisson process is presented in Chapter 5.

interval $[t_1, t_2]$. Before we may find $p_x(\alpha)$, it is necessary to specify some additional assumptions concerning the random phenomenon. These are:

1. The random phenomenon does not change with time, so the probability that $x = k$ in the interval $[\tau, \tau + T]$ is independent of τ; that is,

$$\mathcal{P}\{x = k: 0, T\} = \mathcal{P}\{x = k: \tau, \tau + T\} \tag{3.5-5}$$

   for all τ.
2. The probability that $x = k$ in the interval $[\tau, \tau + T]$ is independent of the number of event points in any nonoverlapping time interval.[1]
3. The probability of the occurrence of an event point is proportional to the length of the observation interval for "short intervals," and the probability of more than one event point in a "short interval" is negligible. In other words,

$$\mathcal{P}\{x = 1: 0, \Delta t\} = \lambda\, \Delta t \tag{3.5-6}$$

   for some $\lambda > 0$ and $\Delta t \approx 0$, and

$$\mathcal{P}\{x = 0: 0, \Delta t\} + \mathcal{P}\{x = 1: 0, \Delta t\} = 1 \tag{3.5-7}$$

   for $\Delta t \approx 0$. The parameter λ is often called the *mean count rate*.

With these three assumptions, we are now ready to derive the probability density function for x. We begin by determining the probability of no event points (that is, $x = 0$) in the interval $[0, t]$. Consider an interval of length $t + \Delta t$, with $\Delta t \sim 0$, that is divided into two nonoverlapping[2] intervals $[0, t]$ and $[t, t + \Delta t]$. Since the probability of an event point occurring during $[t, t + \Delta t]$ is independent of the number of event points in $[0, t]$, by Assumption 2, we may write $\mathcal{P}\{x = 0: 0, t + \Delta t\}$ as

$$\mathcal{P}\{x = 0: 0, t + \Delta t\} = \mathcal{P}\{x = 0: 0, t\}\mathcal{P}\{x = 0: t, t + \Delta t\} \tag{3.5-8}$$

By the use of Eqs. (3.5-5) and (3.5-7), we may write $\mathcal{P}\{x = 0: t, t + \Delta t\}$ as

$$\mathcal{P}\{x = 0: t, t + \Delta t\} = 1 - \mathcal{P}\{x = 1: 0, \Delta t\}$$

and from Eq. (3.5-6) we have

$$\mathcal{P}\{x = 0: t, t + \Delta t\} = 1 - \lambda\, \Delta t \tag{3.5-9}$$

[1] Assumption 2 is often called the independent increment assumption.

[2] The fact that the intervals $[0, t]$ and $[t, t + \Delta t]$ intersect at a single point "t" is not important, since from Eqs. (3.5-5) and (3.5-6) we see that $\mathcal{P}\{x = 1: t, t\} = \mathcal{P}\{x = 1: 0, 0\} = 0$.

Therefore, Eq. (3.5-8) becomes

$$\mathcal{P}\{x = 0: 0, t + \Delta t\} = \mathcal{P}\{x = 0: 0, t\}[1 - \lambda \, \Delta t]$$

or equivalently that

$$\frac{\mathcal{P}\{x = 0: 0, t + \Delta t\} - \mathcal{P}\{x = 0: 0, t\}}{\Delta t} = -\lambda \mathcal{P}\{x = 0: 0, t\}$$

If we take the limit as $\Delta t \longrightarrow 0$, the left-hand side becomes the derivative of $\mathcal{P}\{x = 0: 0, t\}$ with respect to t, so that we have

$$\frac{\partial \mathcal{P}\{x = 0: 0, t\}}{\partial t} = -\lambda \mathcal{P}\{x = 0: 0, t\} \qquad (3.5\text{-}10)$$

By the use of the boundary condition $\mathcal{P}\{x = 0: 0, 0\} = 1$, which is derived from Eqs. (3.5-6) and (3.5-7), we can obtain the solution of Eq. (3.5-10) as

$$\mathcal{P}\{x = 0: 0, t\} = e^{-\lambda t} \qquad (3.5\text{-}11)$$

Thus we have obtained a probability measure as the solution of a differential equation.

Now let us determine the probability that $x = k$ on the time interval $[0, t + \Delta t]$. Again we divide the interval into two nonoverlapping intervals $[0, t]$ and $[t, t + \Delta t]$. There are only two possible ways that x can be equal to k for the interval $[0, t + \Delta t]$: (1) there can be k event points in $[0, t]$ and none in $[t, t + \Delta t]$, or (2) there can be $k - 1$ event points in $[0, t]$ and one event point in $[t, t + \Delta t]$. Hence we may write $\mathcal{P}\{x = k: 0, t + \Delta t\}$ as

$$\begin{aligned}\mathcal{P}\{x = k: 0, t + \Delta t\} = \mathcal{P}\{x = k: 0, t\}\mathcal{P}\{x = 0: t, t + \Delta t\} \\ + \mathcal{P}\{x = k - 1: 0, t\}\mathcal{P}\{x = 1: t, t + \Delta t\}\end{aligned}$$

and by using Eqs. (3.5-5), (3.5-6), and (3.5-9), we have

$$\mathcal{P}\{x = k: 0, t + \Delta t\} = \mathcal{P}\{x = k: 0, t\}(1 - \lambda \, \Delta t) + \mathcal{P}\{x = k - 1: 0, t\}(\lambda \, \Delta t)$$

After some simple rearrangement, this result becomes

$$\begin{aligned}\frac{\mathcal{P}\{x = k: 0, t + \Delta t\} - \mathcal{P}\{x = k: 0, t\}}{\Delta t} + \lambda \mathcal{P}\{x = k: 0, t\} \\ = \lambda \mathcal{P}\{x = k - 1: 0, t\}\end{aligned}$$

and, as $\Delta t \longrightarrow 0$, we obtain the linear nonhomogeneous differential equation for $\mathcal{P}\{x = k: 0, t\}$:

$$\frac{\partial \mathcal{P}\{x = k: 0, t\}}{\partial t} + \lambda \mathcal{P}\{x = k: 0, t\} = \lambda \mathcal{P}\{x = k - 1: 0, t\} \quad (3.5\text{-}12)$$

This first-order differential equation is easy to solve; we use the boundary condition $\mathcal{P}\{x = k: 0, 0\} = 0$ for $k > 0$ and find

$$\mathcal{P}\{x = k: 0, t\} = \lambda e^{-\lambda t} \int_0^t e^{\lambda \tau} \mathcal{P}\{x = k - 1: 0, \tau\} \, d\tau \tag{3.5-13}$$

We may compute $\mathcal{P}\{x = 1: 0, t\}$ by using $\mathcal{P}\{x = 0: 0, t\}$ given by Eq. (3.5-11); this result may then by used to compute $\mathcal{P}\{x = 2: 0, t\}$, and so forth. The general result is easily determined to be

$$\mathcal{P}\{x = k: 0, t\} = \frac{(\lambda t)^k e^{-\lambda t}}{k!}$$

for $k = 0, 1, 2, \ldots$. The desired probability density function for x is therefore

$$p_x(\alpha) = e^{-\lambda t} \sum_{i=0}^{\infty} \frac{(\lambda t)^i}{i!} \delta_D(\alpha - i) \tag{3.5-14}$$

A discrete random variable with this density function is said to have a Poisson distribution with parameter λt. The density function of Eq. (3.5-14) is plotted in Fig. 3.5-2 for several values of λt. Note that, as λt increases, the peak of the density shifts to the right. This is reasonable, since as λt increases one is more likely to find a larger number of event points in the interval $[0, t]$. A tabulation of the Poisson distribution may be found in several places.[1]

Hypergeometric Distribution. Suppose that we have a box containing N objects of two different types. There are M objects of the first type and $N - M$ objects of the second type. Now we draw *without replacement* $n < N$ objects at random from the box. We define a random variable x equal to the number of objects of the first type obtained. Clearly, x is a discrete random variable that takes the values $0, 1, 2, \ldots, n < N$. The probability density function of x is given by

$$p_x(\alpha) = \frac{1}{\binom{N}{n}} \sum_{i=0}^{n} \binom{M}{i} \binom{N - M}{n - i} \delta_D(\alpha - i) \tag{3.5-15}$$

and is known as a hypergeometric distribution with parameters N, M, and n. If M and $N - M$ are large and n is small comparatively, the binomial distribution with $p = M/N$ may be used to approximate the hypergeometric distribution. Tables of the hypergeometric distribution have been tabulated.[2]

[1] See, for example, *Poisson's Exponential Binomial Limit*, E. C. Molina, Van Nostrand Reinhold, Princeton, N.J., 1942. $\lambda = 0.001(0.001)0.01(0.01)0.3(1)1t(1)100$. *Tables of Poisson Distribution*, T. Kitagawa, Baifukan Press, Tokyo, 1952. $\lambda = 0.001(0.001)1.000$, $1.01(0.01)5.00$, $5.01(0.01)10.00$.

[2] See, for example, G. J. Lieberman and D. B. Owen, *Tables of the Hypergeometric Probability Distribution*, Stanford University Press, Stanford, Calif., 1961. $N = 2(1)100$, $M = 1(1)50$.

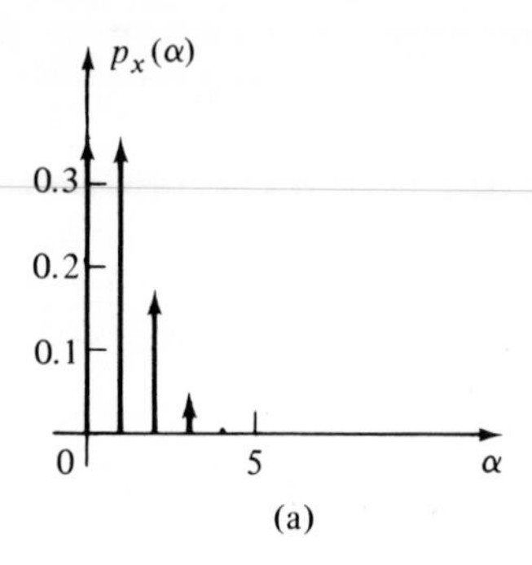

(a)

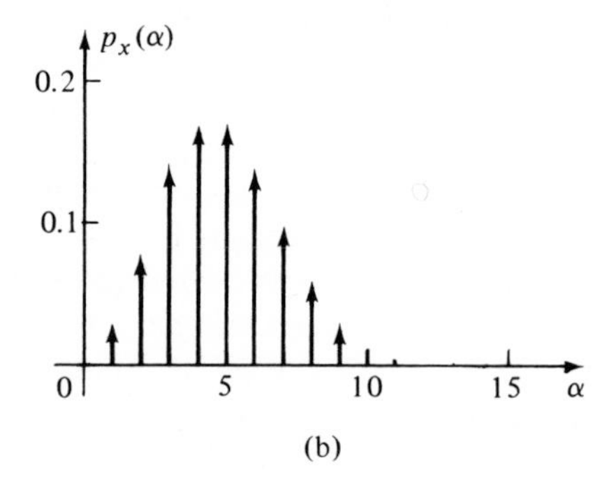

(b)

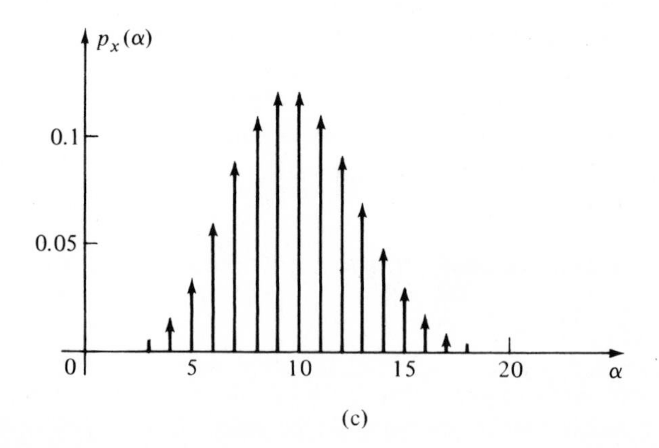

(c)

FIG. 3-5.2. Example of density functions for the Poisson distribution: (a) $\lambda t = 1$, (b) $\lambda t = 5$, (c) $\lambda t = 10$.

Negative Binomial Distribution. We consider independent experiments, each of which has an outcome ξ_0 or ξ_1 with $\mathcal{P}\{\xi_1\} = p$ at each trial. The experiments are run until the kth ξ_1 outcome is observed, where k is a positive integer. We define a random variable x as the number of ξ_0 outcomes observed before the kth ξ_1 outcome. In terms of a physical example, we could consider tossing a coin, with the probability of a head equal to p, until k heads are obtained; then we wish to determine the probability density of x equal to the number of tails observed before the kth head. The discrete random variable x is said to have a negative binomial distribution with a density function given by

$$p_x(\alpha) = p^k \sum_{i=0}^{\infty} \binom{k+i-1}{i} (1-p)^i \delta_D(\alpha - i) \tag{3.5-16}$$

with parameters k and p.

We consider next several continuous distributions.

Continuous Uniform Distribution. Although there are some physical phenomena that are described by the continuous uniform distribution, this distribution is also often used because of its extreme simplicity. Faced with a complete lack of knowledge concerning the distribution of a random phenomenon, one frequently selects a uniform distribution because it is simple and because it does not emphasize any given point. We say that a random variable x has a continuous uniform distribution on $[a, b]$ if

$$p_x(\alpha) = \begin{cases} 0 & \alpha \notin [a, b] \\ \dfrac{1}{b-a} & \alpha \in [a, b] \end{cases} \tag{3.5-17}$$

The associated distribution is easily determined to be

$$F_x(\alpha) = \begin{cases} 0 & \alpha < a \\ \dfrac{\alpha - a}{b - a} & a \le \alpha < b \\ 1 & b \le \alpha \end{cases} \tag{3.5-18}$$

Graphs of the density and distribution functions for a uniform distribution are shown in Fig. 3.5-3. An examination of this figure indicates why this distribution is also referred to as a rectangular distribution.

Normal Distribution. One of the most important continuous distributions is the normal or gaussian distribution. This distribution is widely used because it accurately represents a wide variety of physical situations; in addition, it has numerous convenient mathematical properties. Since we shall discuss the gaussian distribution in detail in Chapter 7, we give no derivation here. A continuous random variable x is said to have a normal distribution with parameters μ and σ^2 if its probability density function is given by

$$p_x(\alpha) = \frac{1}{\sqrt{2\pi}\sigma} \exp\left\{-\frac{(\alpha - \mu)^2}{2\sigma^2}\right\} \tag{3.5-19}$$

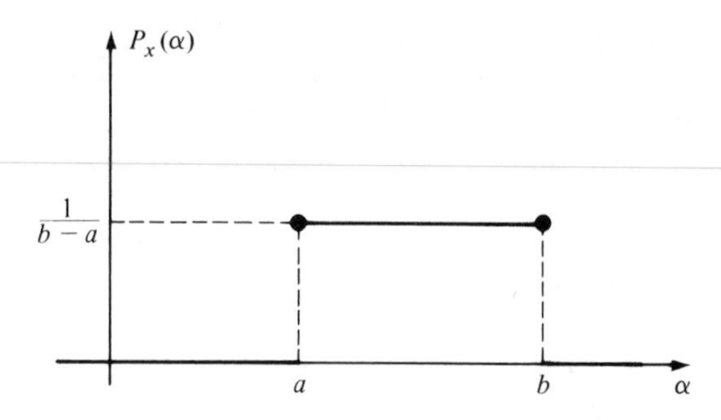

FIG. 3.5-3. Example of density function for a uniform distribution.

The associated distribution function is written as

$$F_x(\alpha) = \frac{1}{2} + \operatorname{erf}\left[\frac{\alpha - \mu}{\sigma}\right] \tag{3.5-20}$$

where erf(x) is the *error function*[1] defined as

$$\operatorname{erf}(x) = \frac{1}{\sqrt{2\pi}} \int_0^x e^{-y^2/2}\, dy \tag{3.5-21}$$

and is well tabulated. Because of the extensive use that we shall make of the normal distribution, we shall use the notation "x is $N(\mu, \sigma^2)$" to indicate that x is normally distributed with parameters μ and σ^2. A graph of the density function for a typical normal distribution is shown in Fig. 3.5-4.

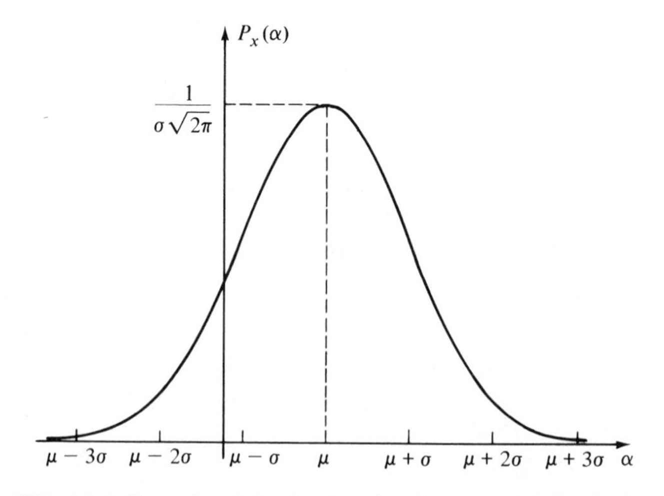

FIG. 3.5-4. Example of density function for a normal distribution.

[1] There are several different definitions of the error function, all essentially equivalent; hence one must be careful of which one is being used.

Rayleigh Distribution. We say that a continuous random variable has a Rayleigh distribution with parameter a^2 if its density function is given by

$$p_x(\alpha) = \frac{\alpha}{a^2} e^{-\alpha^2/2a^2} \mathscr{s}(\alpha) \tag{3.5-22}$$

The graph of this density function is shown in Fig. 3.5-5. Note that p_x is only nonzero for $\alpha \geq 0$; for this reason, the Rayleigh distribution is often used to represent the distribution of "magnitude-type" quantities.

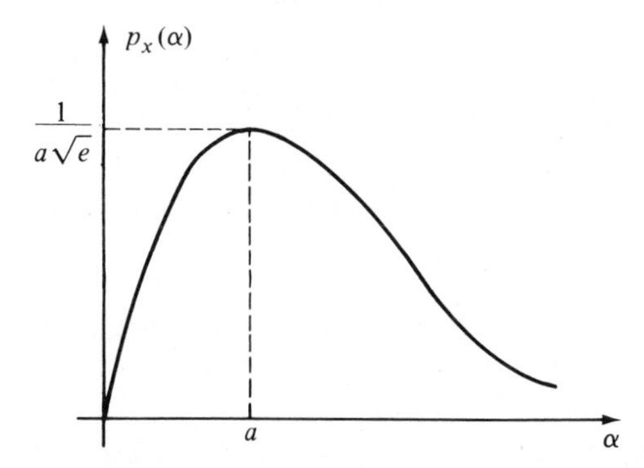

FIG. 3.5-5. Example of density function for a Rayleigh distribution.

Gamma Distributions. The gamma distributions are a two-parameter family of continuous probability density functions described by

$$p_x(\alpha) = \frac{\lambda^n e^{-\lambda\alpha} \alpha^{n-1}}{\Gamma(n)} \mathscr{s}(\alpha) \tag{3.5-23}$$

where $\lambda, n > 0$ and $\Gamma(x)$ is the *gamma function* defined by

$$\Gamma(x) = \int_0^\infty e^{-\gamma} \gamma^{x-1} \, d\gamma \qquad \text{for } x > 0 \tag{3.5-24}$$

If $x > 1$, we may integrate by parts once to obtain the recursive expression

$$\Gamma(x) = (x-1)\Gamma(x-1) \tag{3.5-25}$$

and since

$$\Gamma(1) = \int_0^\infty e^{-\gamma} \, d\gamma = 1 \tag{3.5-26}$$

we find that for x a positive integer

$$\Gamma(x) = (x-1)! \tag{3.5-27}$$

There are two special cases of the general gamma distribution that are particularly useful. If we set $n = 1$, we obtain the *exponential distribution* with parameter λ given by

$$p_x(\alpha) = \lambda e^{-\lambda\alpha} u(\alpha) \tag{3.5-28}$$

The exponential distribution is particularly useful in reliability studies; the reason for its name is obvious.

If we set $\lambda = \frac{1}{2}$ and $n = m/2$, where m is a positive integer, we obtain the χ^2 (*chi-squared*) *distribution* with m degrees of freedom:

$$p_x(\alpha) = \frac{e^{-\alpha/2}\alpha^{(m/2)-1}}{2^{m/2}\Gamma(m/2)} u(\alpha) \tag{3.5-29}$$

This distribution, which is widely used in statistical studies, is extensively tabulated in numerous places. We shall discuss this distribution in more detail in Chapter 7.

Beta Distributions. The beta distributions are also a two-parameter family of probability density functions of the form

$$p_x(\alpha) = \frac{\alpha^{a-1}(1-\alpha)^{b-1}}{B(a, b)}[u(\alpha) - u(\alpha - 1)] \tag{3.5-30}$$

where $a, b > 0$ and $B(a, b)$ is the *Beta function* defined as

$$B(a, b) = \frac{\Gamma(a)\Gamma(b)}{\Gamma(a + b)} \tag{3.5-31}$$

If $a = b = 1$, we obtain the uniform distribution on [0, 1]. Two other interesting special cases are when $a = 1$, $b = 2$, and p_x becomes

$$p_x(\alpha) = 2(1 - \alpha)[u(\alpha) - u(\alpha - 1)] \tag{3.5-32}$$

or when $a = 2$, $b = 1$, and p_x is

$$p_x(\alpha) = 2\alpha[u(\alpha) - u(\alpha - 1)] \tag{3.5-33}$$

These two distributions are known as *triangular distributions* for obvious reasons indicated by a simple sketch.

Cauchy Distribution. A continuous random variable is said to have a Cauchy distribution with parameter λ if its probability density function is given by

$$p_x(\alpha) = \frac{\lambda/\pi}{\lambda^2 + \alpha^2} \tag{3.5-34}$$

The Cauchy distribution has some very strange mathematical properties, which we shall explore in more detail in Chapter 4.

For ease of future reference, the distributions discussed in this section have been summarized in Table 3.5-1.

TABLE 3.5-1. Examples of Density Functions

Name	Parameters	Density Functions
Discrete uniform	n; $\alpha_i, i = 1, 2, \ldots, n$	$p_x(\alpha) = \frac{1}{n}\sum_{i=1}^{n} \delta_D(\alpha - \alpha_i)$
Binomial	$n; p$	$p_x(\alpha) = \sum_{i=0}^{n} \binom{n}{i} p^i(1-p)^{n-i}\delta_D(\alpha - i)$
Poisson	λt	$p_x(\alpha) = e^{-\lambda t}\sum_{i=0}^{n} \frac{(\lambda t)^i}{i!}\delta_D(\alpha - i)$
Hypergeometric	$N; M; n$	$p_x(\alpha) = \frac{1}{\binom{N}{n}}\sum_{i=0}^{n} \binom{M}{i}\binom{N-M}{n-i}\delta_D(\alpha - i)$
Negative binomial	$r; p$	$p_x(\alpha) = p^r \sum_{i=0}^{\infty} \binom{r+i-1}{i}(1-p)^i\delta_D(\alpha - i)$
Continuous uniform	$a; b$	$p_x(\alpha) = \frac{1}{b-a}[\mathscr{u}(\alpha - a) - \mathscr{u}(\alpha - b)]$
Normal (gaussian)	$\mu; \sigma^2$	$p_x(\alpha) = \frac{1}{\sqrt{2\pi}\sigma}\exp\left\{-\frac{(\alpha-\mu)^2}{2\sigma^2}\right\}$
Rayleigh	a^2	$p_x(\alpha) = \frac{\alpha}{a^2}e^{-\alpha^2/2a^2}\mathscr{u}(\alpha)$
Exponential	λ	$p_x(\alpha) = e^{-\lambda\alpha}\mathscr{u}(\alpha)$
Chi-squared	m	$p_x(\alpha) = \frac{e^{-\alpha/2}\alpha^{(m/2-1)}}{2^{m/2}\Gamma(m/2)}\mathscr{u}(\alpha)$
Beta	$a; b$	$p_x(\alpha) = \frac{\alpha^{a-1}(1-\alpha)^{b-1}}{B(a, b)}[\mathscr{u}(\alpha) - \mathscr{u}(\alpha - 1)]$
Cauchy	λ	$p_x(\alpha) = \frac{\lambda/\pi}{\lambda^2 + \alpha^2}$
Gamma	$\lambda; n$	$p_x(\alpha) = \frac{\lambda^n e^{-\lambda\alpha}\alpha^{n-1}}{\Gamma(n)}\mathscr{u}(\alpha)$

3.6. *Joint and marginal distribution and density functions*

In the preceding sections we restricted the treatment to scalar random variables for simplicity. In this section we wish to expand the study to multidimensional random vectors, and in doing so we introduce some new concepts. We shall begin by considering the case of two random variables (or equivalently a two-dimensional random vector); the extension to the general case will follow trivially.

Let us suppose that two random variables x_1 and x_2 are defined on the same physical experiment $(\mathcal{S}, \mathcal{F}, \mathcal{P})$. We use the notation $\{x_1 \leq \alpha_1, x_2 \leq \alpha_2\}$ to represent the set of outcomes $\{\xi: x_1(\xi) \leq \alpha_1 \text{ and } x_2(\xi) \leq \alpha_2\}$ so that the set $\{x_1 \leq \alpha_1, x_2 \leq \alpha_2\}$ is the intersection of the set $\{x_1 \leq \alpha_1\}$ with the set $\{x_2 \leq \alpha_2\}$; that is,

$$\{x_1 \leq \alpha_1, x_2 \leq \alpha_2\} = \{x_1 \leq \alpha_1\}\{x_2 \leq \alpha_2\} \tag{3.6-1}$$

By the definition of a random variable, the sets $\{x_1 \leq \alpha_1\}$ and $\{x_2 \leq \alpha_2\}$ are events from $\mathcal{F}$. Therefore, $\{x_1 \leq \alpha_1, x_2 \leq \alpha_2\}$ is also an event, since it is an intersection of events which must also be an event for every α_1 and $\alpha_2 \in \mathcal{R}$. Hence, $\mathcal{P}\{x_1 \leq \alpha_1, x_2 \leq \alpha_2\}$ is well defined for every $\alpha_1, \alpha_2 \in \mathcal{R}$, and we define the *joint probability distribution* function $F_{x_1x_2}$ as

$$F_{x_1x_2}(\alpha_1, \alpha_2) = \mathcal{P}\{x_1 \leq \alpha_1, x_2 \leq \alpha_2\} \tag{3.6-2}$$

As before, we shall often find it convenient to use the notation $F(x_1, x_2)$ to represent $F_{x_1x_2}$. It is quite easy to show that the joint distribution function has the following properties, which are analogous to the scalar case:

1. $F_{x_1x_2}(+\infty, +\infty) = 1$.
2. $F_{x_1x_2}(-\infty, -\infty) = F_{x_1x_2}(-\infty, \alpha_2) = F_{x_1x_2}(\alpha_1, -\infty) = 0$.
3. $F_{x_1x_2}(\alpha_1, \alpha_1) \leq F_{x_1x_2}(\alpha_1, \alpha_2 + \epsilon)$ for any $\epsilon \geq 0$.
4. $F_{x_1x_2}(\alpha_1, \alpha_2) \leq F_{x_1x_2}(\alpha_1 + \epsilon, \alpha_2)$ for any $\epsilon \geq 0$.

The proof of these properties, which closely follows the arguments of Section 3.3, is left to the reader as an exercise.

One might logically question whether there is any relationship between the distribution functions of the individual random variables F_{x_1} and F_{x_2}, often referred to as *marginal distribution functions*, and the joint distribution $F_{x_1x_2}$. In general, the marginal distributions F_{x_1} and F_{x_2} can be determined from $F_{x_1x_2}$, but the converse is not true; that is, $F_{x_1x_2}$ cannot in general be obtained from F_{x_1} and F_{x_2}. The fact that $F_{x_1x_2}$ cannot be derived from the marginal distributions F_{x_1} and F_{x_2} is easy to understand, since the marginal distributions contain no information about the relationship between the random variables x_1 and x_2 but only information about the individual random variables.

To obtain the marginal distribution F_{x_1} from the joint distribution $F_{x_1x_2}$, we need only recognize that the event $\{x_1 \leq \alpha_1, x_2 \leq +\infty\}$ is identical to the event $\{x_1 \leq \alpha_1\}$. This fact is easily established by reference to Eq. (3.6-1), so that $\{x_1 \leq \alpha_1, x_2 \leq +\infty\}$ becomes

$$\{x_1 \leq \alpha_1, x_2 \leq +\infty\} = \{x_1 \leq \alpha_1\}\{x_2 \leq +\infty\} \tag{3.6-3}$$

However, $\{x_2 \leq +\infty\} = \mathcal{S}$, so Eq. (3.6-3) becomes

$$\{x_1 \leq \alpha_1, x_2 \leq +\infty\} = \{x_1 \leq \alpha_1\}\mathcal{S} = \{x_1 \leq \alpha_1\} \tag{3.6-4}$$

which is the desired result. By the definition of the distribution function and Eq. (3.6-4), we have

$$F_{x_1}(\alpha) = \mathcal{P}\{x_1 \leq \alpha\} = \mathcal{P}\{x_1 \leq \alpha, x_2 \leq +\infty\}$$

and using Eq. (3.6-2) gives

$$F_{x_1}(\alpha) = F_{x_1x_2}(\alpha, +\infty) \tag{3.6-5}$$

In a similar manner, we could show that

$$F_{x_2}(\alpha) = F_{x_1x_2}(+\infty, \alpha) \tag{3.6-6}$$

Associated with the joint distribution function $F_{x_1x_2}$ is the *joint density function* $p_{x_1x_2}$ defined by the integral relation

$$F_{x_1x_2}(\alpha_1, \alpha_2) = \int_{-\infty}^{\alpha_1} d\beta_1 \int_{-\infty}^{\alpha_2} d\beta_2\, p_{x_1x_2}(\beta_1, \beta_2) \tag{3.6-7}$$

with the requirement that $p_{x_1x_2}(\alpha_1, \alpha_2) \geq 0$ for all $\alpha_1, \alpha_2 \in \mathcal{R}$. Now if $\mathfrak{D}$ is any region in the $x_1 - x_2$ plane that is a finite sum or product of sets of the form $\{x_1 \leq \alpha_1, x_2 \leq \alpha_2\}$, the probability that $(x_1, x_2) \in \mathfrak{D}$ is given by

$$\mathcal{P}\{(x_1, x_2) \in \mathfrak{D} \subseteq \mathcal{R}^2\} = \iint_{\mathfrak{D}} p_{x_1x_2}(\alpha_1, \alpha_2)\, d\alpha_1\, d\alpha_2 \tag{3.6-8}$$

In particular, if $\mathfrak{D} = \mathcal{R}^2$, we have

$$\mathcal{P}\{(x_1, x_2) \in \mathcal{R}^2\} = \mathcal{P}\{\mathcal{S}\} = \int_{-\infty}^{\infty} \int_{-\infty}^{\infty} p_{x_1x_2}(\alpha_1, \alpha_2)\, d\alpha_1\, d\alpha_2 = 1 \tag{3.6-9}$$

which says that the total volume under the joint probability density function must be 1.

Just as the marginal distributions can be obtained from the joint distribution function, the marginal density functions p_{x_1} and p_{x_2} can be obtained from the joint density $p_{x_1x_2}$. Let us consider Eq. (3.6-5) again; using the definition of $p_{x_1x_2}$ given in Eq. (3.6-7), we can write Eq. (3.6-5) as

$$F_{x_1}(\alpha) = \int_{-\infty}^{\alpha} d\beta_1 \int_{-\infty}^{\infty} d\beta_2\, p_{x_1x_2}(\beta_1, \beta_2) \tag{3.6-10}$$

A comparison of Eq. (3.6-10) with the original definition of a probability density function [Eq. (3.4-1)] shows that $p_{x_1}(\beta_1)$ is given by

$$p_{x_1}(\beta_1) = \int_{-\infty}^{\infty} p_{x_1x_2}(\beta_1, \beta_2)\, d\beta_2 \tag{3.6-11}$$

since this expression will satisfy Eq. (3.4-1). Often one says that "the variable x_2 has been integrated out" to indicate that we have removed x_2 from the joint density function by integrating over its entire range so that its value is no longer of importance. In a similar fashion, we can show that

$$p_{x_2}(\beta_2) = \int_{-\infty}^{\infty} p_{x_1x_2}(\beta_1, \beta_2)\, d\beta_1 \tag{3.6.12}$$

Hence we can easily obtain the marginal densities p_{x_1} and p_{x_2} from the joint density function; but once again it is, in general, not possible to obtain $p_{x_1x_2}$ from p_{x_1} and p_{x_2}.

The generalization of the above concepts of joint distribution and density functions to the n-dimensional case is quite straightforward. Let us suppose that we have n random variables $x_1, x_2, \ldots, x_n$ defined on the same experiment $(\mathcal{S}, \mathcal{F}, \mathcal{P})$. Then the set $\{x_1 \leq \alpha_1, x_2 \leq \alpha_2, \ldots, x_n \leq \alpha_n\}$ is always an event in $\mathcal{F}$ for every set of $\alpha_i \in \mathcal{R}$, $i = 1, 2, \ldots, n$, and its probability is well defined. The joint probability distribution function of the n random variables $x_1, x_2, \ldots, x_n$ is defined as

$$F_{x_1x_2\cdots x_n}(\alpha_1, \alpha_2, \ldots, \alpha_n) = \mathcal{P}\{x_1 \leq \alpha_1, \ldots, x_n \leq \alpha_n\} \tag{3.6-13}$$

The notation can be considerably simplified if we treat the n random variables as the components of an n-dimensional random vector $\mathbf{x}$. The general definition of a joint distribution function takes the following form:

Definition 3.6-1. ***Joint Distribution Function:*** The joint distribution function $F_{\mathbf{x}}$ of an n-dimensional random vector $\mathbf{x}$ is defined by

$$F_{\mathbf{x}}(\boldsymbol{\alpha}) = \mathcal{P}\{\mathbf{x} \leq \boldsymbol{\alpha}\} = \mathcal{P}\{\xi : \mathbf{x}(\xi) \leq \boldsymbol{\alpha}\}$$

for all $\boldsymbol{\alpha} \in \mathcal{R}^n$.

The concepts of marginal and joint distributions become a bit confusing in the general case. Consider for example the case of three random variables x_1, x_2, and x_3. Here $F_{x_1x_2x_3}$ is clearly the joint distribution for the three random variables, but $F_{x_1x_2}$ might be considered as a joint distribution of x_1 and x_2 or as a marginal distribution of x_1 and x_2 relative to the set of three random variables. Hence we shall typically avoid the use of the terminology "marginal" and "joint" unless the meaning is clear and the use is helpful.

From the distribution $F_{\mathbf{x}}$, we can obtain the distribution of any subset of the elements of $\mathbf{x}$, such as x_1, x_2 by simply setting the value associated with the variables we wish to delete equal to $+\infty$. For example, if we have $F_{x_1x_2x_3x_4}$, we may obtain $F_{x_1x_3}$ as

$$F_{x_1x_3}(\alpha_1, \alpha_3) = F_{x_1x_2x_3x_4}(\alpha_1, +\infty, \alpha_3, +\infty) \tag{3.6-14}$$

or we would find $F_{x_2x_4}$ as

$$F_{x_2x_4}(\alpha_2, \alpha_4) = F_{x_1x_2x_3x_4}(+\infty, \alpha_2, +\infty, \alpha_4)$$

If we wished to obtain $F_{x_3}(\alpha_3)$, we would simply set $\alpha_1 = \alpha_2 = \alpha_4 = \infty$, so that we have

$$F_{x_3}(\alpha_3) = F_{x_1x_2x_3x_4}(+\infty, +\infty, \alpha_3, +\infty) \tag{3.6-15}$$

It might be noted here that the vector notation is not of much assistance when we wish to work with specific elements in $\mathbf{x}$, as in Eqs. (3.6-14) and (3.6-15).

The general case of a density function for n random variables follows directly from the two-dimensional case. We say that $p_{x_1x_2\cdots x_n}$ is the density function associated with $x_1, x_2, \ldots, x_n$ if

$$F_{x_1x_2\cdots x_n}(\alpha_1, \alpha_2, \ldots, \alpha_n) = \int_{-\infty}^{\alpha_1} d\beta_1 \int_{-\infty}^{\alpha_2} d\beta_2 \cdots \int_{-\infty}^{\alpha_n} d\beta_n\, p_{x_1x_2\cdots x_n}(\beta_1, \beta_2, \ldots, \beta_n) \tag{3.6-16}$$

and $p_{x_1x_2\cdots x_n}(\beta_1, \beta_2, \ldots, \beta_n) \geq 0$ for all $\beta_i \in \mathcal{R}$, $i = 1, 2, \ldots, n$. Once again we may simplify the expressions by the use of vector notation so that the definition of the density function for an n-dimensional random vector becomes

Definition 3.6-2. ***Joint Density Function:*** The joint probability density function for an n-dimensional random vector $\mathbf{x}$ having a joint distribution function $F_{\mathbf{x}}$ is any function $p_{\mathbf{x}}$ defined on $\mathcal{R}^n$ and such that

$$F_{\mathbf{x}}(\boldsymbol{\alpha}) = \int_{-\infty}^{\boldsymbol{\alpha}} p_{\mathbf{x}}(\boldsymbol{\gamma})\, d\boldsymbol{\gamma} \tag{3.6-17}$$

for all $\boldsymbol{\alpha} \in \mathcal{R}^n$ and $p_{\mathbf{x}}(\boldsymbol{\alpha}) \geq 0$ for all $\boldsymbol{\alpha} \in \mathcal{R}^n$.

In Eq. (3.6-17) the vector limits on the integral are used to represent the multiple integral, and $d\boldsymbol{\gamma}$ denotes the scalar $d\gamma_1 d\gamma_2 \ldots, d\gamma_n$, which is contrasted to the vector differential $\mathbf{d}\boldsymbol{\gamma} = [d\gamma_1\, d\gamma_2\, d\gamma_3 \cdots d\gamma_n]^T$.

From the joint density function $p_{\mathbf{x}}$ for $\mathbf{x}$ we may obtain the density for any subset of the elements of $\mathbf{x}$ by simply "integrating out" the variables that are not needed. If, for example, $\mathbf{x}$ were a four-dimensional vector and we desired $p_{x_1x_3}$, we would "integrate out" x_2 and x_4 and obtain

$$p_{x_1x_3}(\alpha_1, \alpha_3) = \int_{-\infty}^{\infty} \int_{-\infty}^{\infty} p_{x_1x_2x_3x_4}(\alpha_1, \alpha_2, \alpha_3, \alpha_4)\, d\alpha_2\, d\alpha_4 \tag{3.6-18}$$

Similarly, if we wanted p_{x_2}, we would "integrate out" x_1, x_3, and x_4 so that we would have

$$p_{x_2}(\alpha_2) = \iiint_{-\infty}^{\infty} p_{x_1x_2x_3x_4}(\alpha_1, \alpha_2, \alpha_3, \alpha_4)\, d\alpha_1\, d\alpha_3\, d\alpha_4 \tag{3.6-19}$$

The reader may question why one would obtain the joint density and then use it to obtain the desired marginal density, rather than just compute the marginal density directly. In some cases, one might want to follow this procedure; however, in other cases it may be significantly easier to find the joint density and then compute the marginal density from it. If one wanted to compute a number of different marginal densities, this latter procedure might be preferable since one would need to compute the joint density only once.

So far we have seen that from a given joint distribution or density any desired marginal distribution or density can be found. However, we cannot, in general, obtain the joint distribution or density from the knowledge of marginal distributions or densities. Once again, the basic difficulty is the need for information about the relationships between the elements of $\mathbf{x}$. One way of specifying this information is by means of conditional distributions or densities discussed in Section 3.1.

EXERCISES 3.6

3.6-1. Establish the following properties of the two-dimensional joint distribution function:

(a) $F_{x_1x_2}(+\infty, +\infty) = 1$.

(b) $F_{x_1x_2}(-\infty, -\infty) = F_{x_1x_2}(-\infty, \alpha_2) = F_{x_1x_2}(\alpha_1, -\infty) = 0$.

(c) $F_{x_1x_2}(\alpha_1, \alpha_2) \leq F_{x_1x_2}(\alpha_1, \alpha_2 + \epsilon)$ for all $\epsilon \geq 0$.

(d) $F_{x_1x_2}(\alpha_1, \alpha_2) \leq F_{x_1x_2}(\alpha_1 + \epsilon, \alpha_2)$ for all $\epsilon \geq 0$.

3.6-2. If the joint probability density of x and y is given by

$$p_{xy}(\alpha, \beta) = \begin{cases} \dfrac{1}{\pi} & \alpha^2 + \beta^2 \leq 1 \\ 0 & \text{otherwise} \end{cases}$$

find $p_x(\alpha)$.

Answer:

$$p_x(\alpha) = \begin{cases} \dfrac{2}{\pi}\sqrt{1-\alpha^2} & |\alpha| \leq 1 \\ 0 & \text{otherwise} \end{cases}$$

3.7. *Conditional distribution and density functions*

In Section 2.6 we introduced the concept of conditional probability. We wish to consider the development of conditional distribution and density functions in this section. From Section 2.6 we know that the conditional probability of an event $\mathcal{B}_1$ given an event $\mathcal{B}_2$ is

$$\mathcal{P}\{\mathcal{B}_1 \mid \mathcal{B}_2\} = \frac{\mathcal{P}\{\mathcal{B}_1\mathcal{B}_2\}}{\mathcal{P}\{\mathcal{B}_2\}} \tag{3.7-1}$$

where it is assumed that $\mathcal{P}\{\mathcal{B}_2\} \neq 0$.

Now that we have introduced the concept of a random variable, we can define the events $\mathcal{B}_1$ and $\mathcal{B}_2$ in terms of random variables. In particular, let us assume that $\mathcal{B}_1 = \{x_1 \in \mathcal{A}_1\}$; the conditioning event $\mathcal{B}_2$ may be described in terms of the random variable x_1 or some other random variable x_2. For generality we shall assume that $\mathcal{B}_2$ is defined in terms of x_2, which may be equal to x_1, as $\mathcal{B}_2 = \{x_2 \in \mathcal{A}_2\}$. In terms of these descriptions of the events $\mathcal{B}_1$ and $\mathcal{B}_2$, Eq. (3.7-1) becomes

$$\mathcal{P}\{x_1 \in \mathcal{A}_1 \mid x_2 \in \mathcal{A}_2\} = \frac{\mathcal{P}\{x_1 \in \mathcal{A}_1, x_2 \in \mathcal{A}_2\}}{\mathcal{P}\{x_2 \in \mathcal{A}_2\}} \tag{3.7-2}$$

We may express this result in terms of the joint probability density $p_{x_1x_2}$ and the marginal density p_{x_2} as

$$\mathcal{P}\{x_1 \in \mathcal{A}_1 \mid x_2 \in \mathcal{A}_2\} = \frac{\int_{\mathcal{A}_1} d\alpha_1 \int_{\mathcal{A}_2} d\alpha_2\, p_{x_1x_2}(\alpha_1, \alpha_2)}{\int_{\mathcal{A}_2} p_{x_2}(\alpha_2)\, d\alpha_2}$$

Now let us consider the case when $\mathcal{A}_1 = [-\infty, \alpha_1]$; then we may define the conditional distribution of x_1, given the event $\mathcal{B}_2 = \{x_2 \in \mathcal{A}_2\}$, as

$$F_{x_1|\mathcal{B}_2}(\alpha_1 \mid \mathcal{B}_2) \triangleq \mathcal{P}\{x_1 \leq \alpha_1 \mid \mathcal{B}_2\} = \frac{\mathcal{P}\{x_1 \leq \alpha_1, \mathcal{B}_2\}}{\mathcal{P}\{\mathcal{B}_2\}} \tag{3.7-3}$$

Note that $F_{x_1|\mathcal{B}_2}$ satisfies all the properties of a distribution function; in particular

$$F_{x_1|\mathcal{B}_2}(-\infty \mid \mathcal{B}_2) = \frac{\mathcal{P}\{x_1 \leq -\infty, \mathcal{B}_2\}}{\mathcal{P}\{\mathcal{B}_2\}} = 0$$

and

$$F_{x_1|\mathcal{B}_2}(+\infty \mid \mathcal{B}_2) = \frac{\mathcal{P}\{x_1 \leq +\infty, \mathcal{B}_2\}}{\mathcal{P}\{\mathcal{B}_2\}} = 1$$

Because $F_{x_1|\mathcal{B}_2}$ is a distribution function, we may define a conditional density function $p_{x_1|\mathcal{B}_2}$ of x_1, given $\mathcal{B}_2$ in the usual fashion, as

$$F_{x_1|\mathcal{B}_2}(\alpha_1 \mid \mathcal{B}_2) = \int_{-\infty}^{\alpha_1} p_{x_1|\mathcal{B}_2}(\gamma \mid \mathcal{B}_2)\, d\gamma \tag{3.7-4}$$

Although the distribution function $F_{x_1|\mathcal{B}_2}$ and the density function $p_{x_1|\mathcal{B}_2}$ satisfy all their associated properties, they are not the conditional probability functions that we generally use. To obtain a more useful form, let us assume that $\mathcal{B}_2$ is defined by $\{x_2 \leq \alpha_2\}$; then we define the conditional distribution of x_1, given the event $\{x_2 \leq \alpha_2\}$, as

$$F_{x_1|x_2}(\alpha_1 \mid \alpha_2) \triangleq \mathcal{P}\{x_1 \leq \alpha_1 \mid x_2 \leq \alpha_2\} = \frac{\mathcal{P}\{x_1 \leq \alpha_1, x_2 \leq \alpha_2\}}{\mathcal{P}\{x_2 \leq \alpha_2\}} \tag{3.7-5}$$

In Eq. (3.7-5) we recognize $\mathcal{P}\{x_1 \leq \alpha_1, x_2 \leq \alpha_2\}$ as the joint distribution of x_1 and x_2, $F_{x_1x_2}(\alpha_1, \alpha_2)$, and $\mathcal{P}\{x_2 \leq \alpha_2\}$ as the marginal distribution of x_2, $F_{x_2}(\alpha_2)$, so that Eq. (3.7-5) may be written as

$$F_{x_1|x_2}(\alpha_1 \mid \alpha_2) = \frac{F_{x_1x_2}(\alpha_1, \alpha_2)}{F_{x_2}(\alpha_2)} \tag{3.7.6}$$

Note the similarity of form of Eq. (3.7-6) and Eq. (3.7-1) or (3.7-2). For this reason, $F_{x_1|x_2}$ is the usual form of conditional distribution that we shall use. Once again, a density function may be associated with the distribution function $F_{x_1|x_2}$; however, its use is limited, and hence we will not bother to define it.

To develop a meaningful conditional density, we need to return to Eq. (3.7-3) and let $\mathcal{B}_2 = \{\alpha_2 < x_2 \leq \alpha_2 + d\alpha_2\}$. For simplicity we shall assume that x_2 is a continuous random variable. Now Eq. (3.7-3) becomes

$$\begin{aligned} F_{x_1|\mathcal{B}_2}(\alpha_1 \mid \mathcal{B}_2) &= \frac{\mathcal{P}\{x_1 \leq \alpha_1, \alpha_2 < x_2 \leq \alpha_2 + d\alpha_2\}}{\mathcal{P}\{\alpha_2 < x_2 \leq \alpha_2 + d\alpha_2\}} \\ &= \frac{\mathcal{P}\{x_1 \leq \alpha_1, x_2 \leq \alpha_2 + d\alpha_2\} - \mathcal{P}\{x_1 \leq \alpha_1, x_2 \leq \alpha_2\}}{\mathcal{P}\{x_2 \leq \alpha_2 + d\alpha_2\} - \mathcal{P}\{x_2 \leq \alpha_2\}} \end{aligned} \tag{3.7-7}$$

because of the rule for the probability of the union of two disjoint events $\{x_2 \leq \alpha_2\}$ and $\{\alpha_2 < x_2 \leq \alpha_2 + d\alpha_2\}$.

By using the distribution function equivalent of the probabilities in the numerator and denominator of this expression, we obtain

$$F_{x_1|\mathcal{B}_2}(\alpha_1 \mid \mathcal{B}_2) = \frac{F_{x_1x_2}(\alpha_1, \alpha_2 + d\alpha_2) - F_{x_1x_2}(\alpha_1, \alpha_2)}{F_{x_2}(\alpha_2 + d\alpha_2) - F_{x_2}(\alpha_2)}$$

If we multiply and divide this expression by $d\alpha_2$, we have

$$F_{x_1|\mathcal{B}_2}(\alpha_1 \mid \mathcal{B}_2) = \frac{[F_{x_1x_2}(\alpha_1, \alpha_2 + d\alpha_2) - F_{x_1x_2}(\alpha_1, \alpha_2)]/d\alpha_2}{[F_{x_2}(\alpha_2 + d\alpha_2) - F_{x_2}(\alpha_2)]/d\alpha_2}$$

and the limit as $d\alpha_2 \longrightarrow 0$ of the foregoing becomes

$$F_{x_1|x_2=\alpha_2}(\alpha_1 \mid x_2 = \alpha_2) = \frac{\partial F_{x_1x_2}(\alpha_1, \alpha_2)/\partial\alpha_2}{\partial F_{x_2}(\alpha_2)/\partial\alpha_2} \tag{3.7-8}$$

provided the required derivatives exist.[1] Equation (3.7-8) may be written in terms of density functions as

$$F_{x_1|x_2=\alpha_2}(\alpha_1 \mid x_2 = \alpha_2) = \frac{\int_{-\infty}^{\alpha_1} p_{x_1x_2}(\gamma_1, \alpha_2)\, d\gamma_1}{p_{x_2}(\alpha_2)} \tag{3.7-9}$$

The conditional density function $p_{x_1|x_2}$ defined by

$$F_{x_1|x_2=\alpha_2}(\alpha_1 \mid x_2 = \alpha_2) = \int_{-\infty}^{\alpha_1} p_{x_1|x_2}(\gamma_1 \mid \alpha_2)\, d\gamma_1$$

is obviously given by

$$p_{x_1|x_2}(\alpha_1 \mid \alpha_2) = \frac{p_{x_1x_2}(\alpha_1, \alpha_2)}{p_{x_2}(\alpha_2)} \tag{3.7-10}$$

Once again, we note the similarity in form of this result to Eqs. (3.7-1) and (3.7-2) or (3.7-6). It is this form of conditional probability density that will be of most use to us. It should be emphasized that our principal form of conditional distribution as given by Eq. (3.7-6) and the conditional density of Eq. (3.7-10) are not related by the usual integral relationship; that is,

$$F_{x_1|x_2}(\alpha_1 \mid \alpha_2) \neq \int_{-\infty}^{\alpha_1} p_{x_1|x_2}(\gamma_1 \mid \alpha_2)\, d\gamma_1$$

[1] Since the derivatives will exist except at a countable number of points, we may assign the derivative any positive value at these points since they will not affect the final result.

The basic reason is that in $F_{x_1|x_2}$ the conditioning event is $\{x_2 \leq \alpha_2\}$, whereas in $p_{x_1|x_2}$ the conditioning event is $\{x_2 = \alpha_2\}$. There is no simple expression relating $F_{x_1|x_2}$ and $p_{x_1|x_2}$. The relationship of Eq. (3.7-10) is only valid when x_2 is a continuous random variable. For the treatment of the cases when x_2 is discrete or mixed, it is probably best to revert to basic probability spaces definitions or the use of the conditional distribution of Eq. (3.7-6).

We could carry out a completely parallel treatment for the conditional probability of β_2 given β_1. In particular, we would find that the counterpart of Eq. (3.7-6) is

$$F_{x_2|x_1}(\alpha_2 | \alpha_1) = \frac{F_{x_1x_2}(\alpha_1, \alpha_2)}{F_{x_1}(\alpha_1)} \tag{3.7-11}$$

whereas Eq. (3.7-10) would become

$$p_{x_2|x_1}(\alpha_2 | \alpha_1) = \frac{p_{x_1x_2}(\alpha_1, \alpha_2)}{p_{x_1}(\alpha_1)} \tag{3.7-12}$$

The extension of the development to the case when either or both x_1 and x_2 are random vectors is straightforward. The general definition of conditional distribution takes the following form:

Definition 3.7-1. ***Conditional Distribution Function:*** The conditional distribution function $F_{\mathbf{x}_1|\mathbf{x}_2}$ of a random vector $\mathbf{x}_1$, given the event $\{\mathbf{x}_2 \leq \boldsymbol{\alpha}_2\}$, is defined by

$$F_{\mathbf{x}_1|\mathbf{x}_2}(\boldsymbol{\alpha}_1 | \boldsymbol{\alpha}_2) \triangleq \mathcal{P}\{\mathbf{x}_1 \leq \boldsymbol{\alpha}_1 | \mathbf{x}_2 \leq \boldsymbol{\alpha}_2\} = \frac{F_{\mathbf{x}_1\mathbf{x}_2}(\boldsymbol{\alpha}_1, \boldsymbol{\alpha}_2)}{F_{\mathbf{x}_2}(\boldsymbol{\alpha}_2)} \tag{3.7-13}$$

for all $\boldsymbol{\alpha}_1 \in \mathcal{R}^n$ and $\boldsymbol{\alpha}_2 \in \mathcal{R}^m$.

The general definition of a conditional probability density function is given by

Definition 3.7-2. ***Conditional Density Function:*** The conditional density function $p_{\mathbf{x}_1|\mathbf{x}_2}$ of a random vector $\mathbf{x}_1$, given the event $\{\mathbf{x}_2 = \boldsymbol{\alpha}_2\}$, is defined by

$$p_{\mathbf{x}_1|\mathbf{x}_2}(\boldsymbol{\alpha}_1 | \boldsymbol{\alpha}_2) = \frac{p_{\mathbf{x}_1\mathbf{x}_2}(\boldsymbol{\alpha}_1, \boldsymbol{\alpha}_2)}{p_{\mathbf{x}_2}(\boldsymbol{\alpha}_2)} \tag{3.7-14}$$

or

$$p_{\mathbf{x}_1|\mathbf{x}_2}(\boldsymbol{\alpha}_1 | \boldsymbol{\alpha}_2)\, d\boldsymbol{\alpha}_1 = \mathcal{P}(\boldsymbol{\alpha}_1 < \mathbf{x}_1 \leq \boldsymbol{\alpha}_1 + \mathbf{d}\boldsymbol{\alpha}_1 | \mathbf{x}_2 = \boldsymbol{\alpha}_2)$$

for all $\boldsymbol{\alpha}_1 \in \mathcal{R}^n$ and $\boldsymbol{\alpha}_2 \in \mathcal{R}^m$.

For simplicity, we shall normally refer to $p_{\mathbf{x}_1|\mathbf{x}_2}$ as the conditional (probability) density of $\mathbf{x}_1$ given $\mathbf{x}_2$. Once again, we may use the notation $p(\mathbf{x}_1 \mid \mathbf{x}_2)$, where it is more convenient, to represent $p_{\mathbf{x}_1|\mathbf{x}_2}$.

Conditional probability densities play an important role in almost all applications of random process theory, including statistical communication and stochastic control theory. Their use often permits one to more clearly state a result or to manipulate known information. There are several special rules involving conditional densities that are often quite helpful; we consider some of the more important ones in the following development.

EXERCISES 3.7

3.7-1. If the joint density of x and y is given by

$$p_{xy}(\alpha, \beta) = \begin{cases} \dfrac{1}{\pi} & \alpha^2 + \beta_2 \leq 1 \\ 0 & \text{otherwise} \end{cases}$$

find $p_{x|y}(\alpha \mid \beta)$.

Answer:

$$p_{x|y}(\alpha \mid \beta) = \begin{cases} \dfrac{1}{2\sqrt{1 - \beta^2}} & \alpha^2 + \beta^2 \leq 1 \\ 0 & \text{otherwise} \end{cases}$$

3.7-2. If the joint density of x, y, and z is given by

$$p_{xyz}(\alpha, \beta, \gamma) = \tfrac{1}{8}e^{-|\alpha|}e^{-|\beta|}e^{-|\gamma|}$$

find

(a) $p_{xy|z}(\alpha, \beta \mid \gamma)$.
(b) $p_{x|yz}(\alpha \mid \beta, \gamma)$.

Answers:

(a) $p_{xy|z}(\alpha, \beta \mid \gamma) = \tfrac{1}{4}e^{-|\alpha|}e^{-|\beta|}$.
(b) $p_{x|yz}(\alpha \mid \beta, \gamma) = \tfrac{1}{2}e^{-|\alpha|}$.

3.8. *Special rules*

One of the most useful relations involving conditional probability density functions is known as *Bayes' rule.* To develop this relation, we consider the conditional probability density of $\mathbf{x}_1$ given $\mathbf{x}_2$ as

$$p(\mathbf{x}_1 \mid \mathbf{x}_2) = \frac{p(\mathbf{x}_1, \mathbf{x}_2)}{p(\mathbf{x}_2)} \tag{3.8-1}$$

and the conditional density of $\mathbf{x}_2$ given $\mathbf{x}_1$ as

$$p(\mathbf{x}_2 \mid \mathbf{x}_1) = \frac{p(\mathbf{x}_1, \mathbf{x}_2)}{p(\mathbf{x}_1)} \tag{3.8-2}$$

If we multiply both sides of Eq. (3.8-1) by $p(\mathbf{x}_2)$ and both sides of Eq. (3.8-2) by $p(\mathbf{x}_1)$, we obtain two different expressions for the joint probability density:

$$p(\mathbf{x}_1, \mathbf{x}_2) = p(\mathbf{x}_1 \mid \mathbf{x}_2)p(\mathbf{x}_2) \tag{3.8-3}$$

and

$$p(\mathbf{x}_1, \mathbf{x}_2) = p(\mathbf{x}_2 \mid \mathbf{x}_1)p(\mathbf{x}_1) \tag{3.8-4}$$

By equating these two expressions and dividing by $p(\mathbf{x}_2)$, we can obtain *Bayes' rule:*

$$p(\mathbf{x}_1 \mid \mathbf{x}_2) = \frac{p(\mathbf{x}_2 \mid \mathbf{x}_1)p(\mathbf{x}_1)}{p(\mathbf{x}_2)} \tag{3.8-5}$$

This result allows us to express the conditional density of $\mathbf{x}_1$ given $\mathbf{x}_2$ in terms of the conditional density of $\mathbf{x}_2$ given $\mathbf{x}_1$ and the marginal densities of $\mathbf{x}_1$ and $\mathbf{x}_2$. The usefulness of this result rests in the fact that there are many situations in which it is easier to obtain one conditional density than the other. By multiplying both sides of Eq. (3.8-5) by $p(\mathbf{x}_2)/p(\mathbf{x}_1)$, we can obtain an alternative form of Bayes' rule given by

$$p(\mathbf{x}_2 \mid \mathbf{x}_1) = \frac{p(\mathbf{x}_1 \mid \mathbf{x}_2)p(\mathbf{x}_2)}{p(\mathbf{x}_1)} \tag{3.8-6}$$

Another important rule, which we shall use often, is known as the *chain rule* of probability because of the similarity of the form to the chain rule of differentiation. To illustrate this concept, let us consider the joint probability density $p(\mathbf{x}_1, \mathbf{x}_2, \mathbf{x}_3)$. We may group the random vectors $\mathbf{x}_2$ and $\mathbf{x}_3$ into an augmented[1] random vector $\mathbf{y}$ given by

$$\mathbf{y} = \begin{bmatrix} \mathbf{x}_2 \\ \cdots \\ \mathbf{x}_3 \end{bmatrix}$$

so that the joint density becomes $p(\mathbf{x}_1, \mathbf{y})$. The definition of conditional probability density allows us to write

$$p(\mathbf{x}_1 \mid \mathbf{y}) = \frac{p(\mathbf{x}_1, \mathbf{y})}{p(\mathbf{y})}$$

[1] In effect we define the event $\{\mathbf{y} \leq \boldsymbol{\beta}\}$ as $\{\mathbf{x}_2 \leq \boldsymbol{\alpha}_2\}$ and $\{\mathbf{x}_3 \leq \boldsymbol{\alpha}_3\}$.

or, equivalently,

$$p(\mathbf{x}_1 \mid \mathbf{x}_2, \mathbf{x}_3) = \frac{p(\mathbf{x}_1, \mathbf{x}_2, \mathbf{x}_3)}{p(\mathbf{x}_2, \mathbf{x}_3)} \tag{3.8-7}$$

Now by multiplying both sides of Eq. (3.8-7) by $p(\mathbf{x}_2, \mathbf{x}_3)$ we obtain

$$p(\mathbf{x}_1, \mathbf{x}_2, \mathbf{x}_3) = p(\mathbf{x}_1 \mid \mathbf{x}_2, \mathbf{x}_3)p(\mathbf{x}_2, \mathbf{x}_3) \tag{3.8-8}$$

In a similar manner, we can write $p(\mathbf{x}_2, \mathbf{x}_3)$ as

$$p(\mathbf{x}_2, \mathbf{x}_3) = p(\mathbf{x}_2 \mid \mathbf{x}_3)p(\mathbf{x}_3)$$

so that Eq. (3.8-8) becomes

$$p(\mathbf{x}_1, \mathbf{x}_2, \mathbf{x}_3) = p(\mathbf{x}_1 \mid \mathbf{x}_2, \mathbf{x}_3)p(\mathbf{x}_2 \mid \mathbf{x}_3)p(\mathbf{x}_3) \tag{3.8-9}$$

In general, if we had the joint density of n random vectors, we could use the chain rule to write it as

$$p(\mathbf{x}_1, \mathbf{x}_2, \dots, \mathbf{x}_n) = p(\mathbf{x}_n) \prod_{i=1}^{n-1} p(\mathbf{x}_i \mid \mathbf{x}_{i+1}, \dots, \mathbf{x}_n) \tag{3.8-10}$$

The result will be very useful when we study Markov processes in Chapter 8.

The chain rule, as given by Eq. (3.8-10), and Bayes' rule, given by Eq. (3.8-5), are special cases of two general manipulative rules for moving arguments across the conditioning "bar" or line in the density function. There are two rules, one dealing with the movement of arguments from the right to the left of the conditioning line and the other dealing with movement from left to right.

> **Rule 3.8-1.** ***Right-to-Left Movement:*** To move a variable(s) from the right to the left of the conditioning line, we must multiply by the conditional density of this variable(s) with respect to the remaining right (that is, conditioning) variables.
>
> **Rule 3.8-2.** ***Left-to-Right Movement:*** To move a variable(s) from the left to the right of the conditioning line, we must divide by the conditional density of this variable(s) with respect to the original right (that is, conditioning) variables.

It is quite easy to establish these rules, by the use of the conditional probability definition, following an approach similar to that employed in developing Eq. (3.8-10).

To illustrate the use of these rules, let us consider the density $p(\mathbf{x}_1, \mathbf{x}_2 \mid \mathbf{x}_3, \mathbf{x}_4)$. Suppose that we wish to move $\mathbf{x}_2$ to the right of the conditioning line to obtain $p(\mathbf{x}_1 \mid \mathbf{x}_2, \mathbf{x}_3, \mathbf{x}_4)$; Rule 3.8-2 indicates that to do this we must divide by the conditional density of $\mathbf{x}_2$ with respect to the original right variables $\mathbf{x}_3$ and $\mathbf{x}_4$, that is, $p(\mathbf{x}_2 \mid \mathbf{x}_3, \mathbf{x}_4)$. Therefore, $p(\mathbf{x}_1 \mid \mathbf{x}_2, \mathbf{x}_3, \mathbf{x}_4)$ is given by

$$p(\mathbf{x}_1 \mid \mathbf{x}_2, \mathbf{x}_3, \mathbf{x}_4) = \frac{p(\mathbf{x}_1, \mathbf{x}_2 \mid \mathbf{x}_3, \mathbf{x}_4)}{p(\mathbf{x}_2 \mid \mathbf{x}_3, \mathbf{x}_4)} \tag{3.8-11}$$

On the other hand, let us suppose that we wished to move $\mathbf{x}_4$ to the left in our original density to obtain $p(\mathbf{x}_1, \mathbf{x}_2, \mathbf{x}_4 \mid \mathbf{x}_3)$. Now Rule 3.8-1 indicates that we must multiply by the conditional density of $\mathbf{x}_4$ given the remaining left variables, that is, $\mathbf{x}_3$, so that $p(\mathbf{x}_1, \mathbf{x}_2, \mathbf{x}_4 \mid \mathbf{x}_3)$ is

$$p(\mathbf{x}_1, \mathbf{x}_2, \mathbf{x}_4 \mid \mathbf{x}_3) = p(\mathbf{x}_1, \mathbf{x}_2 \mid \mathbf{x}_3, \mathbf{x}_4)p(\mathbf{x}_4 \mid \mathbf{x}_3) \tag{3.8-12}$$

By repeated use of these two rules in various combinations, one can obtain a wide variety of different density functions from the original density function. Alternatively, one may solve for the original density and thereby obtain alternative and equivalent forms for it. For example, from Eq. (3.8-11) we see that $p(\mathbf{x}_1, \mathbf{x}_2 \mid \mathbf{x}_3, \mathbf{x}_4)$ can be written as

$$p(\mathbf{x}_1, \mathbf{x}_2 \mid \mathbf{x}_3, \mathbf{x}_4) = p(\mathbf{x}_1 \mid \mathbf{x}_2, \mathbf{x}_3, \mathbf{x}_4)p(\mathbf{x}_2 \mid \mathbf{x}_3, \mathbf{x}_4)$$

whereas from Eq. (3.8-12) it becomes

$$p(\mathbf{x}_1, \mathbf{x}_2 \mid \mathbf{x}_3, \mathbf{x}_4) = \frac{p(\mathbf{x}_1, \mathbf{x}_2, \mathbf{x}_4 \mid \mathbf{x}_3)}{p(\mathbf{x}_4 \mid \mathbf{x}_3)}$$

If we were to apply left-to-right movement of $\mathbf{x}_2$ in Eq. (3.8-12), we would find that $p(\mathbf{x}_1, \mathbf{x}_4 \mid \mathbf{x}_2, \mathbf{x}_3)$ is

$$p(\mathbf{x}_1, \mathbf{x}_4 \mid \mathbf{x}_2, \mathbf{x}_3) = \frac{p(\mathbf{x}_1, \mathbf{x}_2, \mathbf{x}_4 \mid \mathbf{x}_3)}{p(\mathbf{x}_2 \mid \mathbf{x}_3)} = \frac{p(\mathbf{x}_1, \mathbf{x}_2 \mid \mathbf{x}_3, \mathbf{x}_4)p(\mathbf{x}_4 \mid \mathbf{x}_3)}{p(\mathbf{x}_2 \mid \mathbf{x}_3)}$$

so we can also express $p(\mathbf{x}_1, \mathbf{x}_2 \mid \mathbf{x}_3, \mathbf{x}_4)$ as

$$p(\mathbf{x}_1, \mathbf{x}_2 \mid \mathbf{x}_3, \mathbf{x}_4) = \frac{p(\mathbf{x}_1, \mathbf{x}_4 \mid \mathbf{x}_2, \mathbf{x}_3)p(\mathbf{x}_2 \mid \mathbf{x}_3)}{p(\mathbf{x}_4 \mid \mathbf{x}_3)}$$

We can now see that Bayes' rule, as given by Eq. (3.8-5), is nothing more than a right-to-left movement of $\mathbf{x}_1$ in $p(\mathbf{x}_2 \mid \mathbf{x}_1)$ followed by a left-to-right movement of $\mathbf{x}_2$.

Earlier in this section we discussed the procedure for obtaining marginal density from joint density. This process is a special case of the more general problem concerning the removal of variables from density functions.

Once again, we have two general rules depending on whether we wish to remove a right or left variable(s).

Rule 3.8-3. ***Right Removal:*** To remove a right variable(s), we multiply by the conditional density of this variable(s) given the remaining right variables and then integrate over the entire range of this variable(s).

Rule 3.8-4. ***Left Removal:*** To remove a left variable(s), we simply integrate over the entire range of this variable(s).

Rule 3.8-4 is simply a restatement of the earlier development concerning the determination of marginal densities from joint densities. The right removal rule is a combination of right-to-left movement (Rule 3.8-1) and left removal.

Let us use the density function $p(\mathbf{x}_1, \mathbf{x}_2 \,|\, \mathbf{x}_3, \mathbf{x}_4)$ to illustrate these two rules. Suppose that we wish to remove the right variable $\mathbf{x}_3$. Rule 3.8-3 indicates that we must multiply by $p(\mathbf{x}_3 \,|\, \mathbf{x}_4)$ and then integrate over the range of $\mathbf{x}_3$ so that $p(\mathbf{x}_1, \mathbf{x}_2 \,|\, \mathbf{x}_4)$ becomes

$$p(\mathbf{x}_1, \mathbf{x}_2 \,|\, \mathbf{x}_4) = \int_{-\infty}^{\infty} p(\mathbf{x}_1, \mathbf{x}_2 \,|\, \mathbf{x}_3, \mathbf{x}_4) p(\mathbf{x}_3 \,|\, \mathbf{x}_4) \, d\mathbf{x}_3$$

To remove a left variable such as $\mathbf{x}_2$, we need only integrate over the range of this variable. Therefore, $p(\mathbf{x}_1 \,|\mathbf{x}_3, \mathbf{x}_4)$ is given by

$$p(\mathbf{x}_1 \,|\, \mathbf{x}_3, \mathbf{x}_4) = \int_{-\infty}^{\infty} p(\mathbf{x}_1, \mathbf{x}_2 \,|\, \mathbf{x}_3, \mathbf{x}_4) \, d\mathbf{x}_2$$

The combination of the removal of $\mathbf{x}_2$ and $\mathbf{x}_3$ would give

$$p(\mathbf{x}_1 \,|\, \mathbf{x}_4) = \int_{-\infty}^{\infty} \int_{-\infty}^{\infty} p(\mathbf{x}_1, \mathbf{x}_2 \,|\, \mathbf{x}_3, \mathbf{x}_4) p(\mathbf{x}_3 \,|\, \mathbf{x}_4) \, d\mathbf{x}_2 \, d\mathbf{x}_3$$

There are also two rules describing how to add variables to density functions, depending on whether one wishes to add a right or left variable. These rules in a sense are the converses of Rules 3.8-3 and 3.8-4.

Rule 3.8-5. ***Right Addition:*** To add a right variable(s), we must multiply by the conditional density of this variable(s) with respect to both the left and right variables in the original density and divide by the conditional density of the variable(s) to be added, given the right variables in the original density, if any. If there were no original conditioning arguments, then one must divide by the marginal density of the variable(s) being added.

Rule 3.8-6. ***Left Addition:*** To add a left variable(s), we must multiply by the conditional density of this variable(s) with respect to both the left and right variables in the original density.

The left-addition rule (Rule 3.8-6) is easily established by the use of Rule 3.8-1 on right-to-left movement. Rule 3.8-5 is then developed by the use of Rule 3.8-6 and Rule 3.8-2 on left-to-right movement.

Let us consider the density $p(\mathbf{x}_1 | \mathbf{x}_2)$ to illustrate the above rules. Suppose first that we wish to add a right variable $\mathbf{x}_3$. The use of Rule 3.8-5 indicates that we must multiply by the density $p(\mathbf{x}_3 | \mathbf{x}_1, \mathbf{x}_2)$ and divide by $p(\mathbf{x}_3 | \mathbf{x}_2)$. Therefore, the desired density $p(\mathbf{x}_1 | \mathbf{x}_2, \mathbf{x}_3)$ is given by

$$p(\mathbf{x}_1 | \mathbf{x}_2, \mathbf{x}_3) = \frac{p(\mathbf{x}_1 | \mathbf{x}_2) p(\mathbf{x}_3 | \mathbf{x}_1, \mathbf{x}_2)}{p(\mathbf{x}_3 | \mathbf{x}_2)} \tag{3.8-13}$$

If we were to use Rule 3.8-3 to remove the right variable $\mathbf{x}_3$, we would expect to obtain the original density $p(\mathbf{x}_1 | \mathbf{x}_2)$. The use of Rule 3.8-3 indicates that we obtain $p(\mathbf{x}_1 | \mathbf{x}_2)$ from $p(\mathbf{x}_1 | \mathbf{x}_2, \mathbf{x}_3)$ by multiplying by $p(\mathbf{x}_3 | \mathbf{x}_2)$ and integrating over the range of $\mathbf{x}_3$. The desired density $p(\mathbf{x}_1 | \mathbf{x}_2)$ is therefore

$$p(\mathbf{x}_1 | \mathbf{x}_2) = \int_{-\infty}^{\infty} p(\mathbf{x}_1 | \mathbf{x}_2, \mathbf{x}_3) p(\mathbf{x}_3 | \mathbf{x}_2) \, d\mathbf{x}_3$$

and substituting from Eq. (3.8-13) gives

$$p(\mathbf{x}_1 | \mathbf{x}_2) = \int_{-\infty}^{\infty} p(\mathbf{x}_1 | \mathbf{x}_2) p(\mathbf{x}_3 | \mathbf{x}_1, \mathbf{x}_2) \, d\mathbf{x}_3$$

But $p(\mathbf{x}_1 | \mathbf{x}_2)$ is not a function of $\mathbf{x}_3$, so we have

$$p(\mathbf{x}_1 | \mathbf{x}_2) = p(\mathbf{x}_1 | \mathbf{x}_2) \int_{-\infty}^{\infty} p(\mathbf{x}_3 | \mathbf{x}_1, \mathbf{x}_2) \, d\mathbf{x}_3$$

and since the integral is unity, we have obtained the desired result.

Let us suppose that we wished to add $\mathbf{x}_3$ as a left variable to $p(\mathbf{x}_1 | \mathbf{x}_2)$. The application of Rule 3.8-6 indicates that we should multiply by $p(\mathbf{x}_3 | \mathbf{x}_1, \mathbf{x}_2)$ so that we have

$$p(\mathbf{x}_1, \mathbf{x}_3 | \mathbf{x}_2) = p(\mathbf{x}_1 | \mathbf{x}_2) p(\mathbf{x}_3 | \mathbf{x}_1, \mathbf{x}_2) \tag{3.8-14}$$

Note that if we begin with $p(\mathbf{x}_1, \mathbf{x}_3 | \mathbf{x}_2)$ and use Rule 3.8-2 to move $\mathbf{x}_1$ from the left to the right of the conditioning line, we directly obtain Eq. (3.8-14).

The use of Rule 3.8-5 gives a very simple development of Bayes' rule as given by Eq. (3.8-5). Consider the density $p(\mathbf{x}_1)$ and suppose that we wish to add a right variable $\mathbf{x}_2$. Rule 3.8-5 indicates that it is necessary to multiply

$p(\mathbf{x}_1)$ by $p(\mathbf{x}_2 \mid \mathbf{x}_1)$ and divide by $p(\mathbf{x}_2)$ so that $p(\mathbf{x}_1 \mid \mathbf{x}_2)$ is given by

$$p(\mathbf{x}_1 \mid \mathbf{x}_2) = \frac{p(\mathbf{x}_1)p(\mathbf{x}_2 \mid \mathbf{x}_1)}{p(\mathbf{x}_2)}$$

This result is identical to Bayes' rule.

The six rules just developed are powerful tools in the application of probability theory to many practical problems. Their use is somewhat of an art that requires an intuitive feeling for the nature of the expressions desired. The advantage of their use is in the ability to find a desired probability density in terms of more easily computed densities. The exact reasons why some densities are easier to determine than others may not be completely clear at this time, but will become more obvious as we continue the study of random phenomena. To illustrate how one might apply some of the above rules in combination to compute some desired densities, let us consider a simple example.

Example *3.8-1*. We wish to compute the conditional probability density $p(\mathbf{x}_1 \mid \mathbf{x}_2)$; however, let us assume that we can only conveniently determine the conditional density $p(\mathbf{x}_2 \mid \mathbf{x}_1)$ and $p(\mathbf{x}_1)$. The application of Bayes' rule allows us to write $p(\mathbf{x}_1 \mid \mathbf{x}_2)$ as

$$p(\mathbf{x}_1 \mid \mathbf{x}_2) = \frac{p(\mathbf{x}_2 \mid \mathbf{x}_1)p(\mathbf{x}_1)}{p(\mathbf{x}_2)}$$

The numerator of this expression is acceptable since we know both $p(\mathbf{x}_2 \mid \mathbf{x}_1)$ and $p(\mathbf{x}_1)$; the denominator, however, poses a problem since we do not know $p(\mathbf{x}_2)$. By the use of Rule 3.8-4 we know that we can write $p(\mathbf{x}_2)$ as

$$p(\mathbf{x}_2) = \int_{-\infty}^{\infty} p(\mathbf{x}_1, \mathbf{x}_2)\, d\mathbf{x}_1$$

but since we do not know $p(\mathbf{x}_1, \mathbf{x}_2)$, we still have a problem. Since we do know $p(\mathbf{x}_2 \mid \mathbf{x}_1)$, we might try to use Rule 3.8-2 to write $p(\mathbf{x}_2, \mathbf{x}_1)$ as

$$p(\mathbf{x}_2, \mathbf{x}_1) = p(\mathbf{x}_2 \mid \mathbf{x}_1)p(\mathbf{x}_1)$$

so that $p(\mathbf{x}_2)$ becomes

$$p(\mathbf{x}_2) = \int_{-\infty}^{\infty} p(\mathbf{x}_2 \mid \mathbf{x}_1)p(\mathbf{x}_1)\, d\mathbf{x}_1$$

This expression is fine since it only involves $p(\mathbf{x}_2 \mid \mathbf{x}_1)$ and $p(\mathbf{x}_1)$, as desired. The final form for $p(\mathbf{x}_1 | \mathbf{x}_2)$ is therefore

$$p(\mathbf{x}_1 | \mathbf{x}_2) = \frac{p(\mathbf{x}_2 | \mathbf{x}_1) p(\mathbf{x}_1)}{\int_{-\infty}^{\infty} p(\mathbf{x}_2 | \mathbf{x}_1) p(\mathbf{x}_1)\, d\mathbf{x}_1}$$

Let us suppose now that $\mathbf{x}_1$ and $\mathbf{x}_2$ are scalars and that $p(x_2 | x_1)$ is $N(x_1, 1)$ and $p(x_1)$ is $N(0, 1)$. Let us compute the integral expression in the denominator first.

$$\int_{-\infty}^{\infty} p(x_2 | x_1) p(x_1)\, dx_1 = \frac{1}{2\pi} \int_{-\infty}^{\infty} e^{-1/2(x_2 - x_1)^2} e^{1/2(x_1)^2}\, dx_1$$
$$= \frac{1}{2\pi} \int_{-\infty}^{\infty} e^{-1/2(x_2^2 - 2x_1x_2 + 2x_1^2)}\, dx_1$$

If we multiply and divide this expression by $e^{-1/2(x_2^2/2)}$, we can write it as

$$\int_{-\infty}^{\infty} p(x_2 | x_1) p(x_1)\, dx_1 = \frac{2^{-\frac{1}{2}}}{\sqrt{2\pi}} e^{-1/2(x_2^2/2)} \int_{-\infty}^{\infty} \frac{\sqrt{2}}{\sqrt{2\pi}} \times \exp\left\{\frac{1}{2}\left[\frac{(x_1 - x_2)^2}{1/2}\right]\right\} dx_1$$

We see that the expression within the integral is a normal density with mean x_2 and variance 1/2; the integral over the range of x_1 must therefore be unity. Therefore, we may write $p(x_1 | x_2)$ as

$$p(x_1 | x_2) = \frac{p(x_2 | x_1) p(x_1)}{\int_{-\infty}^{\infty} p(x_2 | x_1) p(x_1)\, dx_1} = \frac{\dfrac{1}{\sqrt{2\pi}} e^{-1/2(x_2^2 - 2x_1x_2 + x_1^2)}}{\dfrac{2^{-\frac{1}{2}}}{\sqrt{2\pi}} e^{-1/2(x_2^2/2)}}$$
$$= \frac{\sqrt{2}}{\sqrt{2\pi}} \exp\left\{\frac{1}{2}\left[\frac{(x_1 - x_2/2)^2}{1/2}\right]\right\}$$

and we see that $p(x_1 | x_2)$ is $N(x_2/2, 1/2)$.

3.9. *Independence of random variables*

In Section 2.6 we introduced the concept of independent events; here we wish to extend this concept to the independence of random variables. The two events β_1 and β_2 were said to be independent if and only if $\mathcal{P}\{\beta_1, \beta_2\} = \mathcal{P}\{\beta_1\}\mathcal{P}\{\beta_2\}$. We shall base the definition of independent random variables on the independence of the basic events $\{\mathbf{x}_i \leq \boldsymbol{\alpha}_i\}$.

Definition 3.9-1. ***Independence of Two Random Vectors:*** Two random vectors $\mathbf{x}_1$ and $\mathbf{x}_2$ are independent if the events $\{\mathbf{x}_1 \leq \boldsymbol{\alpha}_1\}$ and $\{\mathbf{x}_2 \leq \boldsymbol{\alpha}_2\}$ are independent for any $\boldsymbol{\alpha}_1$ and $\boldsymbol{\alpha}_2$; that is, if

$$\mathcal{P}\{\mathbf{x}_1 \leq \boldsymbol{\alpha}_1, \mathbf{x}_2 \leq \boldsymbol{\alpha}_2\} = \mathcal{P}\{\mathbf{x}_1 \leq \boldsymbol{\alpha}_1\}\mathcal{P}\{\mathbf{x}_2 \leq \boldsymbol{\alpha}_2\} \tag{3.9-1}$$

By the use of the definition of a distribution function, Eq. (3.9-1) can be written as

$$F_{\mathbf{x}_1\mathbf{x}_2}(\boldsymbol{\alpha}_1, \boldsymbol{\alpha}_2) = F_{\mathbf{x}_1}(\boldsymbol{\alpha}_1)F_{\mathbf{x}_2}(\boldsymbol{\alpha}_2) \tag{3.9-2}$$

If we express Eq. (3.9-2) in terms of the associated density functions, we have

$$\begin{aligned}\int_{-\infty}^{\boldsymbol{\alpha}_1} d\boldsymbol{\gamma}_1 \int_{-\infty}^{\boldsymbol{\alpha}_2} d\boldsymbol{\gamma}_2\, p_{\mathbf{x}_1\mathbf{x}_2}(\boldsymbol{\gamma}_1, \boldsymbol{\gamma}_2) &= \int_{-\infty}^{\boldsymbol{\alpha}_1} p_{\mathbf{x}_1}(\boldsymbol{\gamma}_1)\, d\boldsymbol{\gamma}_1 \int_{-\infty}^{\boldsymbol{\alpha}_2} p_{\mathbf{x}_2}(\boldsymbol{\gamma}_2)\, d\boldsymbol{\gamma}_2 \\ &= \int_{-\infty}^{\boldsymbol{\alpha}_1} d\boldsymbol{\gamma}_1 \int_{-\infty}^{\boldsymbol{\alpha}_2} d\boldsymbol{\gamma}_2\, p_{\mathbf{x}_1}(\boldsymbol{\gamma}_1)p_{\mathbf{x}_2}(\boldsymbol{\gamma}_2)\end{aligned}$$

Since this expression must be valid for every $\boldsymbol{\alpha}_1$ and $\boldsymbol{\alpha}_2$, it is necessary that

$$p_{\mathbf{x}_1\mathbf{x}_2}(\boldsymbol{\alpha}_1, \boldsymbol{\alpha}_2) = p_{\mathbf{x}_1}(\boldsymbol{\alpha}_1)p_{\mathbf{x}_2}(\boldsymbol{\alpha}_2) \tag{3.9-3}$$

for every $\boldsymbol{\alpha}_1$ and $\boldsymbol{\alpha}_2$. Equations (3.9-2) and (3.9-3) are completely equivalent to Definition 3.9-1 and could be and often are used as the definition of independence of two random variables.

Equation (3.9-3) is probably the simplest test to use to determine whether two random variables are independent if $p_{\mathbf{x}_1}$, $p_{\mathbf{x}_2}$, and $p_{\mathbf{x}_1\mathbf{x}_2}$ are known. If one multiplies the two marginal densities $p_{\mathbf{x}_1}$ and $p_{\mathbf{x}_2}$ and finds that the product equals $p_{\mathbf{x}_1\mathbf{x}_2}$, then the variables are independent; otherwise they are dependent. Conversely, if one can determine in some other way that two random variables are independent, then Eq. (3.9-3) may be used to determine $p_{\mathbf{x}_1\mathbf{x}_2}$ from the marginal densities. In the case of independent random variables, not only may one determine the marginal densities from the joint densities, but he may also determine the joint density from the marginal densities. This is an important fact; if two random variables are independent, their joint density is the product of their marginal densities.

Consider any two admissible subsets of $\mathcal{R}^n$ given by $\mathcal{A}_1$ and $\mathcal{A}_2$. It is easy to show that if $\mathbf{x}_1$ and $\mathbf{x}_2$ are independent, then

$$\mathcal{P}\{\mathbf{x}_1 \in \mathcal{A}_1, \mathbf{x}_2 \in \mathcal{A}_2\} = \mathcal{P}\{\mathbf{x}_1 \in \mathcal{A}_1\}\mathcal{P}\{\mathbf{x}_2 \in \mathcal{A}_2\} \tag{3.9-4}$$

This result is easily established by the use of Eq. (3.9-3). By the use of the joint density $p_{\mathbf{x}_1\mathbf{x}_2}$, we may write $\mathcal{P}\{\mathbf{x}_1 \in \mathcal{A}_1, \mathbf{x}_2 \in \mathcal{A}_2\}$ as

$$\mathcal{P}\{\mathbf{x}_1 \in \mathcal{A}_1, \mathbf{x}_2 \in \mathcal{A}_2\} = \int_{\mathcal{A}_1} d\boldsymbol{\gamma}_1 \int_{\mathcal{A}_2} d\boldsymbol{\gamma}_2\, p_{\mathbf{x}_1\mathbf{x}_2}(\boldsymbol{\gamma}_1, \boldsymbol{\gamma}_2)$$

Now substituing from Eq. (3.9-3), we have

$$\mathcal{P}\{\mathbf{x}_1 \in \mathcal{A}_1, \mathbf{x}_2 \in \mathcal{A}_2\} = \int_{\mathcal{A}_1} d\boldsymbol{\gamma}_1 \int_{\mathcal{A}_2} d\boldsymbol{\gamma}_2\, p_{\mathbf{x}_1}(\boldsymbol{\gamma}_1) p_{\mathbf{x}_2}(\boldsymbol{\gamma}_2)$$
$$= \left[\int_{\mathcal{A}_1} d\boldsymbol{\gamma}_1\, p_{\mathbf{x}_1}(\boldsymbol{\gamma}_1)\right] \int_{\mathcal{A}_2} d\boldsymbol{\gamma}_2\, p_{\mathbf{x}_2}(\boldsymbol{\gamma}_2)$$

so that finally we obtain

$$\mathcal{P}\{\mathbf{x}_1 \in \mathcal{A}_1, \mathbf{x}_2 \in \mathcal{A}_2\} = \mathcal{P}\{\mathbf{x}_1 \in \mathcal{A}_1\}\mathcal{P}\{\mathbf{x}_2 \in \mathcal{A}_2\}$$

which is the desired result.

Hence we see that although the definition may seem to only relate to the basic events $\{\mathbf{x} \leq \boldsymbol{\alpha}\}$, it does in fact encompass the independence of all events in $\mathfrak{F}$. Once again we see the importance of the basic event $\{\mathbf{x} \leq \boldsymbol{\alpha}\}$.

By the use of conditional densities, we may write $p_{\mathbf{x}_1\mathbf{x}_2}$ as $p_{\mathbf{x}_1 | \mathbf{x}_2} p_{\mathbf{x}_2}$; therefore, Eq. (3.9-3) becomes

$$p_{\mathbf{x}_1 | \mathbf{x}_2}(\boldsymbol{\alpha}_1 \mid \boldsymbol{\alpha}_2) p_{\mathbf{x}_2}(\boldsymbol{\alpha}_2) = p_{\mathbf{x}_1}(\boldsymbol{\alpha}_1) p_{\mathbf{x}_2}(\boldsymbol{\alpha}_2)$$

or, assuming that $p_{\mathbf{x}_2}(\boldsymbol{\alpha}_2) \neq 0$, we have

$$p_{\mathbf{x}_1 | \mathbf{x}_2}(\boldsymbol{\alpha}_1 \mid \boldsymbol{\alpha}_2) = p_{\mathbf{x}_1}(\boldsymbol{\alpha}_1) \tag{3.9-5}$$

This expression is equivalent to Eq. (3.9-3) and is another way of determining if two random variables are independent. Of course, we could also show that Eq. (3.9-3) can be written as

$$p_{\mathbf{x}_2 | \mathbf{x}_1}(\boldsymbol{\alpha}_2 \mid \boldsymbol{\alpha}_1) = p_{\mathbf{x}_2}(\boldsymbol{\alpha}_2) \tag{3.9-6}$$

To summarize, if two random variables are independent, their joint density is equal to the product of their marginal densities, and their conditional densities are equal to their marginal densities.

We can also consider the independence of N random variables. The definition is identical in form to Definition 3.9-1.

Definition 3.9-2. ***Independence of N Random Vectors:*** N random vectors $\mathbf{x}_1, \mathbf{x}_2, \ldots, \mathbf{x}_N$ are independent if the N events $\{\mathbf{x}_1 \leq \boldsymbol{\alpha}_1\}$, $\{\mathbf{x}_2 \leq \boldsymbol{\alpha}_2\}, \ldots, \{\mathbf{x}_N \leq \boldsymbol{\alpha}_N\}$ are independent for any $\boldsymbol{\alpha}_1, \boldsymbol{\alpha}_2, \ldots, \boldsymbol{\alpha}_N$, that is, if

$$\mathcal{P}\{\mathbf{x}_1 \leq \boldsymbol{\alpha}_1, \mathbf{x}_2 \leq \boldsymbol{\alpha}_2, \ldots, \mathbf{x}_N \leq \boldsymbol{\alpha}_N\}$$
$$= \mathcal{P}\{\mathbf{x}_1 \leq \boldsymbol{\alpha}_1\}\mathcal{P}\{\mathbf{x}_2 \leq \boldsymbol{\alpha}_2\}\cdots\mathcal{P}\{\mathbf{x}_N \leq \boldsymbol{\alpha}_N\} \tag{3.9-7}$$

Once again, it is quite easy to establish that if $\mathbf{x}_1, \mathbf{x}_2, \ldots, \mathbf{x}_N$ are independent, then

$$F_{\mathbf{x}_1\mathbf{x}_2\cdots\mathbf{x}_N}(\alpha_1, \alpha_2, \ldots, \alpha_N) = F_{\mathbf{x}_1}(\alpha_1)F_{\mathbf{x}_2}(\alpha_2)\cdots F_{\mathbf{x}_N}(\alpha_N) \tag{3.9-8}$$

and that

$$p_{\mathbf{x}_1\mathbf{x}_2\cdots\mathbf{x}_N}(\alpha_1, \alpha_2, \ldots, \alpha_N) = p_{\mathbf{x}_1}(\alpha_1)p_{\mathbf{x}_2}(\alpha_2)\cdots p_{\mathbf{x}_N}(\alpha_N) \tag{3.9-9}$$

A very important property is that if N random variables are independent, any subset of these N random variables is also independent. In other words, if $\mathbf{x}_1, \mathbf{x}_2, \ldots, \mathbf{x}_N$ are independent, so is $\mathbf{x}_1, \mathbf{x}_2$ or $\mathbf{x}_1, \mathbf{x}_N$ or $\mathbf{x}_1, \mathbf{x}_2, \mathbf{x}_N$, and so forth. This property is quite easy to establish; we need only integrate on both sides in Eq. (3.9-9) over the range of the variables that we wish to delete. For example, let us suppose that we have four independent random variables $\mathbf{x}_1, \mathbf{x}_2, \mathbf{x}_3, \mathbf{x}_4$; we wish to show that the set $\mathbf{x}_1, \mathbf{x}_3$ is also independent. From Eq. (3.9-9) we have

$$p_{\mathbf{x}_1\mathbf{x}_2\mathbf{x}_3\mathbf{x}_4}(\alpha_1, \alpha_2, \alpha_3, \alpha_4) = p_{\mathbf{x}_1}(\alpha_1)p_{\mathbf{x}_2}(\alpha_2)p_{\mathbf{x}_3}(\alpha_3)p_{\mathbf{x}_4}(\alpha_4) \tag{3.9-10}$$

and since we wish to show independence of $\mathbf{x}_1$ and $\mathbf{x}_3$, we integrate over $\mathbf{x}_2$ and $\mathbf{x}_3$ in Eq. (3.9-10) to obtain

$$\begin{aligned} \iint_{-\infty}^{\infty} p_{\mathbf{x}_1\mathbf{x}_2\mathbf{x}_3\mathbf{x}_4}(\alpha_1, \alpha_2, \alpha_3, \alpha_4,)\, d\alpha_2\, d\alpha_4 &= \iint_{-\infty}^{\infty} p_{\mathbf{x}_1}(\alpha_1)p_{\mathbf{x}_2}(\alpha_2)p_{\mathbf{x}_3}(\alpha_3)p_{\mathbf{x}_4}(\alpha_4)\, d\alpha_2 d\alpha_4 \\ &= p_{\mathbf{x}_1}(\alpha_1)p_{\mathbf{x}_3}(\alpha_3)\int_{-\infty}^{\infty} p_{\mathbf{x}_2}(\alpha_2)\, d\alpha_2 \int_{-\infty}^{\infty} p_{\mathbf{x}_4}(\alpha_4)\, d\alpha_4 \end{aligned} \tag{3.9-11}$$

We recognize the integral on the left side of this expression as $p_{\mathbf{x}_1\mathbf{x}_3}(\alpha_1, \alpha_3)$, and the two integrals on the right side as equal to unity; so we have

$$p_{\mathbf{x}_1\mathbf{x}_3}(\alpha_1, \alpha_3) = p_{\mathbf{x}_1}(\alpha_1)p_{\mathbf{x}_3}(\alpha_3)$$

Therefore, we can conclude that $\mathbf{x}_1$ and $\mathbf{x}_3$ are independent. In a similar manner, we could show that any subset of $\mathbf{x}_1, \mathbf{x}_2, \mathbf{x}_3, \mathbf{x}_4$ is independent.

The reader is cautioned against concluding that the converse of the above statement is true. Even if every subset of N random variables (assuming that no subset contains all N variables) is independent, it cannot be concluded that the N random variables are independent. In other words, suppose that we have three random variables and $p(\mathbf{x}_1, \mathbf{x}_2) = p(\mathbf{x}_1)p(\mathbf{x}_2)$, $p(\mathbf{x}_2, \mathbf{x}_3) = p(\mathbf{x}_2)p(\mathbf{x}_3)$, and $p(\mathbf{x}_1, \mathbf{x}_3) = p(\mathbf{x}_1)p(\mathbf{x}_3)$. We do not know that $p(\mathbf{x}_1, \mathbf{x}_2, \mathbf{x}_3) = p(\mathbf{x}_1)p(\mathbf{x}_2)p(\mathbf{x}_3)$, since there is no way to write $p(\mathbf{x}_1, \mathbf{x}_2, \mathbf{x}_3)$ in terms of just the densities $p(\mathbf{x}_1, \mathbf{x}_2)$, $p(\mathbf{x}_2, \mathbf{x}_3)$, and $p(\mathbf{x}_1, \mathbf{x}_3)$. It is necessary to have at least one density that contains all the variables, which is easy to see from Rule 3.8-5 or 3.8-6.

By using the property that any subset of N independent random variables is also independent, we may establish the fact that the conditional density of any subset of N independent random variables conditioned on any or all of the remaining variables is the product of the marginal (unconditioned) densities of the left variables. Consider again, for example, the case of four independent random variables $\mathbf{x}_1$, $\mathbf{x}_2$, $\mathbf{x}_3$, and $\mathbf{x}_4$. The use of the above property of conditional densities indicates that $p(\mathbf{x}_1, \mathbf{x}_2 | \mathbf{x}_3, \mathbf{x}_4)$ is given by

$$p(\mathbf{x}_1, \mathbf{x}_2 | \mathbf{x}_3, \mathbf{x}_4) = p(\mathbf{x}_1)p(\mathbf{x}_2)$$

and that $p(\mathbf{x}_3, \mathbf{x}_2 | \mathbf{x}_1)$ is

$$p(\mathbf{x}_3, \mathbf{x}_2 | \mathbf{x}_1) = p(\mathbf{x}_3)p(\mathbf{x}_2)$$

In other words, one may disregard any conditioning arguments and write the resulting joint density as the product of marginal densities.

It should be noted that just because $\mathbf{x}_1$ and $\mathbf{x}_2$ are independent, it is not possible to conclude that

$$p(\mathbf{x}_1, \mathbf{x}_2 | \mathbf{x}_3) = p(\mathbf{x}_1 | \mathbf{x}_3)p(\mathbf{x}_2 | \mathbf{x}_3) \tag{3.9-12}$$

unless $\mathbf{x}_1$, $\mathbf{x}_2$, and $\mathbf{x}_3$ are independent. To illustrate this fact, let us begin with the joint density $p(\mathbf{x}_1, \mathbf{x}_2)$, which, owing to the independence of $\mathbf{x}_1$ and $\mathbf{x}_2$, we may write as

$$p(\mathbf{x}_1, \mathbf{x}_2) = p(\mathbf{x}_1)p(\mathbf{x}_2) \tag{3.9-13}$$

Using Rule 3.8-5, let us add $\mathbf{x}_3$ as a conditioning argument to $p(\mathbf{x}_1, \mathbf{x}_2)$ so that we have

$$p(\mathbf{x}_1, \mathbf{x}_2 | \mathbf{x}_3) = \frac{p(\mathbf{x}_1, \mathbf{x}_2)p(\mathbf{x}_3 | \mathbf{x}_1, \mathbf{x}_2)}{p(\mathbf{x}_3)}$$

and, upon substituting from Eq. (3.9-13), we obtain

$$p(\mathbf{x}_1, \mathbf{x}_2 | \mathbf{x}_3) = \frac{p(\mathbf{x}_1)p(\mathbf{x}_2)p(\mathbf{x}_3 | \mathbf{x}_1, \mathbf{x}_2)}{p(\mathbf{x}_3)} \tag{3.9-14}$$

Let us use Rule 3.8-5 two more times to add the conditioning argument $\mathbf{x}_3$ to both $p(\mathbf{x}_1)$ and $p(\mathbf{x}_2)$. Equation (3.9-14) is now

$$p(\mathbf{x}_1, \mathbf{x}_2 | \mathbf{x}_3) = p(\mathbf{x}_1 | \mathbf{x}_3)p(\mathbf{x}_2 | \mathbf{x}_3)\left[\frac{p(\mathbf{x}_3)p(\mathbf{x}_3 | \mathbf{x}_1, \mathbf{x}_2)}{p(\mathbf{x}_3 | \mathbf{x}_2)p(\mathbf{x}_3 | \mathbf{x}_1)}\right]$$

For Eq. (3.9-12) to be satisfied, it is necessary that

$$p(\mathbf{x}_3)p(\mathbf{x}_3 \mid \mathbf{x}_1, \mathbf{x}_2) = p(\mathbf{x}_3 \mid \mathbf{x}_2)p(\mathbf{x}_3 \mid \mathbf{x}_1) \tag{3.9-15}$$

This equation will only be valid if $\mathbf{x}_1$, $\mathbf{x}_2$, and $\mathbf{x}_3$ are independent.

In this section we have extended the important concept of statistical independence to random variables. In particular we have expressed the independence of random variables in terms of the relation of their marginal and joint probability density functions. The concept of independent random variables is important in a wide variety of applications of random-variable theory.

EXERCISES 3.9

3.9-1. If the joint density of x and y is given by

$$p_{xy}(\alpha, \beta) = \begin{cases} \dfrac{1}{\pi} & \alpha^2 + \beta^2 \leq 1 \\ 0 & \text{otherwise} \end{cases}$$

are x and y independent?

Answer: No.

3.9-2. Construct a joint density for three random variables x_1, x_2, and x_3 such that they are independent in pairs but such that x_1, x_2, and x_3 are not independent.

3.9-3. Consider the two-dimensional gaussian density given by

$$p_{xy}(\alpha, \beta) = \frac{1}{2\pi\sigma_1\sigma_2\sqrt{1-\rho^2}} \exp\left\{-\frac{1}{2(1-\rho^2)}\left[\frac{\alpha^2}{\sigma_1^2} - \frac{2\rho\alpha\beta}{\sigma_1\sigma_2} + \frac{\beta^2}{\sigma_2^2}\right]\right\}$$

Show that x and y are independent if and only if $\rho = 0$.

3.10. *Conclusions*

The purpose of this chapter was to develop a more convenient mathematical framework for dealing with the probability concepts of the preceding chapter. The random variable was introduced to relate experimental outcomes to real numbers and events to subsets of the real line. The probability measure concept was translated into probability distribution and density functions defined on a random variable. These basic concepts were generalized and extended to conditional probability, leading to the development of a collection of several manipulative rules. Finally, the concept of independent events was related to random variables and expressed in terms of their density functions.

In Chapter 4 we shall concentrate on the problems of manipulating random variables and examine special techniques for treating functions of random variables. In particular, we shall introduce the important concept of expectation, which plays a key role in many applications of probability theory.

3.11. *References*

(1) Chung, Kai Lai, *A Course in Probability Theory*, Harcourt Brace, New York, 1968.

(2) Dubes, R. C., *The Theory of Applied Probability*, Prentice-Hall, Englewood Cliffs, N.J., 1968.

(3) Harris, Bernard, *Theory of Probability*, Addison-Wesley, Reading, Mass., 1966.

(4) Loève, M., *Probability Theory*, 3rd ed., Van Nostrand Reinhold, New York, 1963.

(5) Papoulis, Athanasios, *Probability, Random Variables and Stochastic Processes*, McGraw-Hill, New York, 1964.

3.12. *Problems*

3.12-1. Consider the probability density function $p_{xy}(\alpha, \beta)$ that is zero everywhere except over the shaded area shown. Over the shaded area, it has constant height C.

(a) Find C such that p_{xy} is a valid density.

(b) Find $p_x(\alpha)$.

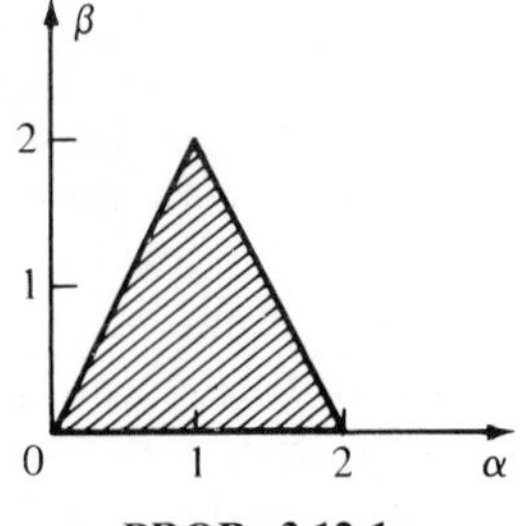

PROB. 3.12-1.

(c) Plot $p_{y|x}(\beta \mid \alpha)$ for $\alpha = 0.2, 0.4, 0.6, 0.8$, and 1.0.
Define the event $\mathcal{A} = \{|x - 1| \leq \frac{1}{2}\}$ and
(d) Find $\mathcal{P}\{\mathcal{A}\}$.
(e) Find $p_y(\beta \mid \mathcal{A})$.

3.12-2. The time in minutes required by a lady to travel from her home to the airport is a random phenomenon obeying a uniform probability law over the interval 25 to 30 minutes. If she leaves her home at exactly 9: 00 A.M., what is the probability that she will catch a plane which leaves at 9:27 A.M.?

3.12-3. The length of time in hours that a certain part remains operational is a numerical-valued random phenomenon with a probability density function specified by

$$p_x(\alpha) = \begin{cases} Ae^{-\alpha/5} & \alpha > 0 \\ 0 & \text{otherwise} \end{cases}$$

(a) Find A such that this is a valid probability density function and plot the resulting density function.
(b) What is the probability that the length of time the part lasts is
 (1) more than 10 hours (hr)?
 (2) less than 5 hr?
 (3) greater than 5 but less than 10 hr?
(c) For any real number τ, let $\mathcal{E}(\tau)$ denote the event that the part lasts longer than τ hours. Find $\mathcal{P}\{\mathcal{E}(\tau)\}$. Show that, for $\tau_1 > 0$ and $\tau_2 > 0$,

$$\mathcal{P}\{\mathcal{E}(\tau_1 + \tau_2) \mid \mathcal{E}(\tau_1)\} = \mathcal{P}\{\mathcal{E}(\tau_2)\}$$

Give a word description of the physical meaning of the above event probability.

3.12-4. You are given a random variable with the distribution function

$$F_x(\alpha) = \begin{cases} 0 & \alpha \leq 0 \\ \frac{1}{2}\alpha & 0 \leq \alpha \leq 1 \\ \frac{1}{2} & 1 \leq \alpha \leq 2 \\ \frac{1}{4}\alpha & 2 \leq \alpha \leq 4 \\ 1 & 4 \leq \alpha < \infty \end{cases}$$

(a) Sketch the probability distribution function.
(b) Sketch the probability density function.
(c) What is the probability that the observed value of the random variable will be (1) greater than 3, (2) less than 1, (3) between 1 and 3?
(d) What is the conditioned probability that the observed value of the random variable will be (1) more than 3, given that it is more than 1, (2) less than 2, given that it is more than 1?

3.12-5. A resistor is selected from one of two urns, A or B, with equal probability. If the resistor comes from urn A, it will fail prior to time τ with probability $1 - e^{-a\tau}$. If the resistor comes from urn B it will fail prior to time τ with probability $1 - e^{-b\tau}$.
(a) Find the probability that a resistor will fail before time τ.

(b) Suppose that the resistor is known to fail some time between τ_1 and τ_2. What is the probability that the resistor came from urn A?

3.12-6. A printer may set any letter or character in a book erroneously, but the fact that a letter is wrong does not affect the probability that any other letter is incorrect. There are 200,000 letters in a book and the probability of a single letter being in error is 10^{-5}.

(a) What is the probability of at least k errors?

(b) What is the probability of at most k errors?

(c) What is the probability of exactly k errors?

(d) Can you suggest an efficient approximation to the above results for small k?

3.12-7. A machine makes parts with an average of 1 per cent defective. What is the probability of there being α defective parts in a sample of 10? If one or more defective parts are found in the sample of 10, production is stopped. What is the probability that production is stopped?

3.12-8. The joint density function of two random variables is

$$p_{xy}(\alpha, \beta) = \begin{cases} e^{-\alpha}e^{-\beta} & \alpha \geq 0,\ \beta \geq 0 \\ 0 & \text{otherwise} \end{cases}$$

(a) Find the densities $p_x(\alpha)$ and $p_y(\beta)$.

(b) Find the densities $p_{x|y}(\alpha \mid \beta)$ and $p_{y|x}(\beta \mid \alpha)$.

(c) Are x and y independent?

3.12-9. Repeat Problem 3.12-8 if

$$p_{xy}(\alpha, \beta) = \begin{cases} a\alpha\beta & 0 \leq \alpha \leq \beta,\ 0 \leq \beta \leq 2 \\ 0 & \text{otherwise} \end{cases}$$

√ ***3.12-10.*** If

$$p_{xy}(\alpha, \beta) = \begin{cases} a(2 + \alpha\beta) & 0 \leq \alpha \leq 1,\ 0 \leq \beta \leq 1 \\ 0 & \text{otherwise} \end{cases}$$

find

(a) $F_{xy}(0.2, 0.5)$

(b) $p_{x|y}(\alpha \mid \beta)$

3.12-11. Repeat Problem 3.12-10 if

$$p_{xy}(\alpha, \beta) = \begin{cases} a(\alpha + 2\beta) & 0 \leq \alpha \leq 1,\ 0 \leq \beta \leq \frac{1}{2} \\ 0 & \text{otherwise} \end{cases}$$

3.12-12. A joint probability density function is

$$p_{x_1x_2}(\alpha_1, \alpha_2) = \begin{cases} \alpha_1\alpha_2 e^{-0.5(\alpha_1^2+\alpha_2^2)} & \alpha_1 \geq 0 \\ & \alpha_2 \geq 0 \\ 0 & \text{otherwise} \end{cases}$$

What is

(a) $p_{x_1}(\alpha_1)$

(b) $\mathcal{P}\{x_1 \leq 1, x_2 \leq 1\}$
(c) $\mathcal{P}\{x_1 + x_2 \leq 1\}$
(d) $\mathcal{P}\{x_1 \leq 1 | x_2 \leq 1\}$

3.12-13. What value of A makes $p_x(\alpha)$ a valid probability density function?

$$p_x(\alpha) = \begin{cases} 0 & -\infty \leq \alpha < 0 \\ A\alpha & 0 \leq \alpha < 4 \\ A(8 - \alpha) & 4 \leq \alpha < 8 \\ 0 & 8 \leq \alpha \end{cases}$$

What is the probability that $x \geq 6$?

3.12-14. Two random variables x and y have the joint density function

$$p_{xy}(\alpha, \beta) = \begin{cases} Ae^{-(\alpha+\beta)} & \alpha \geq 0, \ \beta \geq 0 \\ 0 & \text{otherwise} \end{cases}$$

(a) What is A such that this is a valid probability density function?
(b) What is the probability that $x \leq 2, y \leq 1$?
(c) Are x and y independent random variables?

4. *Functions of Random Variables*

4.1. Introduction

In the preceding chapter we introduced the concept of a random variable and developed two means of describing a random variable—distribution and density functions. In addition, several manipulative rules were developed for treating density functions, especially conditional density functions.

In this chapter we wish to examine several different types of operations on random variables. These techniques, which greatly extend the usefulness of the concept of random variables, are invaluable in the study of such subjects as communication theory, detection theory, and stochastic control theory. First we consider the problem of determining the density and distribution functions for a random variable that is defined as the function of another random variable. Next we consider the expectation operator and develop the concept of statistical moments as a third means of describing a random variable. A fourth method of describing the distribution of a random variable is also developed; this method is the characteristic function method and is essentially the Fourier transform of the density function. The final topic examined concerns sequences of random variables and associated convergence properties.

4.2. Functions of one random variable

One of the most important and useful concepts in the application of probability theory is that of mapping one set of random variables into another set of random variables by means of some functional relationship; this is the subject of this section. We shall begin the treatment by examining the case of a scalar function of a scalar random variable. Once the basic concepts have been mastered, the extension to the general case is direct.

Let us suppose that we have a function[1] f which maps $\mathcal{R}$ into $\mathcal{R}$ and a random variable x, which we remember is a function mapping $\mathcal{S}$ into $\mathcal{R}$. We assume that f is at least defined on the range of x; that is, $\{x(\xi): \xi \in \mathcal{S}\}$. Normally, f will be defined for all of $\mathcal{R}$. For any outcome $\xi \in \mathcal{S}$, we define $y(\xi)$ as

$$y(\xi) = f[x(\xi)] \tag{4.2-1}$$

which defines a new mapping of $\mathcal{S}$ into $\mathcal{R}$. It must be noted that y is, in general, not a random variable unless we place some mild restrictions on f. In particular, we remember that the set of outcomes $\{y \leq \beta\}$ must be an event for all β, and the events $\{y = +\infty\}$ and $\{y = -\infty\}$ must have zero probability for y to be a random variable. To relate these requirements to requirements on f, we shall make use of the following notation. Let $\mathcal{A}_\beta$ be the preimage of the set $\{y: y \leq \beta\}$; that is, $\mathcal{A}_\beta$ is a set of x such that $f(x) \leq \beta$ or

$$\mathcal{A}_\beta \triangleq \{x: f(x) \leq \beta\} \tag{4.2-2}$$

Now we note that the set of outcomes $\{\xi: y(\xi) \leq \beta\}$ must be equal to $\{\xi: x(\xi) \in \mathcal{A}_\beta\}$, or

$$\{\xi: y(\xi) \leq \beta\} = \{\xi: x(\xi) \in \mathcal{A}_\beta\} \tag{4.2-3}$$

which may be written more compactly as

$$\{y \leq \beta\} = \{x \in \mathcal{A}_\beta\} \tag{4.2-4}$$

For $\{y \leq \beta\}$ to be an event, it is necessary that f be defined so that $\{x \in \mathcal{A}_\beta\}$ is an event for every $\beta \in \mathcal{R}$. The function f must also be defined such that the events $\{y = +\infty\}$ and $\{y = -\infty\}$ have zero probability. We shall assume that f satisfies these two requirements so that y is a random variable.

If y is a random variable, it must have a distribution and a density function. We can obviously go back to the basic definitions and find these functions from the statement of the random experiment $[\mathcal{S}, \mathcal{F}, \mathcal{P}]$. However, if the distribution of x is known, it would appear possible to find the distribution of y using the distribution of x and the function f; this is the approach that we wish to explore.

It is convenient to divide the treatment into two parts and treat discrete and continuous random variables separately. One may write a mixed random

[1] See Section 3.2 for a brief discussion of function theory.

variable as the weighted sum of a discrete and a continuous random variable and treat the two parts by their appropriate rules, as in Section 3.4.

Discrete Random Variables. If x is a discrete random variable, it assumes only a countable number of values and its density function takes the form

$$p_x(\alpha) = \sum_{i=1}^{n} p_i \delta_D(\alpha - \alpha_i) \tag{4.2-5}$$

where n is finite or infinite and $p_i = \mathcal{P}\{x = \alpha_i\}$. Now we define a new random variable by the relation

$$y = f(x)$$

where f is assumed to be one to one. Then y must also be a discrete random variable, since it can only take a countable number of values because a countable number of points can only map into a countable number of points, owing to the single-valued nature of the function f. The values that y can assume will be given by $\beta_1, \beta_2, \ldots$, where $\beta_i = f(\alpha_i)$, so the density function of y must take the form

$$p_y(\beta) = \sum_{i=1}^{n} \phi_i \delta_D(\beta - \beta_i) \tag{4.2-6}$$

where $\phi_i = \mathcal{P}\{y = \beta_i\}$. But $\mathcal{P}\{y = \beta_i\} = \mathcal{P}\{x = \alpha_i\}$, so that $\phi_i = p_i$, and Eq. (4.2-6) becomes

$$p_y(\beta) = \sum_{i=1}^{n} p_i \delta_D(\beta - \beta_i) \tag{4.2-7}$$

which gives a very simple way of transforming p_x into p_y.

If f is not one to one, there may be more than one α_i such that $f(\alpha_i) = \beta_j$. Now $\mathcal{P}\{y = \beta_j\} = \mathcal{P}\{x = \alpha_i : f(\alpha_i) = \beta_j\}$, so ϕ_j in Eq. (4.2-6) is given by

$$\phi_j = \sum_{i:\, f(\alpha_i) = \beta_j} p_i \tag{4.2-8}$$

In other words, we sum the probability of all the $\{x = \alpha_i\}$ for which $f(\alpha_i) = \beta_j$ to obtain the probability of $\{y = \beta_i\}$. This simply indicates that all the outcomes ξ, such that $x(\xi) = \alpha_i$ with $f(\alpha_i) = \beta_j$, now yield $y(\xi) = \beta_j$. With the use of Eqs. (4.2-6) and (4.2-8), we may handle the transformation of any discrete random variable. Let us illustrate the use of this result by considering a simple example.

Example *4.2-1.* Let x be a discrete random variable defined on an experiment with 10 equally likely outcomes $\xi_1, \xi_2, \ldots, \xi_{10}$, as $x(\xi_i) =$

i. Then x has a discrete uniform distribution with the density function given by

$$p_x(\alpha) = \tfrac{1}{10} \sum_{i=1}^{10} \delta_D(\alpha - i)$$

so that $\alpha_i = i$ and $p_i = \frac{1}{10}$, $i = 1, 2, \ldots, 10$. The graph of the density function is shown in Fig. 4.2-1a. Let us define the function f by the mapping

$$y = f(x) = \begin{cases} 0 & \text{for } 2n \le x < 2n + 1 \\ 1 & \text{for } 2n - 1 \le x < 2n \end{cases} \qquad n = 0, 1, \ldots$$

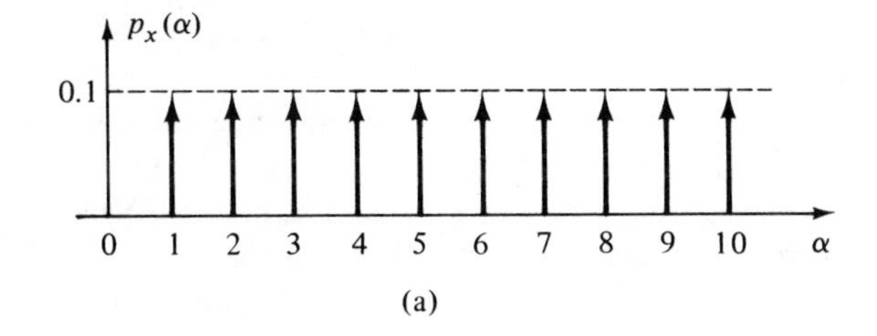

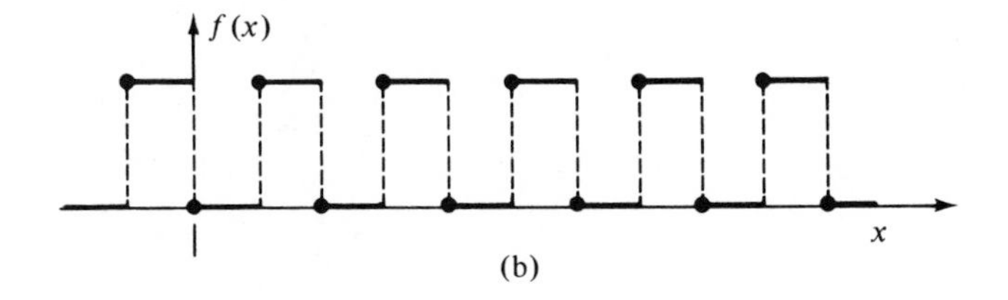

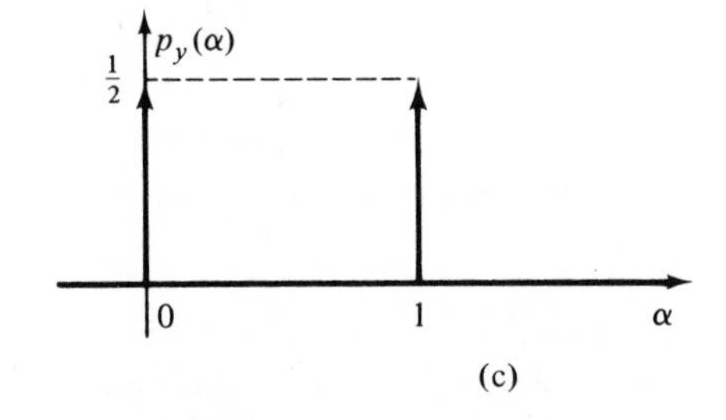

FIG. 4.2-1. Example 4.2-1: (a) density of x, (b) mapping function, (c) density of y.

The graph of this function is shown in Fig. 4.2-1b. Note that y takes on only two values, so it is also a discrete random variable, as predicted with $\beta_1 = 0$ and $\beta_2 = 1$.

In this problem the set of α_i's such that $f(\alpha_i) = \beta_1 = 0$ are $\{\alpha_2, \alpha_4, \alpha_6, \alpha_8, \alpha_{10}\}$, whereas $\{x_i : f(\alpha_i) = \beta_2 = 1\} = \{\alpha_1, \alpha_3, \alpha_5, \alpha_7,$

$\alpha_9\}$. Therefore, the application of Eq. (4.2-8) gives

$$\phi_1 = \sum_{i=2,4,6,8,10} p_i = \tfrac{1}{2}$$

$$\phi_2 = \sum_{i=1,3,5,7,9} p_i = \tfrac{1}{2}$$

and p_y is

$$p_y(\beta) = \tfrac{1}{2}\delta_D(\beta) + \tfrac{1}{2}\delta_D(\beta - 1)$$

This density function is illustrated in Fig. 4.2-1c.

For comparison and clarification, let us determine the distribution of y by the use of the basic definition. For $\beta < 0$, we see that $\{y \leq \beta\} = \varnothing$ since there are no outcomes such that $y(\xi) < 0$. For $\beta \geq 1$, we see that $\{y \leq \beta\} = \mathcal{S}$, since for every outcome $\xi \in \mathcal{S}$, $y(\xi) = f[x(\xi)] \leq 1$. Now for $0 \leq \beta < 1$, we find that $\{y \leq \beta\} = \{\xi_2, \xi_4, \xi_6, \xi_8, \xi_{10}\}$ since ξ_i, $i = 2, 4, 6, 8, 10$, yield $x = 2, 4, 6, 8, 10$, and consequently $y = 0$; on the other hand, ξ_i, $i = 1, 3, 5, 7, 9$, yield $y(\xi) = 1$. Now, using the basic definition of the density function, we have

$$F_y(\beta) = \begin{cases} \mathcal{P}\{\varnothing\} = 0 & \text{for } \beta < 0 \\ \mathcal{P}\{\xi_2, \xi_4, \xi_6, \xi_8, \xi_{10}\} = \frac{1}{2} & \text{for } 0 \leq \beta < 1 \\ \mathcal{P}\{\mathcal{S}\} = 1 & \text{for } \beta \geq 1 \end{cases}$$

so that $p_y(\beta)$ is

$$p_y(\beta) = \tfrac{1}{2}\delta_D(\beta) + \tfrac{1}{2}\delta_D(\beta - 1)$$

which agrees with our previous result. In this simple example there is little advantage in using Eqs. (4.2-6) and (4.2-8). In more complicated problems the use of the basic definitions would normally require prohibitive effort.

Continuous Random Variables. Unfortunately, the treatment for continuous random variables is not as simple as that for discrete random variables. Since the distribution and density functions of y are defined as

$$F_y(\beta) = \int_{-\infty}^{\beta} p_y(\gamma)\, d\gamma = \mathcal{P}\{y \leq \beta\} \tag{4.2-9}$$

we may use Eq. (4.2-4) to write this expression in the form

$$F_y(\beta) = \int_{-\infty}^{\beta} p_y(\gamma)\, d\gamma = \mathcal{P}\{x \in \mathcal{A}_\beta\} \tag{4.2-10}$$

In terms of the density function of x, Eq. (4.2-10) may be written as

$$F_y(\beta) = \int_{-\infty}^{\beta} p_y(\gamma)\, d\gamma = \int_{\mathcal{A}_\beta} p_x(\lambda)\, d\lambda \tag{4.2-11}$$

Although Eq. (4.2-11) expresses in very general form a method of determining the distribution of y from the distribution of x, the result is not too useful in this form, because the general determination of $\mathcal{A}_\beta$ is not trivial.

To develop some useful techniques for dealing with Eq. (4.2-11), we shall begin by severely restricting the class of mapping functions f; then we shall remove these restrictions one by one until we have allowed a very general class of functions to be treated quite easily. Initially, we shall assume that f is strictly monotone increasing, continuous, and differentiable, except at a countable number of points,[1] and "onto" such that the range of f is $\mathcal{R}$. Note that the domain and range of f is $\mathcal{R}$. Because f is strictly monotone and onto, there exists a unique inverse function f^{-1} such that for every y there exists an $x = f^{-1}(y)$ with $f(x) = y$. The pre-image set $\mathcal{A}_\beta$ for $\{y: y \leq \beta\}$ is given by $\{x: x \leq f^{-1}(\beta)\}$. The use of Eq. (4.2-11) therefore yields

$$F_y(\beta) = \int_{-\infty}^{\beta} p_y(\gamma)\, d\gamma = \int_{-\infty}^{f^{-1}(\beta)} p_x(\lambda)\, d\lambda = F_x[f^{-1}(\beta)] \tag{4.2-12}$$

which gives a very simple expression for the distribution function of y in terms of F_x. Let us make a change of variables in the second integral by letting $\tau = f(\lambda)$; then we have

$$\int_{-\infty}^{f^{-1}(\beta)} p_x(\lambda)\, d\lambda = \int_{-\infty}^{\beta} p_x[f^{-1}(\tau)] \frac{df^{-1}(\tau)}{d\tau}\, d\tau \tag{4.2-13}$$

The derivative $df(\tau)/d\tau$, and hence $df^{-1}(\tau)/d\tau$, exists, by assumption, except at a countable number of points. At these points we may assign any arbitrary nonnegative value to $df^{-1}(\tau)/d\tau$ without changing the value of the integral. Hence we may think of $df^{-1}(\tau)/d\tau$ as existing for all $\tau \in \mathcal{R}$.

Substituting Eq. (4.2-13) into Eq. (4.2-12) gives the following relation:

$$\int_{-\infty}^{\beta} p_y(\gamma)\, d\gamma = \int_{-\infty}^{\beta} p_x[f^{-1}(\tau)] \frac{df^{-1}(\tau)}{d\tau}\, d\tau \tag{4.2-14}$$

Since this expression must be valid for every value of $\beta \in \mathcal{R}$, we see that one possible solution for $p_y(\gamma)$ is

$$p_y(\beta) = p_x[f^{-1}(\beta)] \frac{df^{-1}(\beta)}{d\beta} \tag{4.2-15}$$

[1] We need only require that the set of points where f is not differentiable have zero measure.

Let us consider some simple examples to illustrate the use of this result.

Example *4.2-2*. We wish to determine the probability density function of the random variable y defined as

$$y = f(x) = 2x$$

where x is a continuous random variable with a probability density function given by

$$p_x(\alpha) = \frac{1}{\sqrt{2\pi}} e^{-\alpha^2/2}$$

Here f is clearly strictly monotone increasing, continuous, differentiable everywhere, and onto so that Eq. (4.2-15) is applicable. The inverse function $f^{-1}(y)$ is given by

$$f^{-1}(y) = \frac{y}{2}$$

and

$$\frac{df^{-1}(y)}{dy} = \frac{1}{2}$$

Substituting these results into Eq. (4.2-15) yields the general result

$$p_y(\beta) = \tfrac{1}{2} p_x(\beta/2)$$

For the assumed density of x, we obtain

$$p_y(\beta) = p_x[f^{-1}(\alpha)]\frac{df^{-1}(\alpha)}{d\alpha} = \frac{1}{\sqrt{2\pi}} e^{-(1/2)(y/2)^2} \frac{1}{2} = \frac{1}{2\sqrt{2\pi}} e^{-(1/2)(y^2/4)}$$

and we see that y is $N(0, 4)$.

Example *4.2-3*. Let us assume that p_x is uniform on $[-1, 1]$ so that it is given by

$$p_x(\alpha) = \tfrac{1}{2}[\mathscr{s}(\alpha + 1) - \mathscr{s}(\alpha - 1)]$$

and that the function f is

$$f(x) = \begin{cases} x & x \leq 0 \\ \dfrac{x}{2} & x > 0 \end{cases}$$

This function is continuous, strictly monotone increasing, differentiable except at $x = 0$, and onto; hence Eq. (4.2-15) is applicable. The inverse function can be easily found as

$$f^{-1}(y) = \begin{cases} y & y \leq 0 \\ 2y & y > 0 \end{cases}$$

which we may write as

$$f^{-1}(y) = y + y\mathscr{u}(y)$$

The derivative $df^{-1}(y)/dy$ is

$$\frac{df^{-1}(y)}{dy} = \begin{cases} 1 & y < 0 \\ 2 & y > 0 \end{cases}$$

We shall let $df^{-1}(y)/dy = 2$ at $y = 0$ so that

$$\frac{df^{-1}(y)}{dy} = 1 + \mathscr{u}(y)$$

The use of Eq. (4.2-15) yields the following result for p_y:

$$\begin{aligned} p_x(\alpha) = p_x[f^{-1}(\alpha)]\frac{df^{-1}(\alpha)}{d\alpha} &= \frac{1}{2}\{\mathscr{u}[\alpha + \alpha\mathscr{u}(\alpha) + 1] \\ &- \mathscr{u}[\alpha + \mathscr{u}(\alpha) - 1]\}[1 + \mathscr{u}(\alpha)] \end{aligned}$$

It is easy to show by plotting the arguments that

$$\begin{aligned} \mathscr{u}[\alpha + \alpha\mathscr{u}(\alpha) + 1] &= \mathscr{u}(\alpha + 1) \\ \mathscr{u}[\alpha + \alpha\mathscr{u}(\alpha) - 1] &= \mathscr{u}(2\alpha - 1) \end{aligned}$$

so that $p_y(\alpha)$ is now

$$\begin{aligned} p_y(\alpha) &= \tfrac{1}{2}[\mathscr{u}(\alpha + 1) - \mathscr{u}(2\alpha - 1)][1 + \mathscr{u}(\alpha)] \\ &= \tfrac{1}{2}\mathscr{u}(\alpha + 1) + \tfrac{1}{2}\mathscr{u}(\alpha) - \mathscr{u}(2\alpha - 1) \end{aligned}$$

which is the final desired result. The reader is urged to repeat this problem and use Eq. (4.2-12) to determine F_y and then to find p_y.

Before we relax some of the restrictions on the function $f(x)$, let us briefly examine another way of viewing Eq. (4.2-15). Consider the graph of a general strictly monotone increasing function shown in Fig. 4.2-2. Since the events $\{\alpha < x \leq \alpha + d\alpha\}$ and $\{\beta < y \leq \beta + d\beta\}$ are equal, their probabi-

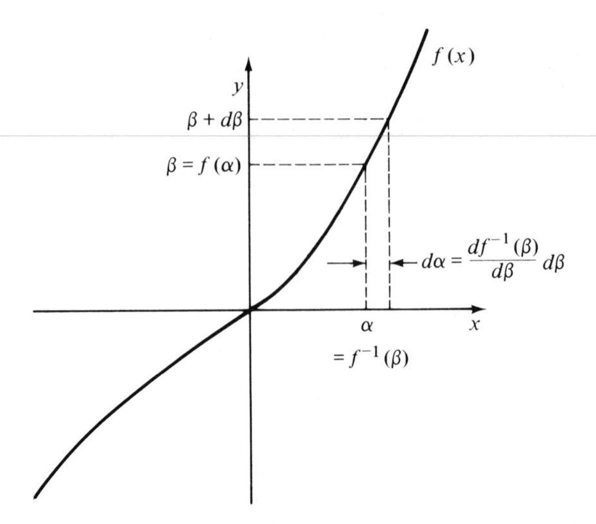

FIG. 4.2-2. Illustration of Eq. (4.2-15).

lities must be equal so that

$$\mathcal{P}\{\beta < y \leq \beta + d\beta\} = \mathcal{P}\{\alpha < x \leq \alpha + d\alpha\} \tag{4.2-16}$$

If we make $d\beta$ and hence $d\alpha$ small enough, Eq. (4.2-16) may be written as

$$p_y(\beta)\, d\beta = p_x(\alpha)\, d\alpha$$

Since $\alpha = f^{-1}(\beta)$ and $d\alpha = [df^{-1}(\beta)/d\beta]\, d\beta$, this result becomes

$$p_y(\beta)\, d\beta = p_x[f^{-1}(\beta)]\frac{df^{-1}(\beta)}{d\beta}\, d\beta$$

or, equivalently,

$$p_y(\beta) = p_x[f^{-1}(\beta)]\frac{df^{-1}(\beta)}{d\beta}$$

which is identical to Eq. (4.2-15). This approach of equating probabilities is a helpful and fundamental approach and should be remembered.

Let us return now to the basic problem and begin to remove some of the restrictions on the function f. As a first step, we consider what happens if f is monotone decreasing instead of monotone increasing. In this case, a unique inverse function f^{-1} still exists, but now the pre-image set for $\{y: y \leq \beta\}$ is $\mathcal{A}_\beta = \{x: f^{-1}(\beta) \leq x\}$, since as β increases $f^{-1}(\beta)$ decreases. The use of Eq. (4.2-11) now gives the following result for F_y:

$$F_y(\beta) = \int_{-\infty}^{\beta} p_y(\gamma)\, d\gamma = \int_{f^{-1}(\beta)}^{\infty} p_x(\lambda)\, d\lambda = 1 - F_x[f^{-1}(\beta)] \tag{4.2-17}$$

If we use the change of variables $\tau = f(\lambda)$, the second integral in Eq. (4.2-17) becomes

$$\int_{f^{-1}(\beta)}^{\infty} p_x(\lambda)\, d\lambda = -\int_{-\infty}^{\beta} [f^{-1}(\tau)] \frac{df^{-1}(\tau)}{d\tau}\, d\tau$$

Substituting this result into Eq. (4.2-17) yields the following expression:

$$\int_{-\infty}^{\beta} p_y(\gamma)\, d\gamma = -\int_{-\infty}^{\beta} p_x[f^{-1}(\tau)] \frac{df^{-1}(\tau)}{d\tau}\, d\tau$$

and if we take the derivative with respect to β we obtain

$$p_y(\beta) = -p_x[f^{-1}(\beta)] \frac{df^{-1}(\beta)}{d\beta} \tag{4.2-18}$$

This expression may look somewhat strange because of the negative sign. However, since f is strictly monotone decreasing, $df^{-1}(\beta)/d\beta$ is always less than 0, so $p_y(\beta)$ remains positive.

A comparison of Eqs. (4.2-18) and (4.2-15) reveals that we may combine these results into one equation of the form

$$p_y(\beta) = p_x[f^{-1}(\beta)] \left| \frac{df^{-1}(\beta)}{d\beta} \right| \tag{4.2-19}$$

which is valid for either strictly monotone increasing or decreasing functions. Equation (4.2-19) represents a most useful expression for working with functions of a random variable.

One additional comment on Eq. (4.2-19) may be appropriate at this time. The derivative $df^{-1}(\beta)/d\beta$ can also be written as

$$\frac{df^{-1}(\beta)}{d\beta} = \left[\frac{df(\alpha)}{d\alpha}\right]^{-1} \Bigg|_{\alpha = f^{-1}(\beta)} \tag{4.2-20}$$

This equivalence is easily established, since

$$\frac{df^{-1}(\beta)}{d\beta} = \frac{d\alpha(\beta)}{d\beta} = \frac{1}{d\beta(\alpha)/d\alpha} \Bigg|_{\alpha = f^{-1}(\beta)}$$

The right-hand side of Eq. (4.2-20) is sometimes simpler to evaluate than the left. In addition, one also finds Eq. (4.2-19) written as

$$p_y(\beta) = p_x[f^{-1}(\beta)] \left| \left[\frac{df(\alpha)}{d\alpha}\right]^{-1} \Bigg|_{\alpha = f^{-1}(\beta)} \right| \tag{4.2-21}$$

We are now able to treat functions that are strictly monotone, continuous and differentiable almost everywhere, and onto. We wish now to remove the requirement that f be onto; note that this requirement rules out such functions as $f(x) = \arctan(x)$ since the range of f is $(-\pi/2, \pi/2)$ and not $\mathcal{R}$. The only way in which we used the property that f was onto was in assuming that $f^{-1}(y)$ existed for all $y \in \mathcal{R}$. If f is not onto, there will exist values of y for which $f^{-1}(y)$ does not exist. In particular, since we have assumed that f is monotone and continuous, then the range of f must be an interval (a, b). Since $\mathcal{P}\{y \in [a, b]\} = \mathcal{P}\{x \in \mathcal{R}\} = \mathcal{P}\{\mathcal{S}\} = 1$, we see that $\mathcal{P}\{y \in [-\infty, a)\} = \mathcal{P}\{y \in (b, +\infty]\} = 0$. In other words, there are no outcomes ξ such that $y \in [-\infty, a)$ or $y \in (b, +\infty]$ since there is no $x \in \mathcal{R}$ such that $f(x) \in [-\infty, a)$ or $f(x) \in (b, +\infty]$. One possible solution would be to let $p_y(\beta) = 0$ for $\beta \in [-\infty, a)$ and $\beta \in (b, +\infty]$, and let $p_y(\beta)$ be given by Eq. (4.2-19) for $\beta \in [a, b]$. Of course, we can always let p_y take arbitrary finite positive values at isolated points in $[-\infty, a)$ and $(b, +\infty]$. In terms of the distribution function, we would have $F_y(\beta) = 0$ for $\beta \in [-\infty, a)$, and $F_y(\beta) = 1$ for $\beta \in (b, +\infty]$.

A similar approach may be used to handle discontinuous functions. We shall now assume that f is strictly monotone, piecewise continuous, and differentiable almost everywhere. For simplicity let us begin by considering the case when $f(x)$ has just one discontinuity at $x = \alpha_0$. Then there are two numbers β_0^+ and β_0^- defined by

$$\beta_0^+ = \lim_{\alpha \downarrow \alpha_0} f(\alpha)$$

$$\beta_0^- = \lim_{\alpha \uparrow \alpha_0} f(\alpha)$$

We shall assume that f is monotone increasing so that $\beta_0^- < \beta_0^+$; if f is monotone decreasing the following arguments still hold if we reverse the roles of β_0^+ and β_0^-. The only event that will yield $y \in (\beta_0^-, \beta_0^+)$ is $x = \alpha_0$, and since x is continuous, $\mathcal{P}\{x = \alpha_0\} = 0$ so that $\mathcal{P}\{y \in (\beta_0^-, \beta_0^+)\} = 0$. Therefore, we can set $p_y(\beta) = 0$ for $\beta \in (\beta_0^-, \beta_0^+)$. If there is more than one discontinuity, we repeat this process for each discontinuity. Note that since $p_y(\beta) = 0$ for $\beta \in (\beta_0^-, \beta_0^+)$, then $F_y(\beta)$ must be constant for $\beta \in (\beta_0^-, \beta_0^+)$.

We can summarize the two preceding developments in a simple manner. We use Eq. (4.2-9) to define $p_y(\beta)$ for β in the range space of f and set $p_y(\beta) = 0$ otherwise. Hence we are now able to handle the class of functions that are strictly monotone, piecewise continuous, and differentiable almost everywhere. Before we continue to broaden the class of allowable functions, let us consider some examples to illustrate the theory to this point.

Example *4.2-4*. Suppose that f is a general linear transformation plus translation given by

$$y = f(x) = ax + b$$

Then f^{-1} is defined for all $y \in \mathcal{R}$ and is given by

$$x = f^{-1}(y) = \frac{y - b}{a}$$

so that $df^{-1}(y)/dy$ is

$$df^{-1}(y)/dy = \frac{1}{a}$$

The application of Eq. (4.2-19) gives the density of y in terms of the density of x as

$$p_y(\beta) = \frac{1}{|a|} p_x\left[\frac{\beta - b}{a}\right]$$

Let us suppose that x is $N(0, \sigma^2)$; then $p_x(\alpha)$ is given by

$$p_x(\alpha) = \frac{1}{\sqrt{2\pi}\,\sigma} e^{-(1/2)(\alpha^2/\sigma^2)}$$

and $p_y(\beta)$ is therefore

$$p_y(\beta) = \frac{1}{\sqrt{2\pi}\,a\sigma} \exp\left\{-\frac{1}{2}\left[\frac{(\beta - b)^2}{(a\sigma)^2}\right]\right\}$$

We see that y is $N(b, a^2\sigma^2)$ and note that a linear transformation plus translation of a gaussian random variable yields another gaussian random variable.

Example 4.2-5. Consider the case when x has an exponential distribution with parameter $\lambda = 1$ so that

$$p_x(\alpha) = e^{-\alpha}\mathcal{s}(\alpha)$$

and f is given by

$$y = f(x) = \arctan x$$

Here the range of f is $(-\pi/2, \pi/2)$, so we set $p_y(\beta) = 0$ for $\beta \notin (-\pi/2, \pi/2)$. The inverse function f^{-1} is given by

$$f^{-1}(y) = \tan y \qquad y \in (-\pi/2, \pi/2)$$

and $df^{-1}(y)/dy$ is equal to

$$\frac{df^{-1}(y)}{dy} = \sec^2 y$$

Equation (4.2-19) may now be used to find $p_y(\beta)$ for $\beta \in (-\pi/2, \pi/2)$ as

$$p_y(\beta) = (\sec^2 \beta)e^{-\tan\beta}\mathscr{s}(\tan \beta) = (\sec^2 \beta)e^{-\tan\beta}\mathscr{s}(\beta) \qquad \text{for } \beta \in (-\pi/2, \pi/2)$$

Since $p_y(\beta) = 0$ for $\beta \in (-\pi/2, \pi/2)$, the final expression for $p_y(\beta)$ is

$$p_y(\beta) = \sec^2 \beta e^{-\tan\beta}[\mathscr{s}(\beta) - \mathscr{s}(\beta - \pi/2)]$$

Example *4.2-6*. Let us consider the application of this technique for f given by

$$f(x) = x + \mathscr{s}(x)$$

We shall assume that the probability density of x is uniform on $[-2, 2]$ so that it is given by

$$p_x(\alpha) = \tfrac{1}{4}[\mathscr{s}(\alpha + 2) - \mathscr{s}(\alpha - 2)]$$

For this problem, $f(x)$ is strictly monotone but is discontinuous at $x = 0$. The range of f does not include the interval $[0, 1)$, so $p_y(\beta) = 0$ for $\beta \in [0, 1)$. We may find $p_y(\beta)$ for $\beta \notin [0, 1)$ by the use of Eq. (4.2-19). The inverse function f^{-1} is given by

$$f^{-1}(y) = y - \mathscr{s}(y - 1) \qquad \text{for } y \notin [0, 1)$$

and $df^{-1}(y)/dy$ is

$$\frac{df^{-1}(y)}{dy} = 1 \qquad \text{for } y \notin [0, 1)$$

The use of Eq. (4.2-19) gives the following result for $p_y(\beta)$, $\beta \notin [0, 1)$:

$$p_y(\beta) = \tfrac{1}{4}[\mathscr{s}(\beta + 2) - \mathscr{s}(\beta - 3)] \qquad \beta \notin [0, 1)$$

And since $p_y(\beta) = 0$ for $\beta \in [0, 1)$, we find that $p_y(\beta)$ is given by

$$p_y(\beta) = \tfrac{1}{4}[\mathscr{s}(\beta + 2) - \mathscr{s}(\beta) + \mathscr{s}(\beta - 1) - \mathscr{s}(\beta - 3)]$$

The graph of $p_y(\beta)$ is shown in Fig. 4.2-3.

Let us return now to the task of generalizing the class of admissible functions. As the next step, we wish to remove the requirement that f be strictly monotonic and just assume that it is monotone. In other words, we now allow $f(x)$ to assume a constant value for a finite interval of x. Suppose,

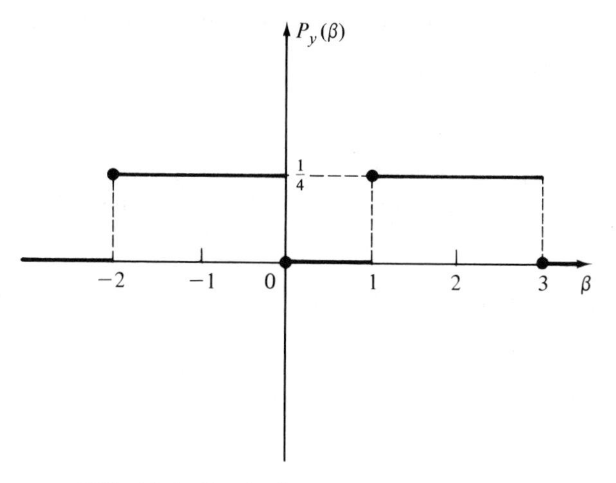

FIG. 4.2-3. Graph of $p_y(\beta)$ for example 4.2-6.

for example, that $f(x) = \beta_0$ for $x \in [\alpha_0, \alpha_1]$. There is clearly no $f^{-1}(\beta)$ for $\beta = \beta_0$ since there is no single value x such that $f(x) = \beta_0$. The event $\{y = \beta_0\}$ is equal to the event $\{x \in [\alpha_0, \alpha_1]\}$; hence

$$\mathcal{P}\{y = \beta_0\} = \mathcal{P}\{x \in [\alpha_0, \alpha_1]\} = \int_{\alpha_0}^{\alpha_1} p_x(\gamma)\, d\gamma = F_x(\alpha_1) - F_x(\alpha_0) \quad (4.2\text{-}22)$$

In general, we see that $\mathcal{P}\{y = \beta_0\} \neq 0$; for this to be true, we must have a delta function in $p_y(\beta)$ at $\beta = \beta_0$ whose area is $F_x(\alpha_1) - F_x(\alpha_0)$.

We can proceed in this fashion to handle any interval where f is constant. This result indicates that y may be a mixed random variable even though we assumed that x was continuous. In fact, if f is a piecewise constant function, then y will be a discrete random variable. Once the intervals where f is constant have been handled by the introduction of appropriate delta functions, then $p_y(\beta)$ may be determined in the remainder of the range of f by the use of Eq. (4.2-19), since f^{-1} will be unique in this region. Again we set $p_y(\beta) = 0$ for β not in the range of f. It may be advantageous to consider an example to illustrate the above concept.

Example *4.2-7*. We assume that f represents a "hard" saturation device with

$$f(x) = \begin{cases} -1 & x < -1 \\ x & -1 \leq x \leq 1 \\ 1 & 1 < x \end{cases}$$

Here f is constant at $f(x) = -1$ for $x \in [-\infty, -1)$, and $f(x) = +1$ for $x \in (1, +\infty]$. Hence $p_y(\beta)$ will have a delta function at $\beta = -1$

and $\beta = +1$. For $\beta \in [-1, 1]$ we may use Eq. (4.2-19) to determine that $p_y(\beta)$ is given by

$$p_y(\beta) = p_x(\beta) \qquad \text{for } \beta \in [-1, 1]$$

In addition, we set $p_y(\beta) = 0$ for $\beta > 1$ and $\beta < -1$ since these regions are not included in the range of f. The final form for $p_y(\beta)$ is therefore

$$p_y(\beta) = p_x(\beta)[\mathscr{s}(\beta + 1) - \mathscr{s}(\beta - 1)] + \rho_1\delta_D(\beta + 1) + \rho_2\delta_D(\beta - 1)$$

where

$$\rho_1 = \int_{-\infty}^{-1} p_x(\gamma)\, d\gamma$$

and

$$\rho_2 = \int_{1}^{\infty} p_x(\gamma)\, d\gamma$$

For a specific case, let us assume that x is uniformly distributed on $[-2, 2]$ so that p_x is

$$p_x(\alpha) = \tfrac{1}{4}[\mathscr{s}(\alpha + 2) - \mathscr{s}(\alpha - 2)]$$

Now ρ_1 is

$$\rho_1 = \int_{-\infty}^{-1} p_x(\gamma)\, d\gamma = \tfrac{1}{4}\int_{-2}^{-1} d\gamma = \tfrac{1}{4}$$

and ρ_2 is similarly determined to be $\frac{1}{4}$. Therefore, the final form of p_y is

$$p_y(\beta) = \tfrac{1}{4}\delta_D(\beta + 1) + \tfrac{1}{4}\delta_D(\beta - 1) + \tfrac{1}{4}[\mathscr{s}(\beta + 1) - \mathscr{s}(\beta - 1)]$$

The last step in the development is to remove the requirement that f be monotonic. If f is not monotonic, an inverse function f^{-1} does not exist, since more than one x may map to the same y. The approach will be to divide the domain space into regions in which f is monotonic. Then Eq. (4.2-19) along with the methods developed above may be applied to each region and the results added to find the density of y.

Let us summarize the general procedure that we have developed for finding the density function p_y for $y = f(x)$ from p_x.

1. Divide the domain space of f into N regions (intervals) where f is either constant or strictly monotone.

2. If on the ith interval, given by $x \in [\alpha_i, \alpha_{i+1}]$, f has a constant value β_i, then the contribution to p_y from this interval is equal to

$$p_{y_i}(\beta) = \rho_i \delta_D(\beta - \beta_i) \tag{4.2-23}$$

where

$$\rho_i = \int_{\alpha_i}^{\alpha_{i+1}} p_x(\gamma)\, d\gamma \tag{4.2-24}$$

3. If on the jth interval f is a monotonic function f_j, use Eq. (4.2-19) with $f = f_j$ to find $p_{y_j}(\beta)$ on the range of f_j, and set $p_{y_j}(\beta) = 0$ for all β not in the range of f_y.
4. The density function of y is given by

$$p_y(\beta) = \sum_{i=1}^{N} p_{y_i}(\beta) \tag{4.2-25}$$

It should be noted that if there are intervals in x such that p_x is zero, one may wish to define such an interval as a region, since it will have no effect on p_y and may be ignored. Let us illustrate the use of this procedure by considering some simple examples.

Example *4.2-8*. Consider the case when x is uniformly distributed on $[-2, 2]$ so that p_x is

$$p_x(\alpha) = \tfrac{1}{4}[\mathcal{s}(\alpha + 2) - \mathcal{s}(\alpha - 2)]$$

and f is given by

$$f(x) = \begin{cases} -x & x < -1 \\ 0 & -1 \leq x \leq 1 \\ x & 1 < x \end{cases}$$

The graphs of the density function of x and f are shown in Fig. 4.2-4. It is easy to see that f is not monotone, so it is necessary to use the general procedure outlined above. We divide the domain of f ($\mathcal{R}$ in this case) into three regions:

Region 1: $x \in [-\infty, -1)$
Region 2: $x \in [-1, 1]$
Region 3: $x \in (1, +\infty]$

In Region 1, f is monotone decreasing; in Region 2, f is constant at 0; and in Region 3, f is monotone increasing. Hence in Regions 1 and 3

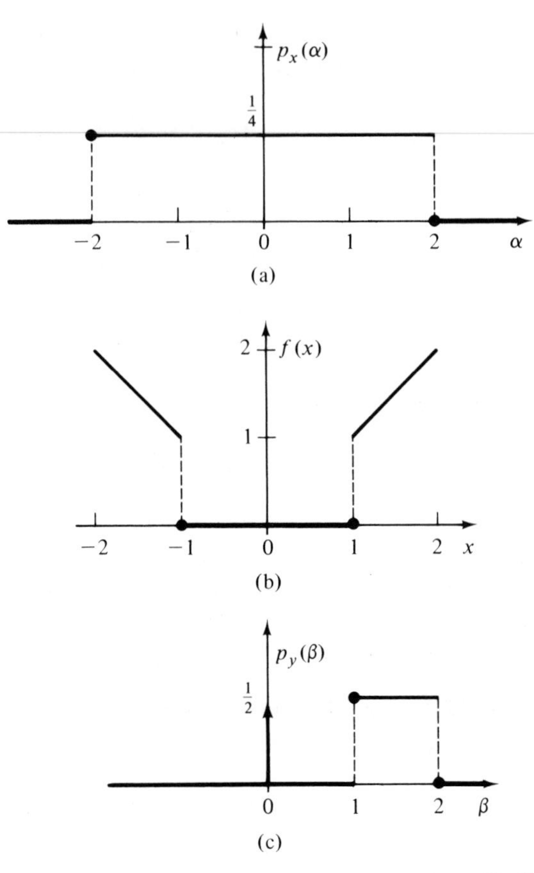

FIG. 4.2-4. Example 4.2-8: (a) density function of x, (b) $f(x)$, (c) density function of y.

we apply Eq. (4.2-19); in Region 2 we must use Eqs. (4.2-23) and (4.2-24) to determine contribution to p_y in each region.

Let us consider Region 1 first. The range of f in Region 1 is $(1, +\infty]$, so we begin by setting $p_{y_1}(\beta) = 0$ for $\beta \in [-\infty, 1)$. The inverse function f^{-1} is given by

$$f^{-1}(y) = -y \qquad \text{for } y \in (1, +\infty]$$

and $df^{-1}(y)/dy$ is equal to -1. Hence Eq. (4.2-19) gives

$$p_{y_1}(\beta) = \tfrac{1}{4}[\mathscr{s}(-\beta + 2) - \mathscr{s}(-\beta - 2)] \qquad \beta \in (1, +\infty]$$

or

$$p_{y_1}(\beta) = \tfrac{1}{4}[\mathscr{s}(2 - \beta)] \qquad \beta \in (1, +\infty]$$

If we combine this result with the fact that $p_{y_1}(\beta) = 0$ for $\beta \in [-\infty, 1)$, we have

$$p_{y_1}(\beta) = \tfrac{1}{4}[\mathscr{u}(\beta - 1) - \mathscr{u}(\beta - 2)]$$

In a similar fashion, we can show that $p_{y_3}(\beta)$ is

$$p_{y_3}(\beta) = \tfrac{1}{4}[\mathscr{u}(\beta - 1) - \mathscr{u}(\beta - 2)]$$

By using Eqs. (4.2-23) and (4.2-24) with $\beta_2 = 0$, $\alpha_2 = -1$, and $\alpha_3 = 1$, $p_{y_2}(\beta)$ becomes

$$p_{y_2}(\beta) = \rho_2 \delta_D(\beta)$$

with

$$\rho_2 = \int_{-1}^{1} p_x(\gamma)\, d\gamma = \tfrac{1}{4} \int_{-1}^{1} d\gamma = \tfrac{1}{2}$$

so that $p_{y_2}(\beta)$ is

$$p_{y_2}(\beta) = \tfrac{1}{2}\delta_D(\beta)$$

Now we use Eq. (4.2-25) to find the complete density function of y given by

$$\begin{aligned} p_y(\beta) &= p_{y_1}(\beta) + p_{y_2}(\beta) + p_{y_3}(\beta) \\ &= \tfrac{1}{2}[\mathscr{u}(\beta - 1) - \mathscr{u}(\beta - 2)] + \tfrac{1}{2}\delta_D(\beta) \end{aligned}$$

This density function is graphed in Fig. 4.2-4c.

In this section we have examined in detail the problem of transforming one random variable into another random variable. In particular, attention has been directed to the determination of the probability density function of the new random variable. In Section 4.3 we extend these concepts to the case of many functions of many random variables.

EXERCISES 4.2

4.2-1. Let x be a scalar random variable and let y be a scalar quantity such that $y = x^2$. If

$$p_x(\alpha) = \frac{1}{\sqrt{2\pi}\sigma} \exp\left[\frac{-\alpha^2}{2\sigma^2}\right]$$

show that

$$p_y(\beta) = \begin{cases} \dfrac{1}{\sqrt{2\pi\beta}\sigma} \exp\left[\dfrac{-\beta}{2\sigma^2}\right] & \beta \geq 0 \\ 0 & \beta < 0 \end{cases}$$

4.2-2. If

$$y = \begin{cases} x & x > 0 \\ 0 & x \leq 0 \end{cases}$$

find $p_y(\beta)$ in terms of a given $p_x(\alpha)$. What are these results if

$$p_x(\alpha) = \frac{1}{(2\pi)^{1/2}} e^{-\alpha^2/2}$$

Answer:

$$p_y(\beta) = \rho\delta_D(\beta) + p_x(\beta)\mathscr{u}(\beta)$$

$$\rho = \int_{-\infty}^{0} p_x(\alpha)\, d\beta$$

4.2-3. If $y = e^x$ and x is uniformly distributed in the interval 0 to 1, find $p_y(\beta)$.

Answer:

$$p_y(\beta) = \begin{cases} 1/\beta & 0 \leq \beta \leq e \\ 0 & \text{otherwise} \end{cases}$$

4.3. *Functions of many random variables*

Since the extension of the methods of the preceding section to functions of many random variables is so direct, the treatment here will be considerably shorter than that of the preceding section. We shall emphasize only the major features of the methods here; the reader is urged to fill in the details following the arguments given in Section 4.2. It should also be noted that the case of functions of many variables is in general considerably more difficult to handle in terms of actual computations, primarily because of the problem of defining the regions discussed in Section 4.2.

Let us consider an m-dimensional random vector $\mathbf{y}$ defined in terms of an n-dimensional random vector $\mathbf{x}$ by the functional relation

$$\mathbf{y} = \mathbf{f}(\mathbf{x}) \tag{4.3-1}$$

We assume that $\mathbf{f}$ is defined so that the set $\{\mathbf{y} \leq \boldsymbol{\beta}\}$ is an event for all $\boldsymbol{\beta} \in \mathcal{R}^m$ and $\mathcal{P}\{\mathbf{y} = +\infty\} = \mathcal{P}\{\mathbf{y} = -\infty\} = 0$, and $\mathbf{y}$ is, in fact, a random vector. It is desired to obtain the probability density function of $\mathbf{y}$ from the density function of $\mathbf{x}$ and the mapping function $\mathbf{f}$. We shall only consider the situation when $\mathbf{x}$ is a continuous random variable. If $\mathbf{x}$ is discrete, then the method presented in the preceding section can be directly applied to obtain $p_{\mathbf{y}}$.

Now let $\mathcal{A}_{\boldsymbol{\beta}}$ be the pre-image set of the set $\{\mathbf{y} : \mathbf{y} \leq \boldsymbol{\beta}\}$; that is, $\mathcal{A}_{\boldsymbol{\beta}} =$

$\{\mathbf{x}: \mathbf{f}(\mathbf{x}) \leq \boldsymbol{\beta}\}$. With this definition it is clear that the event $\{\mathbf{y} \leq \boldsymbol{\beta}\} = \{\xi: \mathbf{y}(\xi) = \mathbf{f}[\mathbf{x}(\xi)] \leq \boldsymbol{\beta}\}$ is equal to the event $\{\mathbf{x} \in \mathcal{A}_{\boldsymbol{\beta}}\} = \{\xi: \mathbf{x}(\xi) \in \mathcal{A}_{\boldsymbol{\beta}}\}$, so their probabilities must also be equal, or

$$\mathcal{P}\{\mathbf{y} \leq \boldsymbol{\beta}\} = \mathcal{P}\{\mathbf{x} \in \mathcal{A}_{\boldsymbol{\beta}}\} \tag{4.3-2}$$

In terms of the distribution and density function of $\mathbf{y}$ and the density function of $\mathbf{x}$, Eq. (4.3-2) may be written as

$$F_{\mathbf{y}}(\boldsymbol{\beta}) = \int_{-\infty}^{\boldsymbol{\beta}} p_{\mathbf{y}}(\boldsymbol{\gamma})\, d\boldsymbol{\gamma} = \int_{\mathcal{A}_{\boldsymbol{\beta}}} p_{\mathbf{x}}(\boldsymbol{\lambda})\, d\boldsymbol{\lambda} \tag{4.3-3}$$

This expression, which is the multidimensional version of Eq. (4.2-11), represents the fundamental relation of $F_{\mathbf{y}}$ and $p_{\mathbf{y}}$ with $p_{\mathbf{x}}$. Once again, the difficult aspect of using this result is the determination of the pre-image set $\mathcal{A}_{\boldsymbol{\beta}}$.

To obtain a more convenient expression for the density function $p_{\mathbf{y}}$, we begin by assuming that the dimension of $\mathbf{y}$ is equal to the of $\mathbf{x}$, that is, $m = n$, and that $\mathbf{f}$ has a unique inverse[1] $\mathbf{f}^{-1}(\mathbf{y})$ for all $\mathbf{y} \in \mathcal{R}^n$. A unique inverse will exist whenever the *Jacobian* $J(\mathbf{x})$ *of the transformation* defined by

$$J(\mathbf{x}) = \det\left[\frac{\partial \mathbf{f}(\mathbf{x})}{\partial \mathbf{x}}\right] \tag{4.3-4}$$

is continuous and does not change sign for all $\mathbf{x} \in \mathcal{R}^n$. Here the ijth element of the matrix $\partial \mathbf{f}(\mathbf{x})/\partial \mathbf{x}$ is defined as

$$\left[\frac{\partial \mathbf{f}(\mathbf{x})}{\partial \mathbf{x}}\right]_{ij} = \frac{\partial f_i(\mathbf{x})}{\partial x_j} \tag{4.3-5}$$

The *jacobian of the inverse transformation* is defined as

$$J_I(\mathbf{y}) = \det\left[\frac{\partial \mathbf{f}^{-1}(\mathbf{y})}{\partial \mathbf{y}}\right] \tag{4.3-6}$$

and can be related to $J(\mathbf{x})$ by the relationship

$$J_I(\mathbf{y}) = \left.\frac{1}{J(\mathbf{x})}\right|_{\mathbf{x}=\mathbf{f}^{-1}(\mathbf{y})} \tag{4.3-7}$$

In the case of a scalar function discussed in Section 4.2, $J(\mathbf{x})$ becomes

$$J(\mathbf{x}) = \frac{\partial f(x)}{\partial x} = \frac{df(x)}{dx}$$

[1] Here it is assumed that the range of $\mathbf{f}$ is $\mathcal{R}^n$.

Hence the requirement that $J(\mathbf{x})$ be continuous and not change sign implies, in the one-dimensional case, that f is monotone and continuously differentiable. Note that Eq. (4.2-20) is the counterpart of Eq. (4.3-7) in this case.

Consider now the probability that $\mathbf{y} \in \mathcal{A}$, where $\mathcal{A}$ is any subset[1] of $\mathcal{R}^n$. Let $\mathcal{B}$ be the pre-image of $\mathcal{A}$ so that $\mathcal{B} = \{\mathbf{x}: \mathbf{f}(\mathbf{x}) \in \mathcal{A}\} = \{\mathbf{x}: \mathbf{x} = \mathbf{f}^{-1}(\mathbf{y}), \mathbf{y} \in \mathcal{A}\}$. The probability that $\mathbf{y} \in \mathcal{A}$ must be equal to the probability that $\mathbf{x} \in \mathcal{B}$, since these two events must correspond to the same set of outcomes; so we have

$$\mathcal{P}\{\mathbf{y} \in \mathcal{A}\} = \mathcal{P}\{\mathbf{x} \in \mathcal{B}\} \tag{4.3-8}$$

In terms of the probability density functions of $\mathbf{y}$ and $\mathbf{x}$, this expression may be written as

$$\int_{\mathcal{A}} p_{\mathbf{y}}(\boldsymbol{\gamma})\, d\boldsymbol{\gamma} = \int_{\mathcal{B}} p_{\mathbf{x}}(\boldsymbol{\eta})\, d\boldsymbol{\eta} \tag{4.3-9}$$

If we make the transformation of variables $\mathbf{v} = \mathbf{f}(\boldsymbol{\eta})$ in the right-hand integral, Eq. (4.3-9) becomes

$$\int_{\mathcal{A}} p_{\mathbf{y}}(\boldsymbol{\gamma})\, d\boldsymbol{\gamma} = \int_{\mathcal{A}} p_{\mathbf{x}}[\mathbf{f}^{-1}(\mathbf{v})]\, |J_I(\mathbf{v})|\, d\mathbf{v} \tag{4.3-10}$$

where we have used the fact from advanced calculus that the volume element $d\boldsymbol{\eta}$ becomes $|J_I(\mathbf{v})|\, d\mathbf{v}$. Here $J_I(\mathbf{v})$ is the Jacobian of the inverse transformation defined in Eq. (4.3-6). Since Eq. (4.3-10) must be valid for every $\mathcal{A}$, one possible way to satisfy this expression is to let

$$p_{\mathbf{y}}(\boldsymbol{\beta}) = p_{\mathbf{x}}[\mathbf{f}^{-1}(\boldsymbol{\beta})]\, |J_I(\boldsymbol{\beta})| \tag{4.3-11}$$

This result gives a direct method for determining $p_{\mathbf{y}}$ from $p_{\mathbf{x}}$ and $\mathbf{f}$.

Equation (4.3-11) can also be obtained in another manner. Consider Fig. 4.3-1; on the left is shown a point and associated volume element in the $\mathbf{y}$ plane. On the right is shown the corresponding point and volume element on the $\mathbf{x}$ plane. Clearly, the probability that $\mathbf{y}$ will lie in the shaded area on the left is equal to the probability that $\mathbf{x}$ will lie in the shaded area on the right. Hence we have

$$p_{\mathbf{y}}(\boldsymbol{\beta})\, d\boldsymbol{\beta} = p_{\mathbf{x}}[\mathbf{f}^{-1}(\boldsymbol{\beta})]\, |J_I(\boldsymbol{\beta})|\, d\boldsymbol{\beta} \tag{4.3-12}$$

which we see is equivalent to Eq. (4.3-11). By the use of Eq. (4.3-7), we can also write Eq. (4.3-11) as

$$p_{\mathbf{y}}(\boldsymbol{\beta}) = p_{\mathbf{x}}[\mathbf{f}^{-1}(\boldsymbol{\beta})][|J[\mathbf{f}^{-1}(\boldsymbol{\beta})]|]^{-1} \tag{4.3-13}$$

[1] We tacitly assume that $\mathcal{A}$ can be expressed as a finite sum or product of sets of the form $\{\mathbf{y}: \mathbf{y} \leq \boldsymbol{\alpha}\}$.

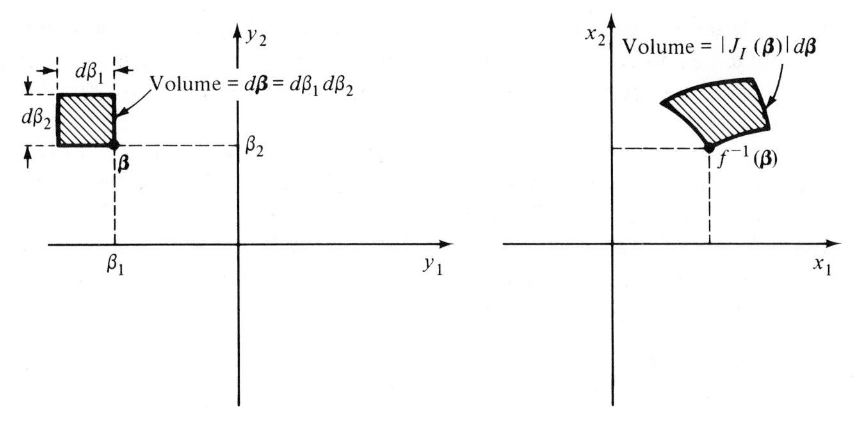

FIG. 4.3-1. Transformation of volume elements.

Example *4.3-1.* To illustrate the use of Eq. (4.3-11), let us suppose that $\mathbf{f}(\mathbf{x})$ is a simple linear transformation given by

$$\mathbf{y} = \mathbf{f}(\mathbf{x}) = \mathbf{A}\mathbf{x}$$

where $\mathbf{A}$ is an $n \times n$ nonsingular matrix. In this case, the inverse function is easily determined as

$$\mathbf{x} = \mathbf{f}^{-1}(\mathbf{y}) = \mathbf{A}^{-1}\mathbf{y}$$

and the jacobian of the inverse transformation is

$$J_I(\mathbf{y}) = \det\left[\frac{\partial \mathbf{f}^{-1}(\mathbf{y})}{\partial \mathbf{y}}\right] = \det \mathbf{A}^{-1} = (\det \mathbf{A})^{-1}$$

Therefore, $p_{\mathbf{y}}$ is given by

$$p_{\mathbf{y}}(\boldsymbol{\beta}) = |(\det \mathbf{A})^{-1}|\, p_{\mathbf{x}}(\mathbf{A}^{-1}\boldsymbol{\beta})$$

Let us suppose that $\mathbf{x}$ has an n-dimensional gaussian distribution[1] with parameters $\boldsymbol{\eta}$ and $\mathbf{V}$ given by

$$p_{\mathbf{x}}(\boldsymbol{\alpha}) = \frac{1}{[(2\pi)^n \det \mathbf{V}]^{1/2}} \exp\left\{-\frac{1}{2}[(\boldsymbol{\alpha} - \boldsymbol{\mu})^T \mathbf{V}^{-1}(\boldsymbol{\alpha} - \boldsymbol{\mu})]\right\}$$

Then $p_{\mathbf{y}}$ is given by

$$\begin{aligned} p_{\mathbf{y}}(\boldsymbol{\beta}) &= \frac{|(\det \mathbf{A})^{-1}|}{[(2\pi)^n \det \mathbf{V}]^{1/2}} \exp\left\{-\frac{1}{2}(\mathbf{A}^{-1}\boldsymbol{\beta} - \boldsymbol{\mu})^T \mathbf{V}^{-1}(\mathbf{A}^{-1}\boldsymbol{\beta} - \boldsymbol{\mu})\right\} \\ &= [(2\pi)^n \det(\mathbf{A}\mathbf{V}\mathbf{A}^T)]^{-1/2} \exp\left\{-\frac{1}{2}(\boldsymbol{\beta} - \mathbf{A}\boldsymbol{\mu})^T(\mathbf{A}\mathbf{V}\mathbf{A}^T)^{-1}(\boldsymbol{\beta} - \mathbf{A}\boldsymbol{\mu})\right\} \end{aligned}$$

[1] See Chapter 7.

Hence we see that $\mathbf{y}$ is also gaussianly distributed with parameters $\mathbf{A\mu}$ and $\mathbf{AVA}^T$. In other words, the gaussian distribution is preserved under a general nonsingular linear transformation.

If the range of $\mathbf{f}$ does not cover all $\mathcal{R}^n$, it is clear that we must set $p_{\mathbf{y}}(\boldsymbol{\beta})$ equal to zero for all $\boldsymbol{\beta}$ not in the range of $\mathbf{f}$. Suppose that there is a region $\mathcal{A}_0$ with nonzero area in the domain of $\mathbf{f}$ for which $\mathbf{f}$ has a constant value $\boldsymbol{\beta}_0$; that is, $\mathbf{f}(\mathbf{x}) = \boldsymbol{\beta}_0$ for all $\mathbf{x} \in \mathcal{A}_0$. In this case, $p_{\mathbf{y}}$ will contain a delta function at $\boldsymbol{\beta} = \boldsymbol{\beta}_0$ whose "area" is equal to the integral of $p_{\mathbf{x}}$ over $\mathcal{A}_0$. Therefore, the contribution to $p_{\mathbf{y}}$ due to this region is

$$p_{\mathbf{y}_0}(\boldsymbol{\rho}) = \rho_0 \delta_D(\boldsymbol{\beta} - \boldsymbol{\beta}_0) \tag{4.3-14}$$

where

$$\rho_0 = \int_{\mathcal{A}_0} p_{\mathbf{x}}(\boldsymbol{\gamma})\, d\boldsymbol{\gamma} \tag{4.3-15}$$

In the n-dimensional case, another situation may occur that has no counterpart in the one-dimensional case of Section 4.2. It is possible that a region $\mathcal{A}_0$ in the domain of $\mathbf{f}$ may exist on which part of the random vector $\mathbf{y}$ is constant while the remainder varies. Consider, for example, that $\mathbf{y}$ is divided into a k vector $\mathbf{y}_1$ and a $n - k$ vector $\mathbf{y}_2$ such that for $\mathbf{x} \in \mathcal{A}_0$, $\mathbf{y}_1 = \boldsymbol{\beta}_{0_1}$ while $\mathbf{y}_2$ varies. In this case, the contribution to $p_{\mathbf{y}}$ associated with this region is

$$p_{\mathbf{y}_0}(\beta) = \rho_0 \delta_D(\boldsymbol{\beta}_1 - \boldsymbol{\beta}_{0_1}) p_{\mathbf{y}_2}(\boldsymbol{\beta}_2) \tag{4.3-16}$$

where

$$\rho_0 = \int_{\mathcal{A}_0} p_{\mathbf{x}}(\boldsymbol{\gamma})\, d\boldsymbol{\gamma} \tag{4.3-17}$$

Here the density $p_{\mathbf{y}_2}$ of the $n - k$ vector $\mathbf{y}_2$ is computed as if $\mathbf{y}_1$ did not exist. Note that Eq. (4.3-11) does not apply here because $\mathbf{y}_2$ is not n dimensional. We shall discuss the approach to be used for computing $p_{\mathbf{y}_2}$ later. In regions in which $\mathbf{y}$ or some subset of $\mathbf{y}$ is constant, the jacobian will be equal to zero.

Example *4.3-2*. Let us consider a second-order problem in which x_1 and x_2 are independent gaussian processes with zero mean and variance σ^2 so that $p_{\mathbf{x}}$ becomes

$$\begin{aligned} p_{\mathbf{x}}(\boldsymbol{\alpha}) = p_{x_1}(\alpha_1)p_{x_2}(\alpha_2) &= \left[\frac{1}{\sqrt{2\pi}\sigma} e^{-(1/2)(\alpha_1{}^2/\sigma^2)}\right]\left[\frac{1}{\sqrt{2\pi}\sigma} e^{-(1/2)(\alpha_1{}^2/\sigma^2)}\right] \\ &= \frac{1}{2\pi\sigma^2} \exp\left[-\frac{1}{2}\frac{\alpha_1^2 + \alpha_2^2}{\sigma^2}\right] \end{aligned}$$

Here the transformation f is given by

$$y_1 = f_1(x) = \tan^{-1}\left(\frac{x_2}{x_1}\right)$$

and

$$y_2 = f_2(x) = \sqrt{x_1^2 + x_2^2}$$

In other words, y_1 and y_2 represent the polar coordinate representation (phase and magnitude) of the vector $\mathbf{x}$.

The range of $\mathbf{f}$ is $-\pi/2 \leq y_1 \leq \pi/2$ (assuming we use the principal value of the arctangent), and $y_2 \geq 0$. Hence we see that $p_{\mathbf{y}}(\boldsymbol{\beta}) = 0$ for $\beta_1 \notin [-\pi/2, \pi/2]$ and $\beta_2 < 0$. We see y_1 and y_2 can equal a constant only for a single point, and y_1 equals a constant only on the line $x_1 = kx_2$, which has zero area; similarly, y_2 can equal a constant only for $x_1^2 + x_2^2 = C^2$, which also has zero area. Therefore, there are no regions of finite area for which $\mathbf{y}$ or any subset of $\mathbf{y}$ is constant.

The jacobian of the transformation is given by

$$J(\mathbf{x}) = \det\begin{bmatrix} \dfrac{-x_2/x_1^2}{1 + (x_2/x_1)^2} & \dfrac{1/x_1}{1 + (x_2/x_1)^2} \\ \dfrac{x_1}{\sqrt{x_1^2 + x_2^2}} & \dfrac{x_2}{\sqrt{x_1^2 + x_2^2}} \end{bmatrix}$$

$$= \frac{-1}{\sqrt{x_1^2 + x_2^2}}$$

so that $J(\mathbf{x}) < 0$ for all $\mathbf{x} \neq \mathbf{0}$. The point at $\mathbf{x} = \mathbf{0}$ causes no difficulty since it has zero area.

The inverse transform in this case is easily found to be

$$x_1 = y_2 \cos y_1$$

$$x_2 = y_2 \sin y_1$$

If we use Eq. (4.3-13) to find $p_y(\beta)$, we obtain

$$p_{\mathbf{y}}(\boldsymbol{\beta}) = \frac{\beta_2}{2\pi\sigma^2} \exp\left[-\frac{1}{2}\frac{\beta_2^2}{\sigma^2}\right] \qquad \text{for } \beta_1 \in [-\pi/2, \pi/2],\ \beta_2 > 0$$

It is interesting to compute the marginal density of β_2, which is given by

$$p_{y_2}(\beta_2) = \int_{-\pi/2}^{\pi/2} p_{\mathbf{y}}(\boldsymbol{\beta})\, d\beta_1 = \frac{\beta_2}{\sigma^2} e^{-\beta_2^2/2\sigma^2} \mathscr{A}(\beta_2)$$

Hence we see that β_2 has a Rayleigh distribution with parameter σ^2. It is also easy to show that $p_{y_1}(\beta_1)$ is uniform on $[-\pi/2, \pi/2]$ so that y_1 and y_2 are independent. In other words, we have transformed two identically distributed independent gaussian processes into polar coordinates, and we find that the magnitude has a Rayleigh distribution with parameter σ^2, the phase angle is uniform on $[-\pi/2, \pi/2]$, and the phase and magnitude are independent.

If in the domain of $\mathbf{f}$ the jacobian may change signs, it is necessary to divide the domain of $\mathbf{f}$ into regions in which J has a constant sign. In these regions, Eq. (4.3-11) may be used to determine the contribution to $p_{\mathbf{y}}$ from each region. The general procedure takes the following form:

1. Divide the domain of $\mathbf{f}$ into regions in which J is either zero or of a constant sign.
2. If $J = 0$ in a given region, Eqs. (4.3-14) and (4.3-15) or (4.3-16) and (4.3-17) must be used to compute the contributions to $p_{\mathbf{y}}$ from this region, depending on whether $\mathbf{y}$ or a subset of $\mathbf{y}$ is constant.
3. If J has a constant sign in a given region, Eq. (4.3-11) is used to compute $p_{\mathbf{y}}(\boldsymbol{\beta})$ for $\boldsymbol{\beta}$ in the range of $\mathbf{f}$ defined on the given region. For $\boldsymbol{\beta}$ not in the range of $\mathbf{f}$, $p_{\mathbf{y}}(\boldsymbol{\beta})$ is set equal to zero.
4. The complete density $p_{\mathbf{y}}$ is determined by summing the contribution from each region.

Example *4.3-3*. Let us examine a simple two-dimensional problem to illustrate the use of the above techniques. We assume that $\mathbf{x}$ is uniformly distributed on $-2 \leq x_1 \leq 2$ and $-2 \leq x_2 \leq 2$ so that $p_{\mathbf{x}}$ is

$$p_{\mathbf{x}}(\boldsymbol{\beta}) = \begin{cases} \frac{1}{16} & \text{for } -2 \leq \beta_1 \leq 2,\ -2 \leq \beta_2 \leq 2 \\ 0 & \text{otherwise} \end{cases}$$

Suppose now that $\mathbf{y}$ is described by

$$y_1 = \begin{cases} |x_1| & \text{for } |x_1| > 1 \\ 0 & \text{for } |x_1| \leq 1 \end{cases}$$

and

$$y_2 = x_2$$

Here we must divide the domain of $\mathbf{f}$ ($\mathcal{R}^2$ in this case) into three regions:

$$\begin{aligned} &\text{Region 1:} && x_1 < -1 \\ &\text{Region 2:} && |x_1| \leq 1 \\ &\text{Region 3:} && x_1 > 1 \end{aligned}$$

In Regions 1 and 2, **f** is invertible, whereas in Region 2, y_1 is constant and y_2 varies so that Eqs. (4.3-16) and (4.3-17) must be used.

Let us consider Region 1 first; in this region f is described by

$$\begin{aligned} y_1 &= -x_1 \\ y_2 &= x_2 \end{aligned} \qquad x_1 < -1$$

so that the inverse transformation is

$$\begin{aligned} x_1 &= -y_1 \\ x_2 &= y_2 \end{aligned}$$

The range defined on Region 1 is $y_1 > 1$, so $p_{\mathbf{y}_1}(\boldsymbol{\beta}) = 0$ for $\beta_1 \leq 1$. On Region 1 the jacobian is given by -1, so $p_{\mathbf{y}_1}(\boldsymbol{\beta})$ for $\beta_1 > 1$ is given by

$$p_{\mathbf{y}_1}(\boldsymbol{\beta}) = p_{\mathbf{x}}(-\beta_1, \beta_2) \qquad \text{for } \beta_1 > 1$$

and by substituting for $p_{\mathbf{x}}$ we obtain

$$p_{\mathbf{y}_1}(\boldsymbol{\beta}) = \begin{cases} \frac{1}{16} & \text{for } 1 < \beta_1 \leq 2,\ -2 \leq \beta_2 \leq 2 \\ 0 & \text{otherwise} \end{cases}$$

In a similar manner, one can find that the contribution to $p_{\mathbf{y}}$ from Region 3 is given by

$$p_{\mathbf{y}_3}(\boldsymbol{\beta}) = \begin{cases} \frac{1}{16} & \text{for } 1 < \beta_1 \leq 2,\ -2 \leq \beta_2 \leq 2 \\ 0 & \text{otherwise} \end{cases}$$

For Region 2 we must make use of Eq. (4.3-16) to write $p_{\mathbf{y}_2}$ as

$$p_{\mathbf{y}_2}(\beta) = \rho_0 \delta_D(\beta_1 - 0) p_{y_2}(\beta_2)$$

where

$$\rho_0 = \int_{-1}^{1} d\alpha_1 \int_{-\infty}^{\infty} d\alpha_2 p_x(\alpha_1, \alpha_2) = \frac{1}{16} \int_{-1}^{1} d\alpha_1 \int_{-2}^{2} d\alpha_2 = \frac{1}{2}$$

The density p_{y_2} can be easily obtained in this case, since $y_2 = x_2$, $p_{y_2} = p_{x_2}$, and

$$p_{y_2}(\beta_2) = p_{x_2}(\beta_2) = \begin{cases} \frac{1}{16} \int_{-2}^{2} d\alpha_1 = \frac{1}{4} & -2 \leq \beta_2 \leq 2 \\ 0 & \text{otherwise} \end{cases}$$

Therefore, $p_{\mathbf{y}_2}(\boldsymbol{\beta})$ becomes

$$p_{\mathbf{y}_2}(\boldsymbol{\beta}) = \begin{cases} \frac{1}{8} \delta_D(\beta_1) & -2 \leq \beta_2 \leq 2 \\ 0 & \text{otherwise} \end{cases}$$

The final complete form for $p_{\mathbf{y}}(\boldsymbol{\beta})$ is given by

$$\begin{aligned} p_{\mathbf{y}}(\boldsymbol{\beta}) &= p_{\mathbf{y}_1}(\boldsymbol{\beta}) + p_{\mathbf{y}_2}(\boldsymbol{\beta}) + p_{\mathbf{y}_3}(\boldsymbol{\beta}) \\ &= \tfrac{1}{8}\delta_D(\beta_1)[\mathscr{s}(\beta_2 + 2) - \mathscr{s}(\beta_2 - 2)] \\ &\quad + \tfrac{1}{8}[\mathscr{s}(\beta_1 - 1) - \mathscr{s}(\beta_1 - 2)][\mathscr{s}(\beta_2 + 2) - \mathscr{s}(\beta_2 - 2)] \end{aligned}$$

So far we have only considered the case when the dimension of $\mathbf{y}$ is equal to the dimension of $\mathbf{x}$. We noted, however, in Eq. (4.3-16) the need to determine the probability density for a vector of smaller dimension than $\mathbf{x}$. In addition, there are many interesting and meaningful applications when $\mathbf{y}$ has a smaller dimension than $\mathbf{x}$. In particular, the case when $\mathbf{y}$ is a scalar is quite important.

The determination of the probability density for a vector $\mathbf{y}$ whose dimension is less than $\mathbf{x}$ can be easily handled by a slight extension of the above techniques. Consider, for example, that $\mathbf{y}$ is of dimension $m < n$; then we add $n - m$ auxiliary variables $y_m, y_{m+1}, \ldots, y_n$, which are simple and convenient functions of $\mathbf{x}$, such that an inverse function exists. Now we use the techniques developed above to find the probability density for the n-dimensional augmented random vector $\mathbf{y}^* = [\mathbf{y}^T \mid y_m \, y_{m+1} \cdots y_n]^T$. Once $p_{\mathbf{y}}^*$ has been obtained, we can determine $p_{\mathbf{y}}$ by integrating over the range of the auxiliary variables.

Example *4.3-4*. Let us consider a simple example to illustrate the use of the concept discussed above. Suppose that $\mathbf{x}$ is two dimensional and y is a scalar given by

$$y_1 = x_1 + x_2$$

In other words, y_1 is the sum of x_1 and x_2. We define an auxiliary variable as

$$y_2 = x_2$$

We could also have set $y_2 = x_1$ or any other simple expression. The problem now takes the form of the linear transformation discussed in Example 4.3-1, with $\mathbf{A}$ equal to

$$\mathbf{A} = \begin{bmatrix} 1 & 1 \\ 0 & 1 \end{bmatrix}$$

so that det $\mathbf{A} = 1$ and

$$\mathbf{A}^{-1} = \begin{bmatrix} 1 & -1 \\ 0 & 1 \end{bmatrix}$$

Therefore, the inverse transformation of the function is given by $\mathbf{x} = \mathbf{A}^{-1}\mathbf{y}$ or

$$x_1 = y_1 - y_2$$
$$x_2 = y_2$$

Now, using the result of Example 4.3-1, we find that $p_{y_1y_2}$ is

$$p_{y_1y_2}(\beta_1, \beta_2) = p_{\mathbf{x}}(\beta_1 - \beta_2, \beta_2)$$

To obtain p_{y_1}, we integrate over the range of y_2 so that p_{y_1} becomes

$$p_{y_1}(\beta_1) = \int_{-\infty}^{\infty} p_{\mathbf{x}}(\beta_1 - \beta_2, \beta_2)\, d\beta_2$$

A very interesting case occurs if x_1 and x_2 are independent so that $p_{\mathbf{x}}$ can be written as

$$p_{\mathbf{x}}(\alpha_1, \alpha_2) = p_{x_1}(\alpha_1)p_{x_2}(\alpha_2)$$

If we make this assumption, then p_{y_1} becomes

$$p_{y_1}(\beta_1) = \int_{-\infty}^{\infty} p_{x_1}(\beta_1 - \beta_2)p_{x_2}(\beta_2)\, d\beta_2$$

Hence we see that the probability density for a sum of two independent random variables is given by the convolution of their probability densities.

This result is quite important. There are at least two ancillary methods that we may use to obtain this result and, for the sake of completeness, we shall present them here.

Let us fix x_1 at the value α_1. The events

$$\{\xi : y_1(\xi) \leq \beta_1, x_1(\xi) = \alpha_1\} = \mathcal{A}_1$$
$$\{\xi : x_2(\xi) \leq \beta_1 - \alpha_1, x_1(\xi) = \alpha_1\} = \mathcal{A}_2$$

are equivalent. Thus we have

$$\mathcal{P}\{\mathcal{A}_1\} = \mathcal{P}\{\mathcal{A}_2\}$$

which may be expressed in terms of density functions as

$$\int_{-\infty}^{\beta_1} p_{y_1|x_1}(\beta \mid \alpha_1)\, d\beta \qquad \int_{-\infty}^{\beta_1 - \alpha_1} p_{x_2|x_1}(\alpha \mid \alpha_1)\, d\alpha$$

Since this expression must be valid for all α_1 and β_1, we have

$$p_{y_1|x_1}(\beta_1 | \alpha_1) = p_{x_2|x_1}(\beta_1 - \alpha_1 | \alpha_1)$$

By multiplying both sides of this expression by $p_{x_1}(\alpha_1)$ and making use of the conditional probability law, we obtain

$$p_{y_1x_1}(\beta_1, \alpha_1) = p_{x_2x_1}(\beta_1 - \alpha_1, \alpha_1)$$

We now integrate both expressions over α_1 from $-\infty$ to ∞ and immediately obtain the desired result:

$$p_{y_1}(\beta_1) = \int_{-\infty}^{\infty} p_{y_1x_1}(\beta_1, \alpha_1)\, d\alpha_1 = \int_{-\infty}^{\infty} p_{x_2x_1}(\beta_1 - \alpha_1, \alpha_1) d\alpha_1$$

Figure 4.3-2 also presents us with a way of obtaining the desired result. The probability that y_1 is less than some value β_1 is the same as the probability that all values of x_1 and x_2 be to the left of the line $x_1 + x_2 = \beta_1$, as illustrated in Fig. 4.3-2. We imagine a strip of width $\Delta\alpha_1$ from $x_2 = -\infty$ to $x_2 = \beta_1 - \alpha_1$. The probability associated with this strip is

$$\int_{-\infty}^{\beta_1 - \alpha_1} p_{x_1x_2}(\alpha_1, \alpha_2)\, d\alpha_2\, \Delta\alpha_1$$

$F_{y_1}(\beta_1)$ is the distribution function associated with the entire area, so we let $\Delta\alpha_1$ become small and sum (integrate) over α_1 from $-\infty$ to ∞

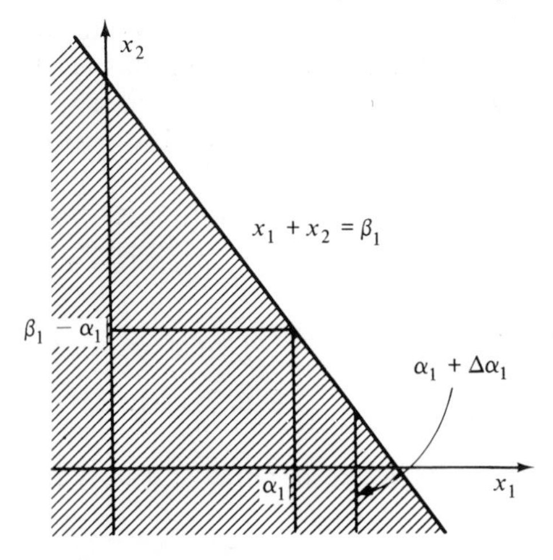

FIG. 4.3-2. Transformation $y_1 = x_1 + x_2$ for Example 4.3-4.

to obtain the distribution function

$$F_{y_1}(\beta_1) = \int_{-\infty}^{\infty} d\alpha_1 \int_{-\infty}^{\beta_1-\alpha_1} p_{x_1x_2}(\alpha_1, \alpha_2)\, d\alpha_2$$

By differentiating this expression with respect to β_1 we immediately obtain the desired result:

$$p_{y_1}(\beta_1) = \int_{-\infty}^{\infty} p_{x_1x_2}(\alpha_1, \beta_1 - \alpha_1)\, d\alpha_1$$

The three methods used to resolve this example are quite important and indicate the many ways in which transformation of random variables may be viewed.

The case in which the dimension of $\mathbf{y}$ is greater than $\mathbf{x}$ is occasionally encountered in practice. It may be easily handled by a simple extension of our general techniques. From the $m > n$ variables in $\mathbf{y}$ we select n, which we designate by $\mathbf{y}_n$, such that a unique transformation $\mathbf{f}_n^{-1}$ exists[1] such that $\mathbf{x} = \mathbf{f}_n^{-1}(\mathbf{y}_n)$. Now the remaining $m - n$ variables, which we designate as $\mathbf{y}_{m-n}$, can be written as functions of $\mathbf{y}_n$ since

$$\mathbf{y}_{m-n} = \mathbf{f}_{m-n}(\mathbf{x}) = \mathbf{f}_{m-n}[\mathbf{f}_n^{-1}(\mathbf{y}_n)] \tag{4.3-18}$$

If we write $p_{\mathbf{y}}$ as

$$p_{\mathbf{y}}(\boldsymbol{\beta}) = p_{\mathbf{y}_{m-n}|\mathbf{y}_n}(\boldsymbol{\beta}_{m-n} \mid \boldsymbol{\beta}_n) p_{\mathbf{y}_n}(\boldsymbol{\beta}_n) \tag{4.3-19}$$

then $p_{\mathbf{y}_n}$ can be determined by the techniques discussed previously and $p_{\mathbf{y}_{m-n}|\mathbf{y}_n}$ is given by

$$p_{\mathbf{y}_{m-n}|\mathbf{y}_n}(\boldsymbol{\beta}_{m-n} \mid \boldsymbol{\beta}_n) = \delta_D\{\boldsymbol{\beta}_{m-n} - \mathbf{f}_{m-n}[\mathbf{f}_n^{-1}(\boldsymbol{\beta}_n)]\} \tag{4.3-20}$$

Example *4.3-5*. Suppose that we have a case in which $\mathbf{x}$ is a scalar and $\mathbf{y}$ is two dimensional with

$$y_1 = x_1$$
$$y_2 = x_1^2$$

In this situation, we can let $y_n = y_1$ and $y_{m-n} = y_2$. Using these definitions, we find that Eq. (4.3-18) becomes

$$y_2 = y_1^2 \tag{4.3-21}$$

[1] In an actual problem we may have to define $\mathbf{y}_n$ differently for different regions of the domain of $\mathbf{f}$.

The application of Eq. (4.3-20) therefore yields

$$p_{y_2|y_2}(\beta_2 \mid \beta_1) = \delta_D(\beta_2 - \beta_1^2)$$

whereas p_{y_1} is easily determined to be

$$p_{y_1}(\beta_1) = p_x(\beta_1)$$

and the final result is therefore

$$p_{y_1y_2}(\beta_1, \beta_2) = \delta_D(\beta_2 - \beta_1^2)p_x(\beta_1)$$

for $\beta_2 > 0$ and zero otherwise.

Suppose now that we integrate over β_1 to obtain $p_{y_2}(\beta_2)$ as

$$p_{y_2}(\beta_2) = \int_{-\infty}^{\infty} p_{y_1y_2}(\beta_1, \beta_2)\, d\beta_1 = \int_{-\infty}^{\infty} \delta_D(\beta_2 - \beta_1^2)p_x(\beta_1)\, d\beta_1$$

If we make the change of variable $\gamma = \beta_1^2$, then $p_{y_2}(\beta_2)$ becomes

$$p_{y_2}(\beta_2) = \frac{1}{2}\int_0^{\infty} \delta_D(\beta_2 - \gamma)p_x(\sqrt{\gamma})\frac{1}{\sqrt{\gamma}}\, d\gamma + \frac{1}{2}\int_0^{\infty} \delta_D(\beta_2 - \gamma)p_x(\sqrt{\gamma})\frac{1}{\sqrt{\gamma}}\, d\gamma$$

which gives

$$p_{y_2}(\beta_2) = \frac{1}{2\sqrt{\beta_2}}[p_x(\sqrt{\beta_2}) + p_x(-\sqrt{\beta_2})] \qquad \beta_2 \geq 0$$

Hence we have obtained the density of $y_2 = x^2$ in a simple and direct manner. This technique suggests an interesting way to obtain complex functions. We can introduce new variables that represent intermediate steps in obtaining the desired density. For example, if we desired to obtain $y = |\cos x|$, we could write this as

$$y_1 = \cos x$$
$$y_2 = |\cos x| = |y_1|$$

and use the technique developed above.

Example *4.3-6*. A case that often occurs is the situation in which

$$\mathbf{y} = \mathbf{A}\mathbf{x}$$

and the dimension of $\mathbf{y}$ is greater than that of $\mathbf{x}$. This transformation may be reposed as one in which

$$\mathbf{y}_1 = \mathbf{A}_1\mathbf{x}$$
$$\mathbf{y}_2 = \mathbf{B}\mathbf{y}_1 = \mathbf{B}\mathbf{A}_1\mathbf{x} = \mathbf{A}_2\mathbf{x}$$

The dimension of $\mathbf{y}_1$ is the same as the of $\mathbf{x}$, and the $\mathbf{A}_1$ matrix is nonsingular. The $\mathbf{B}$ and $\mathbf{A}_2$ matrices may be nonsquare. The complete $\mathbf{y}$ vector is

$$\mathbf{y} = \begin{bmatrix} \mathbf{y}_1 \\ \cdots \\ \mathbf{y}_2 \end{bmatrix} = \begin{bmatrix} \mathbf{A}_1 \\ \cdots \\ \mathbf{A}_2 \end{bmatrix} \mathbf{x}$$

We may easily obtain the density $p_{\mathbf{y}_1}$ as a result of Example 4.3-1:

$$p_{\mathbf{y}_1}(\boldsymbol{\beta}_1) = |(\det \mathbf{A}_1)^{-1}| p_{\mathbf{x}}(\mathbf{A}_1^{-1}\boldsymbol{\beta}_1)$$

If we assume that $\mathbf{y}_1$ is known, $\mathbf{y}_1 = \boldsymbol{\beta}_1$, then $\mathbf{y}_2$ is deterministic if we condition it upon the known $\mathbf{y}_1$, and

$$p_{\mathbf{y}_2|\mathbf{y}_1}(\boldsymbol{\beta}_2 | \boldsymbol{\beta}_1) = \delta_D(\boldsymbol{\beta}_2 - \mathbf{B}\boldsymbol{\beta}_1)$$

We thus have from the foregoing two equations

$$\begin{aligned} p_{\mathbf{y}_2\mathbf{y}_1}(\boldsymbol{\beta}_2, \boldsymbol{\beta}_1) = p_{\mathbf{y}}(\boldsymbol{\beta}) &= p_{\mathbf{y}_2|\mathbf{y}_1}(\boldsymbol{\beta}_2 | \boldsymbol{\beta}_1) p_{\mathbf{y}_1}(\boldsymbol{\beta}) \\ &= |(\det \mathbf{A}_1)^{-1}| p_{\mathbf{x}}(\mathbf{A}^{-1}\boldsymbol{\beta}_1)\delta_D(\boldsymbol{\beta}_2 - \mathbf{B}\boldsymbol{\beta}_1) \end{aligned}$$

The need for this expression often arises in cases in which we are making a dynamic transformation of a random vector. In such cases it often happens that $\mathbf{x}$ is gaussian and the foregoing relation thus allows computation of the density function of $\mathbf{y}$. It is reasonable to expect the impulse functions in the probability density functions, since $\mathbf{y}$ is overspecified in that some components of $\mathbf{y}$ are linearly related to other components in $\mathbf{y}$.

In this section we have examined the extension of the techniques developed in the preceding section to functions of many variables. We have considered several examples to illustrate the techniques, but unfortunately there are still many meaningful examples that we have not had the space to treat. Some of these problems are presented in the exercises at the end of this section; others appear as problems at the end of this chapter. The reader is urged to work with these problems until he has gained a facility with the techniques.

EXERCISES 4.3

4.3-1. If

$$p_{x_1x_2}(\alpha_1, \alpha_2) = \begin{cases} 4\alpha_1\alpha_2 & 0 \leq \alpha_1, \alpha_2 \leq 1 \\ 0 & \text{otherwise} \end{cases}$$

find the joint probability density $p_{y_1y_2}(\alpha_1, \alpha_2)$ if $y_1 = x_1^2$ and $y_2 = x_2^2$.

4.3-2. The random variables x and y are independent with densities

$$p_x(\alpha) = \begin{cases} e^{-\alpha} & \alpha \geq 0 \\ 0 & \alpha < 0 \end{cases}$$

$$p_y(\beta) = \begin{cases} e^{-\beta} & \beta \geq 0 \\ 0 & \beta < 0 \end{cases}$$

(a) Show that the density of

$$z = x + y$$

is

$$p_z(\gamma) = \begin{cases} \gamma e^{-\gamma} & \gamma \geq 0 \\ 0 & \gamma < 0 \end{cases}$$

(b) Show that the density of

$$w = \frac{x}{x + y}$$

is uniform in the interval 0 to 1.

4.3-3. If x_1 and x_2 are independent random variables and $y = a_1x_1 + a_2x_2$, where a_1 and a_2 are nonrandom constants, find $p_y(\alpha)$ in terms of p_{x_1} and p_{x_2}.

Answer:
$$p_y(\alpha) = \frac{1}{|a_1a_2|}\int_{-\infty}^{\infty} p_{x_1}\left(\frac{\alpha - \lambda}{a_1}\right)p_{x_2}\left(\frac{\lambda}{a_2}\right)d\lambda$$

4.3-4. Show that if $z = xy$, then

$$p_z(\gamma) = \int_{-\infty}^{\infty} \frac{1}{|\alpha|}p_{xy}\left(\alpha, \frac{\gamma}{\alpha}\right)d\alpha$$

4.3-5. If

$$y_1 = x_1 + x_2$$
$$y_2 = x_1 - x_2$$
$$y_3 = x_1$$

find

$$p_{y_1y_2y_3}(\beta_1, \beta_2, \beta_3)$$

Answer:
$$p_{y_1y_2y_3}(\beta_1, \beta_2, \beta_3) = \tfrac{1}{2}p_{x_1x_2}[\tfrac{1}{2}(\beta_1 + \beta_2), \tfrac{1}{2}(\beta_1 - \beta_2)] \times \delta_D[\beta_3 - \tfrac{1}{2}(\beta_1 + \beta_2)]$$

4.4. *Expected values, moments, fundamental theorem of expectation*

In the two preceding sections we dealt with functions of random variables. Here we wish to consider another operation on random variables known as expectation; we begin by defining the expected value of a random variable.

Definition 4.4-1. ***Expected Value of a Random Variable:*** The expected value, denoted by $\mathcal{E}\{\mathbf{x}\}$, of a random vector $\mathbf{x}$ with a probability density function $p_{\mathbf{x}}$ is defined as

$$\mathcal{E}\{\mathbf{x}\} = \int_{-\infty}^{\infty} \boldsymbol{\alpha} p_{\mathbf{x}}(\boldsymbol{\alpha})\, d\boldsymbol{\alpha} \tag{4.4-1}$$

if the integral exists.

Let us first consider a scalar random variable x. If one thinks of $p_x(\alpha)$ as having mass distribution along the x axis, then $\mathcal{E}\{x\}$ represents the center of gravity. In other words, if we were to accurately cut p_x out of a piece of stiff paper and attempt to balance it on a knife-edge perpendicular to the x axis, we would find that it would balance at $\alpha = \mathcal{E}\{x\}$. The expected value of $\mathbf{x}$ is also referred to as the *mean* or *average value* of $\mathbf{x}$ and will also be denoted by $\boldsymbol{\mu}_{\mathbf{x}}$, or simply $\boldsymbol{\mu}$ if there is no chance of confusion.

It is instructive to consider the relative-frequency interpretation of Eq. (4.4-1). Let us suppose that we conduct an experiment a large number of times, say $\mathfrak{N}$, and observe the outcome. Associated with each of the $\mathfrak{N}$ outcomes $\xi_1, \xi_2, \ldots, \xi_{\mathfrak{N}}$ is a value of the random variable $\mathbf{x}$. Let $\mathfrak{N}(\boldsymbol{\alpha}_k)$ equal the number of outcomes for which $\mathbf{x}(\xi) \in [\boldsymbol{\alpha}_k, \boldsymbol{\alpha}_k + \boldsymbol{\Delta\alpha}]$, where $\boldsymbol{\Delta\alpha}$ is a small increment. The sum of the $\mathbf{x}(\xi_i)$, $i = 1, 2, \ldots, \mathfrak{N}$ can then be written as

$$\sum_{i=1}^{\mathfrak{N}} \mathbf{x}(\xi_i) \simeq \sum_{k=-\infty}^{\infty} \boldsymbol{\alpha}_k \mathfrak{N}(\boldsymbol{\alpha}_k) \tag{4.4-2}$$

As $\mathfrak{N} \rightarrow \infty$, $\mathfrak{N}(\boldsymbol{\alpha}_k)/\mathfrak{N}$ approaches the probability that $\mathbf{x} \in (\boldsymbol{\alpha}_k, \boldsymbol{\alpha}_k + \boldsymbol{\Delta\alpha})$ so that

$$\mathcal{P}\{\boldsymbol{\alpha}_k < \mathbf{x} < \boldsymbol{\alpha}_k + \boldsymbol{\Delta\alpha}\} = \lim_{\mathfrak{N}\rightarrow\infty} \frac{\mathfrak{N}(\boldsymbol{\alpha}_k)}{\mathfrak{N}} \tag{4.4-3}$$

The average of $\mathbf{x}$, which we denote by $\bar{\mathbf{x}}$, is equal to the sum of $\mathbf{x}(\xi_i)$, $i = 1, 2, \ldots, \mathfrak{N}$, divided by $\mathfrak{N}$, so that $\bar{\mathbf{x}}$ is given by

$$\bar{\mathbf{x}} \triangleq \frac{\sum_{i=1}^{\mathfrak{N}} \mathbf{x}(\xi_i)}{\mathfrak{N}}$$

By the use of Eq. (4.4-2), we see that $\bar{\mathbf{x}}$ may be written as

$$\bar{\mathbf{x}} = \sum_{k=-\infty}^{\infty} \boldsymbol{\alpha}_k \frac{\mathfrak{N}(\boldsymbol{\alpha}_k)}{\mathfrak{N}} \tag{4.4-4}$$

But $\mathfrak{N}(\boldsymbol{\alpha}_k)/\mathfrak{N}$ is approximately $\mathcal{P}\{\boldsymbol{\alpha}_k < \mathbf{x} \leq \boldsymbol{\alpha}_k + \Delta\boldsymbol{\alpha}\}$, which can be written as $p_{\mathbf{x}}(\boldsymbol{\alpha}_k)\Delta\boldsymbol{\alpha}$ for $\Delta\boldsymbol{\alpha}$ sufficiently small. Hence Eq. (4.4-4) may be written as

$$\bar{\mathbf{x}} = \sum_{k=-\infty}^{\infty} \boldsymbol{\alpha}_k \, p_{\mathbf{x}}(\boldsymbol{\alpha}_k) \, \Delta\boldsymbol{\alpha}$$

Now as $\mathfrak{N} \longrightarrow \infty$ and $\Delta\boldsymbol{\alpha} \longrightarrow \mathbf{0}$, $\bar{\mathbf{x}}$ becomes

$$\bar{\mathbf{x}} = \int_{-\infty}^{\infty} \boldsymbol{\alpha} p_{\mathbf{x}}(\boldsymbol{\alpha}) \, d\boldsymbol{\alpha} = \mathcal{E}\{\mathbf{x}\} \tag{4.4-5}$$

It should be noted that $\mathcal{E}\{\mathbf{x}\}$ is a single number. It is *not* a random variable; its value does *not* depend on the outcome of a random experiment. The expected value is a single parameter or number associated with a random variable or, more correctly, with the distribution of a random variable. The expected value might be likened to the number of outcomes in $\mathcal{S}$; this value does not change as a function of the specific outcome due to the performance of the random experiment. In a gross sense, we might say that the random nature of $\mathbf{x}$ has been averaged out to obtain $\mathcal{E}\{\mathbf{x}\}$. The expected value is a macroscopic variable associated with an ensemble (or infinite collection) of identical random experiments.[1] Suppose, for example, that we have an experiment which consists of measuring the voltage of a "random" battery and that x is defined as the value of the voltage. Now suppose that we have a very large number of these batteries, and we connect them in series and measure the voltage with a meter that divides by the number of batteries. The value of the voltage is approximately $\mathcal{E}\{x\}$. There is only one voltage and it does not depend on the outcome of any single outcome but on the collective behavior of a large number of outcomes.

If $\mathbf{x}$ is a discrete random variable, then Eq. (4.4-1) takes a particularly simple form. Suppose that $p_{\mathbf{x}}$ is given by

$$p_{\mathbf{x}}(\boldsymbol{\alpha}) = \sum_{i=1}^{n} p_i \delta_D(\boldsymbol{\alpha} - \boldsymbol{\alpha}_i)$$

and let us substitute this density function into Eq. (4.4-1) to obtain

$$\mathcal{E}\{\mathbf{x}\} = \int_{-\infty}^{\infty} \boldsymbol{\alpha} \sum_{i=1}^{n} p_i \delta_D(\boldsymbol{\alpha} - \boldsymbol{\alpha}_i) \, d\boldsymbol{\alpha}$$

[1] Later we shall consider several other descriptions of a random variable.

Upon using the sifting property of the delta function, this expression becomes

$$\mathcal{E}\{\mathbf{x}\} = \sum_{i=1}^{n} \rho_i \boldsymbol{\alpha}_i \tag{4.4-6}$$

But ρ_i equals the probability that $\mathbf{x} = \alpha_i$, so we may also write Eq. (4.4-6) as

$$\mathcal{E}\{\mathbf{x}\} = \sum_{i=1}^{n} \alpha_i \mathcal{P}\{\mathbf{x} = \boldsymbol{\alpha}_i\} \tag{4.4-7}$$

Let us consider the determination of the expected value of a simple discrete random variable to illustrate the use of Eq. (4.4-6).

Example ***4.4-1.*** Suppose that the density function of x is given by

$$p_x(\alpha) = \tfrac{1}{10} \sum_{i=1}^{4} i\delta_D(\alpha - i)$$

The expected value of x can be obtained by the use of Eq. (4.4-6) as

$$\mathcal{E}\{x\} = \sum_{i=1}^{4} i\left[\frac{i}{10}\right] = \frac{1}{10}\sum_{i=1}^{4} i^2 = 3$$

The use of the terminology "expected value" may lead one to conclude that if we conduct a random experiment, the value of $\mathbf{x}$ which we expect to obtain is $\mathcal{E}\{\mathbf{x}\}$. Although this conclusion is valid for many random variables, it is not true for all. In fact, it is quite possible that there may exist no outcome $\xi \in \mathcal{S}$ such that $\mathbf{x}(\xi) = \mathcal{E}\{\mathbf{x}\}$.

Example ***4.4-2.*** Consider, for example, a single Bernoulli trial with $x(\xi_1) = 1$, $x(\xi_0) = 0$, $\mathcal{P}\{\xi_1\} = p$, and $\mathcal{P}\{\xi_0\} = q = 1 - p$. By using Eq. (4.4-7), we easily find that $\mathcal{E}\{x\}$ is given by

$$\mathcal{E}\{x\} = p\cdot 1 + q\cdot 0 = p$$

In this case, x may take only the values 0 and 1, whereas $\mathcal{E}\{x\} = p$, which is generally not equal to 0 or 1. Hence there is no $\xi \in \mathcal{S}$ such that $x(\xi) = \mathcal{E}\{x\}$ unless $p = 0$ or 1.

If the density function $p_{\mathbf{x}}(\boldsymbol{\alpha})$ is symmetric with respect to $\boldsymbol{\alpha} = \mathbf{a}$, then $\mathcal{E}\{\mathbf{x}\} = \mathbf{a}$. By symmetric with respect to $\boldsymbol{\alpha} = \mathbf{a}$, we mean that $p_{\mathbf{x}}(\mathbf{a} + \mathbf{y}) = p_{\mathbf{x}}(\mathbf{a} - \mathbf{y})$ for any $\mathbf{y}$. This fact may be easily established in the following manner. The $\mathcal{E}\{\mathbf{x}\}$ is defined by

$$\mathcal{E}\{\mathbf{x}\} = \int_{-\infty}^{\infty} \boldsymbol{\alpha} p_{\mathbf{x}}(\boldsymbol{\alpha})\, d\boldsymbol{\alpha} \tag{4.4-8}$$

If we make the substituon $\boldsymbol{\beta} = \boldsymbol{\alpha} - \mathbf{a}$ in Eq. (4.4-8), we obtain

$$\mathcal{E}\{\mathbf{x}\} = \int_{-\infty}^{\infty} (\mathbf{a} + \boldsymbol{\beta}) p_{\mathbf{x}}(\mathbf{a} + \boldsymbol{\beta})\, d\boldsymbol{\beta} \tag{4.4-9}$$

and if we make the substitution $\boldsymbol{\beta} = \mathbf{a} - \boldsymbol{\alpha}$, we obtain

$$\mathcal{E}\{\mathbf{x}\} = \int_{-\infty}^{\infty} (\mathbf{a} - \boldsymbol{\beta}) p_{\mathbf{x}}(\mathbf{a} - \boldsymbol{\beta})\, d\boldsymbol{\beta} \tag{4.4-10}$$

By combining Eqs. (4.4-9) and (4.4-10), we may write $\mathcal{E}\{\mathbf{x}\}$ as one half the sum of these two expressions so that we have

$$\mathcal{E}\{\mathbf{x}\} = \tfrac{1}{2} \int_{-\infty}^{\infty} [(\mathbf{a} + \boldsymbol{\beta}) p_{\mathbf{x}}(\mathbf{a} + \boldsymbol{\beta}) + (\mathbf{a} - \boldsymbol{\beta}) p_{\mathbf{x}}(\mathbf{a} - \boldsymbol{\beta})]\, d\boldsymbol{\beta}$$

Since $p_{\mathbf{x}}$ is assumed to be symmetric with respect to $\mathbf{a}$, we have

$$\begin{aligned} \mathcal{E}\{\mathbf{x}\} &= \tfrac{1}{2} \int_{-\infty}^{\infty} [(\mathbf{a} + \boldsymbol{\beta}) + (\mathbf{a} - \boldsymbol{\beta})]\, p_{\mathbf{x}}(\mathbf{a} + \boldsymbol{\beta})\, d\boldsymbol{\beta} \\ &= \mathbf{a} \int_{-\infty}^{\infty} p_{\mathbf{x}}(\mathbf{a} + \boldsymbol{\beta})\, d\boldsymbol{\beta} = \mathbf{a} \end{aligned}$$

which is the desired result. It is important to note that if $p_{\mathbf{x}}(\boldsymbol{\alpha})$ is symmetric about $\boldsymbol{\alpha} = \mathbf{a}$, then $\boldsymbol{\mu}_{\mathbf{x}} = \mathbf{a}$ independent of any other properties of $p_{\mathbf{x}}$.

Suppose that we consider a random variable $\mathbf{y}$ which is a function of $\mathbf{x}$, such as $\mathbf{f}(\mathbf{x})$. Then the expected value of $\mathbf{y}$ is given by

$$\mathcal{E}\{\mathbf{y}\} = \int_{-\infty}^{\infty} \boldsymbol{\beta} p_{\mathbf{y}}(\boldsymbol{\beta})\, d\boldsymbol{\beta} \tag{4.4-11}$$

To obtain this result we must compute the density function of $\mathbf{y}$, which, as we have seen in the preceding sections, is not always easy to do. Fortunately, it is possible to obtain the expected value of $\mathbf{y}$ without obtaining $p_{\mathbf{y}}$. The *fundamental theorem of expectation* indicates that $\mathcal{E}\{\mathbf{y}\}$ may also be written as

$$\mathcal{E}\{\mathbf{y}\} = \mathcal{E}\{\mathbf{f}(\mathbf{x})\} = \int_{-\infty}^{\infty} \mathbf{f}(\boldsymbol{\alpha}) p_{\mathbf{x}}(\boldsymbol{\alpha})\, d\boldsymbol{\alpha} \tag{4.4-12}$$

For simplicity we shall only establish this theorem[1] for the scalar case in which $\mathbf{f}$ is strictly monotone increasing, differentiable almost everywhere, and onto. The more general situations are handled by applying arguments sim-

[1] We could also have defined $\mathcal{E}\{\mathbf{f}(\mathbf{x})\}$ by Eq. (4.4-12), in which case the $\mathcal{E}\{\mathbf{x}\}$ would have been a simple special case.

ilar to those used in the two preceding sections. We wish to show then that the integral

$$I = \int_{-\infty}^{\infty} f(\alpha)p_x(\alpha)\,d\alpha \tag{4.4-13}$$

is equal to $\mathcal{E}\{y\}$ with $y = f(\alpha)$. Let us make a change of variables such that $\beta = f(\alpha)$; then Eq. (4.4-13) becomes

$$I = \int_{-\infty}^{\infty} \beta p_x[f^{-1}(\beta)]\frac{df^{-1}(\beta)}{d\beta}\,d\beta \tag{4.4-14}$$

From Eq. (4.2-15), we know that

$$p_x[f^{-1}(\beta)]\frac{df^{-1}(\beta)}{d\beta} = p_y(\beta)$$

so that we have

$$\int_{-\infty}^{\infty} f(\alpha)p_x(\alpha)\,d\alpha = \int_{-\infty}^{\infty} \beta p_y(\beta)\,d\beta = \mathcal{E}\{y\}$$

which is the desired result.

Example *4.4-3*. Suppose that a random variable x is uniformly distributed on $[0, 2\pi]$ and that $y = \sin x$. With a modest amount of algebra, it is possible to show that the density function of y is

$$p_y(\beta) = \begin{cases} \dfrac{1}{\pi\sqrt{1-\beta^2}} & \text{for } |\beta| \leq 1 \\ 0 & \text{otherwise} \end{cases}$$

The expected value of y is therefore

$$\mathcal{E}\{y\} = \int_{-1}^{1} \frac{\beta\,d\beta}{\pi\sqrt{1-\beta^2}}$$

This integral can be easily evaluated to find that $\mathcal{E}\{y\} = 0$. Of course, we could have concluded that $\mathcal{E}\{y\} = 0$ because p_y is symmetric with respect to 0.

Now let us use Eq. (4.4-12) to find $\mathcal{E}\{y\} = \mathcal{E}\{\sin x\}$ as

$$\mathcal{E}\{\sin x\} = \frac{1}{2\pi}\int_0^{2\pi} \sin x\,dx = -\frac{1}{2\pi}\cos x\Big|_0^{2\pi} = 0$$

As we expect, we obtain the same value as before; however, in this case we did not need to find p_y.

By the use of the fundamental theorem of expectation, we may easily show that expectation is a linear operator in the sense that if

$$\mathbf{f}(\mathbf{x}) = a_1\mathbf{f}_1(\mathbf{x}) + a_2\mathbf{f}_2(\mathbf{x}) \tag{4.4-15}$$

where a_1 and a_2 are known constants, then

$$\mathcal{E}\{\mathbf{f}(\mathbf{x})\} = a_1\mathcal{E}\{\mathbf{f}_1(\mathbf{x})\} + a_2\mathcal{E}\{\mathbf{f}_2(\mathbf{x})\} \tag{4.4-16}$$

To establish this result, we need only write $\mathcal{E}\{\mathbf{f}(\mathbf{x})\}$ as

$$\mathcal{E}\{\mathbf{f}(\mathbf{x})\} = \int_{-\infty}^{\infty} \mathbf{f}(\boldsymbol{\alpha})p_{\mathbf{x}}(\boldsymbol{\alpha})\, d\boldsymbol{\alpha}$$

and then substitute for $\mathbf{f}(\mathbf{x})$ from Eq. (4.4-16) to obtain

$$\begin{aligned}\mathcal{E}\{\mathbf{f}(\mathbf{x})\} &= \int_{-\infty}^{\infty} [a_1\mathbf{f}_1(\boldsymbol{\alpha}) + a_2\mathbf{f}_2(\boldsymbol{\alpha})]p_{\mathbf{x}}(\boldsymbol{\alpha})\, d\boldsymbol{\alpha} \\ &= a_1 \int_{-\infty}^{\infty} \mathbf{f}_1(\boldsymbol{\alpha})p_{\mathbf{x}}(\boldsymbol{\alpha})\, d\boldsymbol{\alpha} + a_2 \int_{-\infty}^{\infty} \mathbf{f}_2(\boldsymbol{\alpha})p_{\mathbf{x}}(\boldsymbol{\alpha})\, d\boldsymbol{\alpha}\end{aligned}$$

which gives the desired result.

Moments. In Chapter 3 we discussed two methods of describing a random variable: the distribution function and the density function. In this section we have introduced another description of a random variable: the expected value. The expected value, however, is not a complete description in the sense that it does not, in general, completely define the distribution. As noted previously, the expected value is equivalent to the "center of gravity" of the density function; it does not give any information about the "moment of inertia" or the "symmetry" of the density function. Consider, for example, two continuous random variables x_1 and x_2 with uniform distributions such that

$$p_{x_1}(\alpha) = \tfrac{1}{2}[\mathscr{s}(\alpha + 1) - \mathscr{s}(\alpha - 1)]$$

and

$$p_{x_2}(\alpha) = \tfrac{1}{4}[\mathscr{s}(\alpha + 2) - \mathscr{s}(\alpha - 2)]$$

It is easy to show that both random variables have a mean value of zero, although they certainly do not have the same distribution.

There are several other methods of describing a random variable. Two other parameters often used are the mode and the median. The *mode* of a random variable $\mathbf{x}$, often referred to as the *most likely value*, is the value of $\boldsymbol{\alpha}$ (there may be more than one) for which $p_{\mathbf{x}}(\boldsymbol{\alpha})$ is maximum. If a density has only one mode, it is said to be *unimodal;* if it has two, it is *bimodal;* and if it has many modes, it is referred to as *multimodal.* For a scalar random variable x, one may also define the *median*[1] (not necessarily unique) of x as the value $\alpha_{1/2}$ such that

$$\mathcal{P}\{x \leq \alpha_{1/2}\} = \tfrac{1}{2} = F_x(\alpha_{1/2}) \tag{4.4-17}$$

or, equivalently,

$$\int_{-\infty}^{\alpha_{1/2}} p_x(\alpha)\, d\alpha = \int_{\alpha_{1/2}}^{\infty} p_x(\alpha)\, d\alpha = \tfrac{1}{2} \tag{4.4-18}$$

There is no direct way to extend the median to vector-valued random variables. If the density function of a scalar random variable is symmetric about $\alpha = a$, then both the mean and the median are equal to a. If the density function is symmetric about $\alpha = a$ and is unimodal, then the mean and the mode are equal. The proof of these statements is quite easy and is left to the reader as an exercise.

None of the above descriptors—the mean, the mode, and the median—provide a complete description of the distribution of a random variable. We wish to consider now another method of describing a random variable, which will normally uniquely specify p_x. For simplicity, the treatment will be restricted to scalar random variables; the extension of the concepts to random vectors will be developed later. The method, known as *statistical moments,* involves the use of the fundamental theorem of expectation[2] with a particular class of functions $f(x)$.

Definition 4.4-2. ***Moments:*** The kth-order moment of a random variable x, $k = 0, 1, 2, \ldots$, is defined by

$$m_{x^k} = \mathcal{E}\{x^k\} = \int_{-\infty}^{\infty} \alpha^k p_x(\alpha)\, d\alpha \qquad k = 0, 1, 2, \ldots \tag{4.4-19}$$

When there is no possibility of confusion we shall use the simpler notation m_k for m_{x^k}. It is important to note that $m_0 = 1$ and $m_1 = \mu = \mathcal{E}\{x\}$. We shall show in Section 4.5 that if the constants m_k, $k = 0, 1, 2, \ldots$, satisfy some rather simple conditions, then they completely describe $p_x(\alpha)$. In many

[1] In general, the quantiles of order n of a scalar random variable x are the $n - 1$ values of α (not necessarily unique) such that

$$F_x(\alpha_{m/n}) = \frac{m}{n} \qquad m = 1, 2, \ldots, n - 1$$

Quantiles are widely used in statistical analysis. The quantile of order 2 is the median.

[2] In Section 4.5 we shall consider another special $f(x)$ that leads to another method of describing a random variable.

situations it is convenient to compute the moments about the mean value μ rather than zero; in this case one obtains central moments.

Definition 4.4-3. ***Central Moments:*** The kth-order central moment of a random variable x, $k = 0, 1, 2, \ldots$, is defined by

$$\eta_{x^k} = \mathcal{E}\{(x - \mu)^k\} = \int_{-\infty}^{\infty} (\alpha - \mu)^k p_x(\alpha)\, d\alpha \qquad k = 0, 1, 2, \ldots \tag{4.4-20}$$

Again, if there is no possibility of confusion, we shall write simply $\eta_k = \eta_{x^k}$. Here we see that $\eta_0 = 1$ and $\eta_1 = 0$. The moments defined by Eq. (4.4-19) are also referred to as *raw moments* to differentiate them from the central moments.

It would appear that the central moments η_k and moments m_k should be related. To develop this relation, we need only use the binomial formula on the definition of η_k to write

$$\eta_k = \mathcal{E}\{(x - \mu)^k\} = \mathcal{E}\left\{\sum_{i=0}^{k} \binom{k}{i} (-1)^i \mu^i x^{k-i}\right\} \tag{4.4-21}$$

Now applying the definition of moments, Eq. (4.4-21) becomes

$$\eta_k = \sum_{i=0}^{k} \binom{k}{i} (-1)^i \mu^i m_{k-i} \tag{4.4-22}$$

In particular, we see that

$$\eta_0 = m_0 \qquad \eta_1 = m_1 - \mu = 0$$

$$\eta_2 = m_2 - 2\mu m_1 + \mu^2 = m_2 - \mu^2 \tag{4.4-23}$$

$$\eta_3 = m_3 - 3m_2\mu + 3m_1\mu^2 - \mu^3 = m_3 - 3m_2\mu + 2\mu^3 \tag{4.4-24}$$

In a similar manner, one can determine m_k in terms of η_i, $i = 0, 1, 2, \ldots, k$, as

$$m_k = \mathcal{E}\{x^k\} = \mathcal{E}\{[x - \mu + \mu]^k\} = \mathcal{E}\left\{\binom{k}{i} \sum_{i=0}^{k} \mu^i (x - \mu)^{k-i}\right\}$$

so that the definition of the central moments gives

$$m_k = \sum_{i=0}^{k} \binom{k}{i} \mu^i \eta_{k-i} \tag{4.4-25}$$

Because of their wide use, the moments m_1 and m_2 and the central moment η_2 are given special names. We have previously defined $m_1 = \mu_x$

or just μ as the mean or expected value. The second-order moment $m_2 = \mathcal{E}\{x^2\}$ will be referred to as the *mean-square value* and will be denoted by P_x or P. The second-order central moment $\eta_2 = \mathcal{E}\{(x - \mu)^2\}$ is called the *variance* and will be denoted by V_x, V, or var$\{x\}$. From Eq. (4.4-23) we see that we need only know μ_x, and P_x or V_x, since

$$V_x = P_x - \mu_x^2 \tag{4.4-26}$$

The square root of V_x is known as the *standard deviation*; $\sigma_x = \sqrt{V_x}$.

It should be noted that μ_x, P_x, and V_x are *not* functions of x; they are *functionals of the distribution* of x and are constants, not random variables. The mean is a measure of the *location* of the distribution, whereas V_x is a measure of the *dispersion* of the distribution. We also see that V_x is equal to the "moment of interia" of the density function p_x. In other words, as V_x increases, the "area" of the density function is more widely dispersed from the mean.

Example *4.4-4*. To illustrate the above concepts, let us determine the mean and variance of the three continuous random variables shown in Fig. 4.4-1. Let us begin by considering a general uniform density

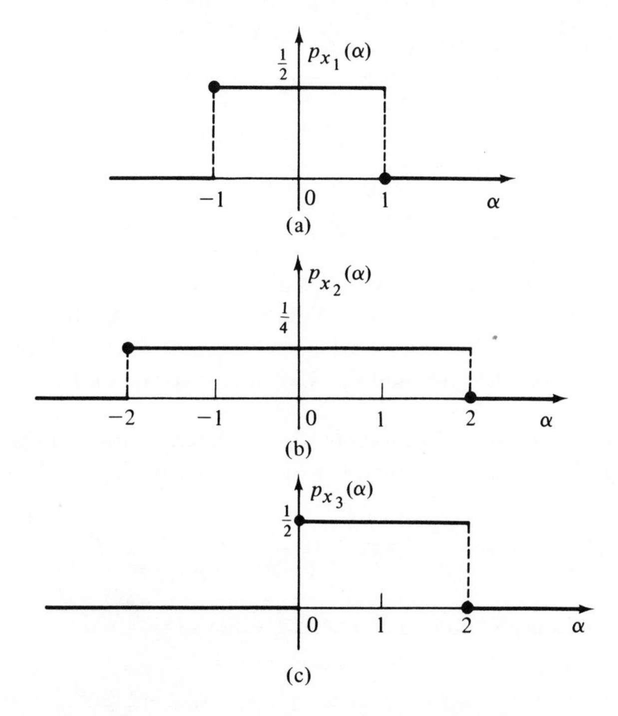

FIG. 4.4-1. Three density functions for Example 4.4-4.

on the interval $[a, b]$ given by

$$p_x(\alpha) = \frac{1}{b-a}[\mathscr{u}(\alpha - a) - \mathscr{u}(\alpha - b)]$$

The expected value of x is given by

$$\mu = \mathcal{E}\{x\} = \int_a^b \frac{1}{b-a}\alpha\, d\alpha = \frac{\alpha^2}{2(b-a)}\bigg|_a^b = \frac{b+a}{2}$$

This result is expected since $p_x(\alpha)$ is symmetric about $\alpha = (b + a)/2$. To find the variance, we shall determine the mean-squared value P and then use Eq. (4.4-26) to find V. The mean-squared value is given by

$$P = \mathcal{E}\{x^2\} = \int_a^b \frac{1}{b-a}\alpha^2\, d\alpha = \frac{\alpha^3}{3(b-a)}\bigg|_a^b = \frac{b^3 - a^3}{3(b-a)}$$

so that the variance is

$$V = P - \mu^2 = \frac{b^3 - a^3}{3(b-a)} - \frac{(b+a)^2}{4} = \frac{(b-a)^2}{12}$$

Using these results, we may easily find the mean and variance for the three random variables given in Fig. 4.4-1.

$$\begin{aligned} \mu_{x_1} &= 0 \qquad & V_{x_1} &= \tfrac{1}{3} \\ \mu_{x_2} &= 0 \qquad & V_{x_2} &= \tfrac{4}{3} \\ \mu_{x_3} &= 1 \qquad & V_{x_3} &= \tfrac{1}{3} \end{aligned}$$

Here we see that x_1 and x_2 have the same means, whereas x_1 and x_3 have the same variance. This result supports the statements made above, since x_1 and x_2 are both "located" at the origin and x_1 and x_3 have the same amount of "spread" about the mean.

Example *4.4-5*. Let us consider the determination of the moments of a gaussian density function with p_x given by

$$p_x(\alpha) = \frac{1}{\sqrt{2\pi}\sigma}\exp\left[-\frac{(\alpha - \mu)^2}{2\sigma^2}\right]$$

We consider first the expected value $\mathcal{E}\{x\}$:

$$\mathcal{E}\{x\} = \frac{1}{\sqrt{2\pi}\sigma}\int_{-\infty}^{\infty} \alpha \exp\left[-\frac{(\alpha - \mu)^2}{2\sigma^2}\right] d\alpha$$

If we make the substitution $z = (\alpha - \mu)/\sigma$, then $\mathcal{E}\{x\}$ becomes

$$\mathcal{E}\{x\} = \frac{\sigma}{\sqrt{2\pi}} \int_{-\infty}^{\infty} z e^{-z^2/2}\, dz + \frac{\mu}{\sqrt{2\pi}} \int_{-\infty}^{\infty} e^{-z^2/2}\, dz$$

The first integral is equal to zero, since $z \exp[-z^2/2]$ is an odd function and the second integral is equal to $\sqrt{2\pi}$; so we have

$$\mathcal{E}\{x\} = \mu$$

Hence we see that our use of the symbol μ for the gaussian density parameter is consistent with the use of μ as the mean.

Let us determine next the variance of x that is given by

$$\begin{aligned} V = \text{var}\{x\} &= \mathcal{E}\{(x - \mu)^2\} \\ &= \frac{1}{\sqrt{2\pi}\sigma} \int_{-\infty}^{\infty} (\alpha - \mu)^2 \exp\left[\frac{-(\alpha - \mu)}{2\sigma^2}\right] d\alpha \end{aligned}$$

If we make the substitution $z = (\alpha - \mu)/\sigma$, then V becomes

$$V = \frac{\sigma^2}{\sqrt{2\pi}} \int_{-\infty}^{\infty} z^2 e^{-z^2/2}\, dz$$

We may use integration by parts on this integral with $u = z$ and $dv = ze^{-z^2/2}\, dz$ to obtain

$$\begin{aligned} V &= \frac{\sigma^2}{\sqrt{2\pi}} \left[-ze^{-z^2/2} \Big|_{-\infty}^{\infty} + \int_{-\infty}^{\infty} e^{-z^2/2}\, dz \right] \\ &= \sigma^2 \end{aligned}$$

Therefore, we may say that a normal distribution $N(\mu, \sigma^2)$ has a mean μ and a variance σ^2.

The general central moments of x are determined in a similar way. It is easy to show that all the odd-order central moments are zero, and by repeated use of integration by parts that the even-order central moments are given by

$$\eta_k = \mathcal{E}\{(x - \mu)^k\} = 1 \cdot 3 \cdots (k - 1)\sigma^k \qquad k = 2, 4, 6, \ldots$$

The raw moments for the gaussian random variable may now be obtained by the use of Eq. (4.4-25), which simplifies somewhat, since $\eta_i = 0$ for i odd. For example, $m_2 = P$ is

$$m_2 = P = \sigma^2 + \mu^2$$

and m_3 becomes

$$m_3 = 3\mu\sigma^2 + \mu^3$$

and m_4 is

$$m_4 = 3\sigma^4 + 6\mu^2\sigma^2 + \mu^4$$

Note that if $\mu = 0$, then all the odd-order raw moments are also zero.

It should be noted that nowhere in the discussion have we stated that any or all of the moments for a given distribution will necessarily exist. In fact, it is quite easy to construct probability distributions for which some or all of the moments do not exist. To illustrate this fact, we shall consider the determination of the mean and variance of the Cauchy distribution.

Example 4.4-6. The reader will remember that the probability density function for the Cauchy distribution is given by

$$p_x(\alpha) = \frac{\lambda/\pi}{\lambda^2 + \alpha^2}$$

where λ is a parameter. The expected value of x is given by

$$\mu = \mathcal{E}\{x\} = \int_{-\infty}^{\infty} \frac{\alpha\lambda/\pi}{\lambda^2 + \alpha^2}\, d\alpha$$

Since the integrand is an odd function, we might conclude that $\mu = 0$. We might also conclude that $\mu = 0$, since $p_x(\alpha)$ is symmetric about $\alpha = 0$. Such a conclusion would be premature and incorrect, as we shall show. The difficulty evolves about the existence of the integral in the above equation. To illustrate the problem, let us divide the integral into two parts so that μ becomes

$$\mu = \int_{0}^{\infty} \frac{\alpha\lambda/\pi}{\alpha^2 + \lambda^2}\, d\alpha + \int_{-\infty}^{0} \frac{\alpha\lambda/\pi}{\alpha^2 + \lambda^2}\, d\alpha$$

Now suppose that we make the substitution $\xi_1 = \alpha^2$ in the first integral and $\xi_2 = \alpha^2$ in the second; then we have

$$\begin{aligned}\mu &= \int_{0}^{\infty} \frac{\lambda/2\pi}{\lambda^2 + \xi_1}\, d\xi_1 - \int_{0}^{\infty} \frac{\lambda/2\pi}{\lambda^2 + \xi_2}\, d\xi_2 \\ &= \frac{\lambda}{2\pi}[\ln(\lambda^2 + \xi_1)|_0^\infty - \ln(\lambda^2 + \xi_3)|_0^\infty]\end{aligned}$$

Here again it may appear that $\mu = 0$; however, the reader is cautioned to remember that $\infty - \infty$ does not necessarily equal zero. To emphasize the problem, let us rewrite the above expression for μ as

$$\mu = \frac{\lambda}{2\pi}[\lim_{\xi_1 \to \infty} \ln(\lambda^2 + \xi_1) - \lim_{\xi_2 \to \infty} \ln(\lambda^2 + \xi_2)]$$

The difference of the two limits only equals the limit of the difference if both initial limits exist. This is not true in this case. If μ is to have a value, the result must be independent of the manner in which ξ_1 and ξ_2 approach infinity. Suppose, for example, that we let $\xi_1 = K\xi_2$, where K is a specified constant; then ξ becomes

$$\mu = \frac{\lambda}{2\pi} \lim_{\mu_2 \to \infty} \ln\left(\frac{\lambda^2 + K\xi_2}{\lambda^2 + \xi_2}\right) = \frac{\lambda}{2\pi} \ln K$$

Hence we see that we may make μ take any value that we wish by proper selection of K. This result indicates that μ does not exist in this case.[1] By similar arguments, we may also show that all the other moments are also undefined for the Cauchy distribution.

Example ***4.4-7.*** As a final example, let us consider the determination of the mean and variance for the Poisson distribution. From Section 3.5 we recall that the Poisson distribution is a discrete distribution with the probability density function given by[2]

$$p_x(\alpha) = e^{-\eta} \sum_{i=0}^{\infty} \frac{\eta^i}{i!} \delta_D(\alpha - i)$$

Therefore, the mean value is given by

$$\mathcal{E}\{x\} = \int_{-\infty}^{\infty} \alpha e^{-\eta} \sum_{i=0}^{\infty} \frac{\eta^i}{i!} \delta_D(\alpha - i)\, d\alpha$$

and using the sifting property of the Dirac delta function, we obtain

$$\mathcal{E}\{x\} = e^{-\eta} \sum_{i=1}^{\infty} i \frac{\eta^i}{i!} = e^{-\eta} \sum_{i=1}^{\infty} \frac{\eta^i}{(i-1)!}$$

We could also have obtained this result directly by the use of Eq. (4.4-7). Now we make the substituion $j = i - 1$ and rewrite the result as

$$\mathcal{E}\{x\} = \eta\left[e^{-\eta} \sum_{j=0}^{\infty} \frac{\eta^j}{j!}\right]$$

[1] The Cauchy principal value of μ is zero.

[2] In Section 3.5 we used λt for the parameter of the Poisson process. For simplicity, we shall use η here.

However

$$e^{-\eta} \sum_{j=0}^{\infty} \frac{\eta^j}{j!} = 1 = \int_{-\infty}^{\infty} p_x(\alpha)\, d\alpha$$

so that $\mathcal{E}\{x\}$ becomes

$$\mu = \mathcal{E}\{x\} = \eta$$

Now the mean-square value is given by

$$\mathcal{E}\{x^2\} = \int_{-\infty}^{\infty} \alpha^2 e^{-\eta} \sum_{i=0}^{\infty} \frac{\eta^i}{i!} \delta_D(\alpha - i)\, d\alpha = e^{-\eta} \sum_{i=1}^{\infty} \frac{\eta^i}{i!} i^2$$
$$= e^{-\eta} \sum_{i=1}^{\infty} \frac{i\eta^i}{(i-1)!}$$

We may write this result in the following form:

$$\mathcal{E}\{x^2\} = \eta e^{-\eta} \sum_{i=1}^{\infty} \frac{\partial}{\partial \eta} \frac{\eta^i}{(i-1)!}$$

which is easily verified by carrying out the indicated operation. If we interchange the summation and differentiation operations,[1] we obtain

$$\mathcal{E}\{x^2\} = \eta e^{-\eta} \frac{\partial}{\partial \eta} \sum_{i=1}^{\infty} \frac{\eta^i}{(i-1)!}$$

If we make the substitution $j = i - 1$, the summation becomes

$$\sum_{i=1}^{\infty} \frac{\eta^i}{(i-1)!} = \sum_{j=0}^{\infty} \frac{\eta^{j+1}}{j!} = \eta e^{\eta} \left[e^{-\eta} \sum_{j=0}^{\infty} \frac{\eta^j}{j!} \right] = \eta e^{\eta}$$

so that $\mathcal{E}\{x^2\}$ is given by

$$P = \mathcal{E}\{x^2\} = \eta e^{-\eta} \frac{\partial}{\partial \eta} [\eta e^{\eta}] = \eta + \eta^2$$

Using Eq. (4.4-26), the variance is easily found to be

$$V = P - \mu^2 = \eta + \eta^2 - \eta^2 = \eta$$

In this case, we find that the mean and variance are equal.

[1] The interchange of these operations is justified in this case because both series are uniformly convergent.

For easy reference the mean and variance of several important distributions are summarized in Table 4.4-1. The reader is urged to derive those results which have not been presented here.

TABLE 4.4-1. Mean and Variance of Several Distributions

Distribution	Parameters[1]	Mean, μ	Variance, V
Binomial	n, p	np	$np(1-p)$
Poisson	λt	λt	λt
Hypergeometric	N, M, n	$\frac{nM}{N}$	$\frac{nM(N-M)}{N^2}\left(1-\frac{n-1}{N-1}\right)$
Negative binomial	r, p	$r\frac{(1-p)}{p}$	$r\frac{1-p}{p^2}$
Continuous uniform	a, b	$\frac{b-a}{2}$	$\frac{(b-a)^2}{12}$
Normal	μ, σ^2	μ	σ^2
Rayleigh	a^2	a	$2a^2$
Gamma	λ, n	$\frac{n}{\lambda}$	$\frac{n}{\lambda^2}$
Beta	a, b	$\frac{a}{a+b}$	$\frac{ab}{(a+b)^2(a+b+1)}$
Cauchy	λ	Do not exist	

[1] See Section 3.5, in particular Table 3.5-1, for definition of the parameters of the distributions.

The method of statistical moments can be directly extended to a random vector $\mathbf{x}$. The only difficulty occurs in the representation of moments higher than the second order. The first-order raw and central moments are given by

$$\mathbf{m}_1 = \mathcal{E}\{\mathbf{x}\} = \int_{-\infty}^{\infty} \boldsymbol{\alpha} p_{\mathbf{x}}(\boldsymbol{\alpha})\, d\boldsymbol{\alpha} = \boldsymbol{\mu}_{\mathbf{x}} \tag{4.4-27}$$

and

$$\boldsymbol{\eta}_1 = \mathcal{E}\{\mathbf{x}\} = \int_{-\infty}^{\infty} (\boldsymbol{\alpha} - \boldsymbol{\mu}_{\mathbf{x}}) p_{\mathbf{x}}(\boldsymbol{\alpha})\, d\boldsymbol{\alpha} = \mathbf{0} \tag{4.4-28}$$

The second-order raw and central moments are given by

$$\mathbf{m}_2 = \mathcal{E}\{\mathbf{x}\mathbf{x}^T\} = \int_{-\infty}^{\infty} \boldsymbol{\alpha}\boldsymbol{\alpha}^T p_{\mathbf{x}}(\boldsymbol{\alpha})\, d\boldsymbol{\alpha} = \mathbf{P}_{\mathbf{x}} \tag{4.4-29}$$

and

$$\boldsymbol{\eta}_2 = \mathcal{E}\{(\mathbf{x} - \boldsymbol{\mu}_{\mathbf{x}})(\mathbf{x} - \boldsymbol{\mu}_{\mathbf{x}})^T\} = \int_{-\infty}^{\infty} (\boldsymbol{\alpha} - \boldsymbol{\mu}_{\mathbf{x}})(\boldsymbol{\alpha} - \boldsymbol{\mu}_{\mathbf{x}})^T p_{\mathbf{x}}(\boldsymbol{\alpha})\, d\boldsymbol{\alpha} = \mathbf{V}_{\mathbf{x}} \tag{4.4-30}$$

Note that both the mean-square value $\mathbf{P_x}$ and the variance $\mathbf{V_x}$ are $N \times N$ symmetric matrices. It is very easy to show that $\boldsymbol{\mu}_\mathbf{x}$, $\mathbf{P_x}$, and $\mathbf{V_x}$ are related in the following manner:

$$\mathbf{V_x} = \mathbf{P_x} - \boldsymbol{\mu}_\mathbf{x}\boldsymbol{\mu}_\mathbf{x}^T \tag{4.4-31}$$

In general, the nth-order moment requires an n-dimensional array, that is, an array in which each element must have n subscripts. For example, the fourth-order moment requires a four-dimensional array in which the $ijkl$ element is

$$[\mathbf{m}_4]_{ijkl} = \mathcal{E}\{x_i x_j x_k x_l\} = \int_{-\infty}^{\infty} \alpha_i \alpha_j \alpha_k \alpha_l \, p_\mathbf{x}(\alpha_i, \alpha_j, \alpha_k, \alpha_l) \, d\alpha_i \, d\alpha_j \, d\alpha_k \, d\alpha_l$$

Unfortunately, we only have general methods for representing first-order moments, by the use of vectors, and second-order moments, by the use of matrices, unless we make use of concepts such as tensors, which are beyond the scope of this treatment. Fortunately, many practical applications only require the use of the first- and second-order moments. If higher-order moments are needed, one must simply express them in elemental form.

One important use of the second-order moment is in an expression known as the *Tchebycheff inequality*. As noted earlier, the variance is a measure of the dispersion of a random variable $\mathbf{x}$ about its mean $\boldsymbol{\mu}_\mathbf{x}$. The Tchebycheff inequality gives an upper bound on the probability that the distance from $\mathbf{x}$ to $\boldsymbol{\mu}_\mathbf{x}$ will exceed any given amount, independent of the distribution of $\mathbf{x}$, in the following form:

$$\mathcal{P}\{\|\mathbf{x} - \boldsymbol{\mu}_\mathbf{x}\| \geq \epsilon > 0\} \leq \frac{\mathrm{tr}\{\mathbf{V_x}\}}{\epsilon^2} \tag{4.4-32}$$

This result may also be expressed in the alternative form

$$\mathcal{P}\{\|\mathbf{x} - \boldsymbol{\mu}_\mathbf{x}\| \leq \epsilon\} \geq 1 - \frac{\mathrm{tr}\{\mathbf{V_x}\}}{\epsilon^2} \tag{4.4-33}$$

For a scalar random variable x, this result takes the somewhat simpler form of

$$\mathcal{P}\{|x - \mu_x| \leq \epsilon\} \geq 1 - \frac{V_x}{\epsilon^2} \tag{4.4-34}$$

To establish Eq. (4.4-32), we begin with the definition of $\mathbf{V_x}$ given by Eq. (4.4-30) and then take the trace to obtain

$$\begin{aligned}\mathrm{tr}\{\mathbf{V_x}\} &= \int_{-\infty}^{\infty} \mathrm{tr}\{(\boldsymbol{\alpha} - \boldsymbol{\mu}_\mathbf{x})(\boldsymbol{\alpha} - \boldsymbol{\mu}_\mathbf{x})^T\} p_\mathbf{x}(\boldsymbol{\alpha}) \, d\boldsymbol{\alpha} \\ &= \int_{-\infty}^{\infty} \|\boldsymbol{\alpha} - \boldsymbol{\mu}_\mathbf{x}\|^2 \, p_\mathbf{x}(\boldsymbol{\alpha}) \, d\boldsymbol{\alpha}\end{aligned}$$

If we integrate on the right-hand side of the above equation only over the values of $\boldsymbol{\alpha}$ for which $\|\boldsymbol{\alpha} - \boldsymbol{\mu}_{\mathbf{x}}\| \geq \epsilon$, then we cannot increase the value of the integral, since the integrand is nonnegative definite; hence we have

$$\operatorname{tr}\{\mathbf{V}_{\mathbf{x}}\} \geq \int_{\|\boldsymbol{\alpha} - \boldsymbol{\mu}_{\mathbf{x}}\| \geq \epsilon} \|\boldsymbol{\alpha} - \boldsymbol{\mu}_{\mathbf{x}}\|^2 p_{\mathbf{x}}(\boldsymbol{\alpha})\, d\boldsymbol{\alpha}$$

Over the range of $\boldsymbol{\alpha}$ for which $\|\boldsymbol{\alpha} - \boldsymbol{\mu}_{\mathbf{x}}\| \geq \epsilon$, we can replace $\|\boldsymbol{\alpha} - \boldsymbol{\mu}_{\mathbf{x}}\|^2$ by ϵ^2 without increasing the value of the integral, so that now we have

$$\operatorname{tr}\{\mathbf{V}_{\mathbf{x}}\} \geq \epsilon^2 \int_{\|\boldsymbol{\alpha} - \boldsymbol{\mu}_{\mathbf{x}}\| \geq \epsilon} p_{\mathbf{x}}(\boldsymbol{\alpha})\, d\boldsymbol{\alpha} = \epsilon^2 \mathcal{P}\{\|\mathbf{x} - \boldsymbol{\mu}_{\mathbf{x}}\| \geq \epsilon\}$$

and simple division gives the desired result of Eq. (4.4-32).

The main advantage of the Tchebycheff inequality is that it may be computed without any knowledge of the probability distribution of the random variable. This makes it very useful in a situation in which the distribution is either unknown or too complex to compute. Unfortunately, because the expression is independent of the distribution, it is generally too conservative to be of major value for many applications. To demonstrate this fact, let us consider a scalar random variable x with a uniform distribution on [0, 2]. From Table 4.4-1 we can easily find that $\mu_x = 1$ and $V_x = \frac{1}{3}$. Suppose that we wish to find the probability that $|x - \mu_x| \leq \frac{1}{2}$. From the Tchebycheff inequality, Eq. (4.4-34), we obtain

$$\mathcal{P}\left\{|x - \mu_x| \leq \frac{1}{2}\right\} \geq 1 - \frac{1/3}{(1/2)^2} = -\frac{1}{3}$$

This result, while correct, is of little value since we already know that the probability must be greater than zero. The actual probability is easily determined for this case to be $\frac{1}{2}$.

Conditional Expectation. A concept that plays a key role in many practical applications, including estimation and detection, of random variable theory is that of conditional expectation. Conditional expectation is a direct extension of the concepts developed above with the usual probability density replaced by a conditional probability density so that we have the following definition:

Definition 4.4-4. ***Conditional Expectation:*** The conditional expectation of a random variable $\mathbf{x}$, given a random variable $\mathbf{y} = \boldsymbol{\beta}$, is

$$\mathcal{E}\{\mathbf{x} \mid \mathbf{y} = \boldsymbol{\beta}\} = \int_{-\infty}^{\infty} \boldsymbol{\alpha} p_{\mathbf{x}|\mathbf{y}}(\boldsymbol{\alpha} \mid \boldsymbol{\beta})\, d\boldsymbol{\alpha} \qquad (4.4\text{-}35)$$

if the integral exists.

We may also develop a *fundamental theorem of conditional expectation* similar to Eq. (4.4-12) given by

$$\mathcal{E}\{\mathbf{f}(\mathbf{x})\,|\,\mathbf{y}=\boldsymbol{\beta}\} = \int_{-\infty}^{\infty} \mathbf{f}(\boldsymbol{\alpha}) p_{\mathbf{x}|\mathbf{y}}(\boldsymbol{\alpha}\,|\,\boldsymbol{\beta})\, d\boldsymbol{\alpha} \qquad (4.4\text{-}36)$$

This result allows us to define conditional moments of a random variable. Once again we restrict attention to the scalar case for notational simplicity. The nth-order (raw) conditional moment of x given $\mathbf{y} = \boldsymbol{\beta}$ is defined by

$$\mathcal{E}\{x^n\,|\,\mathbf{y}=\boldsymbol{\beta}\} = \int_{-\infty}^{\infty} \alpha^n p_{x|\mathbf{y}}(\alpha\,|\,\boldsymbol{\beta})\, d\alpha \qquad (4.4\text{-}37)$$

whereas the nth-order central conditional moment of x given $\mathbf{y} = \boldsymbol{\beta}$ is

$$\mathcal{E}\{(x - \mathcal{E}\{x\,|\,\mathbf{y}=\boldsymbol{\beta}\})^n\,|\,\mathbf{y}=\boldsymbol{\beta}\} = \int_{-\infty}^{\infty} (\alpha - \mathcal{E}\{x\,|\,\mathbf{y}=\boldsymbol{\beta}\})^n p_{x|\mathbf{y}}(\alpha\,|\,\boldsymbol{\beta})\, d\alpha \quad (4.4\text{-}38)$$

As before the first-order raw moment is called the conditional mean vector $\boldsymbol{\mu}_{\mathbf{x}|\mathbf{y}=\boldsymbol{\beta}}$ given by

$$\boldsymbol{\mu}_{\mathbf{x}|\mathbf{y}=\boldsymbol{\beta}} = \int_{-\infty}^{\infty} \boldsymbol{\alpha} p_{\mathbf{x}|\mathbf{y}}(\boldsymbol{\alpha}\,|\,\boldsymbol{\beta})\, d\boldsymbol{\alpha} \qquad (4.4\text{-}39)$$

whereas the *conditional mean-square-value matrix* is defined as

$$\mathbf{P}_{\mathbf{x}|\mathbf{y}=\boldsymbol{\beta}} = \int_{-\infty}^{\infty} \boldsymbol{\alpha}\boldsymbol{\alpha}^T p_{\mathbf{x}|\mathbf{y}}(\boldsymbol{\alpha}\,|\,\boldsymbol{\beta})\, d\boldsymbol{\alpha} \qquad (4.4\text{-}40)$$

The *conditional variance matrix* is given by

$$\mathbf{V}_{\mathbf{x}|\mathbf{y}=\boldsymbol{\beta}} = \int_{-\infty}^{\infty} (\boldsymbol{\alpha} - \boldsymbol{\mu}_{\mathbf{x}|\mathbf{y}=\boldsymbol{\beta}})(\boldsymbol{\alpha} - \boldsymbol{\mu}_{\mathbf{x}|\mathbf{y}=\boldsymbol{\beta}})^T p_{\mathbf{x}|\mathbf{y}}(\boldsymbol{\alpha}\,|\,\boldsymbol{\beta})\, d\boldsymbol{\alpha} \qquad (4.4\text{-}41)$$

Once again, it is quite easy to show that

$$\mathbf{V}_{\mathbf{x}|\mathbf{y}=\boldsymbol{\beta}} = \mathbf{P}_{\mathbf{x}|\mathbf{y}=\boldsymbol{\beta}} - \boldsymbol{\mu}_{\mathbf{x}|\mathbf{y}=\boldsymbol{\beta}}\boldsymbol{\mu}^T_{\mathbf{x}|\mathbf{y}=\boldsymbol{\beta}} \qquad (4.4\text{-}42)$$

Example *4.4-8*. Let us assume that the conditional density of x given y is described by

$$p_{x|y}(\alpha\,|\,\beta) = \frac{1}{\sqrt{2\pi}\sigma} \exp\left\{-\frac{1}{2\sigma^2}(\alpha - a\beta)^2\right\}$$

Let us determine the conditional mean and variance of x given $y = \beta$. The conditional mean is given by

$$\mu_{x|y=\beta} = \int_{-\infty}^{\infty} \frac{\alpha}{\sqrt{2\pi}\sigma} \exp\left\{-\frac{1}{2\sigma^2}(\alpha - a\beta)^2\right\} d\alpha$$

Let us make the substitution $z = (\alpha - a\beta)/\sigma$ to obtain

$$\mu_{x|y=\beta} = \int_{-\infty}^{\infty} \frac{1}{\sqrt{2\pi}}(\sigma z + a\beta)e^{-z^2/2}\,dz$$

$$= \int_{-\infty}^{\infty} \frac{\sigma}{\sqrt{2\pi}} z e^{-z^2/2}\,dz + a\beta \int_{-\infty}^{\infty} \frac{1}{\sqrt{2\pi}} e^{-z^2/2}\,dz$$

The value of the first integral is zero; the value of the second integral is one; so we have

$$\mu_{x|y=\beta} = a\beta$$

The conditional variance is given by

$$V_{x|y-\beta} = \int_{-\infty}^{\infty} \frac{(\alpha - a\beta)^2}{\sqrt{2\pi}\sigma} \exp\left\{-\frac{1}{2\sigma^2}(\alpha - a\beta)^2\right\} d\alpha$$

Again we make the substituion $z = (\alpha - a\beta)/\sigma$ to obtain

$$V_{x|y=\beta} = \sigma^2 \int_{-\infty}^{\infty} \frac{z^2}{\sqrt{2\pi}} e^{-z^2/2}\,dz$$

If we use integration by parts on this integral, we find that $V_{x|y=\beta}$ becomes just

$$V_{x|y=\beta} = \sigma^2$$

Note that, in general, the conditional expectation of $\mathbf{x}$ given $\mathbf{y} = \boldsymbol{\beta}$ is a function of $\boldsymbol{\beta}$. For a given outcome ξ, $\mathbf{y}$ takes a specific value $\mathbf{y}(\xi) = \boldsymbol{\beta}_i$, corresponding to $\boldsymbol{\beta}_i$ is a value $\mathcal{E}\{\mathbf{x} \mid \mathbf{y}(\xi) = \boldsymbol{\beta}_i\}$. In this manner, we can think of $\mathcal{E}\{\mathbf{x} \mid \mathbf{y}(\xi)\}$ as a random variable[1] that is a (complicated) function of the random variable $\mathbf{y}$. It is easier to understand this concept if one remembers that the expected value of a random variable is a functional of the distribution and *not* a function of the random variable. Hence we may think of the conditional expectation of $\mathbf{x}$ as either a number (when $\mathbf{y}$ is a specific number) or as a random variable. To clearly differentiate which meaning is intended, we shall use the notation $\mathcal{E}\{\mathbf{x} \mid \mathbf{y} = \boldsymbol{\beta}\}$ to imply a numerical value; the notation $\mathcal{E}\{\mathbf{x} \mid \mathbf{y}\}$ will be used to indicate a random variable. We shall define $\mathcal{E}\{\mathbf{x} \mid \mathbf{y}\}$ as an integral expression by using a slight abuse of notation in the following form:

$$\mathcal{E}\{\mathbf{x} \mid \mathbf{y}\} = \int_{-\infty}^{\infty} \alpha p_{\mathbf{x}|\mathbf{y}}(\alpha \mid \mathbf{y})\,d\alpha \tag{4.4-43}$$

[1] It is possible to show that $\mathcal{E}\{\mathbf{x} \mid \mathbf{y}(\xi)\}$ is in fact a random variable; that is, it satisfies Definition 3.2-2. This property follows from the requirement that $p_{\mathbf{x}|\mathbf{y}}$ is a density function.

In a similar manner, we can consider the conditional expectation of a function $\mathbf{f}(\mathbf{x}, \mathbf{y})$ given $\mathbf{y}$ as a random variable written in the form

$$\mathcal{E}\{\mathbf{f}(\mathbf{x}, \mathbf{y}) \mid \mathbf{y}\} = \int_{-\infty}^{\infty} \mathbf{f}(\alpha, \mathbf{y}) p_{\mathbf{x}|\mathbf{y}}(\alpha \mid \mathbf{y})\, d\alpha \tag{4.4-44}$$

A very interesting and often useful property is that $\mathcal{E}\{\mathbf{f}(\mathbf{x}, \mathbf{y})\} = \mathcal{E}\{\mathcal{E}\{\mathbf{f}(\mathbf{x}, \mathbf{y}) \mid \mathbf{y}\}\}$; that is, the expected value of the random variable $\mathcal{E}\{\mathbf{f}(\mathbf{x}, \mathbf{y}) \mid \mathbf{y}\}$ is the expected value of $\mathbf{f}(\mathbf{x}, \mathbf{y})$. Note that in $\mathcal{E}\{\mathbf{f}(\mathbf{x}, \mathbf{y})\}$, the expectation is over both $\mathbf{x}$ and $\mathbf{y}$, whereas in $\mathcal{E}\{\mathcal{E}\{\mathbf{f}(\mathbf{x}, \mathbf{y}) \mid \mathbf{y}\}\}$, the outer expectation is over $\mathbf{y}$ and the inner expectation is over $\mathbf{x}$. To emphasize[1] exactly what value the expectation is taken over, we might write the expression as

$$\mathcal{E}_{\mathbf{x},\mathbf{y}}\{\mathbf{f}(\mathbf{x}, \mathbf{y})\} = \mathcal{E}_{\mathbf{y}}\{\mathcal{E}_{\mathbf{x}}\{\mathbf{f}(\mathbf{x}, \mathbf{y}) \mid \mathbf{y}\}\} \tag{4.4-45}$$

where the subscripts are used to indicate the variable(s) over which the expectation is to be taken.

We may establish the validity of Eq. (4.4-45) by writing out the definition of the left side, since $\mathcal{E}_{\mathbf{x}}\{\mathbf{f}(\mathbf{x}, \mathbf{y}) \mid \mathbf{y}\}$ is given by

$$\mathcal{E}_{\mathbf{x}}\{\mathbf{f}(\mathbf{x}, \mathbf{y}) \mid \mathbf{y}\} = \int_{-\infty}^{\infty} \mathbf{f}(\alpha, \mathbf{y}) p_{\mathbf{x}|\mathbf{y}}(\alpha \mid \mathbf{y})\, d\alpha$$

Now the expected value (with respect to $\mathbf{y}$) of this expression gives

$$\begin{aligned}\mathcal{E}_{\mathbf{y}}\{\mathcal{E}_{\mathbf{x}}\{\mathbf{f}(\mathbf{x}, \mathbf{y}) \mid \mathbf{y}\}\} &= \int_{-\infty}^{\infty} \mathcal{E}_{\mathbf{x}}\{\mathbf{f}(\mathbf{x}, \beta) \mid \beta\} p_{\mathbf{y}}(\beta)\, d\beta \\ &= \int_{-\infty}^{\infty} d\beta \int_{-\infty}^{\infty} d\alpha\, \mathbf{f}(\alpha, \beta) p_{\mathbf{x}|\mathbf{y}}(\alpha \mid \beta) p_{\mathbf{y}}(\beta)\end{aligned} \tag{4.4-46}$$

Since $p_{\mathbf{x}|\mathbf{y}}(\alpha \mid \beta) p_{\mathbf{y}}(\beta) = p_{\mathbf{x},\mathbf{y}}(\alpha, \beta)$, Eq. (4.4-46) becomes

$$\begin{aligned}\mathcal{E}_{\mathbf{y}}\{\mathcal{E}_{\mathbf{x}}\{\mathbf{f}(\mathbf{x}, \mathbf{y}) \mid \mathbf{y}\}\} &= \int_{-\infty}^{\infty} d\beta \int_{-\infty}^{\infty} d\alpha\, \mathbf{f}(\alpha, \beta) p_{\mathbf{x}\mathbf{y}}(\alpha, \beta) \\ &= \mathcal{E}_{\mathbf{x},\mathbf{y}}\{\mathbf{f}(\mathbf{x}, \mathbf{y})\}\end{aligned}$$

which is the desired result. Note that if $f(\mathbf{x}, \mathbf{y}) = g_1(\mathbf{x}) g_2(\mathbf{y})$, then $\mathcal{E}\{f(\mathbf{x}, \mathbf{y}) \mid \mathbf{y}\} = \mathcal{E}\{g_1(\mathbf{x}) \mid \mathbf{y}\} g_2(\mathbf{y})$.

Uncorrelated and Orthogonal Random Variable. In Section 3.9 we introduced the concept of independent random variables; at this time we

[1] Note, however, that there is no possible ambiguity in the notation without the subscripts, since we are always taking expectation over all the random variables enclosed. For example, $\mathcal{E}\{\mathbf{f}(\mathbf{x}, \mathbf{y})\}$ is over both $\mathbf{x}$ and $\mathbf{y}$, whereas $\mathcal{E}\{\mathbf{f}(\mathbf{x}, \mathbf{y}) \mid \mathbf{y}\}$ is just over $\mathbf{x}$, since we are assuming that $\mathbf{y}$ is given.

wish to discuss two other concepts—uncorrelated and orthogonal random variables. The definitions of the concepts take the following form:

Definition 4.4-5. ***Uncorrelated Random Variables:*** We say that two random variables $\mathbf{x}$ and $\mathbf{y}$ are uncorrelated if

$$\mathcal{E}\{\mathbf{x}\mathbf{y}^T\} = \mathcal{E}\{\mathbf{x}\}\mathcal{E}\{\mathbf{y}^T\} \tag{4.4-47}$$

Definition 4.4-6. ***Orthogonal Random Variables:*** We say that two N-dimensional random variables $\mathbf{x}$ and $\mathbf{y}$ are orthogonal if

$$\mathcal{E}\{\mathbf{x}^T\mathbf{y}\} = 0 \tag{4.4-48}$$

It should be noted that, unlike independence, the property of being uncorrelated or orthogonal is a binary or pairwise concept. Whenever we speak of a set of N random variables as being uncorrelated or orthogonal, we mean that they are pairwise uncorrelated or orthogonal.

For two random variables $\mathbf{x}$ and $\mathbf{y}$, we shall use the symbols

$$\mathbf{V}_{\mathbf{x}} = \text{var}\{\mathbf{x}\} = \mathcal{E}\{(\mathbf{x} - \boldsymbol{\mu}_{\mathbf{x}})(\mathbf{x} - \boldsymbol{\mu}_{\mathbf{x}})^T\} \tag{4.4-49}$$

and

$$\mathbf{V}_{\mathbf{y}} = \text{var}\{\mathbf{y}\} = \mathcal{E}\{(\mathbf{y} - \boldsymbol{\mu}_{\mathbf{y}})(\mathbf{y} - \boldsymbol{\mu}_{\mathbf{y}})^T\} \tag{4.4-50}$$

to represent the *variance matrices* of $\mathbf{x}$ and $\mathbf{y}$, respectively, whereas

$$\mathbf{V}_{\mathbf{xy}} = \mathbf{V}_{\mathbf{yx}}^T = \text{cov}\{\mathbf{x}, \mathbf{y}\} = \mathcal{E}\{(\mathbf{x} - \boldsymbol{\mu}_{\mathbf{x}})(\mathbf{y} - \boldsymbol{\mu}_{\mathbf{y}})^T\} \tag{4.4-51}$$

will be referred to as the *covariance matrix*. Note that the names are somewhat arbitrary, since the off-diagonal elements of the variances $\mathbf{V}_{\mathbf{x}}$ and $\mathbf{V}_{\mathbf{y}}$ are actually covariances of the form $\text{cov}\{x_i, x_j\} =$ and $\text{cov}\{y_i, y_j\}$, $i \neq j$. The symmetric matrix $\boldsymbol{\Xi}$ defined by

$$\boldsymbol{\Xi}\boldsymbol{\Xi}^T = \mathbf{V}_{\mathbf{x}}^{-1/2}\mathbf{V}_{\mathbf{xy}}\mathbf{V}_{\mathbf{y}}^{-1}\mathbf{V}_{\mathbf{xy}}^T\mathbf{V}_{\mathbf{x}}^{-1/2} \tag{4.4-52}$$

is referred to as the (normalized) correlation coefficient matrix. It is relatively easy to show that $\boldsymbol{\Xi}$ must satisfy the following inequality:

$$-\mathbf{I} \leq \boldsymbol{\Xi} \leq \mathbf{I} \tag{4.4-53}$$

If $\boldsymbol{\Xi} = \pm\mathbf{I}$, $\mathbf{x}$ and $\mathbf{y}$ are said to be perfectly or completely correlated. If $\mathbf{x}$ and $\mathbf{y}$ are related by a nonsingular linear transformation of the form

$$\mathbf{y} = \mathbf{A}\mathbf{x}$$

then we may show that $\boldsymbol{\Xi} = \pm\mathbf{I}$, since $\mathbf{V}_\mathbf{y} = \mathbf{A}\mathbf{V}_\mathbf{x}\mathbf{A}^T$ and $\mathbf{V}_{\mathbf{xy}} = \mathbf{V}_\mathbf{x}\mathbf{A}^T$; so Eq. (4.4-53) becomes

$$\boldsymbol{\Xi}\boldsymbol{\Xi}^T = \mathbf{V}_\mathbf{x}^{-1/2}(\mathbf{V}_\mathbf{x}\mathbf{A}^T)(\mathbf{A}^{-T}\mathbf{V}_\mathbf{x}^{-1}\mathbf{A}^{-1})(\mathbf{A}\mathbf{V}_\mathbf{x})\mathbf{V}_\mathbf{x}^{-1/2} = \mathbf{I}$$

In this case, the sample values of $\mathbf{x}$ and $\mathbf{y}$ would lie on a hyperplane in the $\mathbf{x} - \mathbf{y}$ space.

If $\mathbf{x}$ and $\mathbf{y}$ are uncorrelated, both the covariance matrix $\mathbf{V}_{\mathbf{xy}}$ and the correlation coefficient matrix $\boldsymbol{\Xi}$ will equal zero. To establish this fact, we need to use Eq. (4.4-31) to write

$$\mathbf{V}_{\mathbf{xy}} = \mathcal{E}\{\mathbf{x}\mathbf{y}^T\} - \mathcal{E}\{\mathbf{x}\}\mathcal{E}\{\mathbf{y}^T\}$$

But $\mathcal{E}\{\mathbf{x}\mathbf{y}^T\} = \mathcal{E}\{\mathbf{x}\}\mathcal{E}\{\mathbf{y}^T\}$, since $\mathbf{x}$ and $\mathbf{y}$ are uncorrelated, and we see that $\mathbf{V}_{\mathbf{xy}} = \mathbf{0}$. If $\mathbf{V}_{\mathbf{xy}} = \mathbf{0}$, it is obvious from Eq. (4.4-49) that $\boldsymbol{\Xi}$ is also zero.

In general, two random variables may be uncorrelated but not orthogonal, or orthogonal and not uncorrelated. In other words, one property does not imply the other, although a particular random variable may satisfy both properties. In particular, note that if either $\mathbf{x}$ or $\mathbf{y}$ is zero mean and $\mathbf{x}$ and $\mathbf{y}$ are uncorrelated, then $\mathbf{x}$ and $\mathbf{y}$ are orthogonal. To see this, we need only make use of Eq. (4.4-47) to write

$$\mathcal{E}\{\mathbf{x}\mathbf{y}^T\} = \mathcal{E}\{\mathbf{x}\}\mathcal{E}\{\mathbf{y}^T\} = \boldsymbol{\mu}_\mathbf{x}\boldsymbol{\mu}_\mathbf{y}^T \tag{4.4-54}$$

Now if $\boldsymbol{\mu}_\mathbf{x}$ or $\boldsymbol{\mu}_\mathbf{y}$ or both equal zero, we see that $\mathcal{E}\{\mathbf{x}\mathbf{y}^T\} = \mathbf{0}$. Since $\mathcal{E}\{\mathbf{x}^T\mathbf{y}\}$ may be written as

$$\mathcal{E}\{\mathbf{x}^T\mathbf{y}\} = \operatorname{tr}\mathcal{E}\{\mathbf{x}\mathbf{y}^T\} = \sum_{i=1}^{N} \mathcal{E}\{x_i y_i\} = \boldsymbol{\mu}_\mathbf{x}^T\boldsymbol{\mu}_\mathbf{y} \tag{4.4-55}$$

where tr{ } is the trace operator, we see that if $\mathcal{E}\{\mathbf{x}\mathbf{y}^T\} = \mathbf{0}$, then $\mathcal{E}\{\mathbf{x}^T\mathbf{y}\}$ is also zero. In fact, the condition that $\boldsymbol{\mu}_\mathbf{x}$ or $\boldsymbol{\mu}_\mathbf{y} = \mathbf{0}$ is stronger than necessary; we need only require that $\mathbf{x}$ and $\mathbf{y}$ be uncorrelated and that their means be orthogonal so that $\boldsymbol{\mu}_\mathbf{x}^T\boldsymbol{\mu}_\mathbf{y} = 0$. At the risk of being redundant, let us note again that $\mathbf{x}$ and $\mathbf{y}$ do not need to be uncorrelated for $\mathbf{x}$ and $\mathbf{y}$ to be orthogonal.

It is also possible to show that if $\mathcal{E}_\mathbf{x}\{\mathbf{x} \mid \mathbf{y}\} = \mathcal{E}\{\mathbf{x}\}$ or if $\mathcal{E}_\mathbf{y}\{\mathbf{y} \mid \mathbf{x}\} = \mathcal{E}\{\mathbf{y}\}$, then $\mathbf{x}$ and $\mathbf{y}$ are uncorrelated. Let us write $\mathcal{E}\{\mathbf{x}\mathbf{y}^T\}$ as

$$\mathcal{E}\{\mathbf{x}\mathbf{y}^T\} = \mathcal{E}_\mathbf{y}\{\mathcal{E}_\mathbf{x}\{\mathbf{x}\mathbf{y}^T \mid \mathbf{y}\}\} = \mathcal{E}_\mathbf{y}\{\mathcal{E}_\mathbf{x}\{\mathbf{x} \mid \mathbf{y}\}\mathbf{y}^T\}$$

Now if $\mathcal{E}_\mathbf{x}\{\mathbf{x} \mid \mathbf{y}\} = \mathcal{E}\{\mathbf{x}\}$, then

$$\mathcal{E}\{\mathbf{x}\mathbf{y}^T\} = \mathcal{E}\{\mathbf{x}\}\mathcal{E}\{\mathbf{y}^T\}$$

and $\mathbf{x}$ and $\mathbf{y}$ are uncorrelated. The converse of the above statement is also true, although its proof is more difficult and will be omitted.

It is quite easy to show that if $\mathbf{x}$ and $\mathbf{y}$ are independent, they are uncorrelated, although not necessarily orthogonal. We can establish this statement quite easily; we use the basic definition to write $\mathcal{E}\{\mathbf{x}\mathbf{y}^T\}$ as

$$\mathcal{E}\{\mathbf{x}\mathbf{y}^T\} = \int_{-\infty}^{\infty} d\boldsymbol{\alpha} \int_{-\infty}^{\infty} d\boldsymbol{\beta}\, \boldsymbol{\alpha}\boldsymbol{\beta}^T p_{\mathbf{xy}}(\boldsymbol{\alpha}, \boldsymbol{\beta})$$

If $\mathbf{x}$ and $\mathbf{y}$ are independent, then $p_{\mathbf{xy}}(\boldsymbol{\alpha}, \boldsymbol{\beta}) = p_{\mathbf{x}}(\boldsymbol{\alpha})p_{\mathbf{y}}(\boldsymbol{\beta})$ so that the above expression becomes

$$\begin{aligned}\mathcal{E}\{\mathbf{x}\mathbf{y}^T\} &= \int_{-\infty}^{\infty} \boldsymbol{\alpha} p_{\mathbf{x}}(\boldsymbol{\alpha})\, d\boldsymbol{\alpha} \int_{-\infty}^{\infty} \boldsymbol{\beta}^T p_{\mathbf{y}}(\boldsymbol{\beta})\, d\boldsymbol{\beta} \\ &= \mathcal{E}\{\mathbf{x}\}\mathcal{E}\{\mathbf{y}^T\}\end{aligned}$$

and we see that $\mathbf{x}$ and $\mathbf{y}$ are uncorrelated.

The reader is cautioned against concluding that the converse of the above statement is true. A pair of random variables need *not* be independent to be uncorrelated. In fact, there are very few cases in which the fact that two random variables are uncorrelated allows one to conclude that the random variables are independent. The only important exception to this statement is the gaussian distribution. If two random variables are jointly gaussian and uncorrelated, they are also independent. We defer the proof of this statement until Chapter 7.

The relationship of the properties of independence, correlation, orthogonality, and zero mean are represented graphically in Fig. 4.4-2. In

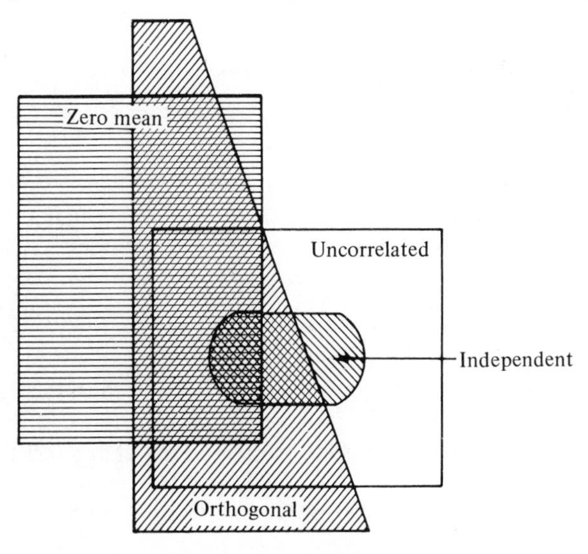

FIG. 4.4-2. Relationship of the properties of independence correlation, zero mean, and orthogonality.

Section 4.5 we shall develop another means of representing a random variable based on the concept of expectation.

EXERCISES 4.4

4.4-1. If $\mathcal{E}\{x\} = 1$ and $\mathcal{E}\{x^2\} = 4$, find the mean and variance of $y = 2x - 3$.

Answer: $\mu_y = -1$, $V_y = 12$.

4.4-2. If $z = x_1 + x_2$, where x_1 and x_2 are independent random variables uniformly distributed on [0, 1], find $\mathcal{E}\{z^k\}$ for $k = 1, 2$, and 3.

Answer: $\mathcal{E}\{z\} = 1$, $\mathcal{E}\{z^2\} = \frac{7}{6}$, $\mathcal{E}\{z^3\} = \frac{3}{2}$.

4.4-3. (a) Show that the variance of a sum of independent random variables equals the sum of the variances.

(b) Show that, for any nonrandom matrix $\mathbf{A}$, $\text{var}\{\mathbf{Ax}\} = \mathbf{A}\,\text{var}\{\mathbf{x}\}\mathbf{A}^T$.

4.4-4. Show that

$$\mathcal{E}\{g_1(x) + g_2(x)\} = \mathcal{E}\{g_1(x)\} + \mathcal{E}\{g_2(x)\}$$

Can you use this relation to show that, under appropriate circumstances,

$$\text{var}\{x_1 + x_2\} = \text{var}\{x_1\} + \text{var}\{x_2\}$$

What are the circumstances?

4.4-5. If

$$p_{x_1x_2}(\alpha_1, \alpha_2) = \frac{1}{2\pi}\exp\left(-\alpha_1^2 + \alpha_1\alpha_2 - \frac{\alpha_2^2}{2}\right)$$

find the conditional mean and variance of x_1 given x_2.

Answer: $\mu_{x_1|x_2} = \frac{1}{2}x_2$, $V_{x_1|x_2} = \frac{1}{2}$.

4.5. *Characteristic functions*

In Sections 3.3 and 3.4 we introduced the density and distribution functions as methods of describing a random variable. The method of statistical moments was developed in Section 4.4 as a third means of describing a random variable. In this section we wish to discuss yet another method of characterizing a random variable, known as the characteristic function.[1]

[1] It is also possible to define a characteristic function of $\mathbf{f}(\mathbf{x})$, a function of $\mathbf{x}$, as

$$\Phi_{\mathbf{f}(\mathbf{x})}(\boldsymbol{\omega}) = \mathcal{E}\{e^{j\boldsymbol{\omega}^T\mathbf{f}(\mathbf{x})}\}$$

However, we shall have little use for this form and shall treat the simpler case when $\mathbf{f}(\mathbf{x}) = \mathbf{x}$.

Definition 4.5-1. ***Characteristic Function:*** The characteristic function $\Phi_{\mathbf{x}}(\boldsymbol{\omega})$ for an N-dimensional random vector $\mathbf{x}$ is defined by

$$\Phi_{\mathbf{x}}(\boldsymbol{\omega}) = \mathcal{E}\{e^{j\boldsymbol{\omega}^T\mathbf{x}}\} \tag{4.5-1}$$

where $\boldsymbol{\omega}$ is a real N vector.

The characteristic function of $\mathbf{x}$ is essentially the Fourier transform of the density function $p_{\mathbf{x}}$. The reader familiar with Fourier transforms may question the existence of $\Phi_{\mathbf{x}}(\boldsymbol{\omega})$ for all random variables, since the Fourier transform does not exist for all functions. The requirements on the definition of a random variable, and hence its density function, however, ensure the existence of $\Phi_{\mathbf{x}}(\boldsymbol{\omega})$ for all random variables, which can be easily demonstrated in the following manner. By definition, $\Phi_{\mathbf{x}}(\boldsymbol{\omega})$ is given by

$$\Phi_{\mathbf{x}}(\boldsymbol{\omega}) = \int_{-\infty}^{\infty} e^{j\boldsymbol{\omega}^T\boldsymbol{\alpha}} p_{\mathbf{x}}(\boldsymbol{\alpha})\, d\boldsymbol{\alpha} \tag{4.5-2}$$

If we take the absolute value of both sides of this relation, we have

$$|\Phi_{\mathbf{x}}(\boldsymbol{\omega})| = \left| \int_{-\infty}^{\infty} e^{j\boldsymbol{\omega}^T\boldsymbol{\alpha}} p_{\mathbf{x}}(\boldsymbol{\alpha})\, d\boldsymbol{\alpha} \right| \leq \int_{-\infty}^{\infty} |e^{j\boldsymbol{\omega}^T\boldsymbol{\alpha}}||p_{\mathbf{x}}(\boldsymbol{\alpha})|\, d\boldsymbol{\alpha}$$

But since $|e^{j\boldsymbol{\omega}^T\boldsymbol{\alpha}}| = 1$ for all real $\boldsymbol{\omega}$ and $\boldsymbol{\alpha}$, and $p_{\mathbf{x}}(\boldsymbol{\alpha}) \geq 0$, we have

$$|\Phi_{\mathbf{x}}(\boldsymbol{\omega})| \leq \int_{-\infty}^{\infty} p_{\mathbf{x}}(\boldsymbol{\alpha})\, d\boldsymbol{\alpha} = 1$$

for all real $\boldsymbol{\omega}$. Hence we see that $\Phi_{\mathbf{x}}(\boldsymbol{\omega})$ exists for all real $\boldsymbol{\omega}$ for all random variables. In this way, the characteristic function has an advantage over the method of statistical moments, since the moments do not exist for all random variables. Before developing additional properties and uses of the characteristic function, let us pause briefly to consider a few examples.

Example ***4.5-1.*** Suppose that we have a random variable which represents a Bernoulli trial such that

$$p_x(\alpha) = p\delta_D(\alpha) + (1 - p)\delta_D(\alpha - 1)$$

The characteristic function for this random variable is given by

$$\Phi_x(\omega)\mathcal{E}\{e^{-j\omega x}\} = \int_{-\infty}^{\infty} e^{j\omega\alpha}[p\delta_D(\alpha) + (1 - p)\delta_D(\alpha - 1)]\, d\alpha$$

so that

$$\Phi_x(\omega) = p + (1 - p)e^{j\omega}$$

Example 4.5-2. As a second example, let us consider a random variable that has a continuous uniform distribution on $[a, b]$ so that

$$p_x(\alpha) = \frac{1}{b-a}[u(\alpha - a) - u(\alpha - b)]$$

The characteristic function is given by

$$\Phi_x(\omega) = \int_{-\infty}^{\infty} e^{j\omega\alpha} \frac{1}{b-a}[u(\alpha - a) - u(\alpha - b)]\, d\alpha$$

$$= \frac{1}{b-a}\int_a^b e^{j\omega\alpha}\, d\alpha = \frac{1}{b-a}\frac{e^{j\omega\alpha}}{j\omega}\bigg|_a^b$$

so that we obtain

$$\Phi_x(\omega) = \frac{1}{b-a}\frac{e^{j\omega b} - e^{j\omega a}}{j\omega} = \frac{2}{\omega(b-a)} e^{j\omega(b+a)/2} \sin\frac{\omega(b-a)}{2}$$

Example 4.5-3. Let us determine next the characteristic function associated with a gaussian random variable whose density function is

$$p_x(\alpha) = \frac{1}{\sqrt{2\pi}\sigma}\exp\left\{-\frac{(\alpha-\mu)^2}{2\sigma^2}\right\}$$

In this case, the characteristic function is given by

$$\Phi_x(\omega) = \frac{1}{\sqrt{2\pi}\sigma}\int_{-\infty}^{\infty} e^{j\omega\alpha}\exp\left\{-\frac{(\alpha-\mu)^2}{2\sigma^2}\right\} d\alpha$$

$$= \frac{1}{\sqrt{2\pi}\sigma}\int_{-\infty}^{\infty} \exp\left\{-\frac{[\alpha^2 - 2a(j\omega\sigma^2 + \mu) + \mu^2]}{2\sigma^2}\right\} d\alpha$$

To complete the square in the exponent, we may add and subtract $-\omega^2\sigma^4 - 2j\omega\sigma^2\mu$ so that we have

$$\Phi_x(\omega) = \exp\left\{j\omega\mu - \frac{\omega^2\sigma^2}{2}\right\}\frac{1}{\sqrt{2\pi}\sigma}\int_{-\infty}^{\infty}\exp\left\{-\frac{[\alpha + (j\omega\sigma^2 + \mu)]^2}{2\sigma^2}\right\} d\alpha$$

The integrand is a gaussian density form so that

$$\Phi_x(\omega) = e^{+j\omega\mu - \omega^2\sigma^2/2}$$

which preserves the e^{-a^2} form of the original gaussian distribution.

Example 4.5-4. In the preceding discussion we noted that the characteristic function exists for all random variables, whereas some or all of the statistical moments may not. For example, we demonstrated in

Example 4.4-6 that the Cauchy distribution does not have statistical moments. On the other hand, it is easy to show that the characteristic function for the Cauchy distribution is

$$\Phi_x(\omega) = e^{-\lambda|\omega|}$$

From the definition we have

$$\Phi_x(\omega) = \int_{-\infty}^{\infty} e^{j\omega\alpha} \frac{\lambda/\pi}{\lambda^2 + \alpha^2} \, d\alpha$$

If we use the Euler relations to write

$$e^{j\omega\alpha} = \cos \omega\alpha + j \sin \omega\alpha$$

then $\Phi_x(\omega)$ becomes

$$\Phi_x(\omega) = \int_{-\infty}^{\infty} \frac{(\lambda/\pi) \cos \omega\alpha}{\lambda^2 + \alpha^2} \, d\alpha + \int_{-\infty}^{\infty} \frac{(\lambda/\pi) \sin \omega\alpha}{\lambda^2 + \alpha^2} \, d\alpha$$

The second integral is zero since it is the integral of an odd function over a symmetric interval; the first integral is symmetric, so we have

$$\Phi_x(\omega) = \int_{-\infty}^{\infty} \frac{(2\lambda/\pi) \cos \omega\alpha}{\lambda^2 + \alpha^2} \, d\alpha$$

This integral may be evaluated by the use of complex-variable theory to give the result shown above. Hence, although the statistical moments do not exist for the Cauchy distribution, the characteristic function does.

For reference the characteristic functions for several important distributions are given in Table 4.5-1.

A careful study of the properties of the characteristic function is beyond the scope of this book. Two of these properties are easily established and are quite important. First, the density function $p_{\mathbf{x}}(\boldsymbol{\alpha})$ can be recovered from $\Phi_{\mathbf{x}}(\boldsymbol{\omega})$; and, second, all the moments (if they exist) can be obtained from $\Phi_{\mathbf{x}}(\boldsymbol{\omega})$. Let us examine these two facts in more detail.

The inversion formula for obtaining $p_{\mathbf{x}}(\boldsymbol{\alpha})$ from $\Phi_{\mathbf{x}}(\boldsymbol{\omega})$ takes the form:

$$p_{\mathbf{x}}(\boldsymbol{\alpha}) = \frac{1}{2\pi} \int_{-\infty}^{\infty} \Phi_{\mathbf{x}}(\omega) e^{-j\boldsymbol{\omega}^T\boldsymbol{\alpha}} \, d\boldsymbol{\omega} \qquad (4.5\text{-}3)$$

We can establish this result in a nonrigorous fashion by using the fact that

$$\frac{1}{2\pi} \int_{-\infty}^{\infty} e^{j\boldsymbol{\omega}^T(\boldsymbol{\xi}-\boldsymbol{\alpha})} \, d\boldsymbol{\omega} = \delta_D(\boldsymbol{\alpha} - \boldsymbol{\xi}) \qquad (4.5\text{-}4)$$

TABLE 4.5-1. Characteristic Function for Several Distributions

Distribution	Parameters[1]	Characteristic Function $\Phi_x(\omega)$
Binomial	n, p	$(1 - p + pe^{j\omega})^n$
Poisson	λt	$e^{-\lambda t} \exp\{\lambda t e^{j\omega}\}$
Negative binomial	r, p	$\left\{\frac{p}{[1 - (1 - p)e^{j\omega}]}\right\}^r$
Continuous uniform	a, b	$\frac{2}{\omega(b - a)} e^{j\omega(b+a)/2} \sin \frac{\omega(b - a)}{2}$
Normal	μ, σ^2	$e^{j\omega\mu - \omega^2\sigma^2/2}$
Gamma	λ, n	$\left(1 - \frac{j\omega}{\lambda}\right)^{-n}$
Cauchy	λ	$e^{-\lambda\lvert\omega\rvert}$

[1] See Section 3.5, especially Table 3.5-1, for definition of the parameters of the distribution.

If we substitute the definition for $\Phi_{\mathbf{x}}(\boldsymbol{\omega})$ into Eq. (4.5-3), we obtain

$$p_{\mathbf{x}}(\boldsymbol{\alpha}) = \frac{1}{2\pi} \int_{-\infty}^{\infty} e^{-j\boldsymbol{\omega}^T\boldsymbol{\alpha}} \, d\boldsymbol{\omega} \int_{-\infty}^{\infty} d\boldsymbol{\xi} \, p_{\mathbf{x}}(\boldsymbol{\xi}) e^{+j\boldsymbol{\omega}^T\boldsymbol{\xi}}$$

and if we interchange the order of integration, we have

$$\begin{aligned} p_{\mathbf{x}}(\boldsymbol{\alpha}) &= \frac{1}{2\pi} \int_{-\infty}^{\infty} d\boldsymbol{\xi} \, p_{\mathbf{x}}(\boldsymbol{\xi}) \int_{-\infty}^{\infty} e^{j\boldsymbol{\omega}^T(\boldsymbol{\xi} - \boldsymbol{\alpha})} \, d\boldsymbol{\omega} \\ &= \int_{-\infty}^{\infty} d\boldsymbol{\xi} \, p_{\mathbf{x}}(\boldsymbol{\xi}) \delta_D(\boldsymbol{\alpha} - \boldsymbol{\xi}) = p_{\mathbf{x}}(\boldsymbol{\alpha}) \end{aligned}$$

as desired. Note that the above procedure is not valid at the points where $p_{\mathbf{x}}(\boldsymbol{\alpha})$ is discontinuous; hence there may be some question about the unique determination of $p_{\mathbf{x}}(\boldsymbol{\alpha})$ from $\Phi_{\mathbf{x}}(\boldsymbol{\omega})$. It is possible to show, however, that the transformation from $p_{\mathbf{x}}$ to $\Phi_{\mathbf{x}}$ and back to $p'_{\mathbf{x}}$ is unique to the extent that $F_{\mathbf{x}}(\boldsymbol{\alpha})$ and $F'_{\mathbf{x}}(\boldsymbol{\alpha})$ are equal. In other words, any probability computations would be unaffected. In this way, $\Phi_{\mathbf{x}}$ is completely equivalent to $p_{\mathbf{x}}$ as the description of a random experiment.

One of the most useful properties of the characteristic function is that it may be used to find the moments (if they exist) for a random variable.[1] Once again we shall restrict the development to the scalar case for

[1] For this reason, the characteristic function is sometimes referred to as the moment-generating function. More properly, the moment-generating function is given by $\mathcal{E}\{e^{\boldsymbol{\omega}^T\mathbf{x}}\}$, where $\boldsymbol{\omega}$ is a real variable.

notational simplicity. Let us consider the Taylor series expansion of $e^{j\omega x}$ about $x = 0$ given by

$$e^{j\omega x} = \sum_{i=0}^{\infty} \frac{1}{i!} \frac{\partial^i e^{j\omega x}}{\partial x^i}\bigg|_{x=0} x^i$$

which becomes

$$e^{j\omega x} = \sum_{i=0}^{\infty} \frac{(j\omega x)^i}{i!} = 1 + j\omega x + \frac{j^2\omega^2 x^2}{2} + \frac{j^3\omega^3 x^3}{3!} + \cdots \tag{4.5-5}$$

Now if we take the expectation of both sides of Eq. (4.5-5), we obtain

$$\Phi_x(\omega) = \sum_{i=0}^{\infty} \frac{(j\omega)^i}{i!} m_i = 1 + j\omega m_1 + \frac{j^2\omega^2}{2} m_2 + \frac{j^3\omega^3}{3!} m_3 + \cdots \tag{4.5-6}$$

assuming that the moments of all order exist. Hence we see that we can obtain the kth moment from $\Phi_x(\omega)$ as

$$m_k = \frac{1}{j^k} \frac{\partial^k \Phi_x(\omega)}{\partial \omega^k}\bigg|_{\omega=0} \tag{4.5-7}$$

We can also develop this result by considering $\partial^k \Phi_x(\omega)/\partial\omega^k$ as

$$\frac{\partial^k \Phi_x(\omega)}{\partial \omega^k} = \frac{\partial^k}{\partial \omega^k} \int_{-\infty}^{\infty} e^{j\omega\alpha} p_x(\alpha)\, d\alpha$$

Interchanging the partial differentiation and integration operators we have

$$\frac{\partial^k \Phi_x(\omega)}{\partial \omega^k} = j^k \int_{-\infty}^{\infty} \alpha^k e^{j\omega\alpha} p_x(\alpha)\, d\alpha$$

and if we let $\omega = 0$, we obtain

$$\frac{\partial^k \Phi_x(\omega)}{\partial \omega^k}\bigg|_{\omega=0} = j^k \int_{-\infty}^{\infty} \alpha^k p_x(\alpha)\, d\alpha = j^k m_k$$

We can also use this procedure to determine the moments of x about any value ξ. If, for example, $\xi = \mu$, then we have the central moments. To obtain this result, we make the Taylor series expansion of $e^{j\omega x}$ about $x = \xi$ to obtain

$$\begin{aligned} e^{j\omega x} &= \sum_{i=0}^{\infty} \frac{1}{i!} \frac{\partial^i e^{j\omega x}}{\partial x^i}\bigg|_{x=\xi} (x-\xi)^i \\ &= e^{j\omega\xi} \sum_{i=0}^{\infty} \frac{[j\omega(x-\xi)]^i}{i!} \end{aligned}$$

Taking the expected value of both sides of the above equation gives (assuming the indicated expected values exist)

$$\Phi_x(\omega) = e^{j\omega\xi} \sum_{i=0}^{\infty} \frac{(j\omega)^i}{i!} \mathcal{E}\{(x - \xi)^i\} \tag{4.5-8}$$

From this result we see that $\mathcal{E}\{(x - \xi)^i\}$ can be expressed as

$$\mathcal{E}\{(x - \xi)^i\} = \frac{1}{j^i} \frac{\partial^i [\Phi_x(\omega) e^{-j\omega\xi}]}{\partial \omega^i} \bigg|_{\omega=0} \tag{4.5-9}$$

so that the central moments η_i are given by

$$\eta_i = \frac{1}{j^i} \frac{\partial^i [\Phi_x(\omega) e^{-j\omega\mu}]}{\partial \omega^i} \bigg|_{\omega=0} \tag{4.5-10}$$

Since we have demonstrated previously that $\Phi_x(\omega)$ uniquely determines $p_x(\alpha)$, Eq. (4.5-6) or (4.5-8) would appear to indicate that $p_{\mathbf{x}}(\alpha)$ could be determined from the statistical moments of the process. If all the moments exist and if the series of Eq. (4.5-6) or (4.5-8) converges absolutely near $\omega = 0$, then $\Phi_x(\omega)$ and hence $p_x(\alpha)$ can be determined from the moments. It was in this sense that we noted earlier that the statistical moments provide a description of a random experiment.

Example *4.5-5.* Let us use the characteristic function developed in Example 4.5-1 to find the moments for a Bernoulli trial. From Example 4.5-1 we know that the characteristic function is given by

$$\Phi_x(\omega) = p + (1 - p)e^{j\omega}$$

The first-order moment $m_1 = \mu$ is therefore

$$\mu = m_1 = \frac{1}{j} \frac{\partial \Phi_x(\omega)}{\partial \omega} \bigg|_{\omega=0} = \frac{1}{j}(1 - p) j e^{j\omega} \big|_{\omega=0} = 1 - p$$

and the second-order moment $m_2 = P$ is

$$P = m_2 = \frac{1}{j^2} \frac{\partial^2 \Phi_x(\omega)}{\partial \omega^2} \bigg|_{\omega=0} = \frac{1}{j^2}(1 - p) j^2 e^{j\omega} \big|_{\omega=0} = 1 - p$$

and, in fact,

$$m_i = \frac{1}{j^i} \frac{\partial^i \Phi_x(\omega)}{\partial \omega^i} \bigg|_{\omega=0} = 1 - p$$

From these two results we see that the variance of x is given by

$$V = \eta_2 = P - \mu^2 = (1 - p) - (1 - p)^2 = p(1 - p)$$

We can also obtain V directly by using Eq. (4.5-10) as

$$V = \eta_2 = \frac{1}{j^2} \frac{\partial^2\{[p + (1-p)e^{j\omega}]e^{-j\omega\mu}\}}{\partial^2\omega}\bigg|_{\omega=0}$$

$$= \frac{1}{j^2}[j^2(1-p)^2pe^{-j\omega\mu} + j^2(1-p)p^2e^{jp\omega}]\,|_{\omega=0}$$

$$= p(1-p)$$

If we compute the mean directly, we obtain

$$\mu = \mathcal{E}\{x\} = \int_{-\infty}^{\infty} \alpha[p\delta_D(\alpha) + (1-p)\delta_D(\alpha-1)]\,d\alpha = 1-p$$

whereas, in general, m_i is given by

$$m_i = \mathcal{E}\{x^i\} = \int_{-\infty}^{\infty} \alpha^i[p\delta_D(\alpha) + (1-p)\delta_D(\alpha-1)]\,d\alpha = 1-p$$

as before.

Example 4.5-6. The characteristic function for a gaussian process was developed in Example 4.5-3 as

$$\Phi_x(\omega) = e^{j\omega\mu - \omega^2\sigma^2/2}$$

The mean is therefore

$$\mu = \frac{1}{j}\frac{\partial\Phi_x(\omega)}{\partial\omega}\bigg|_{\omega=0} = \frac{1}{j}\frac{\partial}{\partial\omega}e^{j\omega\mu-\omega^2\sigma^2/2}\,|_{\omega=0}$$

$$= \frac{1}{j}[(j\mu - \sigma^2\omega)e^{j\omega\mu-\omega^2\sigma^2/2}]\,|_{\omega=0} = \mu$$

as expected. The second-order moment is

$$P = m_2 = \frac{1}{j^2}\frac{\partial}{\partial\omega^2}e^{j\omega\mu-\omega^2\sigma^2/2}\,|_{\omega=0}$$

$$= \frac{1}{j^2}[-\sigma^2 e^{j\omega\mu-\omega^2\sigma^2/2}$$

$$+ (j^2\mu^2 - j\mu\sigma^2\omega + \sigma^2\omega^2)e^{j\omega\mu-\omega^2\sigma^2/2}]\,|_{\omega=0}$$

so that we have

$$P = m_2 = \sigma^2 + \mu^2$$

Example 4.5-7. As a final example, let us consider the Cauchy distribution. The characteristic function, developed in Example 4.5-4, is given by

$$\Phi_x(\omega) = e^{-\lambda|\omega|}$$

We see that $\Phi_x(\omega)$ has no derivatives at $\omega = 0$, so Eq. (4.5-7) cannot be used. Hence, although the characteristic function exists, the distribution has no moments.

Another important property of the characteristic function involves the sum of two independent random variables. In Section 4.3 we showed that the density of $\mathbf{z} = \mathbf{x} + \mathbf{y}$ is given by

$$p_{\mathbf{z}}(\boldsymbol{\alpha}) = \int_{-\infty}^{\infty} p_{\mathbf{x}}(\boldsymbol{\alpha} - \boldsymbol{\beta}) p_{\mathbf{y}}(\boldsymbol{\beta})\, d\boldsymbol{\beta} \tag{4.5-11}$$

Now let us determine the characteristic function of $\mathbf{z}$. By definition, $\Phi_{\mathbf{z}}(\boldsymbol{\omega})$ is

$$\Phi_{\mathbf{z}}(\boldsymbol{\omega}) = \int_{-\infty}^{\infty} e^{j\boldsymbol{\omega}^T\boldsymbol{\alpha}} p_{\mathbf{z}}(\boldsymbol{\alpha})\, d\boldsymbol{\alpha} = \int_{-\infty}^{\infty} e^{j\boldsymbol{\omega}^T\boldsymbol{\alpha}}\, d\boldsymbol{\alpha} \int_{-\infty}^{\infty} p_{\mathbf{x}}(\boldsymbol{\alpha} - \boldsymbol{\beta}) p_{\mathbf{y}}(\boldsymbol{\beta})\, d\boldsymbol{\beta} \tag{4.5-12}$$

If we interchange the order of integration, we obtain

$$\Phi_{\mathbf{y}}(\boldsymbol{\omega}) = \int_{-\infty}^{\infty} d\boldsymbol{\beta}\, p_{\mathbf{y}}(\boldsymbol{\beta}) \int_{-\infty}^{\infty} d\boldsymbol{\alpha}\, e^{j\boldsymbol{\omega}^T\boldsymbol{\alpha}} p_{\mathbf{x}}(\boldsymbol{\alpha} - \boldsymbol{\beta})$$

and if we write $\exp[j\boldsymbol{\omega}^T\boldsymbol{\alpha}]$ as $\exp[j\boldsymbol{\omega}^T(\boldsymbol{\alpha} - \boldsymbol{\beta})] \exp[j\boldsymbol{\omega}^T\boldsymbol{\beta}]$, we have

$$\Phi_{\mathbf{z}}(\boldsymbol{\omega}) = \int_{-\infty}^{\infty} d\boldsymbol{\beta}\, p_{\mathbf{y}}(\boldsymbol{\beta}) e^{j\boldsymbol{\omega}^T\boldsymbol{\beta}} \int_{-\infty}^{\infty} d\boldsymbol{\alpha}\, e^{j\boldsymbol{\omega}^T(\boldsymbol{\alpha}-\boldsymbol{\beta})} p_{\mathbf{x}}(\boldsymbol{\alpha} - \boldsymbol{\beta})$$

Finally, we make the substitution $\boldsymbol{\gamma} = \boldsymbol{\alpha} - \boldsymbol{\beta}$ so that $\Phi_{\mathbf{z}}(\boldsymbol{\omega})$ becomes

$$\Phi_{\mathbf{z}}(\boldsymbol{\omega}) = \int_{-\infty}^{\infty} d\boldsymbol{\beta}\, p_{\mathbf{y}}(\boldsymbol{\beta}) e^{j\boldsymbol{\omega}^T\boldsymbol{\beta}} \int_{-\infty}^{\infty} d\boldsymbol{\gamma}\, e^{j\boldsymbol{\omega}^T\boldsymbol{\gamma}} p_{\mathbf{x}}(\boldsymbol{\gamma})$$

which equals

$$\Phi_{\mathbf{z}}(\boldsymbol{\omega}) = \Phi_{\mathbf{y}}(\boldsymbol{\omega}) \Phi_{\mathbf{x}}(\boldsymbol{\omega}) \tag{4.5-13}$$

Hence the characteristic function for the sum of two independent random variables is the product of their characteristic functions. This property is useful in the study of limit theorems; we shall make use of it when we develop the central limit theorem in Chapter 7.

Once the characteristic function for an N-dimensional random vector is known, it is simple to obtain the characteristic function for a subset of the components of $\mathbf{x}$. Suppose, for example, that $\mathbf{x}$ is partitioned into two parts, $\mathbf{x}_1$ and $\mathbf{x}_2$, such that

$$\mathbf{x} = \begin{bmatrix} \mathbf{x}_1 \\ --- \\ \mathbf{x}_2 \end{bmatrix}$$

In terms of this partition, the characteristic function of $\mathbf{x}$ can be written as

$$\Phi_{\mathbf{x}}(\boldsymbol{\omega}) = \Phi_{\mathbf{x}_1\mathbf{x}_2}(\boldsymbol{\omega}_1, \boldsymbol{\omega}_2) = \mathcal{E}\{\exp(j\boldsymbol{\omega}_1^T\mathbf{x}_1 + j\boldsymbol{\omega}_2^T\mathbf{x}_2)\} \tag{4.5-14}$$

Now if we let $\boldsymbol{\omega}_2 = \mathbf{0}$, then we have

$$\Phi_{\mathbf{x}_1\mathbf{x}_2}(\boldsymbol{\omega}_1, \mathbf{0}) = \mathcal{E}\{\exp(j\boldsymbol{\omega}_1^T\mathbf{x}_1)\} = \Phi_{\mathbf{x}_1}(\boldsymbol{\omega}_1) \tag{4.5-15}$$

Hence we can obtain the marginal characteristic function for any subset of the components of $\mathbf{x}$ from the joint characteristic function $\Phi_{\mathbf{x}}(\boldsymbol{\omega})$ by simply setting the components of $\boldsymbol{\omega}$ associated with the other components of $\mathbf{x}$ to be equal to zero. If, for example, $\mathbf{x}$ were a four-dimensional vector and we want the characteristic function associated with x_1 and x_4, we would have

$$\Phi_{x_1x_4}(\omega_1, \omega_4) = \Phi_{x_1x_2x_3x_4}(\omega_1, 0, 0, \omega_4)$$

It is also possible to define the conditional characteristic function of $\mathbf{x}$ given $\mathbf{y} = \boldsymbol{\alpha}$.

Definition 4.5-2. ***Conditional Characteristic Function:*** The conditional characteristic function $\Phi_{\mathbf{x}|\mathbf{y}=\boldsymbol{\alpha}}(\boldsymbol{\omega})$ for an N-dimensional random vector $\mathbf{x}$, given $\mathbf{y} = \boldsymbol{\alpha}$, is defined as

$$\begin{aligned}\Phi_{\mathbf{x}|\mathbf{y}=\boldsymbol{\alpha}}(\boldsymbol{\omega}) &= \int_{-\infty}^{\infty} e^{j\boldsymbol{\omega}^T\boldsymbol{\beta}} p_{\mathbf{x}|\mathbf{y}}(\boldsymbol{\beta}\,|\,\boldsymbol{\alpha})\, d\boldsymbol{\beta} \\ &= \mathcal{E}\{e^{j\boldsymbol{\omega}^T\boldsymbol{\beta}}\,|\,\mathbf{y} = \boldsymbol{\alpha}\}\end{aligned}$$

where $\boldsymbol{\omega}$ is a real N vector.

The conditional characteristic function can be used to determine conditional moments.

EXERCISES 4.5

4.5-1. If x and y are independent random variables uniformly distributed on $[0, 1]$, find (a) $\Phi_x(\omega)$, (b) $\Phi_{xy}(\omega_1, \omega_2)$, (c) $\mathcal{E}\{xy\}$, (d) $\mathcal{E}\{x\}$, and (e) $\mathcal{E}\{x^2\}$.

Answers: (a) $\Phi_x(\omega) = (e^{j\omega} - 1)/j\omega$.
(b) $\Phi_{xy}(\omega_1, \omega_2) = -(e^{j\omega_1} - 1)(e^{j\omega_2} - 2)/\omega_1\omega_2$.
(c) $\mathcal{E}\{xy\} = \frac{1}{4}$.
(d) $\mathcal{E}\{x\} = \frac{1}{2}$.
(e) $\mathcal{E}\{x^2\} = \frac{1}{3}$.

4.5-2. If x_1 and x_2 are independent random variables and $y = a_1x_2 + a_2x_2$, where a_1 and a_2 are nonrandom constants, find $\Phi_y(\omega)$ in terms of Φ_{x_1} and Φ_{x_2}.

Answer: $\Phi_y(\omega) = \Phi_{x_2}(a_1\omega)\Phi_{x_2}(a_2\omega)$.

4.6. Sequences of random variables and convergence

In this section we wish to examine briefly the important concept of convergence of random sequences. This material is necessary to put the definition of such concepts as continuity, differentiation, and integration of stochastic processes on a rigorous foundation. In addition, the concept of a random sequence serves as a convenient stepping stone from the random-variable theory of this and the preceding chapter to the stochastic-process theory of the following chapters. Most of the results of this section will be offered without proof, since meaningful proofs, in general, require at least a modest knowledge of measure theory and are hence beyond the scope of this treatment.

From Chapter 2 we recall that a random variable $\mathbf{x}$ is basically a function (with mild restrictions) which assigns to each experimental outcome ξ a number $\mathbf{x}(\xi)$. A random sequence, on the other hand, assigns to each outcome ξ an infinite sequence of numbers $\mathbf{x}_1(\xi), \mathbf{x}_2(\xi), \ldots, \mathbf{x}_i(\xi), \ldots$. We shall use the symbolism $\{\mathbf{x}_i\}$ to represent a random sequence; $\{\mathbf{x}_i(\xi)\}$ will represent a realization of the random sequence for a specific outcome ξ. Here each element $\mathbf{x}_j$ of the sequence $\{\mathbf{x}_i\}$ is assumed to be a random variable. In other words, if we specify ξ, then $\{\mathbf{x}_i\}$ is a sequence of real numbers; if we specify i, then $\{\mathbf{x}_i\}$ is a random variable $\mathbf{x}_i$; and if both ξ and i are specified, then $\mathbf{x}_i(\xi)$ is a simple number.

Suppose now that we have a random variable $\mathbf{x}$. We wish to consider what we mean when we say that $\{\mathbf{x}_i\}$ converges to $\mathbf{x}$. Before we attempt to answer this question, let us briefly review the concept of convergence of a nonrandom sequence of numbers $\{\boldsymbol{\alpha}_i\}$, obtained perhaps by evaluating $\{\mathbf{x}_i\}$ for a specific ξ.

We say that a sequence $\{\boldsymbol{\alpha}_i\}$ converges[1] to a vector $\boldsymbol{\alpha}$ or

$$\lim_{n\to\infty} \boldsymbol{\alpha}_n = \boldsymbol{\alpha} \tag{4.6-1}$$

if for every $\epsilon > 0$, there exists an n_0 such that[2]

$$|\boldsymbol{\alpha}_n - \boldsymbol{\alpha}| < \epsilon \qquad \text{for all } n > n_0 \tag{4.6-2}$$

To use this definition, it is necessary that the limit $\boldsymbol{\alpha}$ be known. We may avoid this difficulty by making use of the fact that a sequence converges if and only if it is a Cauchy sequence. A sequence is said to be a Cauchy sequence if for every $\epsilon > 0$, there exists an n_0 such that

$$|\boldsymbol{\alpha}_n - \boldsymbol{\alpha}_m| < \epsilon \qquad \text{for all } n, m > n_0 \tag{4.6-3}$$

Let us return now to a consideration of the convergence properties of random sequences. If $\{\mathbf{x}_n(\xi)\}$ converges to $\mathbf{x}(\xi)$ for all $\xi \in \mathcal{S}$, we say that $\{\mathbf{x}_n\}$ *converges everywhere* to $\mathbf{x}$.

Definition 4.6-1. ***Convergence Everywhere:*** If for every $\xi \in \mathcal{S}$, $\{\mathbf{x}_n(\xi)\}$ converges to $\mathbf{x}(\xi)$, that is,

$$\lim_{n\to\infty} \mathbf{x}_n(\xi) = \mathbf{x}(\xi) \qquad \text{for every } \xi \in \mathcal{S}$$

then we say that $\{\mathbf{x}_n\}$ converges to $\mathbf{x}$ everywhere.

This form of convergence, which represents the ultimate in the convergence of a random sequence, is almost always too strong for practical use. There are many random sequences that do not converge everywhere but yet possess sufficient convergence properties to make them of practical use. Consider for example, the sequence $\{\mathbf{x}_n\}$ defined by

$$x_i = \left(\frac{-1}{y}\right)^i \tag{4.6-4}$$

where the random variable y is uniformly distributed on $[1, 2]$. For all outcomes ξ such that $y(\xi) > 1$, we see that $\{\mathbf{x}_n\}$ converges to zero. If, however,

[1] The convergence of a sequence $\{\boldsymbol{\alpha}_i\}$ should not be confused with the convergence of the series $\{\boldsymbol{\beta}_i\}$ given by

$$\boldsymbol{\beta}_i = \sum_{j=1}^{i} \boldsymbol{\alpha}_i$$

[2] The reader should remember that the notation $|\boldsymbol{\alpha}_n - \boldsymbol{\alpha}| < \epsilon$ implies that the magnitude of each element of $(\boldsymbol{\alpha}_n - \boldsymbol{\alpha})$ is less than ϵ; that is, $|(\boldsymbol{\alpha}_n - \boldsymbol{\alpha})_i| < \epsilon$ for $i = 1, 2, \ldots, N$.

y is equal to 1, then $\{\mathbf{x}_n\}$ does not converge. Hence for all outcomes ξ such that $y(\xi) = 1$, the sequence $\{\mathbf{x}_n\}$ does not converge and hence $\{x_n\}$ does not converge everywhere. Note, however, that $\mathcal{P}\{y = 1\} = 0$, so the probability that the sequence will converge is 1; sequences of this type are said to converge *almost everywhere* (a.e.) or *with probability 1* (w.p.1).

Definition 4.6-2. ***Convergence with Probability 1:*** If the set of outcomes ξ such that

$$\lim_{n\to\infty} \mathbf{x}_n(\xi) = \mathbf{x}(\xi)$$

has probability 1, then $\{\mathbf{x}_n\}$ is said to converge to $\mathbf{x}$ almost everywhere or with probability 1. We shall use the following notation to indicate convergence with probability 1:

$$\lim_{n\to\infty} \mathbf{x}_n = \mathbf{x} \qquad \text{a.e. or w.p.1}$$

We can also write the above property in the following form:

$$\mathcal{P}\{\xi\colon \lim_{n\to\infty} \mathbf{x}_n(\xi) = \mathbf{x}(\xi)\} = 1$$

Another form of convergence often used is convergence in the mean-square sense, which is defined in the following manner:

Definition 4.6-3. ***Mean-Square Convergence:*** A random sequence $\{\mathbf{x}_n\}$ converges to a random variable $\mathbf{x}$ in the mean-square sense if

$$\lim_{n\to\infty} \mathcal{E}\{\|\mathbf{x}_n - \mathbf{x}\|^2\} = 0$$

We shall use the following notation to indicate mean-square convergence:

$$\lim_{n\to\infty} \mathbf{x}_n = \mathbf{x} \qquad \text{m.s.}$$

or more commonly

$$\operatorname*{l.i.m.}_{n\to\infty} \mathbf{x}_n = \mathbf{x}$$

which is read as "limit in the mean."

Convergence with probability 1 and mean-square convergence present two distinctly different approaches to stochastic convergence. For convergence with probability 1, we require that almost every sequence converge but do not consider the combined behavior of all the sequences. Mean-square convergence, on the other hand, is concerned solely with the aggregate behavior of the entire collection of sequences and not with individual sequences, all of which may diverge! Hence it is completely possible that a random sequence will converge with probability 1 but not in the mean-square sense, or will be mean square convergent but will not converge with probability 1. Of course, there are some sequences, such as the one defined by Eq. (4.6-4), that converge with probability 1 and in the mean-square sense. (See Exercise 4.6-3.)

It is relatively easy to construct a random sequence that converges with probability 1 but does not converge in the mean-square sense. We need only define the random sequence so that almost every sequence converges but the average behavior does not converge. Suppose, for example, that we have an experiment whose outcomes are the positive integers; that is, $\xi_i = i$. Each outcome will be defined as an event with the probability measure $\mathcal{P}$ defined by

$$\mathcal{P}\{\xi_i\} = \frac{6/\pi^2}{i^2} \tag{4.6-5}$$

We shall define a random sequence in the following manner:

$$x_n(\xi_i) = \begin{cases} i & \text{for } n = i \\ 0 & \text{otherwise} \end{cases} \tag{4.6-6}$$

It is easy to see that every sequence converges to zero so that the random sequence converges to zero with probability 1. This sequence, however, is not mean-square convergent, since

$$\mathcal{E}\{(x_n - 0)^2\} = \sum_{i=1}^{\infty} x_n^2(\xi_i)\frac{6/\pi^2}{i^2}$$

However, only $x_n(\xi_n)$ is nonzero, so we have

$$\mathcal{E}\{(x_n - 0)^2\} = n^2\frac{6/\pi^2}{n^2} = \frac{6}{\pi^2}$$

so that

$$\lim_{n\to\infty} \mathcal{E}\{(x_n - 0)^2\} = \frac{6}{\pi^2} \neq 0$$

This sequence does not converge in the mean-square sense because the mean-square value of each element of the random sequence is constant. The single nonzero element in each sequence, even though it becomes increasingly large, does not keep the sequences from converging.

To construct a random sequence that converges in the mean-square sense but not with probability 1, we need only define a random sequence in which each sequence fails to converge but yet the mean-square value converges. Let us suppose that the experimental outcome ξ is a number chosen at random from the unit interval so that

$$\mathcal{P}\{0 \leq \xi \leq \alpha\} = \begin{cases} \alpha & \text{for } 0 \leq \alpha \leq 1 \\ 0 & \text{otherwise} \end{cases}$$

Now we define a random sequence as

$$x_n(\xi) = \begin{cases} 1 & \text{for } \dfrac{\sum_{i=0}^{k} i - n + 1}{k} < \xi \leq \dfrac{\sum_{i=0}^{k} - n}{k} \\ 0 & \text{otherwise} \end{cases} \tag{4.6-7}$$

where k is defined such that $\sum_{i=0}^{k} i < n \leq \sum_{i=0}^{k} i$. None of the sequences converge because there is an infinite number of nonzero values. However, because the density of nonzero values becomes smaller and smaller as n increases, the $\lim_{n \to \infty} \mathcal{E}\{x_n^2\} = 0$. Hence this random sequence is mean-square convergent but not convergent with probability 1.

The two preceding types of stochastic convergence are often stronger than needed for many purposes and, in addition, are often quite difficult to establish. For this reason, two other types of stochastic convergence are often used. We shall present their definitions and then discuss their relation to the two preceding definitions.

Definition 4.6-4. ***Convergence in Probability:*** A random sequence $\{\mathbf{x}_n\}$ converges in probability to a random variable $\mathbf{x}$ if for every $\epsilon > 0$

$$\lim_{n \to \infty} \mathcal{P}\{|\mathbf{x}_n - \mathbf{x}| \geq \epsilon\} = 0$$

We shall use the following notation to indicate convergence in probability:

$$\lim_{n \to \infty} \mathbf{x}_n = \mathbf{x} \qquad \text{in } \mathcal{P}$$

Definition 4.6-5. ***Convergence in Distribution:*** A random sequence $\{\mathbf{x}_n\}$ converges in distribution to a random variable $\mathbf{x}$ if at every point for which $F_{\mathbf{x}}(\boldsymbol{\alpha})$ is continuous

$$\lim_{n \to \infty} F_{\mathbf{x}_n}(\boldsymbol{\alpha}) = F_{\mathbf{x}}(\boldsymbol{\alpha})$$

The following notation will be used to indicate convergence in distribution

$$\lim_{n\to\infty} \mathbf{x}_n = \mathbf{x} \qquad \text{in D}$$

It is possible to show that a random sequence $\{\mathbf{x}_n\}$ converges in distribution to a random varaible $\mathbf{x}$ if and only if

$$\lim_{n\to\infty} \Phi_{\mathbf{x}_n}(\omega) = \Phi_{\mathbf{x}}(\omega) \tag{4.6-8}$$

The advantage of this result, which is stated without proof, is the replacement of convergence at continuity points of $F_{\mathbf{x}}$ as required by Definition 4.6-5 with ordinary pointwise convergence.

It is instructive to examine the relationship between the four types of convergences given by Definitions 4.6-2 to 4.6-4. We consider first the relationship of convergence with probability 1 and convergence in probability. A comparison of Definitions 4.6-2 and 4.6-4 reveals that the definitions are similar except for an interchange of the limit and probability operations. Definition 4.6-2 requires that almost all sequences, except for a set of vanishingly small probability, converge. Definition 4.6-4, on the other hand, requires only that the probability that $|\mathbf{x}_n - \mathbf{x}|$ is greater than or equal to any $\epsilon > 0$ becomes vanishingly small; this does *not* require that the deterministic sequence $\{\mathbf{x}_n(\xi)\}$ converge for any $\xi \in \mathcal{S}$. Consider, for example, the sequence given by Eq. (4.6-7). This sequence converges in probability to zero, even though none of the sequences $\{\mathbf{x}_n(\xi)\}$ converge.

To show that every random sequence which converges with probability 1 also converges in probability, it is convenient to restate the definition of convergence with probability 1. If $\lim_{n\to\infty} \mathbf{x}_n(\xi) = \mathbf{x}(\xi)$, then for every $\epsilon > 0$ there must exist an n_0 such that

$$|\mathbf{x}_n(\xi) - \mathbf{x}(\xi)| < \epsilon \qquad \text{for } n \geq n_0 \tag{4.6-9}$$

The set of outcomes $\xi \in \mathcal{S}$ for which Eq. (4.6-9) is satisfied can be written as

$$\mathcal{A} = \bigcap_{i=n_0}^{\infty} \{\xi : |\mathbf{x}_i(\xi) - \mathbf{x}(\xi)| < \epsilon\} \tag{4.6-10}$$

which is the intersection of the set of outcomes ξ, such that $|\mathbf{x}_{n_0}(\xi) - \mathbf{x}(\xi)| < \epsilon$ with the set of outcomes ξ, such that $|\mathbf{x}_{n_0+1}(\xi) - \mathbf{x}(\xi)| < \epsilon$, and so forth. If $\{\mathbf{x}_n\}$ converges with probability 1, then for every $\epsilon > 0$ and $\eta > 0$, there must exist an n_0 such that

$$\mathcal{P}\left\{\bigcap_{i=n_0}^{\infty} \{\xi : |\mathbf{x}_i(\xi) - \mathbf{x}(\xi)| < \epsilon\}\right\} \geq 1 - \eta \tag{4.6-11}$$

In a similar fashion, we can state that if a sequence $\{\mathbf{x}_n\}$ converges in probability, then for every $\epsilon > 0$ and $\eta > 0$ there must exist an n_0 such that

$$\mathcal{P}\{\xi : |\mathbf{x}_n(\xi) - \mathbf{x}(\xi)| < \epsilon\} \geq 1 - \eta \qquad \text{for } n > n_0 \tag{4.6-12}$$

In other words, convergence in probability requires that the set of probability statements

$$\mathcal{P}\{\xi : |\mathbf{x}_{n_0}(\xi) - \mathbf{x}(\xi)| < \epsilon\} \geq 1 - \eta$$
$$\mathcal{P}\{\xi : |\mathbf{x}_{n_0+1}(\xi) - \mathbf{x}(\xi)| < \epsilon\} \geq 1 - \eta$$

and so forth be satisfied. This is quite different from the single probability statement of Eq. (4.6-11).

Now for any $n \geq n_0$, we know that

$$\{\xi : |\mathbf{x}_n(\xi) - \mathbf{x}(\xi)| < \epsilon\} \supset \bigcap_{i=n_0}^{\infty} \{\xi : |\mathbf{x}_n(\xi) - \mathbf{x}(\xi)| < \epsilon\}$$

so that

$$\mathcal{P}\{\xi : \mathbf{x}_n(\xi) - \mathbf{x}(\xi)| < \epsilon\} \geq \mathcal{P} \bigcap_{i=n_0}^{\infty} \{\xi : |\mathbf{x}_n(\xi) - \mathbf{x}(\xi)| < \epsilon\}$$

Hence we see that if Eq. (4.6-11) is satisfied, Eq. (4.6-12) is surely satisfied. Therefore, if a random sequence $\{\mathbf{x}_n\}$ converges with probability 1 to $\mathbf{x}$, it also converges in probability to $\mathbf{x}$. We have shown above, by an example, that the converse of this statement is *not* true. A random sequence may converge in probability but *not* converge with probability 1.

The fact that mean-square convergence implies convergence in probability may be easily established by the use of the Tchebycheff inequality of Section 4.4. By the use of this result, we may write

$$\mathcal{P}\{\xi : |\mathbf{x}_n(\xi) - \mathbf{x}(\xi)| < \epsilon\} > 1 - \frac{\mathcal{E}\{\|\mathbf{x}_n - x\|^2\}}{\epsilon^2} \tag{4.6-13}$$

If $\{\mathbf{x}_n\}$ converges in the mean-square sense, then for any $\delta > 0$ there exists an n_0 such that

$$\mathcal{E}\{\|\mathbf{x}_n - \mathbf{x}\|^2\} < \delta \qquad \text{for } n \geq n_0$$

and therefore

$$\mathcal{P}\{\xi : |\mathbf{x}_n(\xi) - \mathbf{x}(\xi)| < \epsilon\} > 1 - \frac{\delta}{\epsilon^2}$$

Hence Eq. (4.6-12) can be satisfied for any $\epsilon > 0$, $\eta > 0$, if $\delta = \eta\epsilon^2$, and $\{\mathbf{x}_n\}$ must converge in probability. Therefore, if a random sequence $\{\mathbf{x}_n\}$ converges in the mean-square sense to $\mathbf{x}$, it also converges in probability to $\mathbf{x}$.

It is easy to show that the converse of the above is not true. Consider, for example, the random sequence given by Eqs. (4.6-4) and (4.6-6). This sequence converges with probability 1 and hence also converges in probability; however, we have previously argued that this sequence is not mean square convergent. Hence we see that a sequence may converge in probability but *not* converge in the mean-square sense.

It is also possible to construct a random sequence that converges in probability but does not converge either in the mean-square sense or with probability 1. Let us use the random sequence defined by Eq. (4.6-7) again, but now let $\{\mathbf{x}_n(\xi_i)\}$ be defined by

$$x_n(\xi) = \begin{cases} i & \text{for } \dfrac{\sum_{i=1}^{k} i - n + 1}{k} < \xi \leq \dfrac{\sum_{i=0}^{k} - n}{k} \\ 0 & \text{otherwise} \end{cases} \tag{4.6-14}$$

It is reasonably easy to show that this random sequence does not converge in the mean-square sense or with probability 1 but does converge in probability.

Let us turn our attention now to convergence in distribution. We state without proof the fact that if $\{\mathbf{x}_n\}$ converges in probability to $\mathbf{x}$, then $\{\mathbf{x}_n\}$ also converges in distribution to $\mathbf{x}$. The converse of this statement is not true; that is, a random sequence may converge in distribution but not converge in probability. Let $x(\xi)$ have a normal distribution with zero mean and unit variance, and define $\{x_n\}$ by

$$x_n(\xi) = \begin{cases} x(\xi) & \text{for } n = 1, 3, 5, \ldots \\ -x(\xi) & \text{for } n = 2, 4, 6, \ldots \end{cases}$$

Now x_n is normal with zero mean and unit variance for every n so that $\{x_n\}$ converges in distribution to x. However, for $n = 2, 4, 6, \ldots$, we have

$$|x_n(\xi) - x(\xi)| = 2|x(\xi)|$$

so that

$$\mathcal{P}\{\xi : |x_n(\xi) - x(\xi)| \geq 1\} = \mathcal{P}\{\xi : |x(\xi)| \geq \tfrac{1}{2}\} \sim 0.617$$

Thus if we choose $\epsilon = 1$ and $\eta < 0.617$, Eq. (4.6-12) cannot hold.

Although convergence in distribution does not in general imply convergence in probability, there is a special case in which the statement is valid. A random sequence $\{\mathbf{x}_n\}$ converges in probability to the *constant* limit random variable $\mathbf{x} \equiv \mathbf{c}$ if and only if $\{\mathbf{x}_n\}$ converges in distribution to $\mathbf{x}$.

The preceding discussions concerning the relationship between the various modes of stochastic convergence are summarized graphically in Fig.

4.6-1. All the modes of stochastic convergence can also be stated in the form of Cauchy criteria. For convergence with probability 1, we require that the set of outcomes ξ such that

$$\lim_{m,n\to\infty} [\mathbf{x}_n(\xi) - \mathbf{x}_m(\xi)] = \mathbf{0} \tag{4.6-15}$$

has probability 1. Mean-square convergence takes the form

$$\lim_{n,m\to\infty} \mathcal{E}\{\|\mathbf{x}_n - \mathbf{x}_m\|^2\} = 0 \tag{4.6-16}$$

whereas convergence in probability becomes

$$\lim_{n,m\to\infty} \mathcal{P}\{|\mathbf{x}_n - \mathbf{x}_m| \geq \epsilon\} = 0 \tag{4.6-17}$$

Convergence in distribution can be stated as

$$\lim_{n,m\to\infty} [F_{\mathbf{x}_n}(\boldsymbol{\alpha}) - F_{\mathbf{x}_m}(\boldsymbol{\alpha})] = 0 \tag{4.6-18}$$

These expressions allow one to test for the convergence of a random sequence without knowledge of the limit random variable.

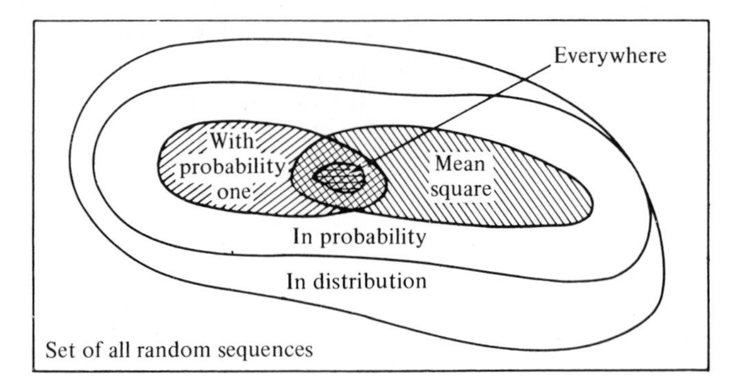

FIG. 4.6-1. Relationship of the various modes of stochastic convergence.

EXERCISES 4.6

4.6-1. A sequence of random variables $\{x_i\}$ converges in the mean-square sense to random variable x; if each x_i has a gaussian distribution, show that x has a gaussian distribution.

4.6-2. Show that the random sequence of Eq. (4.6-14) does not converge in the mean-square sense or with probability 1, but does converge in probability.

4.6-3. Show that the sequence given by Eq. (4.6-4) converges in the mean-square sense and with probability 1.

4.7. Conclusions

In this chapter we examined a number of extensions of the basic random variable theory introduced in Chapter 3 and loosely referred to as functions of random variables. The chapter began with a discussion of algebraic functions of one and then many random variables. Next we examined another type of function, which was really a functional of the distribution, known as the expectation. The study of expectation led to the definition of the characteristic function as another means of describing a random experiment. The final topic in the chapter dealt with the convergence of random sequences. The study of random sequences serves as a transitional device to the study of stochastic process, which will be considered in the remainder of the book.

4.8. Problems

4.8-1. If

$$y = |x|$$

find $p_y(\beta)$ in terms of a given $p_x(\alpha)$. What are the results if

(a) $p_x(\alpha) = \frac{1}{2} \quad |\alpha| \leq 1$?

(b) $p_x(\alpha) = \dfrac{1}{(2\pi)^{1/2}} e^{-\alpha^2/2}$?

(c) $p_x(\alpha) = \frac{1}{3} \quad -1 \leq \alpha \leq 2$?

4.8-2. If

$$y = \begin{cases} x - x^2 & 0 \leq x \leq 1 \\ 0 & \text{otherwise} \end{cases}$$

find $p_y(\beta)$ in terms of a given $p_x(\alpha)$. What are the results if

(a) $p_x(\alpha) = 1 \quad 0 \leq \alpha \leq 1$?

(b) $p_x(\alpha) = \frac{1}{3} \quad -1 \leq \alpha \leq 2$?

4.8-3. If $y = x^2$ and x is gaussian with mean zero, find $p_y(\beta)$.

4.8-4. If

$$y = A \sin \theta$$

where A is a known positive constant and θ is a random variable uniformly distributed in the interval $-\pi/2$ to $\pi/2$, find $p_y(\beta)$.

4.8-5. Show that if the transformation

$$r = y_1 = (x_1^2 + x_2^2)^{1/2}$$

$$\theta = y_2 = \tan^{-1}\left(\frac{x_2}{x_1}\right)$$

is applied to the random vector x, the resulting probability density of y is

$$p_y(\beta) = \beta_1 p_{x_1x_2}(\beta_1 \cos \beta_2, \beta_1 \sin \beta_2)$$

for $\beta_1 \geq 0$, $0 \leq \beta_2 \leq 2\pi$, and $p_y(\beta) = 0$, otherwise.

4.8-6. If

$$y = x_1 - x_2$$

find $p_y(\beta)$ in terms of $p_{x_1x_2}(\alpha_1, \alpha_2)$.

4.8-7. Find the mean value, mean-square value, and variance of

(a) $p_x(\alpha) = \frac{1}{2}e^{-|\alpha|}$

(b) $p_x(\alpha) = \begin{cases} e^{-\alpha} & \alpha \geq 0 \\ 0 & \alpha < 0 \end{cases}$

4.8-8. A joint density function is

$$p_{x_1x_2}(\alpha_1, \alpha_2) = A(\alpha_1^2 + \alpha_2^2) \qquad 0 \leq \alpha_1, \alpha_2 \leq 1$$

What are A, μ_{x_1}, $V_{x_1x_2}$, μ_{x_2}, and V_{x_2}?

4.8-9. For Problem 4.8-8, what is

$$p_{x_1}(\alpha_1) \qquad p_{x_1|x_2}(\alpha_1 \mid \alpha_2)$$

$$V_{x_1|x_2}(\beta_1 \mid \beta_2) \qquad V_{x_1|x_2}(\beta_1 \mid \beta_2)$$

4.8-10. Two independent continuous random variables x_1 and x_2 are uniformly distributed in the interval $[-a, a]$. If $z = x_1 + x_2$, find the density functions $p_z(\gamma)$, $p_{z|x_1}(\gamma \mid \alpha_1)$, and the moments μ_z, V_z, $\mu_{z|x_1}(\alpha_1)$, and $V_{z|x_1}(\alpha_1)$.

4.8-11. If

$$z = \frac{1}{N}\sum_{i=1}^{N} x_i$$

where the x_i are independent and uniformly distributed over the interval $-a$ to $+a$, find the density function $p_z(\gamma)$ and moments μ_z and V_z for $N = 1, 2, 3, 4$. Sketch the density functions and compare with an appropriately chosen gaussian density function curve.

4.8-12. If

$$p_{x_1x_2}(\alpha_1, \alpha_2) = \begin{cases} 4\alpha_1(1 - \alpha_2) & 0 \leq \alpha_1, \alpha_2 \leq 1 \\ 0 & \text{otherwise} \end{cases}$$

find the probability density and distribution function of x_1 given that $x_2 \leq \frac{1}{2}$.

4.8-13. Find the mean and variance of a random variable x with a gamma probability density

$$p_x(\alpha) = \frac{b^a}{\Gamma(a)}\alpha^{a-1}e^{-b\alpha} \qquad \alpha > 0$$

in which a and b are positive constants and Γ is the gamma function

$$\Gamma(n) = \int_0^\infty \eta^n e^{-\eta}\, d\eta$$

$$= (n - 1)! \qquad \text{for } n \text{ an integer}$$

4.8-14. For the conditional gamma density

$$p_{x|y}(\alpha \mid \beta) = \frac{4}{\beta^2}\alpha e^{-\alpha/\beta} \qquad \alpha > 0$$

assume that $\mathcal{E}\{x \mid y\} = \beta$ is a random variable, with an inverted gamma distribution such that $1/\beta$ satisfies (where $z = 1/\beta$)

$$p_z(\gamma) = \frac{b^a}{\Gamma(a)}\gamma^{a-1}e^{-b\gamma}$$

Find the (marginal) density $p_x(\alpha)$ and its mean and variance.

4.8-15. Show that

(a) $\mathcal{E}\{\mathbf{g}(\mathbf{x}) \mid \mathbf{x} = \boldsymbol{\alpha}\} = \mathbf{g}(\boldsymbol{\alpha})$.

(b) $\mathcal{E}\{\mathbf{A}\mathbf{x} \mid \mathbf{y} = \boldsymbol{\beta}\} = \mathbf{A}\mathcal{E}\{\mathbf{x} \mid \mathbf{y} = \boldsymbol{\beta}\}$.

(c) $\mathcal{E}\{\mathbf{G}(\mathbf{y})\mathbf{x} \mid \mathbf{y} = \boldsymbol{\beta}\} = \mathbf{G}(\boldsymbol{\beta})\mathcal{E}\{\mathbf{x} \mid \mathbf{y} = \boldsymbol{\beta}\}$.

(d) $\mathcal{E}\{\mathbf{x} + \mathbf{y} \mid \mathbf{z}\} = \mathcal{E}\{\mathbf{x} \mid \mathbf{z}\} + \mathcal{E}\{\mathbf{y} \mid \mathbf{z}\}$.

4.8-16. The random variables x_1 and x_2 are independent with probability densities

$$p_{x_1}(\alpha_1) = \begin{cases} \dfrac{1}{\pi}\dfrac{1}{(1-\alpha_1^2)^{1/2}} & |a_1| \le 1 \\ 0 & |\alpha_1| > 1 \end{cases}$$

$$p_{x_2}(\alpha_2) = \begin{cases} \dfrac{\alpha_2}{\sigma^2}\exp\left(-\dfrac{\alpha_2^2}{2\sigma^2}\right) & \alpha_2 \ge 0 \\ 0 & \text{otherwise} \end{cases}$$

Show that

$$z = x_1 x_2$$

is gaussian with zero mean and variance $V_z = \sigma^2$.

4.8-17. Determine the mean and variance for the following probability density functions and event probabilities:

(a) $p_x(\alpha) = \begin{cases} |\alpha| & |\alpha| \le 1 \\ 0 & \text{otherwise} \end{cases}$

(b) $p_x(\alpha) = \dfrac{1}{\sqrt{2\pi}\sqrt{2}}\exp\left[-\dfrac{1}{2}\left(\dfrac{x-2}{2}\right)^2\right]$

(c) $\mathcal{P}(a) = \binom{6}{a}\left(\dfrac{2}{3}\right)^a\left(\dfrac{1}{3}\right)^{6-a} \qquad a = 0, 1, 2, \ldots, 6$

4.8-18. A random variable x is uniformly distributed over the interval $-\pi/2$ to $\pi/2$. If $y = A \sin x$, what is the mean and variance of y?

4.8-19. The mean and variance matrix of a three-dimensional random variable is

$$\boldsymbol{\mu}_\mathbf{x} = \begin{bmatrix} \mu_{x_1} \\ \mu_{x_2} \\ \mu_{x^3} \end{bmatrix} \qquad \text{var}\{\mathbf{x}\} = \begin{bmatrix} V_{x_{11}} & 0 & V_{x_{13}} \\ 0 & V_{x_{22}} & 0 \\ V_{x_{31}} & 0 & V_{x_{33}} \end{bmatrix}$$

What is the mean and variance of $\mathbf{y} = A\mathbf{x}$ where

$$\mathbf{A} = \begin{bmatrix} 1 & 0 & 0 \\ 0 & 1 & 0 \\ 2 & 0 & 2 \end{bmatrix}$$

4.8-20. A valid probability density function is given by

$$p_x(\alpha) = 0.5e^{-|\alpha|}$$

Suppose that a nonlinear function

$$y = f(x) = \begin{cases} 2 & x \geq 1 \\ 2x & |x| < 1 \\ -2 & x \leq -1 \end{cases}$$

of the random variable x is formed. Find and sketch $p_y(\beta)$.

4.8-21. Two independent random variables x and y have probability densities

$$p_x(\alpha) = \begin{cases} e^{-\alpha} & \alpha \geq 0 \\ 0 & \alpha \leq 0 \end{cases} \qquad p_y(\beta) = \begin{cases} 2e^{-2\beta} & \beta \geq 0 \\ 0 & \beta \leq 0 \end{cases}$$

What is the probability that a sample taken from the y population is greater than or equal to a sample taken from the x population? That is, the probability that $y \geq x$?

5. *Stochastic Processes*

5.1. *Introduction*

This chapter is concerned with some of the basic concepts in the theory of stochastic processes. As we shall see, stochastic processes are generalizations of random variables, and so almost all the work of the previous chapters can be applied directly to stochastic processes. The consideration of random variables which are functions of time, leads us to the study of stochastic processes. If to every outcome ξ of an experiment, a time function $\mathbf{x}(t, \xi)$ is assigned, we create what is called a *stochastic process*. An interpretative picture of a stochastic process is illustrated in Fig. 5.5-1. A stochastic process is therefore a function of two parameters: the time t and the outcome of an experiment ξ.

Section 5.2 contains a precise definition of a stochastic process along with a simple example. Following this, we shall discuss the minor modifications needed in the work of the previous chapters to consider stochastic processes. After introducing the concepts of covariance, autocorrelation, and cross-correlation functions, we shall discuss errors in actual measurements of these functions. Orthogonal and spectral representations of random processes will be developed, and the concepts of stationarity and ergodicity will be introduced. The extremely important concept of white noise will be developed. Most of the concepts developed in this chapter will be utilized in detail in Chapter 6, which is devoted to a study of the response of linear systems to stochastic processes.

The remainder of this text will also deal with stochastic processes. It is impossible to explain very many physical phenomena that occur in the information, control, or computer sciences without having recourse to the theory of stochastic processes; thus, this topic is vital for further study and deserves a most serious effort.

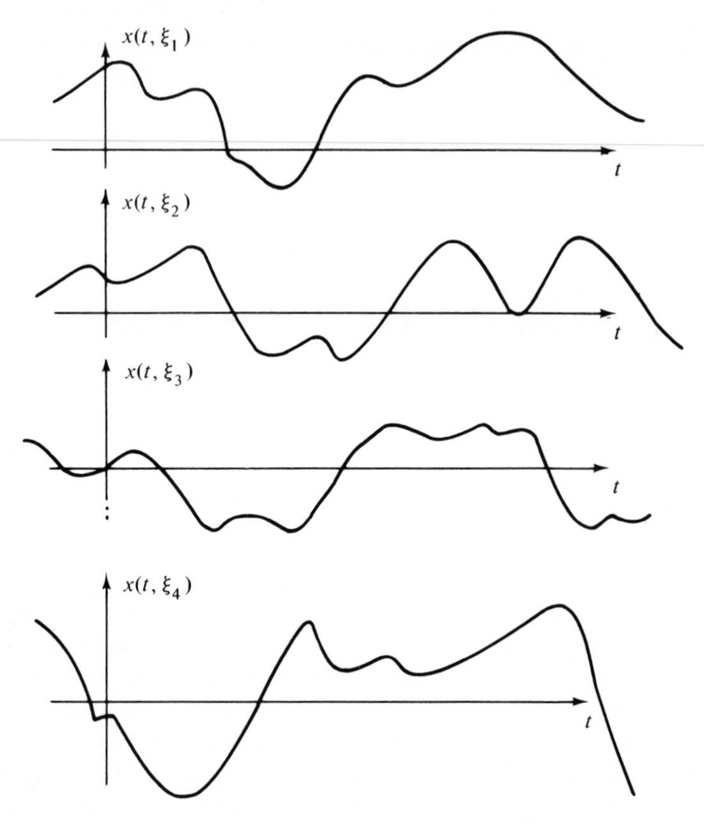

FIG. 5.1-1. Records of a random process.

5.2. *Definition of stochastic processes*

In Section 5.1 a stochastic process was defined as a mapping of outcomes $\xi \in \mathcal{S}$ to time functions $\mathbf{x}(t, \xi)$. Although this definition has a certain intuitive appeal, it has some mathematical drawbacks. Hence we shall use a slightly different approach for the precise definition of a stochastic process.[1]

> **Definition 5.2-1.** ***Stochastic Process:*** A stochastic process $\mathbf{x}(t)$ is a family of random vectors $\{\mathbf{x}(t), t \in \mathfrak{I}\}$ indexed by a parameter t belonging to an index set $\mathfrak{I}$.

The parameter t will normally be interpreted as time, although this is not necessary. No restrictions are placed on the nature of the index set $\mathfrak{I}$;

[1] The terminology *random process* is used interchangeably in this treatment.

there are, however, two important cases. If $\mathfrak{J}$ contains a countable number of values, for example $\mathfrak{J} = \{0, \pm 1, \pm 2, \pm 3, \ldots\}$ or $\mathfrak{J} = \{0, 1, 2, 3, \ldots\}$, then the stochastic process is said to be a *discrete parameter* or *discrete time process*. If, on the other hand, $\mathfrak{J} = \{t: -\infty < t < \infty\}$, $\mathfrak{J} = \{t: t \geq 0\}$, or $\mathfrak{J} = \{t: a \leq t \leq b\}$, then the stochastic process is said to be a *continuous parameter* or *continuous time process*. The random sequences that we treated in Section 4.6 were, in fact, discrete-time stochastic processes.

A stochastic process admits four different interpretations:

1. A deterministic vector when both the outcome ξ_i and the time t_i are given.
2. A random vector when only t_i is specified.
3. A deterministic time function when ξ_i is specified but t_i is not.
4. A stochastic process when neither ξ_i nor t_i is specified.

The notation $\mathbf{x}(t)$ will normally be used to represent a stochastic process. We should more properly use simply the notation $\mathbf{x}$, since $\mathbf{x}(t)$ should imply the evaluation of the stochastic process $\mathbf{x}$ for time t. However, since the use of $\mathbf{x}(t)$ to represent both a general time function and a specific evaluation is common in engineering, we shall use this slight corruption of notation to emphasize the difference between random variables and stochastic processes. To emphasize the evaluation of the stochastic process for a specific outcome, as in cases 1 and 3, we shall use the notation $\mathbf{x}(t, \xi)$ or $\mathbf{x}(t, \xi_i)$.

The reasons why the definition given above is preferable to the concept of assigning time functions to outcomes involve concepts of measure theory that are beyond the scope of this treatement. It is possible, however, to make one simple argument in terms of our level of sophistication. We noted above, for a specific t_i, that $\mathbf{x}(t_i)$ is a random variable. If we use the approach of assigning time functions to outcomes, then although it will be true that $\mathbf{x}(t_i)$ will have a value for each outcome, it will *not* necessarily be true that $\mathbf{x}(t_i)$ will be a random variable. The reader should remember[1] that, in addition to mapping the sample space into $\mathcal{R}^n$, the mapping must be selected so that $\{\mathbf{x} \leq \boldsymbol{\alpha}\}$ is an event for every $\boldsymbol{\alpha} \in \mathcal{R}^n$ and $\mathcal{P}\{\mathbf{x} = \infty\} = \mathcal{P}\{\mathbf{x} = -\infty\} = 0$.

Example ***5.2-1.*** To illustrate the concept of a stochastic process, let us consider the simple scalar linear system

$$\dot{x}(t) = w \qquad \text{with} \qquad x(0) = 0$$

Here w is a random variable that is gaussianly distributed with $\mu_w = 0$ and $V_w = 1$. We may easily solve this differential equation to find that

$$x(t) = wt$$

[1] See Section 3.2.

For each value of t, $t \geq 0$, $x(t)$ is a gaussian random variable with zero mean and variance t^2. Hence $x(t)$ represents a family of random variables with a parameter t and is therefore a stochastic process.

If we let w_i correspond to the outcome ξ_i, we see that $x(t)$ has the four interpretations noted earlier:

1. A deterministic number $x(t_i, \xi_i) = w_i t_i$.
2. A random variable $x(t_i) = wt_i$.
3. A deterministic time function $\mathbf{x}(t, \xi_i) = w_i t$.
4. A stochastic process $x(t) = wt$.

Typical realizations of the stochastic process are illustrated in Fig. 5.2-1.

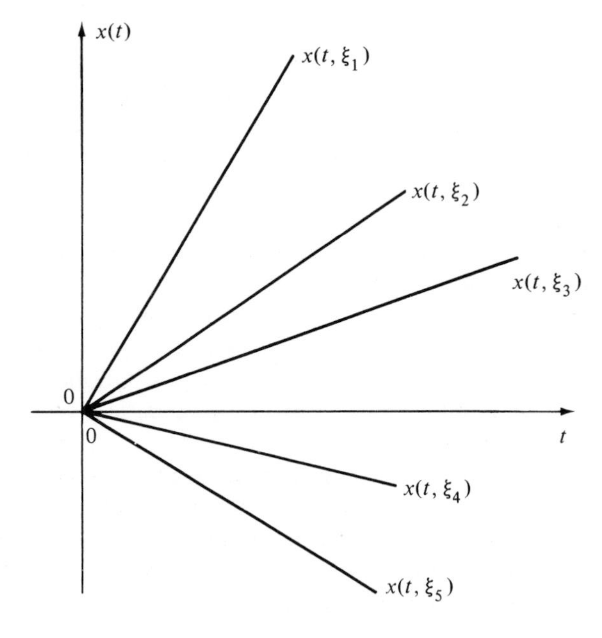

FIG. 5.2-1. Stochastic process $x(t) = wt$.

5.3. Probability expressions for stochastic processes

Probability expressions for stochastic processes are essentially the same as they are for random variables that are not functions of time. The notation $\{\mathbf{x}(t_i) \leq \boldsymbol{\alpha}\}$ will be used to represent the set of outcomes for which

$\mathbf{x}(t)$ at $t = t_i$ is less than or equal to $\boldsymbol{\alpha}$, so that

$$\{\mathbf{x}(t_i) \leq \boldsymbol{\alpha}\} = \{\xi : \mathbf{x}(t_i, \xi) \leq \boldsymbol{\alpha}\} \tag{5.3-1}$$

By definition of a stochastic process, $\mathbf{x}(t_i)$ is a random variable, so $\{\mathbf{x}(t_i) \leq \boldsymbol{\alpha}\}$ is an event for every $t_i \in \mathfrak{I}$ and $\boldsymbol{\alpha} \in \mathcal{R}^n$. The *probability distribution function* of the stochastic process $\mathbf{x}(t)$ is defined as[1]

$$F_{\mathbf{x}(t)}(\boldsymbol{\alpha}, t) = \mathcal{P}\{\mathbf{x}(t) \leq \boldsymbol{\alpha}\} \tag{5.3-2}$$

for all $\boldsymbol{\alpha} \in \mathcal{R}^n$ and $t \in \mathfrak{I}$. Analogous to Eq. (3.6-17), the *probability density function* for $\mathbf{x}(t)$ is any function $p_{\mathbf{x}(t)}$ defined on $\mathcal{R}^n$ and such that

$$F_{\mathbf{x}(t)}(\boldsymbol{\alpha}, t) = \int_{-\infty}^{\boldsymbol{\alpha}} p_{\mathbf{x}(t)}(\boldsymbol{\beta}, t)\, d\boldsymbol{\beta} \tag{5.3-3}$$

for all $\boldsymbol{\alpha} \in \mathcal{R}^n$ and $p_{\mathbf{x}(t)}(\boldsymbol{\beta}, t) \geq 0$ for all $\boldsymbol{\beta} \in \mathcal{R}^n$ and $t \in \mathfrak{I}$. Without exception, all the expressions in Chapters 3 and 4 may be used to express probabilities associated with stochastic processes.

In Section 5.2 we classified a stochastic process as a continuous or discrete-time process depending on the nature of the index set $\mathfrak{I}$. A stochastic process may also be classified in terms of its amplitude distribution. If for every $t_1 \in \mathfrak{I}$, $x(t_1)$ is a continuous random variable, then $x(t)$ is said to be a *continuous-amplitude stochastic process*. If, on the other hand, $\mathbf{x}(t_1)$ is a discrete random variable for every $t_1 \in \mathfrak{I}$, then $\mathbf{x}(t)$ is referred to as a *discrete-amplitude stochastic process*. If $\mathbf{x}(t)$ does not satisfy either of these conditions, it is said to have a *mixed* amplitude. Note that $\mathbf{x}(t)$ may be a mixed-amplitude process because $\mathbf{x}(t_1)$ is a mixed random variable for all $t_1 \in \mathfrak{I}$, or because $\mathbf{x}(t_1)$ is discrete for some t_1, continuous for other t_1, and perhaps mixed for some other t_1. It should be emphasized that the classification of time and amplitude is not related. Hence it is possible for a stochastic process to be a continuous-time discrete-amplitude process, for example.

Second-order statistics are particularly important in discussions of stochastic processes, as we shall soon see. We specify two time instants t_1 and t_2 and then consider the random variables (not stochastic processes, since t_1 and t_2 are fixed) $\mathbf{x}(t_1)$ and $\mathbf{x}(t_2)$. Let $\{\mathbf{x}(t_1) \leq \boldsymbol{\alpha}_1, \mathbf{x}(t_2) \leq \boldsymbol{\alpha}_2\}$ be the set of outcomes (event) that $\mathbf{x}(t)$ is less than or equal to $\boldsymbol{\alpha}_1$ at $t = t_1$, and less than or equal to $\boldsymbol{\alpha}_2$ at $t = t_2$, or

$$\{\mathbf{x}(t_1) \leq \boldsymbol{\alpha}_1, \mathbf{x}(t_2) \leq \boldsymbol{\alpha}_2\} = \{\xi : \mathbf{x}(t_1, \xi) \leq \boldsymbol{\alpha}_1 \text{ and } \mathbf{x}(t_2, \xi) \leq \boldsymbol{\alpha}_2\} \tag{5.3-4}$$

[1] Alternative simpler notations for $F_{\mathbf{x}(t)}(\boldsymbol{\alpha}, t)$ are $F_{\mathbf{x}}(\boldsymbol{\alpha}, t)$ and $F_{\mathbf{x}(t)}(\boldsymbol{\alpha})$. The somewhat longer notation will be used here to emphasize the dependence of the distribution on t.

The joint distribution function of $\mathbf{x}(t_1)$ and $\mathbf{x}(t_2)$ is defined as

$$F_{\mathbf{x}(t_1)\mathbf{x}(t_2)}(\boldsymbol{\alpha}_1, \boldsymbol{\alpha}_2, t_1, t_2) = \mathcal{P}\{\mathbf{x}(t_1) \leq \boldsymbol{\alpha}_1, \mathbf{x}(t_2) \leq \boldsymbol{\alpha}_2\} \tag{5.3-5}$$

for all $\boldsymbol{\alpha}_1, \boldsymbol{\alpha}_2 \in \mathcal{R}^n$ and $t_1, t_2 \in \mathfrak{J}$. This is also referred to as the *second-order distribution function* of $\mathbf{x}(t)$, since it relates the behavior of $\mathbf{x}(t)$ at two time points. The *second-order density function* $p_{\mathbf{x}(t_1)\mathbf{x}(t_2)}$ of $\mathbf{x}(t)$ is given by

$$F_{\mathbf{x}(t_1)\mathbf{x}(t_2)}(\boldsymbol{\alpha}_1, \boldsymbol{\alpha}_2, t_1, t_2) = \int_{-\infty}^{\boldsymbol{\alpha}_1} d\boldsymbol{\beta}_1 \int_{-\infty}^{\boldsymbol{\alpha}_2} d\boldsymbol{\beta}_2 p_{\mathbf{x}(t_1)\mathbf{x}(t_2)}(\boldsymbol{\beta}_1, \boldsymbol{\beta}_2, t_1, t_2) \tag{5.3-6}$$

It is now straightforward to show that

$$p_{\mathbf{x}(t_1)}(\boldsymbol{\alpha}_1, t_1) = \int_{-\infty}^{\infty} p_{\mathbf{x}(t_1)\mathbf{x}(t_2)}(\boldsymbol{\alpha}_1, \boldsymbol{\alpha}_2, t_1, t_2)\, d\boldsymbol{\alpha}_2 \tag{5.3-7}$$

$$p_{\mathbf{x}(t_2)}(\boldsymbol{\alpha}_2, t_2) = \int_{-\infty}^{\infty} p_{\mathbf{x}(t_1)\mathbf{x}(t_2)}(\boldsymbol{\alpha}_1, \boldsymbol{\alpha}_2, t_1, t_2)\, d\boldsymbol{\alpha}_1 \tag{5.3-8}$$

$$F_{\mathbf{x}(t_1)}(\boldsymbol{\alpha}_1, t_1) = F_{\mathbf{x}(t_1)\mathbf{x}(t_2)}(\boldsymbol{\alpha}_1, \infty, t_1, t_2) \tag{5.3-9}$$

$$F_{\mathbf{x}(t_2)}(\boldsymbol{\alpha}_2, t_2) = F_{\mathbf{x}(t_1)\mathbf{x}(t_2)}(\infty, \boldsymbol{\alpha}_2, t_1, t_2) \tag{5.3-10}$$

The conditional probability law and Bayes' theorem hold for stochastic processes just as they do for random variables, so that

$$p_{\mathbf{x}(t_1)|\mathbf{x}(t_2)}(\boldsymbol{\alpha}_1, t_1 \mid \boldsymbol{\alpha}_2, t_2) = \frac{p_{\mathbf{x}(t_1)\mathbf{x}(t_2)}(\boldsymbol{\alpha}_1, \boldsymbol{\alpha}_2, t_1, t_2)}{p_{\mathbf{x}(t_2)}(\boldsymbol{\alpha}_2, t_2)} \tag{5.3-11}$$

$$p_{\mathbf{x}(t_2)|\mathbf{x}(t_1)}(\boldsymbol{\alpha}_2, t_2 \mid \boldsymbol{\alpha}_1, t_1) = \frac{p_{\mathbf{x}(t_1)\mathbf{x}(t_2)}(\boldsymbol{\alpha}_1, \boldsymbol{\alpha}_2, t_1, t_2)}{p_{\mathbf{x}(t_1)}(\boldsymbol{\alpha}_1, t_2)} \tag{5 3-12}$$

$$p_{\mathbf{x}(t_1)|\mathbf{x}(t_2)}(\boldsymbol{\alpha}_1, t_1 \mid \boldsymbol{\alpha}_2, t_2) = \frac{p_{\mathbf{x}(t_2)|\mathbf{x}(t_1)}(\boldsymbol{\alpha}_2, t_2 \mid \boldsymbol{\alpha}_1, t_1) p_{\mathbf{x}(t_1)}(\boldsymbol{\alpha}_1, t_1)}{p_{\mathbf{x}(t_2)}(\boldsymbol{\alpha}_2, t_2)} \tag{5.3-13}$$

as the appropriate relations.

As the next step, we could consider the third-order description of $\mathbf{x}(t)$ at $t = t_1, t_2$, and t_3. The third-order distribution function would give us additional information about the time behavior of $\mathbf{x}(t)$. To completely describe a stochastic process, we must know the distribution functions of all finite orders for all $t_1, t_2, t_3, \ldots, \in T$. Fortunately, the second-order description is sufficient for many engineering applications.

We may also consider the joint distribution and density functions of two stochastic processes defined as

$$F_{\mathbf{x}(t)\mathbf{y}(t)}(\boldsymbol{\alpha}, \boldsymbol{\beta}, t) = \mathcal{P}\{\mathbf{x}(t) \leq \boldsymbol{\alpha}, \mathbf{y}(t) \leq \boldsymbol{\beta}\} \tag{5.3-14}$$

and

$$F_{\mathbf{x}(t)\mathbf{y}(t)}(\boldsymbol{\alpha}, \boldsymbol{\beta}, t) = \int_{-\infty}^{\boldsymbol{\alpha}} d\boldsymbol{\gamma} \int_{-\infty}^{\boldsymbol{\beta}} d\boldsymbol{\lambda}\, p_{\mathbf{x}(t)\mathbf{y}(t)}(\boldsymbol{\gamma}, \boldsymbol{\lambda}, t) \tag{5.3-15}$$

Of course, we may also consider $\mathbf{x}(t)$ and $\mathbf{y}(t)$ at different times to obtain the *second-order joint distribution* function

$$F_{\mathbf{x}(t_1)\mathbf{y}(t_2)}(\boldsymbol{\alpha}, \boldsymbol{\beta}, t_1, t_2) = \mathcal{P}\{\mathbf{x}(t_1) \leq \boldsymbol{\alpha}, \mathbf{y}(t_2) \leq \boldsymbol{\beta}\} \qquad (5.3\text{-}16)$$

and the corresponding definition of the second-order joint density function

$$F_{\mathbf{x}(t_1)\mathbf{y}(t_2)}(\boldsymbol{\alpha}, \boldsymbol{\beta}, t_1, t_2) = \int_{-\infty}^{\boldsymbol{\alpha}} d\boldsymbol{\gamma} \int_{-\infty}^{\boldsymbol{\beta}} d\boldsymbol{\lambda}\, p_{\mathbf{x}(t_1)\mathbf{y}(t_2)}(\boldsymbol{\gamma}, \boldsymbol{\lambda}, t_1, t_2) \qquad (5.3\text{-}17)$$

We may expand the joint distributions of $\mathbf{x}(t)$ and $\mathbf{y}(t)$ to any number of different time points. The combinations are truly limitless.

Example *5.3-1*. Let us determine the density function for the stochastic process of Example 5.2-1. There we stated that for any $t \geq 0$, $x(t)$ given by

$$x(t) = wt$$

was a gaussian variable with mean zero and variance t^2. Hence the density function for $\mathbf{x}(t)$ is

$$p_{x(t)}(\alpha, t) = \frac{1}{\sqrt{2\pi t}} \exp\left\{\frac{-\alpha^2}{2t^2}\right\}$$

We note the dependence of the density function on the parameter t.

Example *5.3-2*. A joint probability function for a single random variable observed at two different times is given by

$$p_{x(t)x(\lambda)}(\alpha, \beta, t, \lambda) = \frac{1}{2\pi\sqrt{1, \rho^2(t, \lambda)}} \exp\left\{\frac{-[\alpha^2 - 2\rho(t, \lambda)\alpha\beta + \beta^2]}{2[1 - \rho^2(t, \lambda)]}\right\}$$

which represents a two-dimensional correlated gaussian density function. We determine the marginal densities in the usual way from Eqs. (5.3-7) and (5.3-8) as

$$p_{x(t)}(\alpha, t) = \int_{-\infty}^{\infty} p_{x(t)x(\lambda)}(\alpha, \beta, t, \lambda)\, d\beta = \frac{1}{\sqrt{2\pi}} e^{-\alpha^2/2}$$

$$p_{x(\lambda)}(\beta, \lambda) = \int_{-\infty}^{\infty} p_{x(t)x(\lambda)}(\alpha, \beta, t, \lambda)\, d\alpha = \frac{1}{\sqrt{2\pi}} e^{-\beta^2/2}$$

Here neither marginal probability density depends upon time.

Example *5.3-3*. A stochastic process is described by

$$x(t) = A \cos(\omega t + \phi)$$

where A and ω are known constants, t is the time variable, and ϕ is a random variable uniformly distributed over the interval $-\pi$ to π:

$$p_\phi(\beta) = \begin{cases} \dfrac{1}{2\pi} & -\pi \leq \beta \leq \pi \\ 0 & \text{otherwise} \end{cases}$$

We desire to determine the density function of the stochastic process $x(t)$.

For a specific value of t, say t_1, $x(t_1)$ is just an algebraic function of the random variable ϕ given by

$$x(t_1) = A\cos(\omega t_1 + \phi)$$

Using the concepts of Section 4.2, it is possible to show that $p_{x(t_1)}$ is given by

$$p_{x(t_1)}(\alpha, t_1) = \begin{cases} \dfrac{1}{\pi(A^2 - \alpha^2)^{1/2}} & |\alpha| < A \\ 0 & \text{otherwise} \end{cases}$$

This particular random process is known as a sine wave with random phase and is a particularly important random process in communication theory. It is interesting to note here that the density function of this random process is not a function of time. This occurs because of the choice of the random variable ϕ. If, for example, ϕ were uniformly distributed only over the interval 0 to $\pi/2$, the density function of the stochastic process $x(t)$ would be a function of time (see Exercise 5.3-1).

If the amplitude A of the sine wave with random phase is not a constant but is itself a random variable independent of ϕ, as in amplitude modulation, we have

$$p_{x(t)}(\alpha, t) = \frac{1}{\pi}\int_{-\infty}^{\infty} \frac{p_A(\gamma)\,d\gamma}{\sqrt{\gamma^2 - \alpha^2}}$$

as can easily be demonstrated.

If the random variable A has a Rayleigh density defined by

$$p_A(\gamma) = \begin{cases} \gamma e^{-\gamma^2/2} & \gamma \geq 0 \\ 0 & \gamma \leq 0 \end{cases}$$

it can be shown (see Exercise 5.3-2) that the random process $x(t)$

formed from

$$x(t) = A \cos(\omega t + \phi)$$

where ϕ is uniformly distributed over the interval $-\pi$ to π, is gaussian.

Example ***5.3-4.*** A very important random process is the Poisson process, which may be used to model such events as the random emission of electrons in a vacuum-tube filament and the time to failure of systems. In Section 3.5 we found that the probability of k events Poisson distributed in the time interval $[0, t]$ was given by

$$\mathcal{P}\{x = k : 0, t\} = \frac{(\lambda t)^k e^{-\lambda t}}{k!}$$

Since this probability depends only on the length of the time interval, we could also write it as

$$\mathcal{P}\{k, T\} = \frac{(\lambda T)^k e^{-\lambda T}}{k!}$$

This expression is the probability that there are k events which occur in a time interval of length T.

Several interesting probabilities can be obtained from this basic expression. One of the most fundamental is the expression for the probability of the number of events that have occurred during the interval T_1 to T_2. This is easily obtained, if N_{T_i} denotes the number of events in the interval 0 to T_i, as

$$\mathcal{P}\{[N_{T_2} - N_{T_1}] = \eta\} = \mathcal{P}[N_{T_2 - T_1} = \eta] = \mathcal{P}(\eta, T_2 - T_1)$$

One very useful random process that is modeled by a Poisson process is the random telegraph signal which can be generated by amplifying and clipping the output of a radioactive source (see Example 5.4-3). This random telegraph process is a binary random process equal to ± 1 with the sign changes generated by the state changes of a Poisson process with mean count rate λ. Several commercially available noise generators are based on this concept.

EXERCISES 5.3

5.3-1. What is the probability density of the sine wave with random phase if the phase is uniformly distributed over the interval
(a) 0 to $\pi/2$?
(b) 0 to π?

5.3-2. Show that the probability density of

$$x(t) = A \sin(\omega t + \phi)$$

is gaussian if A has Rayleigh distribution and ϕ is uniformly distributed on $[-\pi, \pi]$ (see Example 5.3-3).

5.4. *Expectation*

The fundamental theory of expectations developed in Section 4.4 is directly applicable to stochastic processes if we are careful in noting the correct time dependence of the variables. For example, the mean of the stochastic process $\mathbf{x}(t)$ is given by

$$\boldsymbol{\mu}_{\mathbf{x}}(t) = \mathcal{E}\{\mathbf{x}(t)\} = \int_{-\infty}^{\infty} \boldsymbol{\alpha} p_{\mathbf{x}(t)}(\boldsymbol{\alpha}, t)\, d\boldsymbol{\alpha} \tag{5.4-1}$$

We note that the mean of $\mathbf{x}(t)$ is, in general, a deterministic function of t. The mean-square-value matrix of a stochastic process is defined as

$$\mathbf{P}_{\mathbf{x}}(t) \triangleq \mathcal{E}\{\mathbf{x}(t)\mathbf{x}^T(t)\} = \int_{-\infty}^{\infty} \boldsymbol{\alpha}\boldsymbol{\alpha}^T p_{\mathbf{x}(t)}(\boldsymbol{\alpha}, t)\, d\boldsymbol{\alpha} \tag{5.4-2}$$

Again, we note the dependence of the result on the time parameter t. The variance matrix for a stochastic process is defined by

$$\begin{aligned}\mathbf{V}_{\mathbf{x}}(t) &\triangleq \text{var}\{(t)\} = \mathcal{E}\{[\mathbf{x}(t) - \boldsymbol{\mu}_{\mathbf{x}}(t)][\mathbf{x}(t) - \boldsymbol{\mu}_{\mathbf{x}}(t)]^T\} \\ &= \int_{-\infty}^{\infty} [\boldsymbol{\alpha} - \boldsymbol{\mu}_{\mathbf{x}}(t)][\boldsymbol{\alpha} - \boldsymbol{\mu}_{\mathbf{x}}(t)]^T p_{\mathbf{x}(t)}(\alpha, t)\, d\boldsymbol{\alpha}\end{aligned} \tag{5.4-3}$$

It also follows directly that

$$\mathbf{V}_{\mathbf{x}}(t) = \mathbf{P}_{\mathbf{x}}(t) - \boldsymbol{\mu}_{\mathbf{x}}(t)\boldsymbol{\mu}_{\mathbf{x}}^T(t) \tag{5.4-4}$$

Often we shall drop the $\mathbf{x}$ subscript on the mean $\boldsymbol{\mu}$, the variance matrix $\mathbf{V}$, and the mean-square matrix $\mathbf{P}$ whenever no confusion will result. The fundamental theorem of expectation has the following form for stochastic processes:

$$\mathcal{E}\{\mathbf{f}[\mathbf{x}(t)]\} = \int_{-\infty}^{\infty} \mathbf{f}[\boldsymbol{\alpha}]\, p_{\mathbf{x}(t)}(\boldsymbol{\alpha}, t)\, d\boldsymbol{\alpha} \tag{5.4-5}$$

Example *5.4-1*. We desire to determine the mean and variance of the random process $x(t)$, which has the density function

$$p_{x(t)}(\alpha, t) = \begin{cases} r(t) & 0 < \alpha < 1/r(t) \\ 0 & \text{otherwise} \end{cases}$$

Using Eq. (5.4-1), we have

$$\mu_x(t) = \mathcal{E}\{x(t)\} = \int_{-\infty}^{\infty} \alpha p_{x(t)}(\alpha, t)\, d\alpha = \int_0^{1/r(t)} \alpha r(t)\, d\alpha = \frac{r^{-1}(t)}{2}$$

whereas from Eq. (5.4-2), we have

$$P_x(t) = \mathcal{E}\{x^2(t)\} = \int_{-\infty}^{\infty} \alpha^2 p_{x(t)}(\alpha, t)\, d\alpha = \int_0^{1/r(t)} \alpha^2 r(t)\, d\alpha = \frac{r^{-2}(t)}{3}$$

Therefore, the variance is given by

$$V_x(t) = \mathcal{E}\{[x(t) - \mu_x(t)]^2\} = \mathcal{E}\{[x^2(t)]\} - [\mathcal{E}\{x(t)\}]^2 = \frac{r^{-2}(t)}{12}$$

In this example we see that the first two moments are, in general, time varying.

Often we may desire to compute the expected value of the product of two stochastic processes. By use of the fundamental theorem of expectation, we have

$$\mathbf{P}_{\mathbf{xy}}(t) \triangleq \mathcal{E}\{\mathbf{x}(t)\mathbf{y}^T(t)\} = \int_{-\infty}^{\infty} \boldsymbol{\alpha}\boldsymbol{\beta}^T p_{\mathbf{x}(t)\mathbf{y}(t)}(\boldsymbol{\alpha}, \boldsymbol{\beta}, t)\, d\boldsymbol{\alpha}\, d\boldsymbol{\beta} \tag{5.4-6}$$

where $p_{\mathbf{x}(t)\mathbf{y}(t)}(\boldsymbol{\alpha}, \boldsymbol{\beta}, t)$ is the joint density function of the two random processes $\mathbf{x}(t)$ and $\mathbf{y}(t)$. We may also define a *cross-correlation matrix*, which is the expectation of $\mathbf{x}(t)$ and $\mathbf{y}(\tau)$, as

$$\mathbf{P}_{\mathbf{xy}}(t, \tau) = \mathcal{E}\{\mathbf{x}(t)\mathbf{y}^T(\tau)\} = \int_{-\infty}^{\infty} \boldsymbol{\alpha}\boldsymbol{\beta}^T p_{\mathbf{x}(t)\mathbf{y}(\tau)}(\boldsymbol{\alpha}, \boldsymbol{\beta}, t, \tau)\, d\boldsymbol{\alpha}\, d\boldsymbol{\beta} \tag{5.4-7}$$

A special case of this is the *autocorrelation matrix*

$$\mathbf{P}_{\mathbf{x}}(t, \tau) = \mathcal{E}\{\mathbf{x}(t)\mathbf{x}^T(\tau)\} = \int_{-\infty}^{\infty} \boldsymbol{\alpha}\boldsymbol{\beta}^T p_{\mathbf{x}(t)\mathbf{x}(\tau)}(\boldsymbol{\alpha}, \boldsymbol{\beta}, t, \tau)\, d\boldsymbol{\alpha}\, d\boldsymbol{\beta} \tag{5.4-8}$$

Here both t and $\boldsymbol{\tau}$ must belong to $\mathfrak{J}$. In a similar way, we may define covariance and cross-covariance matrices. We define the *covariance matrix* (also called the covariance kernel) as

$$\begin{aligned}\mathbf{V}_{\mathbf{x}}(t, \tau) &= \operatorname{cov}\{\mathbf{x}(t), \mathbf{x}(\tau)\} \triangleq \mathcal{E}\{[\mathbf{x}(t) - \boldsymbol{\mu}_{\mathbf{x}}(t)][\mathbf{x}(\tau) - \boldsymbol{\mu}_{\mathbf{x}}(\tau)]^T\} \\ &= \int_{-\infty}^{\infty} [\boldsymbol{\alpha} - \boldsymbol{\mu}_{\mathbf{x}}(t)][\boldsymbol{\beta} - \boldsymbol{\mu}_{\mathbf{x}}(\tau)]^T p_{\mathbf{x}(t)\mathbf{x}(\tau)}(\boldsymbol{\alpha}, \boldsymbol{\beta}, t, \tau)\, d\boldsymbol{\alpha}\, d\boldsymbol{\beta}\end{aligned} \tag{5.4-9}$$

as the expected value of the product of the random waveform with the mean value removed (at times t and τ). The *variance matrix* is defined as

$$\mathbf{V}_{\mathbf{x}}(t) = \text{var}\{\mathbf{x}(t)\} = \text{cov}\{\mathbf{x}(t), \mathbf{x}(t)\} = \mathbf{V}_{\mathbf{x}}(t, t) \tag{5.4-10}$$

Also, we have the *cross-covariance matrix*

$$\begin{aligned} \mathbf{V}_{\mathbf{xy}}(t, \tau) &= \text{cov}\{\mathbf{x}(t), \mathbf{y}(\tau)\} = \mathcal{E}\{[\mathbf{x}(t) - \boldsymbol{\mu}_{\mathbf{x}}(t)][\mathbf{y}(\tau) - \boldsymbol{\mu}_{\mathbf{y}}(\tau)]^T\} \\ &= \int_{-\infty}^{\infty} [\boldsymbol{\alpha} - \boldsymbol{\mu}_{\mathbf{x}}(t)][\boldsymbol{\beta} - \boldsymbol{\mu}_{\mathbf{y}}(\tau)]^T p_{\mathbf{x}(t)\mathbf{y}(\tau)}(\boldsymbol{\alpha}, \boldsymbol{\beta}, t, \tau)\, d\boldsymbol{\alpha}\, d\boldsymbol{\beta} \end{aligned} \tag{5.4-11}$$

Quite clearly, we have

$$\mathbf{V}_{\mathbf{x}}(t, \tau) = \mathbf{P}_{\mathbf{x}}(t, \tau) - \boldsymbol{\mu}_{\mathbf{x}}(t)\boldsymbol{\mu}_{\mathbf{x}}^T(\tau) \tag{5.4-12}$$

as the relations between mean values, correlation matrices, and covariance matrices, and

$$\mathbf{V}_{\mathbf{xy}}(t, \tau) = \mathbf{P}_{\mathbf{xy}}(t, \tau) - \boldsymbol{\mu}_{\mathbf{x}}(t)\boldsymbol{\mu}_{\mathbf{y}}^T(\tau) \tag{5.4-13}$$

as the relation between mean values, cross-correlation matrices and cross-covariance matrices.

The autocorrelation function and covariance function indicate the relative dependence between the stochastic process $\mathbf{x}(t)$ and the process $\mathbf{x}(t + \tau)$. To illustrate this further, let us consider a zero mean scalar random process $x(t)$ and define

$$y(t) = x(t) - \Xi x(t + \tau) \tag{5.4-14}$$

We desire to determine the value of Ξ that minimizes

$$J = \mathcal{E}\{y^2(t)\} \tag{5.4-15}$$

By determining the value of Ξ, we shall have an indication of how much the process $x(t)$ is contained in $x(t + \tau)$. By combining the foregoing two relations, we obtain

$$J = P_x(t) - 2\Xi P_x(t + \tau, t) + \Xi^2 P_x(t + \tau) \tag{5.4-16}$$

The best value of Ξ to minimize is easily determined as

$$\Xi = \frac{P_x(t + \tau, t)}{P_x(t + \tau)} = \rho(t + \tau, t) \tag{5.4-17}$$

the (normalized) correlation function, or correlation coefficient.

Example 5.4-2. Let us reconsider the Poisson process considered in Example 5.3-4. There we obtained an expression for the probability of n events in a time interval of length T as

$$\mathcal{P}\{n, T\} = \frac{(\lambda T)^n}{n!} e^{-\lambda T}$$

Let us define a stochastic process $x(t)$ as the number of events that have occurred in a time interval of length t. Thus

$$\mathcal{P}\{x(t) = n\} = \mathcal{P}\{n, t\}$$

where it is assumed that $x(0) = 0$, and thus we find that $x(t_2) - x(t_1)$ equals the number of points in the interval (t_2, t_1). Each function of the process is staircase like, as indicated in Fig. 5.4-1, with each step

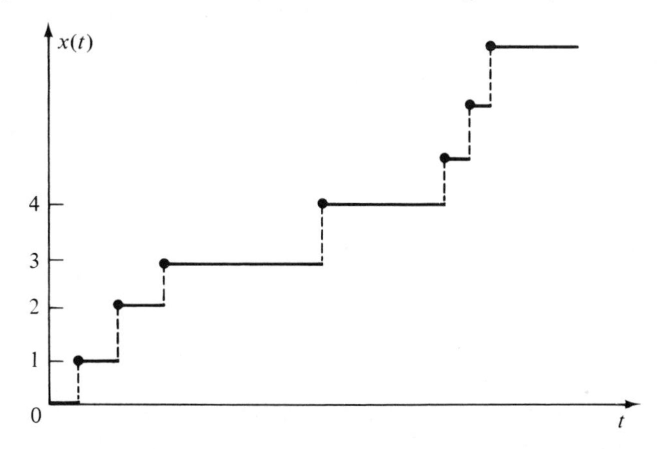

FIG. 5.4-1. Poisson process.

of amplitude $+1$ occurring at random points t_i. We shall say that this Poisson process "has the parameter λt," which is just the mean of the process

$$\mu_x(t) = \mathcal{E}\{x(t)\} = \sum_{n=0}^{\infty} n\mathcal{P}\{x(t) = n\} = \lambda t$$

The mean-square value of the process is determined in straightforward but tedious fashion as

$$P_x(t) = \mathcal{E}\{x^2(t)\} = \sum_{n=0}^{\infty} n^2\mathcal{P}\{x(t) = n\} = (\lambda t)^2 + \lambda t$$

and thus the variance of the Poisson process is

$$V_x(t) = P_x(t) - \mu_x^2(t) = \lambda t$$

The autocorrelation function of this Poisson process can be determined directly. By definition, we have

$$P_x(t_1, t_2) = \mathcal{E}\{x(t_1)x(t_2)\} = \mathcal{E}\{[x(t_1)][x(t_1) + x(t_2) - x(t_1)]\}$$
$$= \mathcal{E}\{x^2(t_1)\} + \mathcal{E}\{[x(t_1)][x(t_2) - x(t_1)]\}$$

The random variables $x(t_1)$ and $x(t_2) - x(t_1)$ are independent (they represent the number of events in different time intervals) for $t_2 \geq t_1$ because of the independent increment property of the Poisson process (see Section 3.5). Thus

$$P_x(t_1, t_2) = \mathcal{E}\{x^2(t_1)\} + \mathcal{E}\{x(t_1)\}\mathcal{E}\{x(t_2) - x(t_1)\}$$

By utilizing the results obtained previously in this example for the mean and mean-square value of the Poisson process, we have for the correlation function, assuming $t_2 \geq t_1$,

$$P_x(t_1, t_2) = (\lambda t_1)^2 + (\lambda t_1) + (\lambda t_1)[\lambda(t_2 - t_1)]$$
$$= \lambda t_1(1 + \lambda t_2) \qquad t_2 \geq t_1$$

If $t_1 \geq t_2$, this result becomes

$$P_x(t_1, t_2) = (\lambda t_2)^2 + (\lambda t_2) + (\lambda t_2)[\lambda(t_1 - t_2)]$$
$$= \lambda t_2(1 + \lambda t_1) \qquad t_1 \geq t_2$$

Example *5.4-3*. The random telegraph process is an important random process that may be generated from the Poisson process. We define the random telegraph process $y(t)$ as follows. If the total number of events in the Poisson process $x(t)$ is even, then $y(t) = 1$. If the total number of events in the Poisson process $x(t)$ is odd, then $y(t) = -1$. Thus $y(t)$ takes on ± 1 values with the axis crossings being determined by a Poisson process, as indicated in Fig. 5.4-2.

The events $\{n$ points in $(0, t)\}$ are mutually exclusive because of the independent increment property of the Poisson process. Thus the probability of an even number of points in the interval $(0, t)$ is just

$$\mathcal{P}\{0, t\} + \mathcal{P}\{2, t\} + \cdots = e^{-\lambda t}\left[1 + \frac{(\lambda t)^2}{2!} + \frac{(\lambda t)^4}{4!} + \cdots\right]$$
$$= e^{-\lambda t} \cosh \lambda t$$

In a similar way, the probability of having an odd number of points in the interval $(0, t)$ is

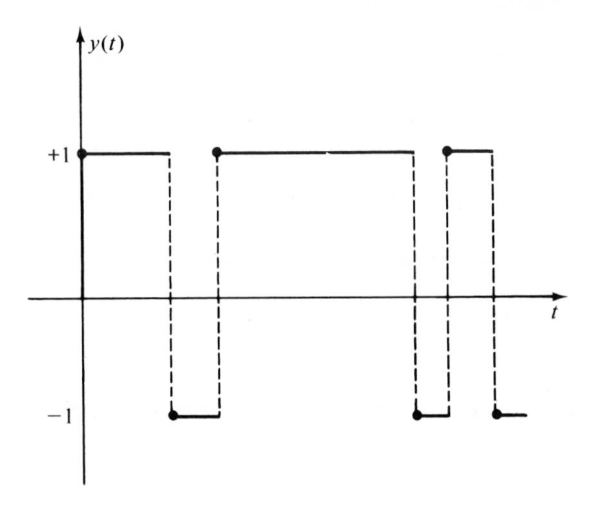

FIG. 5.4-2. Poisson square wave or random telegraph signal.

$$\mathcal{P}\{1, t\} + \mathcal{P}\{3, t\} + \cdots = e^{-\lambda t}\left[\lambda t + \frac{(\lambda t)^3}{3!} + \cdots\right]$$
$$= e^{-\lambda t} \sinh \lambda t$$

Thus we have

$$\mathcal{P}\{y(t) = +1\} = e^{-\lambda t} \cosh \lambda t$$
$$\mathcal{P}\{y(t) = -1\} = e^{-\lambda t} \sinh \lambda t$$

and have for the average value of the random telegraph process, which is here assumed to start at $t = 0$,

$$\mathcal{E}\{y(t)\} = +1\mathcal{P}\{y(t) = +1\} + (-1)\mathcal{P}\{y(t) = -1\} = e^{-2\lambda t}$$

In most actual uses of the random telegraph signal, it is assumed that the random process starts at $t = -\infty$, or sufficiently far in the past such that the mean value is essentially zero over the time duration of interest of the process.

We need to obtain the joint probabilities to obtain the autocorrelation function. These are probably most easily obtained from the conditional probabilities. We let

$$t_2 - t_1 = \tau > 0$$

If $y(t_1) = +1$, then $y(t_2) = -1$ if there is an odd number of events in the interval (t_1, t_2). Thus

$$\mathcal{P}\{y(t_2) = -1 \mid y(t_1) = +1\} = e^{-\lambda t} \sinh \lambda \tau$$

In a similar way,

$$\mathcal{P}\{y(t_2) = +1 \mid y(t_1) = +1\} = e^{-\lambda t} \cosh \lambda\tau$$
$$\mathcal{P}\{y(t_2) = -1 \mid y(t_1) = -1\} = e^{-\lambda t} \cosh \lambda\tau$$
$$\mathcal{P}\{y(t_2) = +1 \mid y(t_1) = -1\} = e^{-\lambda t} \sinh \lambda\tau$$

From the joint probability law, we have

$$\mathcal{P}\{y(t_2) = -1, y(t_1) = +1\} = \mathcal{P}\{y(t_2) = -1 \mid y(t_1) = +1\} \times \mathcal{P}\{y(t_1) = +1\}$$

and three similar relations for the other joint probabilities. Using the definition of the correlation function as the weighted sum of probabilities,

$$P_y(t_1, t_2) = \mathcal{E}\{y(t_1)y(t_2)\} = \sum (\pm 1)(\pm 1) \, \mathcal{P}\{y(t_2) = \pm 1, y(t_1) = \pm 1\}$$

we have, after a modest amount of algebra,

$$P_y(t_1, t_2) = e^{-2\lambda(t_2 - t_1)}$$

If we reverse the roles of t_1 and t_2, we see that under all circumstances we have

$$P_y(t_1, t_2) = e^{-2\lambda|t_2 - t_1|}$$

EXERCISES 5.4

5.4-1. You are given a deterministic periodic function $y(t)$ with period T and a random variable ϕ uniformly distributed in the interval 0 to T. Show, for $x(t) = y(t + \phi)$, that

$$\mathcal{E}\{x(t)x(t + \tau)\} = \frac{1}{T} \int_0^T y(t)y(t + \tau)\, dt$$

5.4-2. The stochastic process $x(t)$ is defined by

$$x(t) = e^{-t} \sin a$$

where a is a random variable uniformly distributed on $[0, \pi]$. Find
(a) $\mu_x(t)$.
(b) $V_x(t, \tau)$.

Answers: (a) $\mu_x(t) = \frac{2}{\pi} e^{-t}$, (b) $V_x(t, \tau) = \left(\frac{1}{2} - \frac{4}{\pi^2}\right) e^{-(t+\tau)}$.

5.4-3. The stochastic process $x(t)$ is defined by

$$x(t) = a \sin t$$

where a is a random variable with

$$p_a(\alpha) = e^{-\alpha} \mathcal{A}(\alpha)$$

Find

(a) $\mu_x(t)$.

(b) $V_x(\tau, t)$.

(c) $p_{x(t)}(\alpha, t)$.

Answers: (a) $\mu_x(t) = \sin t$, (b) $V_x(\tau, t) = \sin \tau \sin t$, (c) $p_{x(t)}(\alpha, t) = e^{-\alpha/\sin t}/ |\sin t| \mathcal{A}[a - \sin t]$.

5.5. Stationary processes

In the preceding sections we classified stochastic processes in terms of the index set $\mathfrak{J}$ as discrete or continuous time, and in terms of amplitude as discrete, continuous, or mixed. We have considered in the previous two sections both deterministic and nondeterministic random processes. Two simple examples of deterministic random processes that we considered are the random ramp, $x(t) = wt$, where w is a gaussian random variable, and the sine wave with random phase, $x(t) = \cos(\omega t + \phi)$. In each of these two processes the future value of any particular sample function $x(t, \xi_i)$ can be predicted *exactly* from a knowledge of past values. Thus these two processes are called *deterministic random processes.* The random telegraph signal or the Poisson process are, however, nondeterministic random processes. For these it is not possible to predict *exactly* future values from past values.

Stochastic processes can also be divided into two broad classes: stationary and nonstationary. A *stationary process* is one in which the joint distribution or density functions are invariant with time.

> **Definition 5.5-1. (*Strictly*) *Stationary Random Process:*** A random process is strictly stationary if $\mathbf{x}(t)$ and $\mathbf{x}(t + \eta)$ have the same statistics of for all η such that t and $t + \eta \in \mathfrak{J}$.

For a stochastic process to be strict-sense stationary, the probability density and distribution functions of all order must be invariant under any finite time translation. In some cases, it is only possible to establish that the probability functions for some finite order are invariant under a time shift; then one speaks of stationarity of finite order.

Definition 5.5-2. ***Stationarity of Order*** k**:** A stochastic process is stationary of order k if the joint distribution or density function of order k is independent under a shift of time such that

$$p_{\mathbf{x}(t_1)\mathbf{x}(t_2)\cdots\mathbf{x}(t_k)}(\alpha, \beta, \ldots, \kappa, t_1, t_2, \ldots, t_k)$$
$$= p_{\mathbf{x}(t_1+\eta)\mathbf{x}(t_2+\eta)\cdots\mathbf{x}(t_k+\eta)}(\alpha, \beta, \ldots, \kappa, t_1+\eta, t_2+\eta, \ldots, t_k+\eta) \tag{5.5-1}$$

for any η such that t_i and $t_i + \eta \in \mathfrak{I}$ for all $i = 1, 2, \ldots, k$.

We note that a process which is stationary of order k is also stationary of order $n \leq k$, since the lower-order densities may be determined from the kth-order density. If a stochastic process is strict-sense stationary, it is stationary of order k for all $k = 1, 2, \ldots$.

If a stochastic process is stationary of order 1 or higher, the first-order density must satisfy

$$p_{\mathbf{x}(t)}(\alpha, t) = p_{\mathbf{x}(t+\eta)}(\alpha, t+\eta) \tag{5.5-2}$$

for all η such that t and $t + \eta \in \mathfrak{I}$. This condition can only be satisfied for arbitrary η if the density is independent of t so that

$$p_{\mathbf{x}(t)}(\boldsymbol{\alpha}, t) = p_{\mathbf{x}(t)}(\boldsymbol{\alpha}) \tag{5.5-3}$$

From this fact we may conclude that the mean as well as all higher-order moments of $\mathbf{x}(t)$, such as the variance and mean-square value, must be independent of t, and hence are constants. We shall say that a stochastic process is *mean value stationary* if the mean is not a function of time such that

$$\boldsymbol{\mu}_{\mathbf{x}}(t) = \mathcal{E}\{\mathbf{x}(t)\} = \int_{-\infty}^{\infty} \boldsymbol{\alpha} p_{\mathbf{x}(t)}(\boldsymbol{\alpha}, t)\, d\boldsymbol{\alpha} = \boldsymbol{\mu}_{\mathbf{x}} \tag{5.5-4}$$

Hence a process that is stationary of order 1 or higher is mean value stationary, although the converse is *not* true.

For a process that is stationary of order 2 or higher, the second-order density must be such that

$$p_{\mathbf{x}(t_2)\mathbf{x}(t_1)}(\boldsymbol{\alpha}_2, \boldsymbol{\alpha}_1, t_2, t_1) = p_{\mathbf{x}(t_2+\eta)\mathbf{x}(t_1+\eta)}(\boldsymbol{\alpha}_2, \boldsymbol{\alpha}_1, t_2+\eta, t_1+\eta) \tag{5.5-5}$$

for all η such that t_1, t_2, $t_2 + \eta$, and $t_1 + \eta$ are in $\mathfrak{I}$. This condition will only be satisfied if the second-order density is only a function of the difference between t_1 and t_2, often called the "age variable" $\tau = t_2 - t_1$. Thus we require

$$p_{\mathbf{x}(t_2)\mathbf{x}(t_1)}(\boldsymbol{\alpha}_2, \boldsymbol{\alpha}_1, t_2, t_1) = p_{\mathbf{x}(t_2)\mathbf{x}(t_1)}(\boldsymbol{\alpha}_2, \boldsymbol{\alpha}_1, \tau) \tag{5.5-6}$$

where $\tau = t_2 - t_1$. In this case, we see that the correlation function

$$\mathbf{P}_{\mathbf{x}}(t_2, t_1) = \int_{-\infty}^{\infty} \boldsymbol{\alpha}\boldsymbol{\beta}^T p_{\mathbf{x}(t_2)\mathbf{x}(t_1)}(\boldsymbol{\alpha}, \boldsymbol{\beta}, \tau)\, d\boldsymbol{\alpha}\, d\boldsymbol{\beta} \tag{5.5-7}$$

is just a function of $\tau = t_2 - t_1$. To emphasize this fact, we shall use the symbolism

$$\mathbf{C}_{\mathbf{x}}(\tau) = \mathcal{E}\{\mathbf{x}(t+\tau)\mathbf{x}^T(t)\} = \mathbf{P}_{\mathbf{x}}(t+\tau, t) \tag{5.5-8}$$

Processes for which Eq. (5.5-8) is valid are referred to as *correlation stationary.* In a similar manner, we can show that if a stochastic process is stationary of order 2 or higher, the covariance function

$$\mathbf{V}_{\mathbf{x}}(t+\tau, t) = \operatorname{cov}\{\mathbf{x}(t+\tau), \mathbf{x}(t)\} \tag{5.5-9}$$

will only be a function of τ. This result will be expressed in the form

$$\mathbf{R}_{\mathbf{x}}(\tau) = \operatorname{cov}\{\mathbf{x}(t+\tau), \mathbf{x}(t)\} = \mathbf{V}_{\mathbf{x}}(t+\tau, t) \tag{5.5-10}$$

If a stochastic process satisfies Eq. (5.5-10), it will be said to be *covariance stationary.*

We note that any stochastic process which is stationary of order 2 or higher will also be correlation and covariance stationary as well as mean value stationary. The converse of this statement is *not* true, however. We remark, in addition, that it is possible for a process to be correlation stationary but not covariance or mean value stationary; similarly, it is possible for a process to be covariance stationary but not correlation or mean value stationary. If a process is both correlation and mean value stationary, the process is also covariance stationary, since

$$\begin{aligned}\mathbf{V}_{\mathbf{x}}(t+\tau, t) &= \mathbf{P}_{\mathbf{x}}(t+\tau, t) - \boldsymbol{\mu}_{\mathbf{x}}(t+\tau)\boldsymbol{\mu}_{\mathbf{x}}^T(t) \\ &= \mathbf{C}_{\mathbf{x}}(\tau) - \boldsymbol{\mu}_{\mathbf{x}}\boldsymbol{\mu}_{\mathbf{x}}^T = \mathbf{R}_{\mathbf{x}}(\tau)\end{aligned} \tag{5.5-11}$$

In an analogous manner, we can establish that if a process is covariance and mean value stationary, it is also correlation stationary.

Autocorrelation and cross-correlation functions for stationary random processes possess many useful properties. Some of these, which the interested reader can easily verify, are

1. $\mathbf{C}_{\mathbf{x}}(0) = \mathbf{P}_{\mathbf{x}}$. The mean-square value of the random process is obtained by setting $\tau = 0$.
2. $\mathbf{C}_{\mathbf{x}}(-\tau) = \mathbf{C}_{\mathbf{x}}^T(\tau)$. For scalar processes, $C_x(-\tau) = C_x(\tau)$ so that autocorrelation function is an even function of τ.
3. $\mathbf{C}_{\mathbf{x}}(\tau = \infty) = \boldsymbol{\mu}_{\mathbf{x}}\boldsymbol{\mu}_{\mathbf{x}}^T$ if $\mathbf{x}(t)$ contains no periodic components. The average

value of the random process acts as a simple bias for the correlation function. Also $\mathbf{R}_{\mathbf{x}}(\tau = \infty) = \mathbf{0}$ if $\mathbf{x}(t)$ contains no periodic components.

4. For a scalar process $|C_x(\tau)| \leq C_x(0) = P_x$. The largest value of the correlation function occurs at $\tau = 0$. We may easily show this by considering the zero-mean processes $x(t)$ and $x(t + \tau)$. Since

$$\mathcal{E}\{[x(t + \tau) - x(t)]^2\} = 2P_x - 2C_x(\tau) \geq 0$$

and

$$\mathcal{E}\{[x(t + \tau) + x(t)]^2\} = 2P_x + 2C_x(\tau) \geq 0$$

the inequality $|C_x(\tau)| \leq C_x(0)$ follows immediately.

5. $\mathbf{C}_{\mathbf{xy}}(\tau) = \mathbf{C}^T_{\mathbf{yx}}(-\tau)$.
6. For a scalar process $|C_{xy}(\tau)| \leq [C_x(0)C_y(0)]^{1/2}$.
7. If $\mathbf{x}(t)$ has a periodic component, then $\mathbf{C}_{\mathbf{x}}(\tau)$ will also have periodic components of the same frequency.

It is often difficult in practice to establish that a stochastic process is strict-sense stationary or even stationary of order k for any k. For this reason a weaker form of stationarity is often used.

Definition 5.5-3. ***Wide-Sense (Weak-Sense) Stationarity:*** A stochastic process $\mathbf{x}(t)$ is stationary in the wide sense if it is mean value and correlation stationary.

If $\mathbf{x}(t)$ is strictly stationary of order 2 or higher, it is wide-sense stationary. However, the definition of wide-sense stationarity involves only the first and second moments and not the probability density function. Thus the converse statement that if a process is wide-sense stationary, it is strictly stationary of order 2 *is not correct*. It turns out that if the random process is gaussian amplitude distributed and stationary in the wide sense, it is stationary in the strict sense (of any order).

Example *5.5-1*. We consider the one-dimensional correlated gaussian density function

$$p_{x(t)x(\tau)}(\alpha, \beta, t, \tau) = \frac{1}{2\pi\sqrt{1 - \rho^2(t - \tau)}} \exp\left[\frac{-[a^2 - 2\rho(t - \tau)ab + b^2]}{2[1 - \rho^2(t - \tau)]}\right]$$

where $a = \alpha - f(t)$ and $b = \beta - f(\tau)$. We obtain the marginal density from Eq. (5.3-7) as

$$p_{x(t)}(\alpha, t) = \int_{-\infty}^{\infty} p_{x(t)x(\tau)}(\alpha, \beta, t, \pi)\, d\beta = \frac{1}{\sqrt{2\pi}} \exp\left\{\frac{-[\alpha - f(t)]^2}{2}\right\}$$

Thus the marginal probability density depends upon time. The process is not strictly stationary of order 2, since the second-order statistics depend upon time. Nor is the process stationary of order 1, since the first-order statistics also depend upon time. We can now compute the mean value, correlation, and covariance function. We obtain

$$\mu_x(t) = \mathcal{E}\{x(t)\} = \int_{\infty}^{\infty} \alpha p_{x(t)}(\alpha, t)\, d\alpha = f(t)$$

$$P_x(t) = \mathcal{E}\{x^2(t)\} = \int_{\infty}^{\infty} \alpha^2 p_{x(t)}(\alpha, t)\, d\alpha = 1 + f^2(t)$$

$$\begin{aligned} P_x(t, \tau) &= \mathcal{E}\{x(t)x(\tau)\} = \int_{-\infty}^{\infty} \alpha\beta p_{x(t)x(\tau)}(\alpha, \beta, t, \tau)\, d\alpha\, d\beta \\ &= \rho(t - \tau) + f^2(t) \\ V_x(t, \tau) &= \mathcal{E}\{[x(t) - \mu_x(t)][x(\tau) - \mu_x(\tau)]\} = \rho(t - \tau) \end{aligned}$$

We see that even though the process is not wide-sense stationary, since $\mu_x(t)$ depends upon t and $P_x(t, \tau)$ is not a function of $t - \tau$ only; it is, however, covariance stationary. If the mean value, $f(t)$, is not actually a function of time, the process is wide-sense stationary and also strict-sense stationary.

Example 5.5-2. Let us consider again the random-phase sine wave of Example 5.3-3 with $A = 1$ given by

$$x(t) = \sin(\omega t + \phi)$$

with

$$p_\phi(\beta) = \begin{cases} \dfrac{1}{2\pi} & -\pi \leq \beta \leq \pi \\ 0 & \text{otherwise} \end{cases}$$

The probability density function for $x(t)$ was determined to be

$$p_{x(t)}(\alpha, t) = \begin{cases} \dfrac{1}{\pi\sqrt{1 - \alpha^2}} & |\alpha| < 1 \\ 0 & \text{otherwise} \end{cases}$$

The mean and mean-square value of the process may be obtained from the foregoing probability density of $x(t)$ as

$$\mu_x(t) = \mathcal{E}\{x(t)\} = \int_{-\infty}^{\infty} \alpha p_{x(t)}(\alpha, t)\, d\alpha = 0$$

$$P_x(t) = \mathcal{E}\{x^2(t)\} = \int_{-\infty}^{\infty} \alpha^2 p_{x(t)}(\alpha, t)\, d\alpha = \tfrac{1}{2}$$

Since $x(t)$ is actually a function of ϕ, we may also use the fundamental theorem of expectation to obtain

$$x = g(y) \qquad \mathcal{E}\{x\} = \int_{-\infty}^{\infty} \alpha p_x(\alpha)\, d\alpha = \int_{-\infty}^{\infty} g(\beta) p_\phi(\beta)\, d\beta$$

where $p_\phi(\beta) = \frac{1}{2}\pi$ for $-\pi \leq \beta \leq \pi$. To obtain $\mu_x(t)$, we set $g(\beta) = \sin(\omega t + \beta)$ such that

$$\mu_x(t) = \frac{1}{2\pi} \int_{-\pi}^{\pi} \sin(\omega t + \beta)\, d\beta = 0$$

To obtain $P_x(t)$, we use $g(\beta) = \sin^2(\omega t + \beta)$ such that

$$P_x(t) = \frac{1}{2\pi} \int_{-\pi}^{\pi} \sin^2(\omega t + \beta) d\beta = \frac{1}{2}$$

It is an easy matter to compute the correlation function using this approach. We set $g(\beta) = \sin(\omega t + \beta) \sin(\omega \tau + \beta)$ and find

$$P_x(t, \tau) = \frac{1}{2\pi} \int_{-\pi}^{\pi} \sin(\omega t + \beta) \sin(\omega \tau + \beta)\, d\beta = \frac{1}{2} \cos[\omega(t - \tau)]$$

The correlation function obtained here is a function of the age variable, $t - \tau$, only; so the process is wide-sense stationary. Also, it is covariance stationary. The correlation function is phase insensitive, since the same results would have been obtained if we assumed $x(t) = \sin(\omega t + \phi + \gamma)$ for any deterministic value of γ.

Example 5.5-3. We again assume a sine wave with random phase, except that we assume the phase is uniformly distributed over the interval 0 to π such that

$$x(t) = \sin(\omega t + \phi)$$

$$p_\phi(\beta) = \begin{cases} \dfrac{1}{\pi} & 0 \leq \beta \leq \pi \\ 0 & \text{otherwise} \end{cases}$$

We again use the fundamental theorem of expectation to obtain

$$\mu_x(t) = \frac{1}{\pi} \int_0^{\pi} \sin(\omega t + \beta)\, d\beta = \frac{2}{\pi} \cos \omega t$$

$$P_x(t) = \frac{1}{\pi} \int_0^{\pi} \sin^2(\omega t + \beta)\, d\beta = \frac{1}{2}$$

$$P_x(t, \tau) = \frac{1}{\pi} \int_0^{\pi} \sin(\omega t + \beta) \sin(\omega \tau + \beta)\, d\beta = \frac{1}{2} \cos[\omega(t - \tau)]$$

$$V_x(t, \tau) = P_x(t, \tau) - \mu_x(t)\mu_x(\tau) = \frac{1}{2} \cos[\omega(t - \tau)] - \frac{4}{\pi^2} \cos^2(\omega t)$$

Thus we see that this process is not stationary in the mean. It is correlation stationary but not covariance stationary. Since the process is not stationary in the mean, it is not wide-sense stationary.

EXERCISES 5.5

5.5-1. Show that the stochastic process $x(t) = A \sin (t + \theta)$ is wide-sense stationary if θ is a random variable whose characteristic function $\Phi_\theta(\omega) = 0$ for $\omega = 1, 2$, and 4.

5.5-2. Consider the random process $x(t)$ defined by

$$x(t) = z(t) + w$$

where $z(t)$ is a stationary random process and w is a random variable independent of $z(t)$. Is $x(t)$ stationary?

Answer: Stationary.

5.5-3. Repeat Exercise 5.5-2 but let

$$x(t) = z(t) + wt$$

Answer: Not stationary.

5.5-4. Show that the covariance of a random process $x(t)$ remains unchanged when an arbitrary deterministic function $d(t)$ is added to it. What happens to the correlation function under these circumstances? Discuss the various types of stationarity of the process $z(t) = x(t) + d(t)$.

5.6. *Spectral density*

In the analysis of deterministic time-invariant systems, Laplace and Fourier transform techniques[1] may often be used to great advantage over time-domain approaches. This also turns out to be the case in the analysis of stochastic processes. In this section we shall discuss the spectral-density characterization of stochastic processes and relations between spectral density and correlation functions. In Chapter 6 we shall discuss, in part, the

[1] See Appendix A for a brief review of transform methods in linear systems.

response of linear systems with stochastic process inputs using spectral density concepts. In this section we shall restrict attention to stochastic processes that are at least wide-sense stationary.

Definition 5.6-1. ***Spectral Density (Continuous Time):*** The spectral density $\mathbf{C}_\mathbf{x}(\omega)$ of a wide-sense-stationary, continuous-time stochastic process $\mathbf{x}(t)$ is the Fourier transform of its autocorrelation $\mathbf{C}_\mathbf{x}(\tau)$:

$$\mathbf{C}_\mathbf{X}(\omega) = \int_{-\infty}^{\infty} e^{-j\omega\tau}\mathbf{C}_\mathbf{x}(\tau)\, d\tau \tag{5.6-1}$$

From the Fourier transform inversion formula, we may express $\mathbf{C}_\mathbf{x}(\tau)$ as

$$\mathbf{C}_\mathbf{x}(\tau) = \frac{1}{2\pi}\int_{-\infty}^{\infty} \mathbf{C}_\mathbf{X}(\omega)e^{j\omega\tau}\, d\omega \tag{5.6-2}$$

The fact that the autocorrelation function and spectral density constitute a Fourier transform pair is known as the *Wiener–Khintchine relationship.* The Wiener–Khintchine relationship may also be expressed by use of the Fourier cosine transformation for the scalar case. For the scalar case, we may easily show that correlation functions and spectral densities are symmetric in that $C_x(\tau) = C_x(-\tau)$ and $C_X(\omega) = C_X(-\omega)$. By using Euler's identity,[1] Eqs. (5.6-1) and (5.6-2) become (where $s = j\omega$)

$$C_X(\omega) = \int_{-\infty}^{\infty} C_x(\tau)\cos\omega\tau\, d\tau = 2\int_{0}^{\infty} C_x(\tau)\cos\omega\tau\, d\tau \tag{5.6-3}$$

$$C_x(\tau) = \frac{1}{2\pi}\int_{\infty}^{\infty} C_X(\omega)\cos\omega\tau\, d\omega = \frac{1}{\pi}\int_{0}^{\infty} C_X(\omega)\cos\omega\tau\, d\omega \tag{5.6-4}$$

Since the spectral density is the Fourier (or bilateral Laplace) transform of the correlation function,

$$\mathbf{C}_\mathbf{X}(\omega) = \int_{-\infty}^{\infty} \mathbf{C}_\mathbf{x}(\tau)e^{-j\omega\tau}\, d\tau \tag{5.6-5}$$

and the correlation function is obtained from the second-order probability density as

$$\mathbf{C}_\mathbf{x}(\tau) = \int_{-\infty}^{\infty}\int_{-\infty}^{\infty} \boldsymbol{\alpha}_1\boldsymbol{\alpha}_2^T\, p_{\mathbf{x}(t+\tau)\mathbf{x}(t)}(\boldsymbol{\alpha}_1, \boldsymbol{\alpha}_2, \tau)\, d\boldsymbol{\alpha}_1\, d\boldsymbol{\alpha}_2 \tag{5.6-6}$$

it is natural to inquire concerning a possible relation between the spectral density and the second-order probability density of $\mathbf{x}(t)$. If we define the

[1] Euler's identity is $e^{j\alpha} = \cos\alpha + j\sin\alpha$.

Fourier transform of the probability density as

$$G_{\mathbf{x}(t+\tau)\mathbf{x}(t)}(\boldsymbol{\alpha}_1, \boldsymbol{\alpha}_2, \omega) = \int_{-\infty}^{\infty} p_{\mathbf{x}(t+\tau)\mathbf{x}(t)}(\boldsymbol{\alpha}_1, \boldsymbol{\alpha}_2, \tau)e^{-j\omega\tau}\, d\tau \tag{5.6-7}$$

it follows directly by interchanging the order of integration in Eqs. (5.6-5) and (5.6-6) that

$$\mathbf{C}_{\mathbf{X}}(\omega) = \int_{-\infty}^{\infty}\int_{-\infty}^{\infty} \boldsymbol{\alpha}_1\boldsymbol{\alpha}_2{}^T G_{\mathbf{x}(t+\tau)\mathbf{x}(t)}(\boldsymbol{\alpha}_1, \boldsymbol{\alpha}_2, \omega)\, d\boldsymbol{\alpha}_1\, d\boldsymbol{\alpha}_2 \tag{5.6-8}$$

The *cross-spectral density* $C_{\mathbf{XY}}(s)$ of the two jointly wide-sense-stationary processes $\mathbf{x}(t)$ and $\mathbf{y}(t)$ is the Fourier transform of their cross-correlation function. Thus we have the transform and inverse-transform relations

$$\mathbf{C}_{\mathbf{XY}}(\omega) = \int_{-\infty}^{\infty} \mathbf{C}_{\mathbf{xy}}(\tau)e^{-j\omega\tau}\, d\tau \tag{5.6-9}$$

$$\mathbf{C}_{\mathbf{xy}}(\tau) = \frac{1}{2\pi}\int_{-\infty}^{\infty} \mathbf{C}_{\mathbf{XY}}(\tau)e^{-j\omega\tau}\, d\omega \tag{5.6-10}$$

Spectral and cross-spectral densities have several characteristics that are worth mentioning here. In each case, the reader can easily verify the following properties.

1. The diagonal element of $\mathbf{C}_{\mathbf{X}}(\omega)$ are *real positive even* functions of ω. For the scalar case, $C_X(\omega) = C_X(-\omega)$.
2. The mean-square-value matrix can be evaluated directly from the spectral density by

$$\mathbf{P}_{\mathbf{x}} = \mathbf{C}_{\mathbf{x}}(0) = \frac{1}{2\pi}\int_{-\infty}^{\infty} \mathbf{C}_{\mathbf{X}}(\omega)\, d\omega \tag{5.6-11}$$

   Since the spectral density $\mathbf{C}_{\mathbf{X}}(\omega)$ will often be a *rational* real positive even function of ω, the integral table given in Appendix A will often be of value in evaluating mean-square values.
3. $\mathbf{C}_{\mathbf{X}}(\omega) = \mathbf{C}_{\mathbf{X}}^T(-\omega)$.
4. $\mathbf{C}_{\mathbf{XY}}(\omega) = \mathbf{C}_{\mathbf{YX}}^T(-\omega)$.
5. For the scalar case, $\text{Re}[C_{XY}(\omega)]$ is an even function of ω.
6. For the scalar case, $\text{Im}[C_{XY}(\omega)]$ is an odd function of ω.

Some useful autocorrelation functions and their associated spectral densities are given in Fig. 5.6-1. Particularly important among these is the *white-noise process*, which has an autocorrelation and spectral density

$$C_x(\tau) = \Psi\delta_D(\tau) \qquad C_X(\omega) = \Psi$$

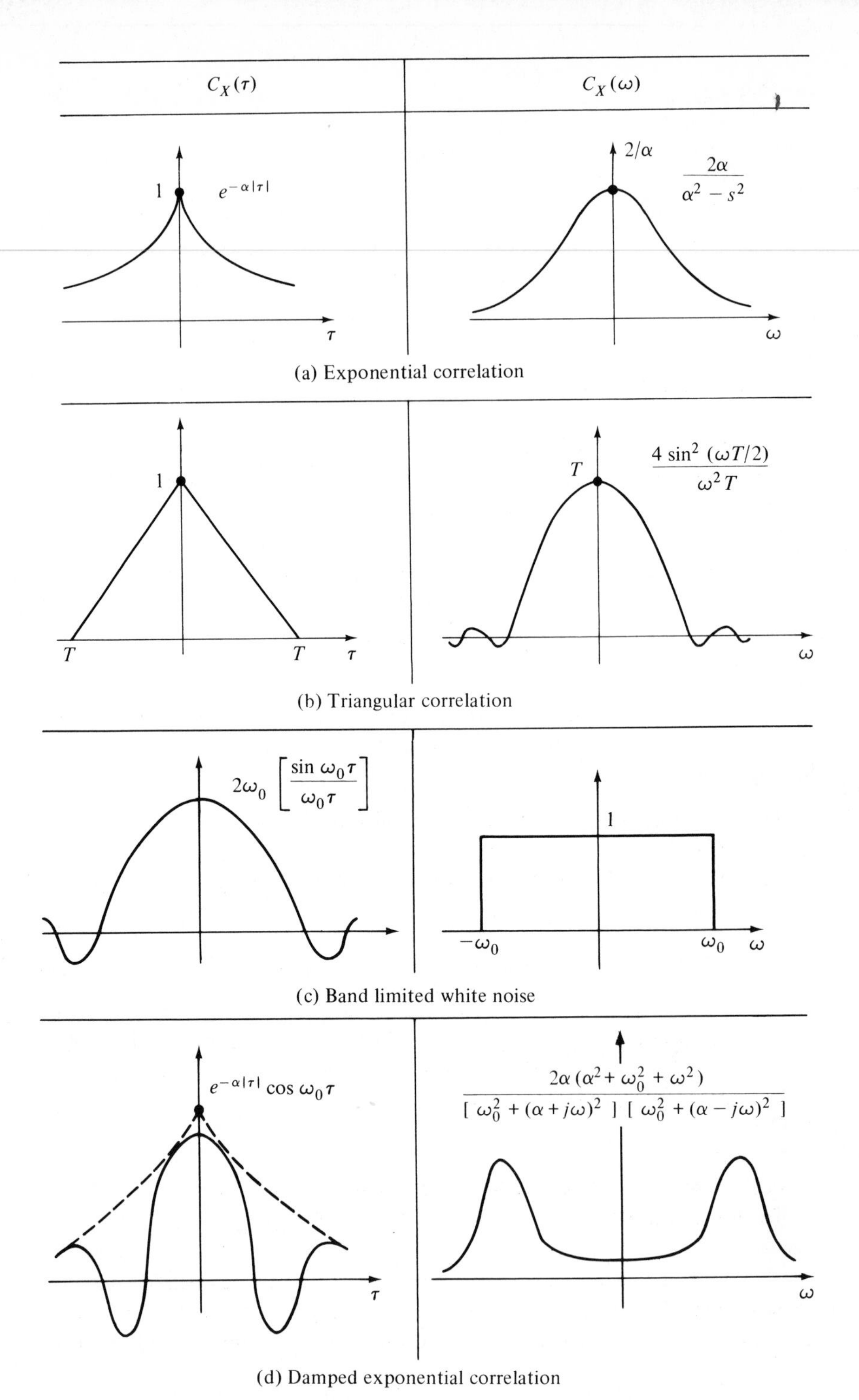

FIG. 5.6-1. Typical correlation functions and associated spectra.

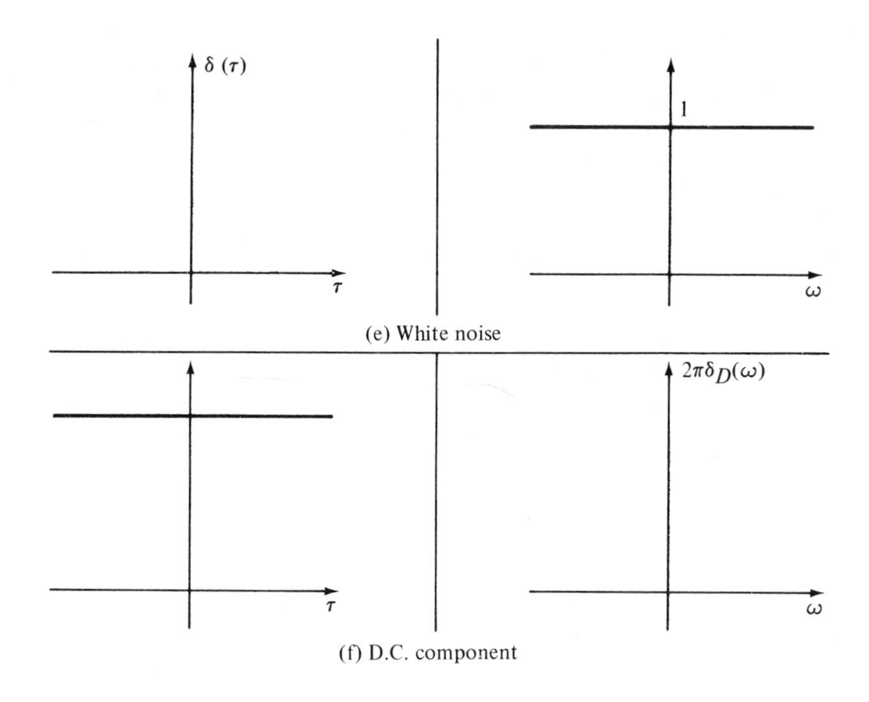

FIG. 5.6-1. *(continued.)*

The fact that the white-noise process has a flat spectral density and possesses equal energy at all frequencies makes it especially useful for test purposes in measurements and for analytical work. We shall see much use for the white-noise process in the remaining three chapters of this text.

We may also define and obtain spectral densities for discrete-time wide-sense-stationary random processes. For the stationary case we may express the correlation function as

$$\mathbf{C}_{\mathbf{x}}(nT) = \mathbf{P}_{\mathbf{x}}(\overline{k + n}T, kT) = \mathcal{E}\{\mathbf{x}(\overline{k + n}T)\mathbf{x}^T(kT)\} \tag{5.6-12}$$

where n takes on only integer values. We shall define discrete spectral density as being the discrete Fourier or bilateral z transform of the correlation function, which is sometimes called the correlation sequence for discrete random processes. Thus, we have the *discrete Wiener–Khintchine relations*

$$\mathbf{C}_{\mathbf{X}}(z) = T \sum_{n=-\infty}^{\infty} \mathbf{C}_{\mathbf{x}}(nT)z^{-n} \qquad \mathbf{C}_{\mathbf{x}}(nT) = \frac{1}{2\pi Tj} \oint \mathbf{C}_{\mathbf{X}}(z)z^{n-1}\,dz \tag{5.6-13}$$

where the integration is over the unit circle centered at $z = 0$. In the presentation to follow we shall make little use of discrete spectral density concepts, although we shall often consider discrete-time random processes.

We can also relate in a nonrigorous manner spectral density to signal power in a random waveform. The average power, or more correctly time average[1] of the waveform, squared in the ith record of a random scalar waveform $x^i(t)$ is

$$P_{AV} = \lim_{T \to \infty} \frac{1}{T} \int_{-T/2}^{T/2} [x^i(t)]^2 \, dt \tag{5.6-14}$$

To resolve existence problems associated with the Fourier transform of a waveform $x_i(t)$ that may have infinite energy, we define a truncated signal

$$x_T^i(t) = \begin{cases} x^i(t) & |t| \leq T/2 \\ 0 & |t| > T/2 \end{cases} \tag{5.6-15}$$

such that we have for the Fourier transform of the truncated signal

$$X_T^i(\omega) = \int_{-\infty}^{\infty} x_T^i(t) e^{-j\omega t} \, dt = \int_{-T/2}^{T/2} x^i(t) e^{-j\omega t} \, dt \tag{5.6-16}$$

The average power in the truncated waveform or the original waveform over the interval $-T/2$ to $T/2$ may be written using Parceval's theorem:[2]

$$\int_{-\infty}^{\infty} f_1(t) f_2(t) \, dt = \frac{1}{2\pi} \int_{-\infty}^{\infty} F_1(\pm\omega) F_2(\mp\omega) \, d\omega \tag{5.6-17}$$

as

$$\int_{-T/2}^{T/2} \frac{[x^i(t)]^2}{T} \, dt = \int_{-\infty}^{\infty} \frac{[x_T^i(t)]^2}{T} \, dt = \frac{1}{2\pi} \int_{-\infty}^{\infty} \frac{X_T^i(\omega) X_T^i(-\omega)}{T} \, d\omega \tag{5.6-18}$$

where $X_T^i(\omega)$ is the Fourier or bilateral Laplace transform of $x_T^i(t)$. It is convenient to call the term $X_T^i(\omega) X_T^i(-\omega)/T$ the power spectral density or simply spectral density of the ith record, $x^i(t)$, in the time interval $-0.5T \leq t \leq 0.5T$, and denote it by $C_T^i(T, \omega)$. We wish to show that the spectral density of the random waveform $x(t)$ is the limit as $T \longrightarrow \infty$ of the expected value of the power spectral density associated with $X_T(t)$, or

$$C_X(\omega) = \lim_{T \to \infty} \frac{1}{T} \mathcal{E}\{X_T(\omega) X_T(-\omega)\} = \lim \frac{1}{T} \mathcal{E}\{C_X(T, \omega)\} \tag{5.6-19}$$

[1] Integrals and time averages of stochastic processes will be discussed in Section 5.8.

[2] See Appendix A.

For this result to be valid we must impose the restrictions that

$$\lim_{T\to\infty} \mathcal{E}[C_X(T, \omega)] = \lim_{T\to\infty} \mathcal{E}\left\{\frac{X_T(\omega)X_T(-\omega)}{T}\right\} = C_X(\omega) \qquad (5.6\text{-}20)$$

and

$$\lim_{T\to\infty} \operatorname{var}\{C_X(T, \omega)\} = 0 \qquad (5.6\text{-}21)$$

We shall only consider the first requirement. We may use the Fourier transform of $x_T(t)$ in Eq. (5.6-20) to obtain

$$\begin{aligned}\mathcal{E}\{C_X(T, \omega)\} &= \mathcal{E}\left\{\frac{1}{T}\int_{-T/2}^{T/2}\int_{-T/2}^{T/2} x(\lambda_1)x(\lambda_2)e^{-j\omega\lambda_1}e^{j\omega\lambda_2}\,d\lambda_1\,d\lambda_2\right\}\\ &= \frac{1}{T}\int_{-T/2}^{T/2}\int_{-T/2}^{T/2} C_x(\lambda_1 - \lambda_2)e^{-i\omega\lambda_1}e^{i\omega\lambda_2}\,d\lambda_1\,d\lambda_2 \qquad (5.6\text{-}22)\end{aligned}$$

Now if we make the change of variable $\tau = \lambda_1 - \lambda_2$, $\eta = \lambda_1 + \lambda_2$, it is a simple matter to show that

$$\begin{aligned}\mathcal{E}\{C_X(t, \omega)\} &= \int_{-T}^{T}\left[1 - \frac{|\tau|}{T}\right]C_x(\tau)e^{-j\omega\tau}\,d\tau\\ &= \int_{-T}^{T} C_x(\tau)e^{-j\omega\tau}\,d\tau - \int_{-T}^{T}\frac{|\tau|}{T}C_x(\tau)e^{-j\omega\tau}\,d\tau \qquad (5.6\text{-}23)\end{aligned}$$

The first expression in the foregoing is the desired result. The second expression must go to zero as $T \longrightarrow \infty$, and this will happen if

$$\int_{-\infty}^{\infty} |\tau C_x(\tau)|\,d\tau < \infty \qquad (5.6\text{-}24)$$

If this restriction is satisfied, Eq. (5.6-23) becomes

$$C_X(\omega) = \lim_{T-\infty}\frac{1}{T}\mathcal{E}\{X_T(\omega)X_T(-\omega)\} = \int_{-\infty}^{\infty} C_x(\tau)e^{-j\omega\tau}\,d\tau \qquad (5.6\text{-}25)$$

which shows that the spectral density [as defined by Eq. (5.6-19)] is the Fourier transformation of the correlation function, as required.

Example *5.6-1.* We again consider the sine wave with random phase distributed uniformly, $-\pi \leq \theta \leq \pi$,

$$x^i(t) = \sin(\omega_1 t + \theta^i)$$

where we use ω_1 rather than ω in order not to confuse the fixed known frequency ω_1 with the Fourier transform variable ω. The Fourier transform of a finite time sample of the ith record is

$$X_T^i(\omega) = \int_{-T/2}^{T/2} \sin(\omega_1 t + \theta^i)e^{-j\omega t}\,dt$$
$$= \frac{\sin[(\omega - \omega_1)T/2]}{j(\omega - \omega_1)}e^{j\theta^i} - \frac{\sin[(\omega + \omega_1)T/2]}{j(\omega + \omega_1)}e^{-j\theta^i}$$

Thus

$$X_T^i(\omega)X_T^i(-\omega) = \frac{\sin^2[(\omega - \omega_1)T/2]}{(\omega - \omega_1)^2} + \frac{\sin^2[(\omega + \omega_1)T/2]}{(\omega + \omega_1)^2}$$
$$+ 2\cos(2\theta^i)\frac{\sin[(\omega - \omega_1)T/2]\sin[(\omega + \omega_1)T/2]}{\omega^2 - \omega_1^2}$$

From Eq. (5.6-12) we see we must take the expectation of the foregoing to obtain the spectral density. The expectation of the last term on the right in the foregoing is zero, and since the first two terms are deterministic, we have

$$C_X(\omega) = \lim_{T\to\infty} \frac{1}{T}\frac{\sin^2[(\omega - \mathrm{A}_1)T/2]}{(\omega - \omega_1)^2} + \frac{\sin^2[(\omega + \omega_1)T/2]}{(\omega + \omega_1)^2}$$

If ω is not equal to $\pm\omega_1$, the limit of this expression as T becomes infinite is clearly zero. If ω is equal to $\pm\omega_1$, the sine expressions are small such that they may be expanded in a Taylor series to give $\sin^2(\alpha T/2) \simeq \alpha^2 T^2/4$, and we see that the limit does not exist. We might guess, therefore, that the spectral density contains impulses at $\pm\omega_1$ and write

$$C_X(\omega) = K[\delta_D(\omega - \omega_1) + \delta_D(\omega + \omega_1)]$$

Occurrence of thei mpulses is not unreasonable, since all the energy in the random wave should exist at the two discrete frequencies $\pm\omega_1$. From Eqs. (5.6-14) and (5.6-18) [or alternatively the definition of Eq. (5.6-1)], we have

$$C_x(0) = \mathcal{E}\{P_{\mathrm{AV}}\} = \frac{1}{2\pi j}\int_{-j\infty}^{j\infty} C_X(\omega)\,d\omega$$

which becomes for our impulsive spectral density

$$\mathcal{E}\{P_{\mathrm{AV}}\} = \frac{1}{2\pi}\int_{-\infty}^{\infty} K[\delta_K(\omega - \omega_1) + \delta_K(\omega + \omega_1)]\,d\omega = \frac{K}{\pi}$$

By physical inspection of the random waveform, we know that its average power is $\frac{1}{2}$. Therefore, we have $K = \pi/2$ and have thus completely specified the spectral density. A much simpler method is to take the Fourier transform of the correlation function that we previously obtained in Example 5.5-3, which leads more directly to the spectral density of the sine wave with random phase:

$$\mathcal{C}_X(\omega) = \pi/2[\delta_D(\omega + \omega_1) + \delta_D(\omega - \omega_1)] \qquad C_x(\tau) = 0.5 \cos \omega\tau$$

EXERCISES 5.6

5.6-1. If the correlation function for the process $x(t)$ is given by

$$C_x(\tau) = \begin{cases} 1 - \dfrac{|\tau|}{T} & |\tau| < T \\ 0 & \text{otherwise} \end{cases}$$

find $\mathcal{C}_X(\omega)$.

Answer: $\mathcal{C}_X(\omega) = \dfrac{-(e^{j\omega T/2} - e^{-j\omega T/2})^2}{\omega^2 T}$

5.6-2. Which of the following are permissible correlation functions and spectral densities? Give reasons for each answer.

(a) $\mathcal{C}_x(\omega) = \cos \omega/(1 + \omega^2)$.
(b) $C_x(\tau) = [1 - 2|\tau|]e^{-|\tau|}$.
(c) $C_x(\tau) = |\tau| e^{-|\tau|}$.
(d) $C_x(\tau) = (1 + \alpha|\tau|)e^{-|\tau|}$

Answer: (a) yes, (b) no, (c) no, (d) $\alpha \geq 1$

5.7. Orthogonal representation

As is well known, a real periodic function $x(t)$ may be expanded in a Fourier series of sines and cosines, or complex exponentials, that represents $x(t)$. This expansion is unique and is given by the well-known Fourier series relations

$$\gamma_n = \frac{1}{T} \int_a^{T+a} x(t)e^{-jn\omega_0 t}\, dt \tag{5.7-1}$$

$$x(t) = \sum_{n=-\infty}^{\infty} \gamma_n e^{jn\omega_0 t} \tag{5.7-2}$$

for the complex representation, or

$$\alpha_n = \frac{1}{T} \int_a^{T+a} x(t) \cos n\omega_0 t \, dt \tag{5.7-3}$$

$$\beta_n = \frac{1}{T} \int_a^{T+a} x(t) \sin n\omega_0 t \, dt \tag{5.7-4}$$

$$x(t) = \alpha_0 + 2 \sum_{n=1}^{\infty} (\alpha_n \cos n\omega_0 t + \beta_n \sin n\omega_0 t) \tag{5.7-5}$$

for the sine–cosine representation. Here T is the period of the periodic process and ω_0 is the fundamental frequency equal to $\omega_0 = 2\pi/T$.

This same representation can be used to represent a real nonperiodic deterministic function $x(t)$ over any *specified finite* interval $a \leq t \leq a + b$. This expansion is nonunique, however, since the series may be chosen to represent any periodic function that represents $x(t)$ for $a \leq t \leq a + b$. This situation is illustrated in Fig. 5.7-1, which shows two periodic waveforms, $f_1(t)$ and $f_2(t)$, each of which represents $x(t)$ over the interval 0 to 1 second(s).

There is no fundamental reason why a stationary random process cannot be expanded in a Fourier series over an interval $a \leq t \leq a + b$.

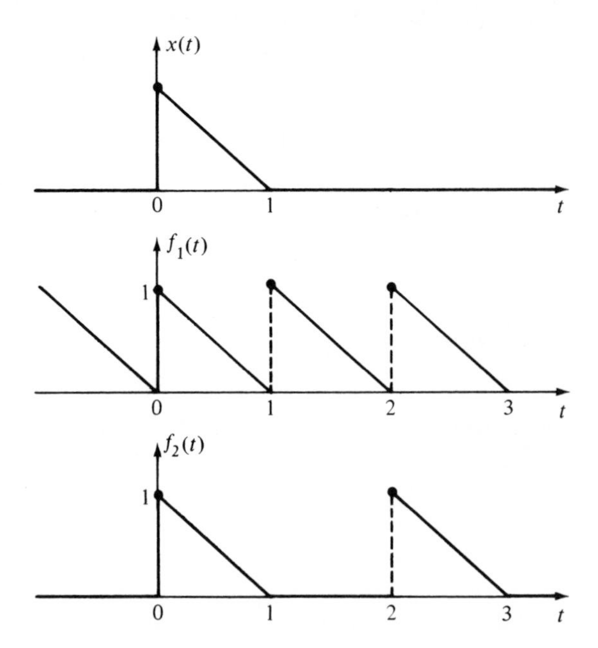

FIG. 5.7-1. Different periodic waveforms, which represent $x(t)$ for $0 \leq t \leq 1$.

A simple way[1] of doing this is to use the expansion of Eqs. (5.7-1) and (5.7-2) or, alternatively, that of Eqs. (5.7-3) through (5.7-5) with $T = b$. Unfortunately, there is, in general, correlation between the coefficients of the series in that

$$\operatorname{cov}\{\gamma_i, \gamma_n\} \neq 0 \qquad i \neq n \tag{5.7-6}$$

and this makes the representation very cumbersome to use on actual problems. It is desirable that the coefficients γ_i be uncorrelated. This can be seen immediately by considering the correlation function of the random process $x(t)$ and use of the series representation of Eq. (5.7-2), which yields

$$C_x(\tau) = \mathcal{E}\{x(t + \tau)x(t)\} = \sum_{i=-\infty}^{\infty} \sum_{n=-\infty}^{\infty} \mathcal{E}\{\gamma_i \gamma_n\} e^{j(\overline{i+n}\omega_0 t + i\omega_0 \tau)} \tag{5.7-7}$$

It turns out that this expression is valid only if t and $t + \tau$ are in the same interval of length b. That γ_i and γ_n are complex quantities imposes no fundamental problems in determining the desired expectation. It would be highly desirable for this expectation to be zero for $i \neq n$, but it turns out that this will happen only for $b \longrightarrow \infty$. Let us now look at a procedure, known as the Karhunen–Loéve expansion, that will force the coefficients to be uncorrelated.

We shall consider a time-limited waveform in which the energy has some finite value

$$\int_{-T/2}^{T/2} x^2(t)\, dt = \int_{-\infty}^{\infty} x_T^2(t)\, dt < \infty \tag{5.7-8}$$

and will for convenience let $a = -T/2$ and $b = T/2$. We inquire concerning the possibility of writing an infinite series expression for $x_T(t)$ such that

$$x_T(t) = \sum_{i=0}^{\infty} \gamma_i \psi_i(t) \tag{5.7-9}$$

where the $\psi_i(t)$ are a set of orthonormal functions. An elementary set of orthonormal function is $\psi_i(t) = \sqrt{2}/T \cos 2\pi i t/T$. This set is orthonormal over the interval $-T/2 < t < T/2$, since the definition of an orthonormal function

$$\int_{-T/2}^{T/2} \psi_i(t)\psi_n(t)\, dt = \delta_K(i - n) \tag{5.7-10}$$

is satisfied. Clearly, we may alter the time region of orthogonality by a different choice of a and b. The interval 0 to T is often used.

[1] Integrals of stochastic processes will be discussed in Section 5.8.

We shall pick the γ_i, the coefficients in the orthogonal series expansion, such that

$$J = \int_{-T/2}^{T/2} \tilde{x}_k^2(t)\, dt = \int_{-T/2}^{T/2} \left[x(t) - \sum_{k=1}^{K} \gamma_k \psi_k(t)\right]^2 dt \qquad (5.7\text{-}11)$$

the integral square error in a finite sum approximation to $x(t)$, is minimized. We set $\partial J/\partial \gamma_i = 0$ for each of the K values of γ, use the orthonormality property of the ψ_i, and obtain

$$\gamma_i = \int_{-T/2}^{T/2} x(t)\psi_i(t)\, dt \qquad (5.7\text{-}12)$$

which can be shown to yield a minimum of J in Eq. (5.7-11). We may also show that $J \longrightarrow 0$ as $K \longrightarrow \infty$, which indicates mean-square convergence, since $\lim_{K\to\infty} \mathcal{E}\{J\} = 0$. An advantage to the orthogonal representation of $x(t)$ is apparent from Eq. (5.7-12) in that the coefficients γ_i may be determined by a single integration rather than having to solve a set of integral equations.

There may be an infinite number of solutions to Eq. (5.7-12), since we have not specified the ψ_i. It is advantageous to choose a set of ψ_i such that we obtain uncorrelated zero mean γ_i, as we have previously indicated. Thus we require

$$\mathcal{E}\{\gamma_i\} = 0 \qquad \mathcal{E}\{\gamma_i\gamma_n\} = \lambda_i\delta_K(k - n) \qquad (5.7\text{-}13)$$

If we substitute the γ_i from Eq. (5.7-12) into Eq. (5.7-13), we obtain

$$\mathcal{E}\{\gamma_i\gamma_n\} = \int_{-T/2}^{T/2}\int_{-T/2}^{T/2} \psi_i(t)P_x(t, \tau)\psi_n(\tau)\, dt\, d\tau = \lambda_i\delta_K(i - n) \qquad (5.7\text{-}14)$$

The foregoing relation will only be satisfied if

$$\lambda_n\psi_n(t) = \int_{-T/2}^{T/2} P_x(t, \tau)\psi_n(\tau)\, d\tau \qquad (5.7\text{-}15)$$

as can easily be verified. It is common practice to denote the nonnegative numbers λ_i as eigenvalues and to call the $\psi_i(t)$ eigenfunctions. This particular expansion is known as the Karhunen–Loéve expansion. Equations (5.7-9), (5.7-12), and (5.7-15) are the defining relations for this expansion.

Example *5.7-1*. We shall now consider one of the most fundamental of stochastic processes, the Wiener process. This process will be discussed in considerable detail in Chapter 7. It is obtained by passing white gaussian noise through an integrator such that we have the model

$$\dot{x} = w(t) \qquad x(0) = 0 \qquad C_w(\tau) = \sigma^2\delta_D(\tau)$$

such that the *Wiener process* $x(t)$ is

$$x(t) = \int_0^t w(\eta)\, d\eta$$

which has the correlation function

$$\begin{aligned} P_x(t, \tau) &= \mathcal{E}\{x(t)x(\tau)\} = \int_0^t \int_0^\tau \mathcal{E}\{w(\eta_1)w(\eta_2)\}\, d\eta_1\, d\eta_2 \\ &= \int_2^t \int_0^\tau \sigma^2 \delta_D(\eta_1 - \eta_2)\, d\eta_2\, d\eta_1 = \sigma^2 \min(t, \tau) \\ &= \begin{cases} \sigma^2 t & t \leq \tau \\ \sigma^2 \tau & \tau \leq t \end{cases} \end{aligned}$$

For this process Eq. (5.7-15) yields, for the region of desired orthogonality, $0 < t < T$,

$$\lambda_n \psi_n(t) = \int_0^T P_x(t, \tau)\psi_n(\tau)\, d\tau = \sigma^2 \int_0^t \tau \psi_n(\tau)\, d\tau + \sigma^2 \int_0^T t \psi_n(\tau)\, d\tau$$

We wish to solve this integral equation for the orthogonal series representation. An effective way to attempt this is to convert the integral equation into a differential equation. If we differentiate the integral equation twice with respect to time, we obtain

$$\ddot{\psi}_n + \frac{\sigma^2}{\lambda_n} \psi_n(t) = 0$$

The solution to this equation (since σ^2 and λ_n are positive) is

$$\psi_n(t) = a_0 \cos b_n t + a_1 \sin b_n t \qquad b_n = \sqrt{\frac{\sigma}{\lambda_n^{1/2}}}$$

If we substitute this solution into the integral equation, we obtain

$$\begin{aligned} \lambda_n a_0 \cos b_n t + \lambda_n a_1 \sin b_n t = \sigma^2 a_0 &\left[\frac{1}{b_n^2} \cos b_n t - \frac{1}{b_n^2} + \frac{t}{b_n} \sin b_n T\right] \\ &+ \sigma^2 a_1 \left[\frac{1}{b_n^2} \sin b_n t - \frac{t}{b_n} \cos b_n T\right] \end{aligned}$$

For this equation to hold for all t and T, it is necessary that $a_0 = 0$ such that the $\sigma^2 a_0 / b_n$ term, which does not depend upon t and T, is zero. Setting $a_0 = 0$ reduces the foregoing equation to

$$\lambda_n \sin b_n t = \sigma^2 \left[\frac{1}{b_n^2} \sin b_n t - \frac{t}{b_n} \cos b_n T\right]$$

For this to be true for all t, we must have

$$\lambda_n = \frac{\sigma^2 T^2}{(n - \frac{1}{2})^2 \pi^2}$$

Since $\psi(t)$ is orthonormal, we must have $a_1 = \sqrt{2/T}$, and thus the orthogonal representation is

$$x(t) = \sum_{n=1}^{\infty} \gamma_n \psi_n(t) \qquad \gamma_n = \int_0^T x(t)\psi_n(t)\, dt$$

where

$$\psi_n(t) = \sqrt{\frac{2}{T}} \sin\left[(n - 0.5)\frac{\pi t}{T}\right]$$

We now generate the orthogonal coefficients as illustrated in Fig. 5.7-2. At first glance it may appear that we have greatly complicated any problem by using this orthogonal series expansion. This is not the case, however, since the first few, often the first one, of the γ_i tell us all we need to know to make an optimum decision or estimation in an actual problem. These few γ_n will constitute what are called a *sufficient statistic*. This is indeed fortunate, since it will not often be a trivial task to compute the orthogonal representation of random processes. It also turns out that many other procedures in use in estimation and decision theory lead to the same general result for the optimum system structure as obtained by use of the Karhunen–Loéve expansion. The

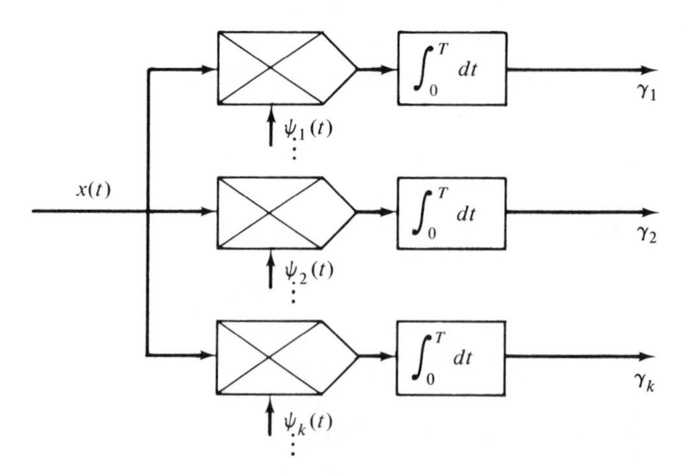

FIG. 5.7-2. Generation of γ_n for the orthogonal series expansion of $x(t)$, Example 5.7-1.

concepts involved in the orthogonal expansion are, however, quite important, even though it may not be often that one is required to actually obtain the solution to a given problem.

EXERCISES 5.7

5.7-1. Consider a stationary random process with

$$C_x(\tau) = 0.25e^{-2|\tau|}$$

It is desired to find an orthogonal expansion for $x(t)$ for $-T/2 < t < T/2$. Show that the eigenfunctions of the Karhunen–Loéve expansion are obtained by solution of the differential equation

$$\ddot{\psi}_n + \frac{2 - \lambda_n}{\lambda_n}\psi_n = 0$$

Show also that the characteristic functions and characteristic values are such that the expansion is

$$x(t) = \sum_{i=1}^{\infty} \gamma_i \psi_i + \bar{\gamma}_i \bar{\psi}_i$$

where

$$\psi_i = \frac{\cos 2\alpha_i t}{2(1 + \alpha_i^2)^{1/2}\left(\frac{T}{2} + \frac{\sin 2\alpha_i T}{4\alpha_i}\right)^{1/2}}$$

$$\bar{\psi}_i = \frac{\sin 2\bar{\alpha}_i t}{2(1 + \alpha_i^2)\left(\frac{T}{2} - \frac{\sin 2\bar{\alpha}_i T}{4\bar{\alpha}_i}\right)^{1/2}}$$

$$\alpha_i \tan \alpha_i T = \bar{\alpha}_i \cot \bar{\alpha}_i T = 1$$

5.7-2. What does Eq. (5.7-7) become for the sine–cosine Fourier series representation?

5.8. Derivatives, integrals, and ergodicity

In Sections 4.2 and 4.3 we examined in detail the problems and techniques associated with functions of random variables. Without exception, these results may be applied to functions of stochastic processes. In the case of stochastic processes, we can consider a broader class of functions, actually operators, which include in particular time derivatives and integrals. The purpose of this section is to explore the definitions and meanings of these operators for stochastic processes.

We note first that derivatives and integrals of stochastic processes, in general, are stochastic processess. As in the case of deterministic time functions, these operators are defined in terms of limits. For stochastic processess, the definitions will involve limits of random sequences, which were discussed in Section 4.6. As noted there, these limits can be defined in several ways, including convergence with probability 1, mean-square convergence, and convergence in probability. We shall consider only convergence in the mean-square sense here.

The definitions of derivatives and integrals of stochastic processes differ from those of ordinary time functions because we wish to ensure convergence of the associated limits for the entire family of functions that comprise the stochastic process. Of course, if the limits exist for every member of the family, there is no problem; this is, however, much too stringent a requirement. For most practical applications, it is sufficient to ensure that the limits exist in the mean-square sense. We shall begin with a brief discussion of the continuity of stochastic process. Next the definition of the integral of a stochastic process will be presented and integrals will be applied to a study of time averages and ergodicity. The final topic considered will be the definition of derivatives of stochastic process. We shall consider only continuous-time stochastic processes in this section, since derivatives and integrals (in the usual sense) are meaningless for discrete-time processes. Of course, there are discrete equivalents, but these are simply algebraic operations that can be handled by the methods of Sections 4.2 and 4.3.

Stochastic Continuity. We say that a stochastic process $x(t)$ is continuous in the mean-square sense at $t = t_1$ if

$$\operatorname*{l.i.m.}_{\epsilon \to 0} \mathbf{x}(t_1 + \epsilon) = \mathbf{x}(t_1) \tag{5.8-1}$$

or, equivalently,

$$\lim_{\epsilon \to 0} \mathcal{E}\{\|\mathbf{x}(t_1 + \epsilon) - \mathbf{x}(t_1)\|^2\} = 0 \tag{5.8-2}$$

We may use the correlation function to express the left-hand side of Eq. (5.8-2) as

$$\begin{aligned} \lim_{\epsilon \to 0} \mathcal{E}\{\|\mathbf{x}(t_1 + \epsilon) - \mathbf{x}(t_1)\|^2\} = \lim_{\epsilon \to 0} \operatorname{tr}\{&\mathbf{P_x}(t_1 + \epsilon, t_1 + \epsilon) \\ &- \mathbf{P_x}(t_1 + \epsilon, t_1) - \mathbf{P_x}(t_1, t_1 + \epsilon) + \mathbf{P_x}(t_1, t_1)\} \end{aligned} \tag{5.8-3}$$

If the correlation function $\mathbf{P_x}(t, \tau)$ (at least the diagonal terms) is continuous in t and τ at $t = \tau = t_1$, then $\mathbf{x}(t)$ is continuous in the mean-square sense at t_1.

It should be noted that continuity in the mean-square sense does not guarantee that all or even one of the time functions which comprise the stochastic process are continuous in the ordinary sense. Consider, for instance, the Poisson process discussed in Example 5.4-2. The correlation function for this process is continuous everywhere, so that the Poisson process is continuous in the mean-square sense for all t. We know, however, that almost every time function of the Poisson process is discontinuous.

For stationary stochastic processes, Eq. (5.8-3) becomes

$$\lim_{\epsilon \to 0} \mathcal{E}\{\| \mathbf{x}(t_1 + \epsilon) - \mathbf{x}(t_1) \|^2 = \lim_{\epsilon \to 0} \operatorname{tr}\{\mathbf{C}_\mathbf{x}(0) - \mathbf{C}_\mathbf{x}(-\epsilon) - \mathbf{C}_\mathbf{x}(\epsilon) + \mathbf{C}_\mathbf{x}(0)\} \tag{5.8-4}$$

Hence, if $\mathbf{C}_\mathbf{x}(\tau)$ is continuous at $\tau = 0$, a stationary stochastic process is continuous in the mean-square sense for all t.

If a stochastic process $\mathbf{x}(t)$ is continuous in the mean-square sense at t_1, then the mean $\boldsymbol{\mu}_\mathbf{x}(t)$ of $\mathbf{x}(t)$ is continuous at t_1 in the ordinary sense, since $\boldsymbol{\mu}_\mathbf{x}(t)$ is a deterministic time function. To prove this result, we note that

$$\begin{aligned} &\mathcal{E}\{\| \mathbf{x}(t_1 + \epsilon) - \mathbf{x}(t_1) \|^2\} = \operatorname{tr}\{\operatorname{var}\{\mathbf{x}(t_1 + \epsilon) - \mathbf{x}(t_1)\}\} \\ &+ \mathcal{E}\{\mathbf{x}^T(t_1 + \epsilon) - \mathbf{x}^T(t_1)\}\mathcal{E}\{\mathbf{x}(t_1 + \epsilon) - \mathbf{x}(t)\} \geq \mathcal{E}\{\mathbf{x}^T(t_1 + \epsilon) - \mathbf{x}^T(t_1)\} \\ &\mathcal{E}\{\mathbf{x}(t_1 + \epsilon) - \mathbf{x}(t_1)\} \end{aligned} \tag{5.8-5}$$

Since $\mathbf{x}(t)$ is mean-square continuous at t_1, the limit of the left-hand side of Eq. (5.8-4), as ϵ approaches zero, is zero. Hence, the limit of the right-hand side must also be zero, so that

$$\lim_{\epsilon \to 0} \mathcal{E}\{\mathbf{x}(t_1 + \epsilon) + \mathbf{x}(t_1)\} = \lim_{\epsilon \to 0} \boldsymbol{\mu}_\mathbf{x}(t_1 + \epsilon) - \boldsymbol{\mu}_\mathbf{x}(t_1) = \mathbf{0} \tag{5.8-6}$$

which establishes the continuity of $\boldsymbol{\mu}_\mathbf{x}(t)$ at t_1 as stated. This result means that for a stochastic process which is continuous in the mean-square sense we may interchange the ordering of the operations of expectation and limiting, since

$$\lim_{\epsilon \to 0} \mathcal{E}\{\mathbf{x}(t_1 + \epsilon)\} = \boldsymbol{\mu}_\mathbf{x}(t_1) = \mathcal{E}\{\mathbf{x}(t_1)\} = \mathcal{E}\{\underset{\epsilon \to 0}{\text{l.i.m.}}\ \mathbf{x}(t + \epsilon)\} \tag{5.8-7}$$

This is an important result that will be of value in the material which follows.

Stochastic Integration. Time integrals of stochastic processes are often of value in the practical application of stochastic process theory. For example, the finite time average defined by

$$\hat{\mathbf{x}}(t_f) = \frac{1}{t_f} \int_0^{t_f} \mathbf{x}(t)\, dt \tag{5.8-8}$$

is used to approximate the mean $\boldsymbol{\mu}_\mathbf{x}$ for a stationary stochastic process. We shall examine the concept of time averages in detail in the sequel. First, however, we consider the meaning of the stochastic integral in Eq. (5.8-8).

The definition of the stochastic integral[1] that we wish to use takes the following form:

$$\int_a^b \mathbf{x}(t)\,dt = \underset{\substack{k\to\infty \\ \delta\to 0}}{\text{l.i.m.}} \sum_{i=1}^{k} \mathbf{x}(t_i)[t_i - t_{i-1}] \tag{5.8-9}$$

where $a = t_0 < t_1 < t_2 < \cdots < t_k = b$, and δ is the maximum of $t_i - t_{i-1}$, $i = 1, 2, \ldots, k$. It can be shown that a necessary and sufficient condition for the existence of the limit of Eq. (5.8-9) is that $\mathbf{P}_\mathbf{x}(t, \tau)$ be Riemann integrable over $a \le t \le b$, $a \le \tau \le b$. This is a very mild restriction on the class of stochastic processes. We noted that if the limit on the right-hand side of Eq. (5.8-9) exists, it will in general be equal to a random variable $\mathbf{y}$. In other words, the result of the integration of the stochastic process $\mathbf{x}(t)$ is to generate a random variable defined as

$$\mathbf{y} = \int_{-\infty}^{\infty} \mathbf{x}(t)\,dt \tag{5.8-10}$$

with the equality to be interpreted in the mean-square sense.

Let us suppose that we have a continuous-time stochastic process $\mathbf{x}(t)$ with finite second moments [mean $\boldsymbol{\mu}_\mathbf{x}(t)$ and variance $\mathbf{V}_\mathbf{x}(t, \tau)$] that are continuous functions of t and τ. Then the integral of Eq. (5.8-9) is well defined in the sense that the limit in the mean exists. In addition, the mean and variance of the integral are given by

$$\mathcal{E}\left\{\int_a^b \mathbf{x}(t)\,dt\right\} = \int_a^b \mathcal{E}\{\mathbf{x}(t)\}\,dt = \int_a^b \boldsymbol{\mu}_\mathbf{x}(t)\,dt \tag{5.8-11}$$

$$\text{var}\left\{\int_a^b \mathbf{x}(t)\,dt\right\} = \int_a^b dt \int_a^b d\tau\, \text{cov}\{\mathbf{x}(t), \mathbf{x}(\tau)\} = \int_a^b dt \int_a^b d\tau\, \mathbf{V}_\mathbf{x}(t, \tau) \tag{5.8-12}$$

Equations (5.8-11) and (5.8-12) are extremely important; they indicate that we may interchange the linear operations of integration and expectation. In the case of a stationary process, we need only require that $\mathbf{R}_\mathbf{x}(\tau)$ be continuous. We shall not attempt to prove this result rigorously here; we do, however, offer a plausibility argument. Since $\mathbf{V}_\mathbf{x}(t, \tau)$ and $\boldsymbol{\mu}_\mathbf{x}(t)$ are continuous in t, τ, $\mathbf{P}_\mathbf{x}(t, \tau)$ must also be continuous and hence Riemann integrable; so our previous statement would guarantee the existence of the integral. From the preceding discussion of continuity we know that if $\mathbf{P}_\mathbf{x}(t, \tau)$ is continuous, then $\mathbf{x}(t)$ will be continuous in the mean-square sense, and we may interchange the expectation and limit operators as indicated by Eq.

[1] Another form of stochastic integral is discussed in detail in Section 8.4.

(5.8-7). The argument is somewhat faulty at this point, however, since Eq. (5.8-7) involves ordinary limits, whereas stochastic integrals require limits in the mean-square sense. From Eqs. (5.8-11) and (5.8-12) we may deduce that the mean-square value of the integral is given by

$$\mathcal{E}\left\{\int_a^b \mathbf{x}(t)\,dt \int_a^b \mathbf{x}^T(\tau)\,dt\right\} = \int_a^b dt \int_a^b d\tau\, \mathcal{E}\{\mathbf{x}(t)\mathbf{x}^T(\tau)\}$$
$$= \int_a^b dt \int_a^b d\tau\, \mathbf{P}_{\mathbf{x}}(t, \tau) \tag{5.8-13}$$

Time Averages and Ergodicity. One of the important uses of stochastic integrals is in the definition of time averages for *stationary* stochastic processes. Time averages are an attempt to deduce from a single observation of a stochastic process the statistics of the process. To illustrate the concept, let us examine the *time average of the mean* of a stationary stochastic process given by

$$\vec{\mathbf{x}} = \lim_{T\to\infty} \frac{1}{T} \int_0^T \mathbf{x}(t)\,dt \tag{5.8-14}$$

We wish to determine the relationship between the time average of the mean $\vec{\mathbf{x}}$ and the actual mean $\boldsymbol{\mu}_{\mathbf{x}}$. We note the requirement that $\mathbf{x}(t)$ be stationary so that the mean is a constant, since $\vec{\mathbf{x}}$ is not a function of t. Since $\vec{\mathbf{x}}$ is a random variable, as noted above, and $\boldsymbol{\mu}_{\mathbf{x}}$ is a constant, it is obvious that $\vec{\mathbf{x}}$ and $\boldsymbol{\mu}_{\mathbf{x}}$ cannot be equal in the ordinary sense. We can, however, show that they are equal with probability 1 if $\mathcal{E}\{\vec{\mathbf{x}}\} = \boldsymbol{\mu}_{\mathbf{x}}$ and $\text{var}\{\vec{\mathbf{x}}\} = \mathbf{0}$.

To establish that $\vec{\mathbf{x}}$ and $\boldsymbol{\mu}_{\mathbf{x}}$ are equal with probability one, we begin by considering the *finite time average of the mean* given by

$$\hat{\mathbf{x}}(t_f) = \frac{1}{t_f} \int_0^{t_f} \mathbf{x}(t)\,dt \tag{5.8-15}$$

If we can establish that

$$\lim_{t_f\to\infty} \mathcal{E}\{\hat{\mathbf{x}}(t_f)\} = \boldsymbol{\mu}_{\mathbf{x}} \tag{5.8-16}$$

and

$$\lim_{t_f\to\infty} \text{var}\{\hat{\mathbf{x}}(t_f)\} = \mathbf{0} \tag{5.8-17}$$

then we shall have established the equality (with probability 1) of $\vec{\mathbf{x}}$ and $\boldsymbol{\mu}_{\mathbf{x}}$ since

$$\vec{\mathbf{x}} = \lim_{t_f\to\infty} \hat{\mathbf{x}}(t_f) \tag{5.8-18}$$

We note that in addition to being used as a step in relating $\vec{\mathbf{x}}$ and $\boldsymbol{\mu}_{\mathbf{x}}$, the finite time average has great practical significance. In an actual application it will not be possible to find the time average of Eq. (5.8-14), since an infinite time integral is required. It is obvious that the finite time average of Eq. (5.8-15) must be employed. In this context, $\hat{\mathbf{x}}(t_f)$ is often referred to as an estimator of $\boldsymbol{\mu}_{\mathbf{x}}$. Hence the question of how accurately the finite time average approximates the infinite time average is as important as the equivalence of the time average and the true mean. We shall see that these questions are intimately related.

We begin by considering the expected value of the finite time average. Assuming the continuity of $\mathbf{R}_{\mathbf{x}}(\tau)$, we may use Eq. (5.8-11) to express $\mathcal{E}\{\hat{\mathbf{x}}(t_f)\}$ as

$$\mathcal{E}\{\hat{\mathbf{x}}(t_f)\} = \frac{1}{t_f}\int_0^{t_f} \mathcal{E}\{\mathbf{x}(t)\}\,dt = \frac{1}{t_f}\int_0^{t_f} \boldsymbol{\mu}_{\mathbf{x}}\,dt = \boldsymbol{\mu}_{\mathbf{x}} \tag{5.8-19}$$

Hence we see that the expected value of $\hat{\mathbf{x}}(t_f)$ is equal to $\boldsymbol{\mu}_{\mathbf{x}}$ for all t_f and not just as t_f approaches infinity, as required. Because of this property, $\hat{\mathbf{x}}(t_f)$ is said to be an *unbiased estimator* of $\boldsymbol{\mu}_{\mathbf{x}}$.

The variance of $\hat{\mathbf{x}}(t_f)$ may be computed by the use of Eq. (5.8-12) as

$$\mathbf{V}_{\hat{\mathbf{x}}}(t_f) = \frac{1}{t_f^2}\int_0^{t_f} dt \int_0^{t_f} d\tau\, \mathbf{R}_{\mathbf{x}}(t-\tau) \tag{5.8-20}$$

This double integral can be simplified by letting $\lambda = t - \tau$; then $\mathbf{R}_{\mathbf{x}}(t-\tau) = \mathbf{R}_{\mathbf{x}}(\lambda) =$ constant along the line of $t - \tau = \lambda$, shown as the shaded region in Fig. 5.8-1. The area of this region is $(t_f - |\lambda|)\,d\lambda$, and λ ranges from $-t_f$ to t_f; so the double integral becomes

$$\mathbf{V}_{\hat{\mathbf{x}}}(t_f) = \frac{1}{t_f^2}\int_{-t_f}^{t_f} (t_f - |\lambda|)\mathbf{R}_{\mathbf{x}}(\lambda)\,d\lambda \tag{5.8-21}$$

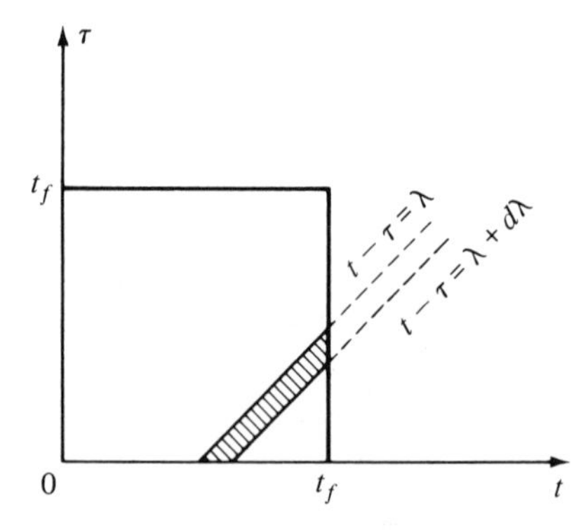

FIG. 5.8-1. Transformation of variables.

Hence a sufficient condition for $\vec{\mathbf{x}}$ to be equal with probability 1 to $\boldsymbol{\mu}_{\mathbf{x}}$ is that

$$\lim_{t_f \to \infty} \frac{1}{t_f^2} \int_{-t_f}^{t_f} (t_f - |\lambda|)\mathbf{R}_{\mathbf{x}}(\lambda)\, d\lambda = \mathbf{0} \tag{5.8-22}$$

Processes for which Eq. (5.8-22) is valid, that is, processes for which $\vec{\mathbf{x}}$ equals $\boldsymbol{\mu}_{\mathbf{x}}$ with probability 1, are referred to as *ergodic in the mean.* For processes that are ergodic in the mean, the finite time average of Eq. (5.8-15) may be used as an estimate of $\boldsymbol{\mu}_{\mathbf{x}}$. In this case, Eq. (5.8-21) may be used with a finite t_f to determine the variance of the error in the measurement of $\boldsymbol{\mu}_{\mathbf{x}}$ by $\hat{\mathbf{x}}(t_f)$. In practice one often selects limits for $\mathbf{V}_{\mathbf{x}}(t_f)$ and then uses Eq. (5.8-21) to find the required integration time t_f.

Example *5.8-1.* Let us consider the use of the above results for a stationary stochastic process in which $\mu_x = 0$ and the covariance $R_x(\tau) = \psi_x \delta_D(\tau)$. This stochastic process is referred to as a zero-mean white-noise process. It is desired to show that this process is ergodic in the mean. From Eq. (5.8-21) we find that $V_{\hat{x}}(t_f)$ is

$$V_{\hat{x}}(t_f) = \frac{1}{t_f^2} \int_{ft}^{t_f} (t_f - |\lambda|)\psi_x \delta_D(\lambda)\, d\lambda = \frac{\psi_x}{t_f}$$

and we see that

$$\lim_{t_f \to \infty} V_{\hat{x}}(t_f) = \lim_{t_f \to \infty} \frac{\psi_x}{t_f} = 0$$

so that this process is ergodic in the mean.

Example *5.8-2.* We desire to show that a stationary stochastic process $x(t)$ with zero mean and covariance

$$R_x(\tau) = ae^{-b|\tau|}$$

is ergodic in the mean. The use of Eq. (5.8-21) gives

$$\begin{aligned} V_{\hat{x}}(t_f) &= \frac{1}{t_f^2} \int_{-t_f}^{t_f} (t_f - |\lambda|)ae^{-b|\lambda|}\, d\lambda \\ &= \frac{1}{t_f^2}\left[\int_{-t_f}^{0} (t_f - |\lambda|)ae^{-b|\lambda|}\, d\lambda + \int_{0}^{t_f} (t_f - |\lambda|)ae^{-b|\lambda|}\, d\lambda\right] \\ &= \frac{2a}{t_f^2} \int_{0}^{t_f} (t_f - \lambda)e^{-b\lambda}\, d\lambda \\ &= \frac{2}{t_f^2}\left[\frac{a}{b}t_f + \frac{a}{b^2}(e^{-bt_f} - 1)\right] \end{aligned}$$

Hence we see that

$$\lim_{t_f \to \infty} V_{\hat{x}}(t_f) = 0$$

and the process is ergodic in the mean. Now we may think of $\hat{x}(t_f)$ as an estimate of μ_x, and $V_{\hat{x}}(t_f)$ as the associated variance of the error between $\hat{x}(t_f)$ and μ_x.

It is of interest to consider the discrete equivalent of Eq. (5.8-15), since a digital computer will often be used to accomplish the required computations needed to measure or estimate the mean value of a random process. Although a discrete time equivalent of Eq. (5.8-15) could take several forms, we shall use

$$\hat{\mathbf{x}}(KT) = \frac{1}{K+1} \sum_{j=0}^{K} \mathbf{x}(jT) \tag{5.8-23}$$

Again we wish to show that the stationary discrete-time process $\mathbf{x}(jT)$ is ergodic in the mean and that $\hat{\mathbf{x}}(KT)$ represents an estimator of $\boldsymbol{\mu}_{\mathbf{x}}$. We do this by showing

$$\lim_{K \to \infty} \mathcal{E}\{\hat{\mathbf{x}}(KT)\} = \boldsymbol{\mu}_{\mathbf{x}} \tag{5.8-24}$$

and

$$\lim_{K \to \infty} \text{var}\{\hat{\mathbf{x}}(KT)\} = \lim_{K \to \infty} \mathbf{V}_{\hat{\mathbf{x}}}(KT) = \mathbf{0} \tag{5.8-25}$$

If we take the expected value of both sides of Eq. (5.8-23), we obtain

$$\mathcal{E}\{\hat{\mathbf{x}}(KT)\} = \frac{1}{k+1} \sum_{j=0}^{K} \boldsymbol{\mu}_{\mathbf{x}} = \boldsymbol{\mu}_{\mathbf{x}} \tag{5.8-26}$$

so that $\hat{\mathbf{x}}(KT)$ is an unbiased estimator of $\boldsymbol{\mu}_{\mathbf{x}}$. The variance of $\hat{\mathbf{x}}(KT)$ is given by

$$\mathbf{V}_{\hat{\mathbf{x}}}(KT) = \frac{1}{(K+1)^2} \sum_{j=0}^{K} \sum_{i=0}^{K} \mathbf{R}_{\mathbf{x}}(\overline{i-j}T) \tag{5.8-27}$$

which is the discrete equivalent of Eq. (5.8-20). By making the substitution $k = i - j$, it is possible to rewrite Eq. (5.8-27) in the form

$$\mathbf{V}_{\hat{\mathbf{x}}}(KT) = \frac{1}{(K+1)^2} \sum_{k=-K}^{K} (K - |k|)\mathbf{R}_{\mathbf{x}}(kT) \tag{5.8-28}$$

A sufficient condition for $\hat{\mathbf{x}}(kT)$ to be ergodic in the mean is that

$$\lim_{K \to \infty} \frac{1}{(K+1)^2} \sum_{k=-K}^{K} (K - |k|)\mathbf{R}_{\mathbf{x}}(kT) = \mathbf{0} \tag{5.8-29}$$

Example *5.8-3.* We wish to consider the discrete counterpart of Example 5.8-1. We consider discrete white noise of zero mean with a correlation function

$$\mathbf{P}_{\mathbf{x}}(jT, kT) = \frac{\boldsymbol{\Psi}_{\mathbf{x}}}{T}\delta_K(\overline{j-k}T)$$

where δ_K is the Kroneker delta function equal to 1 if $j = k$ and equal to zero otherwise. At first glance it may appear unreasonable to replace the Dirac delta or impulse function, δ_D, by the Kroneker delta function in changing from a continuous-time random white waveform to a white discrete-time random waveform.

To show that the procedure is completely reasonable, let us consider the generalization to the continuous-time case of the discrete random process $\mathbf{x}(kT)$ with correlation function

$$\mathbf{P}_{\mathbf{x}}(jT, kT) = \mathbf{C}_{\mathbf{x}}(\overline{j-k}T) = \frac{\boldsymbol{\Psi}_{\mathbf{x}}}{T}\delta_K(\overline{j-k}T)$$

Since

$$\mathbf{x}(t) = \lim_{T\to 0,\, kT\to t} \mathbf{x}(kT) \qquad \mathbf{x}(\tau) = \lim_{T\to 0,\, jT\to \tau} \mathbf{x}(jT)$$

we have

$$\mathbf{P}_{\mathbf{x}}(t, \tau) = \mathcal{E}\{\mathbf{x}(t)\mathbf{x}^T(\tau)\} = \lim_{T\to 0,\, kT\to t,\, jT\to\tau} \mathcal{E}\{\mathbf{x}(kT)\mathbf{x}^T(jT)\}$$

Thus we see that the continuous-time correlation function is

$$\mathbf{P}_{\mathbf{x}}(t, \tau) = \lim_{T\to 0,\, kT\to t,\, jT\to\tau} \frac{\boldsymbol{\Psi}_{\mathbf{x}}}{T}\delta_K(t-\tau) = \boldsymbol{\psi}_{\mathbf{x}}\delta_D(t-\tau)$$

since we may visualize a Kroneker delta function as the limiting case of a Dirac delta function times the sampling period. In other words

$$\delta_D(t-\tau) = \lim_{T\to 0}\frac{1}{T}\left[\mathscr{A}\left(t-\tau+\frac{T}{2}\right) - \mathscr{A}\left(t-\tau-\frac{T}{2}\right)\right]$$

$$\delta_D(t-\tau) = \lim_{T\to 0}\frac{1}{T}\delta_K(\overline{j-k}T)$$

We shall provide ancillary justification for this discretization of white noise in Chapter 6. This concept is a rather vital one.

From Eq. (5.8-28) we have for the variance in the discrete finite time average of the mean-value vector

$$\mathbf{V}_{\hat{\mathbf{x}}}(KT) = \frac{1}{(K+1)^2}\sum_{k=-K}^{K}(K-|k|)\frac{\boldsymbol{\Psi}_{\mathbf{x}}}{T} = \frac{\boldsymbol{\Psi}_{\mathbf{x}}}{(K+1)T}$$

In comparing this result with the result of the continuous case in Example 5.4-2, we see that since $t_f = (K+1)T$, the two error variances are the same, as well they should be. This fact is also justification for the equivalence between discrete- and continuous-time white-noise representations

$$\mathbf{P}_{\mathbf{x}}(t, \tau) = \mathcal{E}\{\mathbf{x}(t)\mathbf{x}^T(\tau)\} = \boldsymbol{\psi}_{\mathbf{x}}\delta_D(t-\tau)$$

$$\mathbf{P}_{\mathbf{x}}(jT, kT) = \mathcal{E}\{\mathbf{x}(jT)\mathbf{x}^T(kT)\} = \frac{\boldsymbol{\psi}_{\mathbf{x}}}{T}\delta_K(\overline{j-k}T)$$

We can also write the time average of the correlation function of a stationary process $\mathbf{x}(t)$ in the form

$$\vec{\mathbf{C}}_{\mathbf{x}}(\tau) = \lim_{T\to\infty} \frac{1}{T}\int_0^T \mathbf{x}(t)\mathbf{x}^T(t-\tau)\,dt \tag{5.8-30}$$

A stochastic process is said to be correlation ergodic if $\vec{\mathbf{C}}_{\mathbf{x}}(\tau)$ equals $\mathbf{C}_{\mathbf{x}}(\tau)$ with probability 1. We can define covariance ergodicity in a similar manner.

In attempting to determine $\vec{\mathbf{C}}_{\mathbf{x}}(\tau)$, it will be necessary to approximate the results with a finite time average. If there is a record of the data of length 0 to T seconds available, we might use the finite time average estimator

$$\hat{\mathbf{C}}_{\mathbf{x}}(\tau) = \frac{1}{T-\tau}\int_\tau^T \mathbf{x}(t)\mathbf{x}^T(t-\tau)\,dt \tag{5.8-31}$$

The limits of integration can only extend from τ to T rather than 0 to T, since no data is assumed available for $t < 0$. The measurements implied by Eq. (5.8-31) are accurate only for $\tau \ll T$. Of course, if a record of longer length than 0 to T is available, integration over the full T seconds is possible. If we were processing the data digitally and had a total of N samples, the discrete approximation

$$\hat{\mathbf{C}}_{\mathbf{x}}(NT) = \frac{1}{N-n}\sum_{k=n}^{N} \mathbf{x}(kT)\mathbf{x}^T(\overline{k-n}T) \tag{5.8-32}$$

would be used.

Example *5.8-4*. We desire to obtain expressions for errors in the measurement of autocorrelation functions. This is a fairly complicated subject, and many papers and books have been published on the subject. We shall consider only a fairly simple version of the problem and shall ignore many of the practical details.

We assume a scalar ergodic process and that the estimate of the correlation function is given by Eq. (5.8-31) and instrumented by the block diagram of Fig. 5.8-2. The error in the estimate is given by

$$\tilde{C}_x(\tau) = C_x(\tau) - \hat{C}_x(\tau) = C_x(\tau) - \frac{1}{T}\int_0^T x(t)x(t-\tau)\,dt$$

where we assume that τ is so small compared with T that it can be neglected in the $T - \tau$ terms of Eq. (5.8-31), or the equivalent assumption that a data record of length $T + \tau$ (or longer) is available.

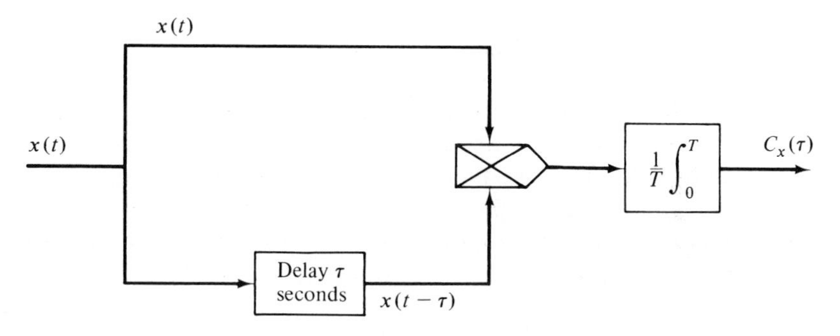

FIG. 5.8-2. Block diagram of correlator.

The estimate of the correlation function

$$\hat{C}_x(\tau) = \frac{1}{T}\int_0^T x(t)x(t-\tau)\,dt$$

is unbiased since

$$\mathcal{E}\{\hat{C}_x(\tau)\} = \frac{1}{T}\int_0^T \mathcal{E}\{x(t)x(t-\tau)\}\,dt = \frac{1}{T}\int_0^T C_x(\tau)\,dt = C_x(\tau)$$

The error variance in the estimate is nonzero, however. We have, since $\mathcal{E}\{\tilde{C}_x(\tau)\} = 0$,

$$\begin{aligned}\text{var}\{\tilde{C}_x(\tau)\} &= \mathcal{E}\{\tilde{C}_x^2(\tau)\} = \mathcal{E}\{C_x^2(\tau) - 2C_x(\tau)\hat{C}_x(\tau) + \hat{C}_x^2(\tau)\}\\ &= \mathcal{E}\{\hat{C}_x^2(\tau)\} - C_x^2(\tau)\end{aligned}$$

To evaluate this error variance expression, we need to obtain an expression for $\mathcal{E}\{\hat{C}_x^2(\tau)\}$ that is, in terms of the expression for the estimate,

$$\mathcal{E}\{\hat{C}_x^2(\tau)\} = \mathcal{E}\left\{\frac{1}{T^2}\int_0^T\int_0^T x(\lambda_1)x(\lambda_1 - \tau)x(\lambda_2)x(\lambda_2 - \tau)\, d\lambda_1\, d\lambda_2\right\}$$

which may be written in terms of the fourth moment of $x(t)$ as

$$\mathcal{E}\{C_x^2(t)\} = \frac{1}{T^2}\int_0^T\int_0^T \mathfrak{F}_x(\lambda_1, \lambda_2, \tau)\, d\lambda_1\, d\lambda_2$$

where

$$\mathfrak{F}_x(\lambda_1, \lambda_2, \tau) = \mathcal{E}\{x(\lambda_1)x(\lambda_2)x(\lambda_1 - \tau)x(\lambda_2 - \tau)\}$$

is a fourth moment of $x(t)$. The precise expression for the fourth moment will depend upon the particular probability density function of the ergodic stochastic process $x(t)$. For the particular case of a gaussian stochastic process with zero mean value, it is shown in Chapter 7 that this fourth moment is given by the fourth-product-moment expansion for a gaussian random variable:

$$\begin{aligned}\mathfrak{F}_x(\lambda_1, \lambda_2, \tau) = C_x^2(\tau) + C_x^2(\lambda_1 - \lambda_2)& \\ + C_x(\lambda_1 - \lambda_2 + \tau)C_x(\lambda_1 - \lambda_2 - \tau)&\end{aligned}$$

The error variance expression thus becomes

$$\begin{aligned}\text{var}\{\tilde{C}_x(\tau)\} = \frac{1}{T^2}\int_0^T\int_0^T [C_x^2(\lambda_1 - \lambda_2) + C_x(\lambda_1 - \lambda_2 + \tau)& \\ \times C_x(\lambda_1 - \lambda_2 - \tau)]\, d\lambda_1\, d\lambda_2&\end{aligned}$$

A simpler version of this expression is desirable. If we make the change of variable $\lambda_3 = \lambda_1 - \lambda_2$, the foregoing expression becomes

$$\text{var}\{\tilde{C}_x(\tau)\} = \frac{1}{T^2}\int_0^T d\lambda_2 \int_{-\lambda_2}^{T-\lambda_2} [C_x^2(\lambda_3) + C_x^2(\lambda_3 + \tau)C_x^2(\lambda_3 - \tau)]\, d\lambda_3$$

The integrand of this expression is an even function of λ_3, and it is convenient to denote it by $f(\lambda_3)$. The region of integration for the integral in the foregoing is shown in Fig. 5.8-3. By careful inspection of Fig. 5.8-3 it is apparent that we may exchange the order of integra-

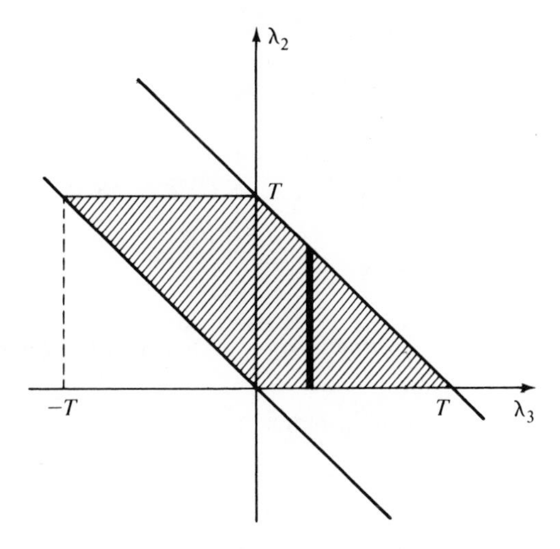

FIG. 5.8-3. Integration area for double integral.

tion if we write the integral in two parts. We obtain

$$\frac{1}{T^2}\int_0^T d\lambda_2 \int_{-\lambda_2}^{T-\lambda_2} f(\lambda_3)\,d\lambda_3$$

$$= \frac{1}{T^2}\int_0^T f(\lambda_3)\,d\lambda_3 \int_0^{T-\lambda_3} d\lambda_2 + \frac{1}{T^2}\int_{-T}^0 f(\lambda_3)\,d\lambda_3 \int_{-\lambda_3}^T d\lambda_2$$

$$= \frac{1}{T^2}\int_0^T (T-\lambda_3)f(\lambda_3)\,d\lambda_3 + \frac{1}{T^2}\int_{-T}^0 (T+\lambda_3)f(\lambda_3)\,d\lambda_3$$

We now let $\lambda_3 = -\lambda_4$ in the last integral and note that $f(\lambda_4) = f(-\lambda_4)$ so that

$$\operatorname{var}\{\tilde{C}_x(\tau)\} = \frac{1}{T^2}\int_0^T (T-\lambda_3)f(\lambda_3)\,d\lambda_3 - \frac{1}{T^2}\int_T^0 (T-\lambda^4)f(\lambda_4)\,d\lambda_4$$

$$= \frac{2}{T^2}\int_0^T (T-\lambda)[C_x^2(\lambda) + C_x(\lambda+\tau)C_x(\lambda-\tau)]\,d\lambda$$

This expression for the error variance in correlation estimation will generally be easier to evaluate than the first obtained one. Still, in practice even this expression may be difficult to evaluate. This is particularly true since it involves the unknown correlation function $C_x(\tau)$. We shall therefore obtain an upper bound for this error variance expression, which is useful in many applications.

We have, for $\lambda < T$,

$$\operatorname{var}\{\tilde{C}_x(\tau)\} \leq \frac{2}{T}\int_0^T \left|1 - \frac{\lambda}{T}\right| |C_x^2(\lambda) + C_x(\lambda+\tau)C_x(\lambda-\tau)|\, d\lambda$$
$$\leq \frac{2}{T}\int_0^T [C_x^2(\lambda) + C_x(\lambda+\tau)C_x(\lambda-\tau)]\, d\lambda$$

Schwartz's inequality

$$\left[\int_c^d a(t)b(t)\, dt\right]^2 \leq \int_c^d a^2(t)\, dt \int_c^d b^2(t)\, dt$$

may be used to obtain from

$$\frac{2}{T}\int_0^T C_x(\lambda+\tau)C_x(\lambda-\tau)\, d\lambda = \frac{1}{T}\int_{-T}^T C_x(\lambda+\tau)C_x(\lambda-\tau)\, d\lambda$$

the inequality

$$\frac{1}{T}\int_{-T}^T C_x(\lambda+\tau)C_x(\lambda-\tau)\, d\tau$$
$$\leq \frac{1}{T}\left[\int_{-T}^T C_x^2(\lambda+\tau)\, d\lambda \int_{-T}^T C_x^2(\lambda-\tau)\, d\lambda\right]$$

But surely

$$\int_{-T}^T C_x^2(\lambda+\tau)\, d\lambda \leq 2\int_0^\infty C_x^2(\lambda)\, d\lambda$$
$$\int_{-T}^T C_x^2(\lambda-\tau)\, d\lambda \leq 2\int_0^\infty C_x^2(\lambda)\, d\lambda$$

as a simple sketch, recalling the fact that $C_x(\tau) = C_x(-\tau)$, will show.

Thus we have, finally, the error bound obtained by collecting the foregoing six equations:

$$\operatorname{var}\{\tilde{C}_x(\tau)\} \leq \frac{4}{T}\int_0^\infty C_x^2(\lambda)\, d\lambda \qquad |\tau| < T$$

To illustrate the dependence of the autocorrelation function estimate upon the age variable τ and the record length T, let us consider the correlation function

$$C_x(\tau) = a^2 e^{-\alpha|\tau|} + \mu_x^2$$

Use of the exact expression for the error variance leads immediately to

$$\operatorname{var}\{\tilde{C}_x(\tau)\} = \frac{a^4}{2(\alpha T)^2}\{2\alpha T - 1 + 2e^{-2\alpha T} + [(2\alpha\tau + 1)(2\alpha T - 1) - 2(a\tau)^2]e^{-2\alpha\tau}\}$$

for $T > \tau > 0$. The smallest error variance occurs when $\tau = 0$. This error variance is

$$\operatorname{var}\{\tilde{C}_x(\tau)\} = \frac{a^4}{(\alpha T)^2}(\alpha T - 1 + e^{-2\alpha T})$$

For reasonable estimation it would be necessary to restrict $\alpha T > 10^2$. In this case, the smallest error variance value may be bounded by

$$\operatorname{var}\{\tilde{C}_x(\tau)\} \geq \frac{a^4}{(\alpha T)}$$

Use of the expression for the upper bound on the error variance leads to the inequality

$$\frac{a^4}{\alpha T} \leq \operatorname{var}\{\tilde{C}_x(\tau)\} \leq \frac{2a^4}{\alpha T}$$

The signal-to-noise ratio for this problem would be the square of the correlation function divided by the error variance in estimation. Thus we have, for the case $\mu_x = 0$,

$$\frac{\alpha T}{2}e^{-\alpha|\tau|} \leq \frac{C_x^2(\tau)}{\operatorname{var}\{\tilde{C}_x(\tau)\}} \leq \alpha T e^{-\alpha|\tau|}$$

and it is seen that difficulties will be encountered in maintaining a high signal-to-noise ratio for large τ, as very large data length records will be required.

Many other factors, such as spurious noise inputs, enter into more practical problems involving correlation function measurement, but they are beyond the scope of the present effort.

So far we have discussed stochastic processes for which the time averages of the mean, covariance, and correlation function were equal with probability 1 to the true mean, covariance, and correlation function as obtained from normal expectation definition. These processes are referred to as mean value, covariance, and correlation ergodic, respectively. For some stochastic processes, the time average of any function $\mathbf{f}[\mathbf{x}(t)]$ is equal with probability 1 to the expected value of $\mathbf{f}[\mathbf{x}(t)]$; these processes are referred to as simply *ergodic.*

Definition 5.8-1. ***Ergodicity:*** A stochastic process $\mathbf{x}(t)$ is ergodic if

$$\lim_{t_f \to \infty} \frac{1}{t_f} \int_0^{t_f} \mathbf{f}[\mathbf{x}(t)]\, dt = \mathcal{E}\{\mathbf{f}[\mathbf{x}(t)]\} \qquad \text{w.p.1} \tag{5.8-33}$$

for any $\mathbf{f}[\mathbf{x}(t)]$ for which the integral exists.

For an ergodic process, the statistical properties of the process can be determined from almost every realization of the process, that is, the time function associated with almost every outcome. We note that it is necessary for a process to be stationary for it to be ergodic. However, the converse is *not* true; not all stationary processes are ergodic.

Example ***5.8-5.*** Let us consider the random-dc-value stochastic process for which

$$x(t) = a \qquad \text{for all } t$$

where a is a random variable with known distribution. To be specific, let us assume that the mean of a is zero. This process is obviously strict sense stationary. However, this process is not even ergodic in the mean. The time average of the mean is given by

$$\vec{x} = \lim_{t_f \to \infty} \frac{1}{t_f} \int_0^{t_f} x(t)\, dt = \lim_{t_f \to \infty} \frac{1}{t_f} \int_0^{t} a\, dt = a$$

The mean of $x(t)$ is

$$\mathcal{E}\{x(t)\} = \mathcal{E}\{a\} = \mu_a = 0$$

Hence we see that in general $\vec{x} \neq \mu_x$.

Stochastic Differentiation. Given a stochastic process $\mathbf{x}(t)$, the time derivative of $\mathbf{x}(t)$ is defined by

$$\frac{d\mathbf{x}(t)}{dt} = \dot{\mathbf{x}}(t) = \underset{\epsilon \to 0}{\text{l.i.m.}} \frac{\mathbf{x}(t+\epsilon) - \mathbf{x}(t)}{\epsilon} \tag{5.8-34}$$

The limit of Eq. (5.8-34) will exist if $\boldsymbol{\mu}_{\mathbf{x}}(t)$ is differentiable and the mixed second partial derivative $\partial^2 \mathbf{V}_{\mathbf{x}}(t, \tau)/\partial t\, \partial \tau$ exists and is continuous. Under these conditions, it is possible to show that

$$\mathcal{E}\{\dot{\mathbf{x}}(t)\} = \frac{d}{dt}\mathcal{E}\{\mathbf{x}(t)\} = \dot{\boldsymbol{\mu}}_{\mathbf{x}}(t) \tag{5.8-35}$$

$$\operatorname{cov}\{\dot{\mathbf{x}}(t), \mathbf{x}(\tau)\} = \frac{d}{dt}\operatorname{cov}\{\mathbf{x}(t), \mathbf{x}(\tau)\} = \frac{\partial}{\partial t}\mathbf{V}_{\mathbf{x}}(t, \tau) \tag{5.8-36}$$

$$\begin{aligned}\operatorname{cov}\{\dot{\mathbf{x}}(t), \dot{\mathbf{x}}(\tau)\} &= \frac{d}{dt}\frac{d}{d\tau}\operatorname{cov}\{\mathbf{x}(t), \mathbf{x}(\tau)\} \\ &= \frac{\partial^2 \mathbf{V}_{\mathbf{x}}(t, \tau)}{\partial t \partial \tau}\end{aligned} \tag{5.8-37}$$

Equations (5.8-35) to (5.8-37) indicate that the operations of differentiation and expectation may be interchanged.

The establishment of the above results is reasonably easy and is left as an exercise for the reader. We note only that it is convenient to make use of the Cauchy criterion in demonstrating the existence of the limit in Eq. (5.8-34). Since we shall make extensive use of stochastic derivatives in the next chapter, we shall not consider them further at this time.

EXERCISES 5.8

5.8-1. Use the Cauchy criterion to show that the limit in Eq. (5.8-34) exists if $\mu_x(t)$ is differentiable and $\partial^2 V_x(t, \tau)/\partial t\, \partial \tau$ exists and is continuous.

5.8-2. Establish the validity of Eqs. (5.8-35), (5.8-36), and (5.8-37).

5.8-3. The process $x(t)$ is stationary with $C_x(\tau) = 1 + e^{-2|\tau|}$. Find the mean, variance, and correlation function of the stochastic process

$$y(t) = \int_0^t x(\tau)\, d\tau$$

Answers: $\mu_y(t) = t$, $C_y(\tau) = \tau^2 + \tau + \frac{1}{2}e^{-2\tau} - \frac{1}{2}$.

5.8-4. Given

$$y(t) = \frac{dx(t)}{dt}$$

with $V_x(t, \tau) = \sin t \cos \tau + 2$, find $V_y(t, \tau)$ and $V_{xy}(t, \tau)$.

Answers: $V_y(t, \tau) = -\cos t \sin \tau$, $V_{xy}(t, \tau) = -\sin t \sin \tau$.

5.8-5. If $y(t)$ is defined by

$$y(t) = \int_0^t x(\tau)\, d\tau$$

with $\mu_x(t) = 0$ and $V_x(t, \tau) = 2\delta_D(t - \tau)$, find $V_y(t, \tau)$. Show that $[y(t_3) - y(t_2)]$ and $[y(t_2) - y(t_1)]$ are uncorrelated if $t_1 < t_2 < t_3$.

Answer: $V_y(t, \tau) = \min(t, \tau)$.

5.9. Summary

In this chapter we studied stochastic processes and related topics. All the efforts of the previous three chapters apply almost without modification to stochastic process. The various moments of a stochastic process were defined, and particular attention was given to the concepts of stationarity and ergodicity. References to stochastic processes abound and only a selection of these are given here. All are exceptionally good references and are at the senior, first-year graduate level and therefore appropriate for the audience for whom this book is intended. The texts by Parzen and Papoulis are especially recommended for their fundamental discussions of stochastic processes and are in addition quite readable. The other texts cited are concerned more with applications of stochastic process theory.

5.10. References

(1) BENDAT, J. S., and A. G. PIERSOL, *Measurement and Analysis of Random Data*, Wiley, New York, 1966.

(2) DAVENPORT, W. D., and W. L. ROOT, *Random Signals and Noise*, McGraw-Hill, New York, 1958.

(3) FREEMAN, J. J., *Principles of Noise*, Wiley, New York, 1958.

(4) HANCOCK, J. C., *An Introduction to the Principles of Communication Theory*, McGraw-Hill, New York, 1961.

(5) JENKINS, G. M., and D. G. WATTS, *Spectral Analysis and Its Applications*, Holden-Day, San Francisco, 1968.

(6) LANING, J. H., and R. H. BATTIN, *Random Processes in Automatic Control*, McGraw-Hill, New York, 1956.

(7) LATHI, B. P., *An Introduction to Random Signals and Communication Theory*, International Textbook Co., Scranton, Pa., 1968.

(8) LEE, Y. W., *Statistical Theory of Communication*, Wiley, New York, 1960.

(9) PAPOULIS, A., *Probability, Random Variables and Stochastic Processes*, McGraw-Hill, New York 1965.

(10) PARZEN, E., *Stochastic Processes*, Holden-Day, San Francisco, 1962.

(11) THOMAS, J. B., *An Introduction to Statistical Communication Theory*, Wiley, New York, 1969.

(12) WOZENCRAFT, J. M., and I. M. JACOBS, *Principles of Communication Engineering*, Wiley, New York, 1965.

5.11. Problems

5.11-1. $x(t)$ and $y(t)$ are independent stationary random processes. Stochastic processes $u(t)$ and $v(t)$ are formed by

$$u(t) = x(t) + y(t)$$
$$v(t) = x(t) - y(t)$$

Find (a) $C_u(\tau)$, (b) $C_v(\tau)$, (c) $C_{uv}(\tau)$, and (d) $C_{vu}(\tau)$ in terms of the correlation functions of $x(t)$ and $y(t)$.

5-11-2. A stochastic process $z(t)$ is defined by

$$z(t) = x(t)y(t)$$

where $x(t)$ and $y(t)$ are independent stationary random processes with

$$C_x(\tau) = \mu_x{}^2 + e^{-\alpha|\tau|}$$
$$C_y(\tau) = \mu_y{}^2 + e^{-\beta|\tau|}$$

What is $C_z(\tau)$?

5.11-3. Suppose that

$$z(t) = x(t) \cos (\omega t + \phi)$$

where $x(t)$ is an ergodic random process independent of the random variable ϕ. Determine the mean and correlation function for $z(t)$ and compare them with the time average of the mean and correlation function. Assume that ϕ is uniformly distributed. What are the requirements for $z(t)$ to be mean and correlation ergodic?

5.11-4. You are given a stationary stochastic process $x(t)$ and attempt to calculate the stochastic derivative $y(t) = dx(t)/dt$. For each of the correlation functions given below

(a) determine whether or not $y(t)$ exists.
(b) determine the correlation functions $C_y(\tau)$ and $C_{xy}(\tau)$.
(c) determine the spectral densities $C_X(s)$, $C_Y(s)$, and $C_{XY}(s)$.

(i) $C_x(\tau) = (1 + \alpha|\tau|)e^{-|\tau|}$
(ii) $C_x(\tau) = (\cos \omega\tau + (\alpha/\omega) \sin \omega|\tau|)e^{-\alpha|\tau|}$
(iii) $C_x(\tau) = e^{-|\tau|}$
(iv) $C_x(\tau) = (\sin \omega\tau)/\tau$

5.11-5. Show that for a stationary stationary process $x(t)$, the nth derivative

$$y(t) = \frac{dx^n(t)}{dt^n}$$

has the correlation function and spectral density

$$C_y(\tau) = (-1)^n \frac{d^{2n}C_x(\tau)}{d\tau^{2n}}$$
$$C_Y(\omega) = (-1)^n(j\omega)^{2n}C_X(\omega)$$

if they exist. What are the existence requirements?

5.11-6. The process $x(t)$ is ergodic with $C_x(\tau) = a^2 e^{-\alpha|\tau|} \cos \omega\tau$. Find the mean, variance, and correlation function of the stochastic process

$$y(t) = \int_0^t x(\tau)\, d\tau$$

5.11-7. A random process is defined by

$$x(t) = \cos(\omega t + \phi) + \sin(3\omega t + \phi)$$

where ϕ is uniformly distributed in the interval 0 to 2π.

(a) Is the process stationary?

(b) Determine $p_{x(t)}(\alpha, t)$.

(c) Determine $P_x(t_1, t_2)$.

5.11-8. An observation of a constant is corrupted by additive white noise

$$z(t) = a + v(t)$$

where $v(t)$ is a zero-mean white random process with $C_v(\tau) = \psi_v \delta_D(\tau)$. An estimator of the form

$$\hat{a}(t) = \frac{1}{t} \int_0^t z(\tau)\, dt$$

turns out to be an optimum estimator. Determine the mean and variance of the estimation error $\tilde{a}(t) = a - \hat{a}(t)$ as a function of ψ_w and t.

5.11-9. An observation of a constant is corrupted by additive white noise

$$z(k) = a + v(k) \qquad k = 1, 2, \ldots$$

where $v(k)$ is a zero-mean white random sequence with $E\{v(k)v(j)\} = V_v \delta_K(k - j)$. An estimator of the form

$$\hat{a}(k) = \frac{1}{k} \sum_{j=1}^{k} z(j)$$

turns out to be an optimum estimator. Determine the mean and variance of the estimation error $\tilde{a}(k) = a - \hat{a}(k)$ as a function of V_v and k.

5.11-10. Find a value for t in Problem 5.11-8 such that we shall have a probability of 0.9 that $|\tilde{a}(t)|$ will be less than or equal to 0.1.

(a) Use the Tchebycheff inequality to approximate t.

(b) Assume that $\tilde{a}(t)$ is gaussian and compute an exact value for t.

6. Linear System Response to Stochastic Processes

6.1. Introduction

Many problems in the information, computer, and control sciences may be posed as problems involving the calculation of the output response of a linear system to a stochastic process input. It is, of course, actually possible to solve the differential equation

$$\dot{\mathbf{x}}(t) = \mathbf{F}(t)\mathbf{x}(t) + \mathbf{G}(t)\mathbf{u}(t)$$

where $\mathbf{u}(t)$ is a stochastic process input. Each time the problem is repeated a different result is obtained, since the input $\mathbf{u}(t)$ would be different. Often we are more interested in the probability density function of the output $p_{\mathbf{x}(t)}(\boldsymbol{\alpha}, t)$, or perhaps the second-order joint density $p_{\mathbf{x}(t_1)\mathbf{x}(t_2)}(\boldsymbol{\alpha}_1, \boldsymbol{\alpha}_2, t_1, t_2)$, or perhaps even higher-order joint densities. For most problems these joint densities would be very difficult to obtain, and so we are often content with a knowledge of the mean of the process $\boldsymbol{\mu}_{\mathbf{x}}(t)$, the variance of the process $\mathbf{V}_{\mathbf{x}}(t)$, the covariance of the process $\mathbf{V}_{\mathbf{x}}(t_1, t_2)$, the correlation function $\mathbf{P}_{\mathbf{x}}(t_1, t_2)$, and perhaps the covariance function $\mathbf{V}_{\mathbf{xu}}(t_1, t_2)$. These are the first- and second-order moments and, although they do not provide a complete description of the process, they do give as much statistical knowledge of the process as is useful in many practical problems. In addition, if the random process $\mathbf{x}(t)$ has a gaussian density function, knowledge of these first- and second-order moments is sufficient to allow a complete statistical description of the process.

Of course, even this complete statistical knowledge is not sufficient to completely characterize $\mathbf{x}(t)$ as a time function.

We shall first give consideration to the state-space representation of the response of linear continuous systems to random inputs. Then we shall consider the frequency-domain approach, which is most useful for single input–single output systems. Finally, discrete systems will be considered, and the equivalence and differences between continuous-time and discrete-time representation of random processes will be explored. From an applications viewpoint, the material of this chapter is most useful, and the results of this chapter form the basis for further work in many applications areas for stochastic processes.

6.2. *State-variable approach to continuous systems*

In previous efforts we have considered algebraic operations on random variables and stochastic processes. Because time is a parameter in a stochastic process, it is also possible to consider dynamic transformations, that is, operations on stochastic processes described by differential or difference equations. In this section we shall consider the response of linear systems to stochastic-process inputs by state-variable techniques.

Because the output of a dynamic system depends on past as well as current inputs, the output tends to be more correlated, and in that sense less random, than the input. This feature of "smoothing" random data by means of a dynamic transformation is, in fact, one of the key concepts of estimation theory.

Since it is an exceptionally difficult problem to attempt to discuss the subject of the response of systems to random inputs, the goals in this chapter will be much less ambitious than to treat the general case. We shall restrict the development to linear systems of equations and shall deal only with the mean and covariances of the stochastic processes. Chapter 8 will extend these results to a fairly general class of nonlinear systems.

Although it may appear overly restrictive to treat only the mean and covariance of the response, we shall see, in Chapter 7, that, for gaussian distributions, specification of the mean and variance is equivalent to a complete statistical description of the process. We shall show this by proving the result of any linear operation on a gaussian distribution is another gaussian distribution! Hence if the input and initial condition for a dynamic linear system are gaussian, the output distribution of the system is completely defined statistically.

The combination of gaussian signals, linear systems, and the study of

the propagation of mean and variance is so natural that the reader is cautioned not to think of these subjects as an inseparable group. The developments that follow are valid for any distribution, not only gaussian distributions. For nongaussian distributions, the mean and variance are generally not sufficient to completely define the distribution. Fortunately, much work done in estimation theory and related subjects requires only these first two moments.

The class of continuous dynamic systems that we consider is given by the first-order, time-varying vector differential equation

$$\dot{\mathbf{x}}(t) = \mathbf{F}(t)\mathbf{x}(t) + \mathbf{G}(t)\mathbf{u}(t) \tag{6.2-1}$$

which has as its input the white-noise process $\mathbf{u}(t)$. The mean and covariance of $\mathbf{u}(t)$ are assumed to be known as

$$\boldsymbol{\mu}_{\mathbf{u}}(t) \triangleq \mathcal{E}\{\mathbf{u}(t)\} \tag{6.2-2}$$

$$\mathbf{V}_{\mathbf{u}}(t_1, t_2) \triangleq \operatorname{cov}\{\mathbf{u}(t_1), \mathbf{u}(t_2)\} = \boldsymbol{\psi}_{\mathbf{u}}(t_1)\delta_D(t_1 - t_2) \tag{6.2-3}$$

Note that we are *not* assuming that $\mathbf{u}(t)$ is a stationary process. The problem is to determine the mean $\boldsymbol{\mu}_{\mathbf{x}}(t)$ and variance $\mathbf{V}_{\mathbf{u}}(t_1, t_2)$ of $\mathbf{x}(t)$ as well as the covariances of $\mathbf{x}(t)$ and $\mathbf{u}(t)$, $\mathbf{V}_{\mathbf{xu}}(t_1, t_2)$, and $\mathbf{V}_{\mathbf{ux}}(t_1, t_2)$.

The solution of Eq. (6.2-1) is given by[1]

$$\mathbf{x}(t) = \boldsymbol{\Phi}(t, t_0)\mathbf{x}(t_0) + \int_{t_0}^{t} \boldsymbol{\Phi}(t, \tau)\mathbf{G}(\tau)\mathbf{u}(\tau)\, d\tau \tag{6.2-4}$$

where the state transition matrix $\boldsymbol{\Phi}(t, t_0)$ is a solution of the homogeneous differential equation (fundamental matrix equation)

$$\dot{\boldsymbol{\Phi}}(t, t_0) = \frac{\partial}{\partial t}\boldsymbol{\Phi}(t, t_0) = \mathbf{F}(t)\boldsymbol{\Phi}(t, t_0) \tag{6.2-5}$$

with the initial condition

$$\boldsymbol{\Phi}(t_0, t_0) = \mathbf{I} \tag{6.2-6}$$

It may seem at first that white noise is not the most general input noise which could be assumed. However, as we shall see, many linear systems driven by nonwhite noise can be represented as another linear system driven by white noise by using the method of state augmentation.

We may easily determine an equation for the propagation of the mean of the state vector $\mathbf{x}(t)$ by taking the expectation of both sides of Eq. (6.2-1),

[1] For a discussion of the state-variable description of linear systems, see Appendix B.

interchanging the operation of differentiation (a time operation), and taking the expectation to obtain

$$\frac{d\boldsymbol{\mu}_{\mathbf{x}}(t)}{dt} = \dot{\boldsymbol{\mu}}_{\mathbf{x}}(t) = \mathbf{F}(t)\boldsymbol{\mu}_{\mathbf{x}}(t) + \mathbf{G}(t)\boldsymbol{\mu}_{\mathbf{u}}(t) \tag{6.2-7}$$

which is solved subject to the initial condition $\boldsymbol{\mu}_{\mathbf{x}}(t_0) = \boldsymbol{\mu}_{\mathbf{x}_0}$. The prior mean of the process $\boldsymbol{\mu}_{\mathbf{x}}(t_0)$ must be given as part of the problem specification. This represents the desired relation for propagation of the mean of $\mathbf{x}(t)$.

The mean of $\mathbf{x}(t)$ may also be easily determined by taking the expected value of both sides of Eq. (6.2-4) to obtain

$$\begin{aligned}\boldsymbol{\mu}_{\mathbf{x}}(t) \triangleq \mathcal{E}\{\mathbf{x}(t)\} &= \mathcal{E}\{\boldsymbol{\Phi}(t, t_0)\mathbf{x}(t_0)\} + \mathcal{E}\left\{\int_{t_0}^{t} \boldsymbol{\Phi}(t, \tau)\mathbf{G}(\tau)\mathbf{u}(\tau)\, d\tau\right\} \\ &= \boldsymbol{\Phi}(t, t_0)\boldsymbol{\mu}_{\mathbf{x}}(t_0) + \mathcal{E}\left\{\int_{t_0}^{t} \boldsymbol{\Phi}(t, \tau)\mathbf{G}(\tau)\mathbf{u}(\tau)\, d\tau\right\}\end{aligned}$$

where $\boldsymbol{\mu}_{\mathbf{x}}(t_0)$ is the prior mean of the $\mathbf{x}(t)$ process. Interchanging the expectation and integration operators, we have

$$\boldsymbol{\mu}_{\mathbf{x}}(t) = \boldsymbol{\Phi}(t, t_0)\boldsymbol{\mu}_{\mathbf{x}}(t_0) + \int_{t_0}^{t} \boldsymbol{\Phi}(t, \tau)\mathbf{G}(\tau)\boldsymbol{\mu}_{\mathbf{u}}(\tau)\, d\tau \tag{6.2-8}$$

which is the desired result. Note that $\boldsymbol{\mu}_{\mathbf{x}}(t)$ consists of an unforced portion which depends on the value of $\boldsymbol{\mu}_{\mathbf{x}}$ at t_0, $\boldsymbol{\mu}_{\mathbf{x}}(t_0)$, and a forced portion given by a convolution-type integral. Hence the solution for $\boldsymbol{\mu}_{\mathbf{x}}(t)$ is identical to the solution for $\mathbf{x}(t)$, except that the stochastic variables are replaced by their expected values in the expression for $\boldsymbol{\mu}_{\mathbf{x}}(t)$. It is a simple task to show that the solution of Eq. (6.2-7) is (6.2-8).

The integral expression for $\boldsymbol{\mu}_{\mathbf{x}}(t)$ given by Eq. (6.2-8) is difficult to use because of the convolution integral and the need to determine the state transition matrix. The convolution integral is particularly troublesome because it is not well suited for either hand or computer evaluation. Therefore, it is desirable to write the expression for $\boldsymbol{\mu}_{\mathbf{x}}(t)$ in an alternative form. In particular, let us differentiate both sides of Eq. (6.2-8) with respect to time to obtain, using Leibnitz's rule[1] on the convolution integral, the expression

$$\dot{\boldsymbol{\mu}}_{\mathbf{x}}(t) = \dot{\boldsymbol{\Phi}}(t, t_0)\boldsymbol{\mu}_{\mathbf{x}}(t_0) + \int_{t_0}^{t} \dot{\boldsymbol{\Phi}}(t, \tau)\mathbf{G}(\tau)\boldsymbol{\mu}_{\mathbf{u}}(\tau)\, d\tau + \boldsymbol{\Phi}(t, t)\mathbf{G}(t)\boldsymbol{\mu}_{\mathbf{u}}(t)$$

[1] Leibnitz's rule, which is proved in most books on advanced calculus, states that if $f(t, \tau)$, $a(t)$, and $b(t)$ are continuous and differentiable with respect to t, then

$$\frac{\partial}{\partial t}\int_{a(t)}^{b(t)} f(t, \tau)\, d\tau = \int_{a(t)}^{b(t)} \frac{\partial f(t, \tau)}{\partial t}\, d\tau - f(t, a)\frac{da(t)}{\partial t} + f(t, b)\frac{db(t)}{\partial t}$$

provided that the integrals exist.

We now replace $\dot{\mathbf{\Phi}}(t, t_0)$ in the foregoing by $\mathbf{F}(t)\mathbf{\Phi}(t, t_0)$, from Eq. (6.2-5), to obtain

$$\dot{\boldsymbol{\mu}}_{\mathbf{x}}(t) = \mathbf{F}(t)\left\{\mathbf{\Phi}(t, t_0)\boldsymbol{\mu}_{\mathbf{x}}(t_0) + \int_{t_0}^{t} \mathbf{\Phi}(t, \tau)\mathbf{G}(\tau)\boldsymbol{\mu}_{\mathbf{u}}(\tau)\, d\tau\right\} + \mathbf{\Phi}(t, t)\mathbf{G}(t)\boldsymbol{\mu}_{\mathbf{u}}(t) \tag{6.2-9}$$

If we recognize the quantity in brackets as simply $\boldsymbol{\mu}_{\mathbf{x}}(t)$ as determined in Eq. (6.2-8), and if we make use of the fact that $\mathbf{\Phi}(t, t) = \mathbf{I}$, we have

$$\dot{\boldsymbol{\mu}}_{\mathbf{x}}(t) = \mathbf{F}(t)\boldsymbol{\mu}_{\mathbf{x}}(t) + \mathbf{G}(t)\boldsymbol{\mu}_{\mathbf{u}}(t) \tag{6.2-10}$$

which is, as it should be, the same as obtained in Eq. (6.2-7).

Now $\boldsymbol{\mu}_{\mathbf{x}}(t)$ may be obtained by solving Eq. (6.2-10) subject to the initial condition $\boldsymbol{\mu}_{\mathbf{x}}(t_0)$. Of course, the general solution of Eq. (6.2-10) is given by Eq. (6.2-8). However, if numerical methods are to be used to find $\boldsymbol{\mu}_{\mathbf{x}}(t)$, which is necessary for all but trivial problems, then the computations of directly integrating Eq. (6.2-10) are considerably easier than those required by Eq. (6.2-8).

Example *6.2-1*. Let us consider the simple first-order (scalar) system

$$\dot{x}(t) = -x(t) + u(t)$$

where $u(t)$ is a white random process with

$$\mathcal{E}\{u(t)\} = \mu_u(t) = 1$$
$$\text{cov}\{u(t), u(\tau)\} = \psi_u(t)\delta_D(t - \tau) = \delta_D(t - \tau)$$

and the initial mean and variance of $x(t)$ are

$$\mu_x(t_0) = \mathcal{E}\{x(t_0)\} = 0$$
$$V_x(t_0) = \text{var}\{x(t_0)\} = 0$$

We desire to obtain expressions for the mean of $x(t)$ by several approaches and note the general features of each approach.

The differential equation for the mean vector, Eq. (6.2-7), becomes

$$\dot{\mu}_x = -\mu_x + \mu_u \qquad \mu_x(0) = 0$$

The solution to this equation may be obtained simply by means of Laplace transforms or simply by "guessing" the result. It follows easily [since $\mu_u(t) = 1$] that

$$\mu_x(t) = 1 - e^{-t} \qquad t \geq 0$$

To use the solution of Eq. (6.2-8), we must first solve the transition matrix equation (6.2-5), which is here

$$\frac{\partial}{\partial t}\Phi(t, 0) = -\Phi(t, 0)$$
$$\Phi(0, 0) = 1$$

This is easily accomplished and we have

$$\Phi(t, 0) = e^{-t} \qquad t > 0$$

or, in general,

$$\Phi(t, \tau) = e^{-(t-\tau)} \qquad t > \tau$$

Equation (6.2-8) becomes

$$\mu_x(t) = \int_0^t e^{-(t-\tau)}\, d\tau = 1 - e^{-t} \qquad t \geq 0$$

For this simple example, either method is of approximately the same complexity. This is not, however, the case for time-varying or higher-order systems.

Example ***6.2-2.*** We desire to repeat Example 6.2-1 with the same random input but with the system differential equation changed to

$$\dot{x}(t) = a(t)x(t) + u(t)$$

where $a(t)$ is a known time function.

Equation (6.2-7) yields

$$\frac{d\mu_x(t)}{dt} = a(t)\mu_x(t) + 1 \qquad t \geq 0$$

Because of the time-varying nature of the system, we cannot use Laplace transforms in any simple fashion, nor are we able to guess a simple solution to the equation as in the time-invariant case.

The analog-computer mechanization to solve the mean-propagation equation is indicated in Fig. 6.2-1. Unless the $a(t)$ function were quite complex, mechanization and solution of this equation would be a simple matter.

The approach must be contrasted with that based on solution of the matrix equation (6.2-5) and use of Eq. (6.2-8). The solution to

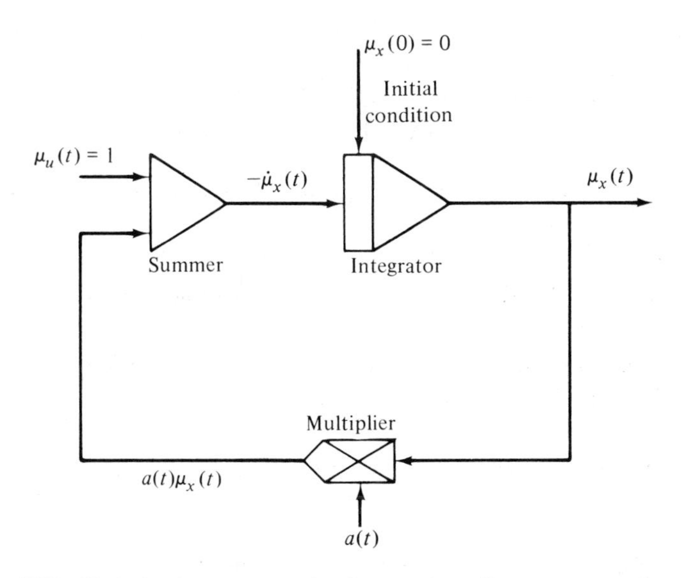

FIG. 6.2-1. Analog-computer implementation of mean-propagation equation for Example 6.2-2.

the fundamental matrix equation for this example

$$\Phi(t, \tau) = e^{\int_\tau^t a(\eta)\, d\eta}$$

cannot be simply obtained unless $a(t)$ is a constant.

We now turn attention to the derivation of an algorithm for propagation of the mean-square value $\mathbf{P}_\mathbf{x}(t)$ for the system of Eq. (6.2-1), where $\mathbf{u}(t)$ is a white-noise input.

The expression for the symmetric $\mathbf{P}_\mathbf{x}(t)$ matrix may be derived directly from the defining relation

$$\mathbf{P}_\mathbf{x}(t) = \mathcal{E}\{\mathbf{x}(t)\mathbf{x}^T(t)\} \tag{6.2-11}$$

By differentiation with respect to time and use of Eq. (6.2-1), there results

$$\begin{aligned}\dot{\mathbf{P}}_\mathbf{x}(t) &= \mathcal{E}\{\dot{\mathbf{x}}(t)\mathbf{x}^T(t)\} + \mathcal{E}\{\mathbf{x}(t)\dot{\mathbf{x}}^T(t)\} \\ &= \mathbf{F}(t)\mathcal{E}\{\mathbf{x}(t)\mathbf{x}^T(t)\} + \mathbf{G}(t)\mathcal{E}\{\mathbf{u}(t)\mathbf{x}^T(t)\} \\ &\quad + \mathcal{E}\{\mathbf{x}(t)\mathbf{x}^T(t)\}\mathbf{F}^T(t) + \mathcal{E}\{\mathbf{x}(t)\mathbf{u}^T(t)\}\mathbf{G}^T(t)\end{aligned}$$

Use of the definition of Eq. (6.2-11) and the definitions

$$\begin{aligned}\mathbf{P}_{\mathbf{ux}}(t) &= \mathcal{E}\{\mathbf{u}(t)\mathbf{x}^T(t)\} \\ &= \mathbf{P}_{\mathbf{xu}}^T(t)\end{aligned} \tag{6.2-12}$$

results in

$$\begin{aligned}\dot{\mathbf{P}}_{\mathbf{x}}(t) &= \mathbf{F}(t)\mathbf{P}_{\mathbf{x}}(t) + \mathbf{P}_{\mathbf{x}}(t)\mathbf{F}^T(t) \\ &\quad + \mathbf{G}(t)\mathbf{P}_{\mathbf{ux}}(t) + \mathbf{P}_{\mathbf{xu}}(t)\mathbf{G}^T(t)\end{aligned} \tag{6.2-13}$$

and we see the need to determine $\mathbf{P}_{\mathbf{ux}}(t)$ and $\mathbf{P}_{\mathbf{xu}}(t)$. From Eqs. (6.2-12) and (6.2-4) there results

$$\mathbf{P}_{\mathbf{xu}}(t) = \mathcal{E}\left\{\left[\mathbf{\Phi}(t, t_0)\mathbf{x}(t_0) + \int_{t_0}^{t} \mathbf{\Phi}(t, \tau)\mathbf{G}(\tau)\mathbf{u}(\tau)\,d\tau\right][\mathbf{u}^T(t)]\right\} \tag{6.2-14}$$

It will be assumed that there is no correlation between $\mathbf{x}(t_0)$ and $\mathbf{u}(t)$ for $t \geq t_0$. Thus the foregoing relation becomes

$$\mathbf{P}_{\mathbf{xu}}(t) = \mathbf{\Phi}(t, t_0)\boldsymbol{\mu}_{\mathbf{x}}(t_0)\boldsymbol{\mu}_{\mathbf{u}}^T(t) + \int_{t_0}^{t} \mathbf{\Phi}(t, \tau)\mathbf{G}(\tau)[\boldsymbol{\mu}_{\mathbf{u}}(\tau)\boldsymbol{\mu}_{\mathbf{u}}^T(t) + \boldsymbol{\psi}_{\mathbf{u}}(t)\delta_D(t - \tau)]\,d\tau \tag{6.2-15}$$

From Eq. (6.2-8) the foregoing relation simplifies to

$$\mathbf{P}_{\mathbf{xu}}(t) = \boldsymbol{\mu}_{\mathbf{x}}(t)\boldsymbol{\mu}_{\mathbf{u}}^T(t) + \int_{t_0}^{t} \mathbf{\Phi}(t, \tau)\mathbf{G}(\tau)\boldsymbol{\psi}_{\mathbf{u}}(t)\delta_D(t - \tau)\,d\tau \tag{6.2-16}$$

Evaluation of the integral relation in the foregoing requires a bit of care. However, since the covariance function $\mathbf{V}_{\mathbf{u}}(t, \tau)$ is symmetric, we shall assume a symmetric impulse function[1] to obtain

$$\mathbf{P}_{\mathbf{xu}}(t) = \boldsymbol{\mu}_{\mathbf{x}}(t)\boldsymbol{\mu}_{\mathbf{u}}^T(t) + \tfrac{1}{2}\mathbf{G}(t)\boldsymbol{\psi}_{\mathbf{u}}(t) \tag{6.2-17}$$

In a similar way we may show that

$$\mathbf{P}_{\mathbf{ux}}(t) = \boldsymbol{\mu}_{\mathbf{u}}(t)\boldsymbol{\mu}_{\mathbf{x}}^T(t) + \tfrac{1}{2}\boldsymbol{\psi}_{\mathbf{u}}(t)\mathbf{G}^T(t) \tag{6.2-18}$$

such that Eq. (6.2-13) becomes

$$\begin{aligned}\dot{\mathbf{P}}_{\mathbf{x}}(t) &= \mathbf{F}(t)\mathbf{P}_{\mathbf{x}}(t) + \mathbf{P}_{\mathbf{x}}(t)\mathbf{F}^T(t) + \mathbf{G}(t)\boldsymbol{\psi}_{\mathbf{u}}(t)\mathbf{G}^T(t) \\ &\quad + \mathbf{G}(t)\boldsymbol{\mu}_{\mathbf{u}}(t)\boldsymbol{\mu}_{\mathbf{x}}^T(t) + \boldsymbol{\mu}_{\mathbf{x}}(t)\boldsymbol{\mu}_{\mathbf{u}}^T(t)\mathbf{G}^T(t)\end{aligned} \tag{6.2-19}$$

[1] The symmetric Dirac delta function is defined by

$$\int_a^b f(\tau)\delta_D(\tau - t)\,d\tau = \begin{cases} 0 & \text{if } t < a \text{ or } t > b \\ f(a)/2 & \text{if } t = a \\ f(b)/2 & \text{if } t = b \\ f(t) & \text{if } a < t < b \end{cases}$$

where $f(\tau)$ is any function continuous on the closed interval $[a, b]$. Reasons for the choice of this particular form for the delta function and why we must be rather careful with stochastic integrals will be discussed in Chapter 8.

This is the desired algorithm whose solution yields the propagation of the mean-square value of $\mathbf{x}(t)$. The initial value for the foregoing is

$$\mathbf{P}_{\mathbf{x}}(t_0) = \boldsymbol{\mu}_{\mathbf{x}}(t_0)\boldsymbol{\mu}_{\mathbf{x}}^T(t_0) + \mathbf{V}_{\mathbf{x}}(t_0) = \mathbf{P}_{\mathbf{x}_0} \tag{6.2-20}$$

The mean-square-propagation algorithm is a bit more complex than we might desire. Often, but by no means always, we are more interested in the variance of the state.

Since we know that

$$\mathbf{V}_{\mathbf{x}}(t) = \mathbf{P}_{\mathbf{x}}(t) - \boldsymbol{\mu}_{\mathbf{x}}(t)\boldsymbol{\mu}_{\mathbf{x}}^T(t) \tag{6.2-21}$$

we have, upon taking the time derivative,

$$\dot{\mathbf{V}}_{\mathbf{x}}(t) = \dot{\mathbf{P}}_{\mathbf{x}}(t) - \dot{\boldsymbol{\mu}}_{\mathbf{x}}(t)\boldsymbol{\mu}_{\mathbf{x}}^T(t) - \boldsymbol{\mu}_{\mathbf{x}}(t)\dot{\boldsymbol{\mu}}_{\mathbf{x}}^T(t) \tag{6.2-22}$$

Use of the mean-propagation-algorithm (6.2-7) relation for $\dot{\boldsymbol{\mu}}_{\mathbf{x}}(t)$ in Eqs. (6.2-22), (6.2-21), and (6.2-19) leads directly to the desired variance-propagation algorithm

$$\dot{\mathbf{V}}_{\mathbf{x}}(t) = \mathbf{F}(t)\mathbf{V}_{\mathbf{x}}(t) + \mathbf{V}_{\mathbf{x}}(t)\mathbf{F}^T(t) + \mathbf{G}(t)\boldsymbol{\psi}_{\mathbf{u}}(t)\mathbf{G}^T(t) \tag{6.2-23}$$

which is solved with the initial condition $\mathbf{V}_{\mathbf{x}}(t_0) = \mathbf{V}_{\mathbf{x}_0}$.

This relation is most important. Of at least equal importance is the method used in the derivation, since derivations of this type are most commonly needed in the information, control, and communication sciences. Now let us look at an ancillary approach. We begin by assuming that $\mathbf{u}$ is not necessarily white.

First, we consider the determination of the covariance $\mathbf{V}_{\mathbf{xu}}(t_1, t_2)$, which becomes, by use of Eq. (6.2-4) and the assumption that $\mathbf{x}(t_0)$ and $\mathbf{u}(t)$ for $t \geq t_0$ are uncorrelated,

$$\begin{aligned}\mathbf{V}_{\mathbf{xu}}(t_1, t_2) \triangleq \operatorname{cov}\{\mathbf{x}(t_1), \mathbf{u}(t_2)\} &= \boldsymbol{\Phi}(t_1, t_0) \operatorname{cov}\{\mathbf{x}(t_0), \mathbf{u}(t_2)\} \\ &\quad + \operatorname{cov}\left\{\int_{t_0}^{t_1} \boldsymbol{\Phi}(t_1, \tau)\mathbf{G}(\tau)\mathbf{u}(\tau)\, d\tau,\ \mathbf{u}(t_2)\right\}\end{aligned} \tag{6.2-24}$$

By interchanging the expectation and integration operations, we obtain

$$\mathbf{V}_{\mathbf{xu}}(t_1, t_2) = \boldsymbol{\Phi}(t_1, t_0)\mathbf{V}_{\mathbf{xu}}(t_0, t_2) + \int_{t_0}^{t_1} \boldsymbol{\Phi}(t_1, \tau)\mathbf{G}(\tau)\mathbf{V}_{\mathbf{u}}(\tau, t_2)\, d\tau \tag{6.2-25}$$

A convolution-type integral is again involved. In this case, the variance $\mathbf{V}_{\mathbf{u}}(\tau, t_2)$ acts as the input. Before manipulating this expression further, let us consider the development of an expression for the variance $\mathbf{V}_{\mathbf{x}}(t_1, t_2)$.

By use of Eq. (6.2-4), the system response equation, we find that

$$\begin{aligned}
\mathbf{V}_{\mathbf{x}}(t_1, t_2) &\triangleq \operatorname{cov}\{\mathbf{x}(t_1), \mathbf{x}(t_2)\} \\
&= \mathbf{\Phi}(t_1, t_0)\operatorname{cov}\{\mathbf{x}(t_0), \mathbf{x}(t_0)\}\mathbf{\Phi}^T(t_2, t_0) \\
&\quad + \mathbf{\Phi}(t_1, t_0)\operatorname{cov}\left\{\mathbf{x}(t_0), \int_{t_0}^{t_2} \mathbf{\Phi}(t_2, \tau)\mathbf{G}(\tau)\mathbf{u}(\tau)\,d\tau\right\} \qquad (6.2\text{-}26) \\
&\quad + \operatorname{cov}\left\{\int_{t_0}^{t_1} \mathbf{\Phi}(t_1, \tau)\mathbf{G}(\tau)\mathbf{u}(\tau)\,d\tau, \mathbf{\Phi}(t_2, t_0)\mathbf{x}(t_0)\right\} \\
&\quad + \operatorname{cov}\left\{\int_{t_0}^{t_1} d\tau_1 \int_{t_0}^{t_2} d\tau_2 \mathbf{\Phi}(t_1, \tau_1)\mathbf{G}(\tau_1)\mathbf{u}(\tau_1), \mathbf{\Phi}(t_2, \tau_2)\mathbf{G}(\tau_2)\mathbf{u}(\tau_2)\right\}
\end{aligned}$$

After slight rearrangement, this result becomes, since $\mathbf{u}(\tau)$, $t_0 \leq \tau \leq t_2$, is uncorrelated with $\mathbf{x}(t_0)$,

$$\begin{aligned}
\mathbf{V}_{\mathbf{x}}(t_1, t_2) &= \mathbf{\Phi}(t_1, t_0)\mathbf{V}_{\mathbf{x}}(t_0)\mathbf{\Phi}^T(t_2, t_0) \\
&\quad + \int_{t_0}^{t_1} d\tau_1 \int_{t_0}^{t_2} d\tau_2 \mathbf{\Phi}(t_1, \tau_1)\mathbf{G}(\tau_1)\mathbf{V}_{\mathbf{u}}(\tau_1, \tau_2)\mathbf{G}^T(\tau_2)\mathbf{\Phi}^T(t_2, \tau_2)
\end{aligned} \qquad (6.2\text{-}27)$$

Equations (6.2-25) and (6.2-27) are the most general expressions that we can obtain for the variances $\mathbf{V}_{\mathbf{xu}}(t_1, t_2)$ and $\mathbf{V}_{\mathbf{x}}(t_1, t_2)$. Unfortunately, the expressions are not convenient or easy to use in our later development. To simplify the expressions, it is convenient to make a restriction on the input signal. We shall assume that $\mathbf{u}(t)$ is white noise so that the relation of Eq. (6.2-3) may be used. By augmenting the state variable, $\mathbf{x}(t)$, it is a trivial matter to consider non-white-noise inputs $\mathbf{u}(t)$. We are in effect, therefore, assuming that the state variable has so been augmented that $\mathbf{u}(t)$ is white.

With the white-noise assumption $\mathbf{V}_{\mathbf{xu}}(t_1, t_2)$ in Eq. (6.2-25) becomes

$$\mathbf{V}_{\mathbf{xu}}(t_1, t_2) = \mathbf{\Phi}(t_1, t_0)\mathbf{V}_{\mathbf{xu}}(t_0, t_2) + \int_{t_0}^{t_1} \mathbf{\Phi}(t_1, \tau)\mathbf{G}(\tau)\boldsymbol{\psi}_{\mathbf{u}}(\tau)\delta_D(\tau - t_2)\,d\tau \qquad (6.2\text{-}28)$$

Using the properties of the Dirac delta function, the foregoing equation reduces to

$$\begin{aligned}
\mathbf{V}_{\mathbf{xu}}(t_1, t_2) &= \mathbf{\Phi}(t_1, t_0)\mathbf{V}_{\mathbf{xu}}(t_0, t_2) \\
&\quad + \begin{cases} \mathbf{0} & \text{if } t_0 < t_1 < t_2 \\ \mathbf{G}(t_2)\boldsymbol{\psi}_{\mathbf{u}}(t_2)/2 & \text{if } t_0 < t_1 = t_2 \\ \mathbf{\Phi}(t_1, t_2)\mathbf{G}(t_2)\boldsymbol{\psi}_{\mathbf{u}}(t_2) & \text{if } t_0 < t_2 < t_1 \end{cases}
\end{aligned} \qquad (6.2\text{-}29)$$

Hence we see that a discontinuity occurs in $\mathbf{V}_{\mathbf{xu}}(t_1, t_2)$ at the point $t_1 = t_2$.

The expression for $\mathbf{V}_{\mathbf{x}}(t_1, t_2)$ in Eq. (6.2-27) becomes, with the white-noise assumption for $\mathbf{V_u}(t_1, t_2)$,

$$\mathbf{V_x}(t_1, t_2) = \mathbf{\Phi}(t_1, t_0)\mathbf{V_x}(t_0)\mathbf{\Phi}^T(t_2, t_0) + \int_{t_0}^{t_1} d\tau_1 \int_{t_0}^{t_2} d\tau_2 \mathbf{\Phi}(t_1, \tau_1)\mathbf{G}(\tau_1)\boldsymbol{\psi}_\mathbf{u}(\tau_1)\mathbf{G}^T(\tau_2)\mathbf{\Phi}^T(t_2\ \tau_2)\delta_D(\tau_1 - \tau_2) \tag{6.2-30}$$

We must be careful in selecting the order in which we carry out this double integration. To easily note the point at which the Dirac delta function will be encountered, we shall integrate first over the variable that has the larger range. In other words, if $t_1 < t_2$, we shall integrate with respect to τ_1 first. Hence the resulting expression for $\mathbf{V_x}(t_1, t_2)$ becomes

$$\mathbf{V_x}(t_1, t_2) = \mathbf{\Phi}(t_1, t_0)\mathbf{V_x}(t_0)\mathbf{\Phi}^T(t_2, t_0) + \int_{t_0}^{\min\{t_1, t_2\}} \mathbf{\Phi}(t_1, \tau)\mathbf{G}(\tau)\boldsymbol{\psi}_\mathbf{u}(\tau)\mathbf{G}^T(\tau)\mathbf{\Phi}^T(t_2, \tau)\, d\tau \tag{6.2-31}$$

where the operation $\min\{t_1, t_2\}$ implies

$$\min\{t_1, t_2\} = \begin{cases} t_1 & \text{if } t_1 \leq t_2 \\ t_2 & \text{if } t_1 > t_2 \end{cases}$$

Equation (6.2-31) is still not easy to use because of the convolution integral that is involved. To simplify this expression, it is convenient to first restrict attention to the case in which $t_1 = t_2 = t$ so that $\mathbf{V_x}(t_1, t_2) = \mathbf{V_x}(t)$. Equation (6.2-31) then becomes

$$\mathbf{V_x}(t) = \mathbf{\Phi}(t, t_0)\mathbf{V_x}(t_0)\mathbf{\Phi}^T(t, t_0) + \int_{t_0}^{t} \mathbf{\Phi}(t, \tau)\mathbf{G}(\tau)\boldsymbol{\psi}_\mathbf{u}(\tau)\mathbf{G}^T(\tau)\mathbf{\Phi}^T(t, \tau)\, d\tau \tag{6.2-32}$$

where $\mathbf{V_x}(t_0)$ is the prior variance of the $\mathbf{x}(t)$ process.

It may appear that this last expression is not really simpler than Eq. (6.2-31). With this form, however, it is possible to transform the integral expression into a differential equation. This transformation may be accomplished by differentiating both sides of Eq. (6.2-32), once again making use of Leibnitz's rule for differentiation under integration, to obtain

$$\begin{aligned}\dot{\mathbf{V}}_\mathbf{x}(t) = {} & \dot{\mathbf{\Phi}}(t, t_0)\mathbf{V_x}(t_0)\mathbf{\Phi}^T(t, t_0) + \mathbf{\Phi}(t, t_0)\mathbf{V_x}(t_0)\dot{\mathbf{\Phi}}^T(t, t_0) \\ & + \int_{t_0}^{t} \dot{\mathbf{\Phi}}(t, \tau)\mathbf{G}(\tau)\boldsymbol{\psi}_\mathbf{u}(\tau)\mathbf{G}^T(\tau)\mathbf{\Phi}^T(t, \tau)\, d\tau \\ & + \int_{t_0}^{t} \mathbf{\Phi}(t, \tau)\mathbf{G}(\tau)\boldsymbol{\psi}_\mathbf{u}(\tau)\mathbf{G}^T(\tau)\dot{\mathbf{\Phi}}^T(t, \tau)\, d\tau \\ & + \mathbf{\Phi}(t, t)\mathbf{G}(t)\boldsymbol{\psi}_\mathbf{u}(t)\mathbf{G}^T(t)\mathbf{\Phi}(t, t)\end{aligned}$$

We now use Eqs. (6.2-5) and (6.2-6) so that the foregoing becomes

$$\dot{\mathbf{V}}_{\mathbf{x}}(t) = \mathbf{F}(t)\left\{\mathbf{\Phi}(t, t_0)\mathbf{V}_{\mathbf{x}}(t_0)\mathbf{\Phi}^T(t, t_0) + \int_{t_0}^{t} \mathbf{\Phi}(t, \tau)\mathbf{G}(\tau)\boldsymbol{\psi}_{\mathbf{u}}(\tau)\mathbf{G}^T(\tau)\mathbf{\Phi}^T(t, \tau)\, d\tau\right\}$$
$$+ \left\{\mathbf{\Phi}(t, t_0)\mathbf{V}_{\mathbf{x}}(t_0)\mathbf{\Phi}^T(t, t_0) + \int_{t_0}^{t} \mathbf{\Phi}(t, \tau)\mathbf{G}(\tau)\boldsymbol{\psi}_{\mathbf{u}}(\tau)\mathbf{G}^T(\tau)\mathbf{\Phi}^T(t, \tau)\, d\tau\right\}\mathbf{F}^T(t)$$
$$+ \mathbf{G}(t)\boldsymbol{\psi}_{\mathbf{u}}(t)\mathbf{G}^T(t) \tag{6.2-33}$$

We compare the foregoing equation with Eq. (6.2-12) and recognize the quantity in brackets as $\mathbf{V}_{\mathbf{x}}(t)$. Thus this equation may be written, finally, as

$$\dot{\mathbf{V}}_{\mathbf{x}}(t) = \mathbf{F}(t)\mathbf{V}_{\mathbf{x}}(t) + \mathbf{V}_{\mathbf{x}}(t)\mathbf{F}^T(t) + \mathbf{G}(t)\boldsymbol{\psi}_{\mathbf{u}}(t)\mathbf{G}^T(t) \tag{6.2-34}$$

Equation (6.2-34) is the desired result and allows us to determine $\mathbf{V}_{\mathbf{x}}(t)$ by solving a linear matrix differential equation with the initial condition $\mathbf{V}_{\mathbf{x}}(t_0)$. We note that explicit knowledge of the transition matrix is not required in Eq. (6.2-34) and that this equation is the same as Eq. (6.2-23), as it should be.

It is now a very simple matter to show that

$$\mathbf{V}_{\mathbf{x}}(t_1, t_2) = \begin{cases} \mathbf{\Phi}(t_1, t_2)\mathbf{V}_{\mathbf{x}}(t_2) & t_1 > t_2 \\ \mathbf{V}_{\mathbf{x}}(t_1)\mathbf{\Phi}^T(t_2, t_1) & t_2 > t_1 \end{cases} \tag{6.2-35}$$

by substitution of the foregoing and Eq. (6.2-32) into Eq. (6.2-31).

The reader is cautioned that Eqs. (6.2-29), (6.2-31), (6.2-34), and (6.2-35) are only valid if $\mathbf{u}(t)$ is white noise. If $\mathbf{u}(t)$ is not white, it is often possible to model $\mathbf{u}(t)$ as the output of a dynamic system of the form of Eq. (6.2-1), which is driven by a white-noise input. By adjoining this dynamic model to the original system equations, an augmented dynamic system is obtained that has a white-noise input and to which these variance relations may be applied. The mean, mean-square, and variance-propagation algorithms of this section are summarized in Table 6.2-1.

Example *6.2-3*. To illustrate the use of the variance algorithms that we have obtained, let us reconsider the simple scalar Example 6.2-1, where the dynamic model was given by

$$\dot{x}(t) = -x(t) + u(t)$$

We shall assume that the state of the system is known at $t = 0$, $x(0) = 0$, and thus the prior mean and variance are $\mu_x(0) = \mathcal{E}\{x(0)\} = 0$ and $V_x(0) = \text{var}\{x(0)\} = 0$. The input noise $u(t)$ is white with unit mean and a unit variance parameter $\mu_u(t) = 1$ and $\psi_u(t) = 1$.

The differential equation for the mean of $x(t)$ was determined in Example 6.2-1 to be

$$\dot{\mu}_x(t) = -\mu_x(t) + 1$$

TABLE 6.2-1. State-Variable Relations for Continuous Linear System Response to White Random Inputs

State equation	$\dot{\mathbf{x}}(t) = \mathbf{F}(t)\mathbf{x}(t) + \mathbf{G}(t)\mathbf{u}(t)$
Moments	$\mathcal{E}\{\mathbf{u}(t)\} = \boldsymbol{\mu}_{\mathbf{u}}(t)$ $\quad$ $\operatorname{cov}\{\mathbf{u}(t), \mathbf{u}(\tau)\} = \boldsymbol{\psi}_{\mathbf{u}}(t)\delta_D(t-\tau)$ $\mathcal{E}\{\mathbf{x}(t_0)\} = \boldsymbol{\mu}_{\mathbf{x}}(t_0)$ $\quad$ $\operatorname{var}\{\mathbf{x}(t_0)\} = \mathbf{V}_{\mathbf{x}}(t_0)$
Mean-value algorithm	$\dot{\boldsymbol{\mu}}_{\mathbf{x}}(t) = \mathbf{F}(t)\boldsymbol{\mu}_{\mathbf{x}}(t) + \mathbf{G}(t)\boldsymbol{\mu}_{\mathbf{u}}(t)$ $\boldsymbol{\mu}_{\mathbf{x}}(t_0) = \boldsymbol{\mu}_{\mathbf{x}_0}$
Variance algorithm	$\dot{\mathbf{V}}_{\mathbf{x}}(t) = \mathbf{F}(t)\mathbf{V}_{\mathbf{x}}(t) + \mathbf{V}_{\mathbf{x}}(t)\mathbf{F}^T(t) + \mathbf{G}(t)\boldsymbol{\psi}_{\mathbf{u}}(t)\mathbf{G}^T(t)$ $\mathbf{V}_{\mathbf{x}}(t_0) = \mathbf{V}_{\mathbf{x}_0}$
Mean-square algorithm	$\mathbf{P}_{\mathbf{x}}(t) = \mathbf{V}_{\mathbf{x}}(t) + \boldsymbol{\mu}_{\mathbf{x}}(t)\boldsymbol{\mu}_{\mathbf{x}}^T(t)$
Alternative mean-square algorithm	$\dot{\mathbf{P}}_{\mathbf{x}}(t) = \mathbf{F}(t)\mathbf{P}_{\mathbf{x}}(t) + \mathbf{P}_{\mathbf{x}}(t)\mathbf{F}^T(t) + \mathbf{G}(t)\boldsymbol{\psi}_{\mathbf{u}}(t)\mathbf{G}^T(t)$ $+ \mathbf{G}(t)\boldsymbol{\mu}_{\mathbf{u}}(t)\boldsymbol{\mu}_{\mathbf{x}}^T(t) + \boldsymbol{\mu}_{\mathbf{x}}(t)\boldsymbol{\mu}_{\mathbf{u}}^T(t)\mathbf{G}^T(t)$ $\mathbf{P}_{\mathbf{x}}(t_0) = \mathbf{V}_{\mathbf{x}_0} + \boldsymbol{\mu}_{\mathbf{x}_0}\boldsymbol{\mu}_{\mathbf{x}_0}^T = \mathbf{P}_{\mathbf{x}_0}$
Covariance algorithm	$\mathbf{V}_{\mathbf{x}}(t_1, t_2) = \begin{cases} \boldsymbol{\Phi}(t_1, t_2)\mathbf{V}_{\mathbf{x}}(t_2) & t_1 \geq t_2 \\ \mathbf{V}_{\mathbf{x}}(t_1)\boldsymbol{\Phi}^T(t_2, t_1) & t_1 \leq t_2 \end{cases}$
Correlation algorithm	$\mathbf{P}_{\mathbf{x}}(t_1, t_2) = \mathbf{V}_{\mathbf{x}}(t_1, t_2) + \boldsymbol{\mu}_{\mathbf{x}}(t_1)\boldsymbol{\mu}_{\mathbf{x}}^T(t_2)$
Transition-matrix algorithm	$\dfrac{\partial \boldsymbol{\Phi}(t_1, t_2)}{\partial t_1} = \mathbf{F}(t)\boldsymbol{\Phi}(t_1, t_2)$ $\boldsymbol{\Phi}(t_2, t_2) = \mathbf{I}$

with the initial condition $\mu_x(0) = 0$. The solution for this equation was easily found to be

$$\mu_x(t) = 1 - e^{-t} \qquad t \geq 0$$

The variance equation is, from Eq. (6.2-34),

$$\dot{V}_x(t) = -2V_x(t) + 1$$

with $V_x(0) = 0$. The solution is easily found to be

$$V_x(t) = \tfrac{1}{2}(1 - e^{-2t})$$

These relations may also be obtained by means of the fundamental system matrix, which is

$$\Phi(t, \tau) = e^{-(t-\tau)} \qquad t \geq \tau$$

for this example. In particular, Eq. (6.2-35) yields for the covariance function

$$V_x(t_1, t_2) = \begin{cases} 0.5e^{-(t_1-t_2)}(1 - e^{-2t_2}) & t_1 > t_2 \\ 0.5e^{-(t_2-t_1)}(1 - e^{-2t_1}) & t_2 > t_1 \end{cases}$$

and it is easily seen that the stochastic process $x(t)$ is nonstationary and therefore nonergodic.

Figure 6.2-2 shows the time history of the mean, variance, and mean-square-value propagation for this example. It is known that

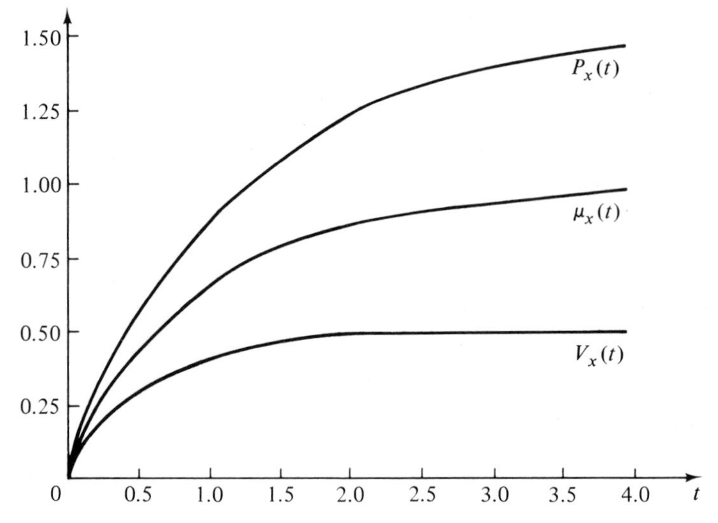

FIG. 6.2-2. Moments of $x(t)$, Example 6.2-3.

the mean of the state is zero at time equals zero. The mean response of the system is precisely the same as that of the differential equation

$$\dot{y}(t) = -y(t) + 1 \qquad y(0) = 0$$

in which no mention need be made of any random inputs. The input $+1$ is simply regarded as deterministic.

The initial state of the system is known precisely, so the initial state variance must be zero. As time increases, the state variance increases because of the increasing ignorance of the value of the state due to the white random input $u(t)$. It is important to recall, however, that the variance of the white-noise input is infinite, whereas the variance of the state output is finite. Figure 6.2-3 shows a sketch of the covariance function $V_x(t_1, t_2)$ for this example.

Clearly, the white-noise input is not physically realizable since

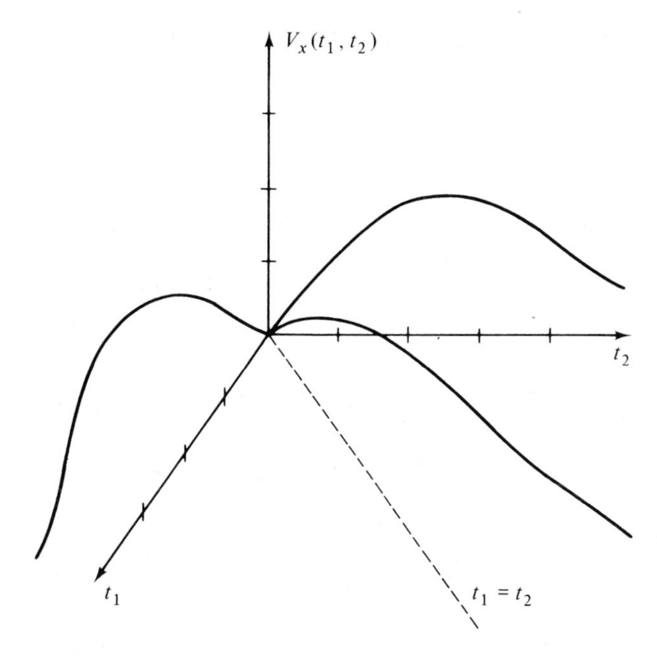

FIG. 6.2-3. Variance $V_x(t_1, t_2)$ for Example 6.2-3.

it has infinite variance. We are assuming that the covariance of the input noise is

$$V_u(t_1, t_2) = \delta_D(t_1 - t_2)$$

If it turns out that the actual input covariance is

$$V_{u_a}(t_1, t_2) = \frac{1}{2\alpha}e^{-|t_1-t_2|/\alpha}$$

the system variance $V_x(t)$ will be essentially the same as for the white-noise input, as long as $\alpha \ll 1$. It will be a simple matter for the reader to show that actual noise $u_a(t)$ is generated from the relationships

$$\dot{u}_a(t) = \frac{-u_a(t)}{\alpha} + \frac{w(t)}{\alpha}$$

with the prior statistics for the initial state $u_a(0)$ and the white-noise input $w(t)$,

$$\mu_{u_a}(0) = 1 \qquad V_{u_a}(0) = \frac{1}{2\alpha}$$

$$\mu_w(t) = 1 \qquad \psi_w(t) = 1$$

We may now augment the state vector by the definition

$$\mathbf{x}(t) = \begin{bmatrix} x(t) \\ u_a(t) \end{bmatrix} \qquad \mathbf{F}(t) = \begin{bmatrix} -1 & 1 \\ 0 & \dfrac{-1}{\alpha} \end{bmatrix} \qquad \mathbf{G}(t) = \begin{bmatrix} 0 \\ \dfrac{1}{\alpha} \end{bmatrix}$$

such that the differential equation for the augmented process is

$$\dot{\mathbf{x}}(t) = \mathbf{F}(t)\mathbf{x}(t) + \mathbf{G}(t)\mathbf{w}(t)$$

The variance equation

$$\dot{\mathbf{V}}_{\mathbf{x}}(t) = \mathbf{F}(t)\mathbf{V}_{\mathbf{x}}(t) + \mathbf{V}_{\mathbf{x}}(t)\mathbf{F}^T(t) + \mathbf{G}(t)\psi_w(t)\mathbf{G}^T(t)$$

and the associated initial conditions become in component form

$$\dot{V}_x(t) = -2V_x(t) + 2V_{xu_a}(t) \qquad V_x(0) = 0$$

$$\dot{V}_{xu_a}(t) = -\left(1 + \frac{1}{\alpha}\right)V_{xu_a}(t) + V_{u_a}(t) \qquad V_{xu_a}(0) = 0$$

$$\dot{V}_{u_a}(t) = -\frac{2}{\alpha}V_{u_a}(t) + \frac{1}{\alpha^2}\psi_w(t) \qquad V_{u_a}(0) = \frac{1}{2\alpha}$$

The solution to these equations is easily determined to be, for $\psi_w(t) = 1$,

$$V_{u_a}(t) = \frac{1}{2\alpha}$$

$$V_{xu_a}(t) = \frac{1}{2(\alpha + 1)}[1 - e^{-(\alpha+1)t/\alpha}]$$

$$V_x(t) = \frac{1}{2(\alpha + 1)} - \frac{1}{2(1 - \alpha)}e^{-2t} + \frac{\alpha}{1 - \alpha^2}e^{-(\alpha+1)t/\alpha}$$

As $\alpha \longrightarrow 0$, we see that the $V_x(t)$ with the non-white-noise input has essentially the same value that was obtained with the white-noise assumption. As indicated in Fig. 6.2-4, the white-noise approximation is quite valid for $\alpha < 0.2$.

Example *6.2-4*. As a somewhat more complicated example, we consider the double integrator system illustrated in Fig. 6.2-5. As is apparent from the figure, the equations of state that describe the system are

$$\dot{x}_1(t) = x_2(t)$$

$$\dot{x}_2(t) = u(t)$$

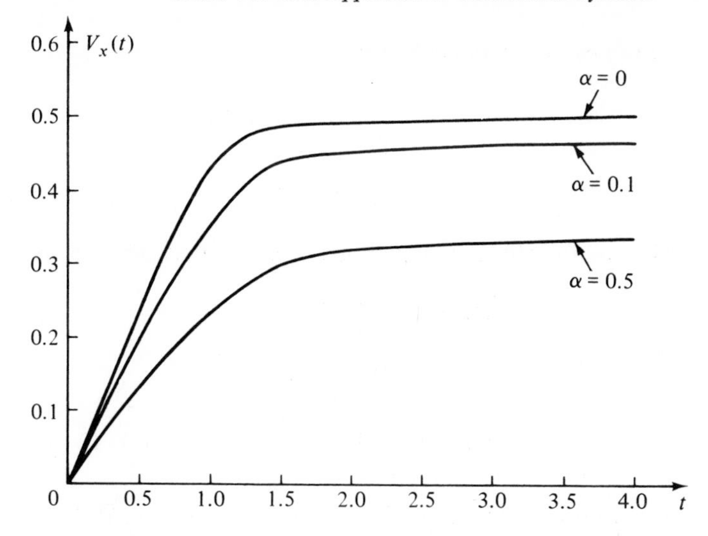

FIG. 6.2-4. $V_x(t)$ for different white-noise approximations.

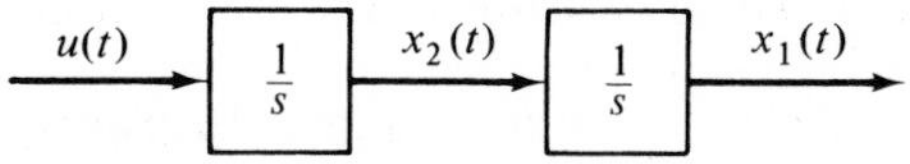

$u(t)$ → $\frac{1}{s}$ → $x_2(t)$ → $\frac{1}{s}$ → $x_1(t)$

FIG. 6.2-5. Double integrator for Example 6.2-4.

where $u(t)$ is white noise with unit mean $\mu_u(t)$ and variance parameter $\psi_u(t)$. Thus the system can be represented by the state equation

$$\dot{\mathbf{x}}(t) = \mathbf{F}\mathbf{x}(t) + \mathbf{G}\mathbf{u}(t)$$

where

$$\mathbf{x}(t) = \begin{bmatrix} x_1(t) \\ x_2(t) \end{bmatrix} \qquad \mathbf{F}(t) = \begin{bmatrix} 0 & 1 \\ 0 & 0 \end{bmatrix} \qquad \mathbf{G}(t) = \begin{bmatrix} 0 \\ 1 \end{bmatrix}$$

The variance algorithm

$$\dot{\mathbf{V}}_{\mathbf{x}}(t) = \mathbf{F}(t)\mathbf{V}_{\mathbf{x}}(t) + \mathbf{V}_{\mathbf{x}}(t)\mathbf{F}^T(t) + \mathbf{G}(t)\boldsymbol{\psi}_{\mathbf{u}}(t)\mathbf{G}^T(t)$$

yields immediately

$$\begin{bmatrix} \dot{V}_{x_{11}}(t) & \dot{V}_{x_{12}}(t) \\ \dot{V}_{x_{12}}(t) & \dot{V}_{x_{22}}(t) \end{bmatrix} = \begin{bmatrix} 0 & 1 \\ 0 & 0 \end{bmatrix} \begin{bmatrix} V_{x_{11}}(t) & V_{x_{12}}(t) \\ V_{x_{12}}(t) & V_{x_{22}}(t) \end{bmatrix}$$
$$+ \begin{bmatrix} V_{x_{11}}(t) & V_{x_{12}}(t) \\ V_{x_{12}}(t) & V_{x_{22}}(t) \end{bmatrix} \begin{bmatrix} 0 & 0 \\ 1 & 0 \end{bmatrix}$$
$$+ \begin{bmatrix} 0 \\ 1 \end{bmatrix} 1 \, [0 \quad 1]$$

which is, in component form,

$$\dot{V}_{x_{11}}(t) = 2V_{x_{12}}(t)$$
$$\dot{V}_{x_{12}}(t) = V_{x_{22}}(t)$$
$$\dot{V}_{x_{22}}(t) = 1$$

The solution to these equations is easily determined to be

$$V_{x_{22}}(t) = t + V_{x_{22}}(0)$$
$$V_{x_{12}}(t) = V_{x_{22}}(0)t + \frac{t^2}{2} + V_{x_{12}}(0)$$
$$V_{x_{11}}(t) = V_{x_{22}}(0)t^2 + \frac{t^3}{3} + 2V_{x_{12}}(0)t + V_{x_{11}}(0)$$

in the order in which analytical solution is obtained. It is possible to determine the mean-square value as a function of time by solution of Eq. (6.2-19), although if the variance propagation has already been obtained, it would be simpler to obtain it from the relation

$$\mathbf{P}_{\mathbf{x}}(t) = \mathbf{V}_{\mathbf{x}}(t) + \boldsymbol{\mu}_{\mathbf{x}}(t)\boldsymbol{\mu}_{\mathbf{x}}^T(t)$$

which requires solution of the mean-propagation equation

$$\dot{\boldsymbol{\mu}}_{\mathbf{x}}(t) = \mathbf{F}(t)\boldsymbol{\mu}_{\mathbf{x}}(t) + \mathbf{G}(t)\boldsymbol{\mu}_{\mathbf{w}}(t)$$

which is here

$$\dot{\mu}_{x_1}(t) = \mu_{x_2}(t)$$
$$\dot{\mu}_{x_2}(t) = \mu_u(t)$$

The solution to these equations is

$$\mu_{x_2}(t) = \mu_{x_2}(0) + t$$
$$\mu_{x_1}(t) = \mu_{x_2}(0)t + \frac{t^2}{2}$$

Example ***6.2-5.*** We consider the message model

$$\dot{\mathbf{y}}(t) = \mathbf{A}(t)\mathbf{y}(t) + \mathbf{B}(t)\mathbf{v}(t)$$

where $\mathbf{v}(t)$ is nonwhite noise that can be modeled by

$$\dot{\boldsymbol{\alpha}}(t) = \mathbf{C}(t)\boldsymbol{\alpha}(t) + \mathbf{D}(t)\mathbf{v}(t)$$
$$\mathbf{v}(t) = \mathbf{E}(t)\boldsymbol{\alpha}(t) + \mathbf{w}(t)$$

where $\mathbf{w}(t)$ and $\mathbf{v}(t)$ are white noises that are assumed uncorrelated. We augment the state by definition of the augmented state vector and white noises

$$\mathbf{x}(t) = \begin{bmatrix} \mathbf{y}(t) \\ \hline \boldsymbol{\alpha}(t) \end{bmatrix} \qquad \mathbf{u}(t) = \begin{bmatrix} \mathbf{v}(t) \\ \hline \mathbf{w}(t) \end{bmatrix}$$

such that the augmented-state differential equation

$$\dot{\mathbf{x}}(t) = \mathbf{F}(t)\mathbf{x}(t) + \mathbf{G}(t)\mathbf{w}(t)$$

where

$$\mathbf{F}(t) = \left[\begin{array}{c|c} \mathbf{A}(t) & \mathbf{B}(t)\mathbf{E}(t) \\ \hline \mathbf{0} & \mathbf{C}(t) \end{array}\right] \qquad \mathbf{G}(t) = \left[\begin{array}{c|c} \mathbf{0} & \mathbf{B}(t) \\ \hline \mathbf{D}(t) & \mathbf{0} \end{array}\right]$$

results. Thus the derived white-noise-input mean- and variance-propagation algorithms may be used. These algorithms are solved with the initial conditions and moments

$$\boldsymbol{\mu}_{\mathbf{x}}(t_0) = \begin{bmatrix} \boldsymbol{\mu}_{\mathbf{y}}(t_0) \\ \hline \boldsymbol{\mu}_{\alpha}(t_0) \end{bmatrix} \qquad \mathbf{V}_{\mathbf{x}}(t_0) = \left[\begin{array}{c|c} \mathbf{V}_{\mathbf{y}}(t_0) & \mathbf{V}_{\mathbf{y}\alpha}(t_0) \\ \hline \mathbf{V}_{\alpha\mathbf{y}}(t_0) & \mathbf{V}_{\alpha}(t_0) \end{array}\right]$$

$$\boldsymbol{\mu}_{\mathbf{u}}(t) = \begin{bmatrix} \boldsymbol{\mu}_{\mathbf{v}}(t) \\ \hline \boldsymbol{\mu}_{\mathbf{w}}(t) \end{bmatrix} \qquad \mathbf{V}_{\mathbf{u}}(t, \tau) = \left[\begin{array}{c|c} \boldsymbol{\psi}_{\mathbf{v}}(t) & \boldsymbol{\psi}_{\mathbf{w}}(t) \\ \hline \mathbf{0} & \mathbf{0} \end{array}\right] \delta_D(t - \tau)$$

We see that we need the initial cross variances $\mathbf{V}_{\alpha\mathbf{y}}(t_0)$ to solve the variance-propagation algorithms. Often this term will be zero in physical problems.

Example *6.2-6*. One major application area for the material in this chapter and in the whole text is that area of information and control science concerned with sequential filter theory. A complete exposition of this subject would require an entire text, and much of the material is beyond the scope of the present efforts. To indicate some of the essential features of sequential filter theory and illustrate uses for the material of this section, we present a somewhat simplified derivation of the Kalman filter equations.

We assume that a state-vector message model is generated by the linear differential equation

$$\dot{\mathbf{x}}(t) = \mathbf{F}(t)\mathbf{x}(t) + \mathbf{G}(t)\mathbf{w}(t)$$

where $\mathbf{w}(t)$ is zero-mean white noise. The first- and second-order statistics of the white plant noise $\mathbf{w}(t)$ and the initial state vector are

$$\mathcal{E}\{\mathbf{w}(t)\} = \mathbf{0} \qquad \mathbf{V}_{\mathbf{w}}(t_1, t_2) = \operatorname{cov}\{\mathbf{w}(t_1), \mathbf{w}(t_2)\}$$
$$= \boldsymbol{\psi}_{\mathbf{w}}(t_1)\delta_D(t_1 - t_2)$$
$$\boldsymbol{\mu}_{\mathbf{x}}(t_0) = \mathcal{E}\{\mathbf{x}(t_0)\} = \boldsymbol{\mu}_{\mathbf{x}_0} \qquad \mathbf{V}_{\mathbf{x}}(t_0) = \operatorname{var}\{\mathbf{x}(t_0)\} = \mathbf{V}_{\mathbf{x}_0}$$
$$\operatorname{cov}\{\mathbf{x}(t_0), \mathbf{w}(t_1)\} = \mathbf{0} \qquad t_1 \geq t_0$$

A noise-corrupted linearly modulated observation of the state vector

$$\mathbf{z}(t) = \mathbf{H}(t)\mathbf{x}(t) + \mathbf{v}(t)$$

is available. $\mathbf{v}(t)$ is zero-mean white noise with

$$\mathcal{E}\{\mathbf{v}(t)\} = \mathbf{0} \qquad \operatorname{cov}\{\mathbf{v}(t_1), \mathbf{v}(t_2)\} = \boldsymbol{\psi}_{\mathbf{v}}(t)\delta_D(t_1 - t_2)$$
$$\operatorname{cov}\{\mathbf{v}(t_1), \mathbf{w}(t_2)\} = \mathbf{0} \qquad \operatorname{cov}\{\mathbf{v}(t_1), \mathbf{x}(t_2)\} = \mathbf{0} \qquad \text{for all } t_1, t_2$$

It is desired to construct a linear, unbiased, minimum error variance, sequential estimator of the state vector $\mathbf{x}(t)$. We postulate a filter of the form

$$\dot{\hat{\mathbf{x}}}(t) = \mathbf{A}(t)\hat{\mathbf{x}}(t) + \mathbf{K}(t)\mathbf{z}(t)$$

which is linear and sequential. To make the filter unconditionally unbiased, we must have

$$\mathcal{E}\{\hat{\mathbf{x}}(t)\} = \mathcal{E}\{\mathbf{x}(t)\} = \boldsymbol{\mu}_{\mathbf{x}}(t)$$

We take the unconditioned expectation of the postulated filter algorithm and obtain

$$\mathcal{E}\{\dot{\hat{\mathbf{x}}}(t)\} = \mathbf{A}(t)\mathcal{E}\{\hat{\mathbf{x}}(t)\} + \mathbf{K}(t)\mathcal{E}\{\mathbf{z}(t)\}$$

Since

$$\mathcal{E}\{\mathbf{z}(t)\} = \mathcal{E}\{\mathbf{H}(t)\mathbf{x}(t) + \mathbf{v}(t)\} = \mathbf{H}(t)\boldsymbol{\mu}_{\mathbf{x}}(t)$$

and since we desire

$$\mathcal{E}\{\hat{\mathbf{x}}(t)\} = \boldsymbol{\mu}_{\mathbf{x}}(t)$$

we have for an equivalent expression for the expectation of the filter algorithm

$$\dot{\boldsymbol{\mu}}_{\mathbf{x}}(t) = [\mathbf{A}(t) + \mathbf{K}(t)\mathbf{H}(t)]\boldsymbol{\mu}_{\mathbf{x}}(t)$$

We know that the mean of the message-model state vector $\mathbf{x}(t)$ propagates according to

$$\dot{\boldsymbol{\mu}}_{\mathbf{x}}(t) = \mathbf{F}(t)\boldsymbol{\mu}_{\mathbf{x}}(t) \qquad \boldsymbol{\mu}_{\mathbf{x}}(t_0) = \boldsymbol{\mu}_{\mathbf{x}_0}$$

Thus we see that the requirement that the filter be unbiased leads us to the restrictions

$$\mathbf{A}(t) = \mathbf{F}(t) - \mathbf{K}(t)\mathbf{H}(t)$$
$$\hat{\mathbf{x}}(t_0) = \boldsymbol{\mu}_{\mathbf{x}}(t_0) = \boldsymbol{\mu}_{\mathbf{x}_0}$$

It is also possible to show that these two requirements are necessary but not sufficient for the filter to be conditionally unbiased in that

$$\mathcal{E}\{\mathbf{x}(t) \mid \boldsymbol{Z}(t)\} = \mathcal{E}\{\hat{\mathbf{x}}(t) \mid \boldsymbol{Z}(t)\}$$

where the conditioning variable $\boldsymbol{Z}(t)$ represents all the data $\mathbf{z}(\tau)$ for $t_0 \leq \tau \leq t$. With the unbiased requirements satisfied, the postulated filter is of the form

$$\dot{\hat{\mathbf{x}}}(t) = \mathbf{F}(t)\hat{\mathbf{x}}(t) + \mathbf{K}(t)[\mathbf{z}(t) - \mathbf{H}(t)\hat{\mathbf{x}}(t)]$$
$$\hat{\mathbf{x}}(t_0) = \boldsymbol{\mu}_{\mathbf{x}}(t_0)$$

The filter algorithm is arranged in this form for convenience. When $\mathbf{K}$ is optimally selected, the quantity $\mathbf{z}(t) - \mathbf{H}(t)\hat{\mathbf{x}}(t)$ is a white-noise stochastic process called the "innovation," because it contains all the *new* information in the observation $\mathbf{z}(t)$. This innovation has many important properties, which are discussed in advanced works on estimation theory.

We now seek to complete the derivation by determining $\mathbf{K}(t)$ such that the filter has minimum error variance. We shall find $\mathbf{K}(t)$ such as to minimize

$$J(t) = \text{tr}\{\text{var}\{\tilde{\mathbf{x}}(t)\}\} = \text{tr}\{\mathbf{V}_{\tilde{\mathbf{x}}}(t)\}$$

where tr $\{\cdot\}$ denotes the trace operation and $\tilde{\mathbf{x}}(t)$ is the error in filtering given by

$$\tilde{\mathbf{x}}(t) = \mathbf{x}(t) - \hat{\mathbf{x}}(t)$$

We now desire to obtain an expression for $\mathbf{V}_{\tilde{\mathbf{x}}}(t)$. From the expression for the message model and the unbiased postulated filter, we have

$$\dot{\tilde{\mathbf{x}}}(t) = \dot{\mathbf{x}}(t) - \dot{\hat{\mathbf{x}}}(t) = \mathbf{F}(t)\mathbf{x}(t) + \mathbf{G}(t)\mathbf{w}(t)$$
$$- \mathbf{F}(t)\hat{\mathbf{x}}(t) - \mathbf{K}(t)[\mathbf{z}(t) - \mathbf{H}(t)\hat{\mathbf{x}}(t)]$$

Letting $\mathbf{z}(t) = \mathbf{H}(t)\mathbf{x}(t) + \mathbf{v}(t)$ and using the error definition, we obtain

$$\dot{\tilde{\mathbf{x}}}(t) = [\mathbf{F}(t) - \mathbf{K}(t)\mathbf{H}(t)]\tilde{\mathbf{x}}(t) + \mathbf{u}(t)$$

where

$$\mathbf{u}(t) = \mathbf{G}(t)\mathbf{w}(t) - \mathbf{K}(t)\mathbf{v}(t)$$

is a zero-mean white-noise term with

$$\text{cov}\{\mathbf{u}(t_1), \mathbf{u}(t_2)\} = [\mathbf{G}(t_1)\boldsymbol{\psi}_{\mathbf{w}}(t_1)\mathbf{G}^T(t_1) + \mathbf{K}(t_1)\boldsymbol{\psi}_{\mathbf{v}}(t_1)\mathbf{K}^T(t_1)]\delta_D(t_1 - t_2)$$
$$\text{cov}\{\tilde{\mathbf{x}}(t_1), \mathbf{u}(t_2)\} = \mathbf{0} \qquad t_1 \geq t_0$$

We may write down by inspection the relation for the propagation of the error variance. We have (from Table 6.2-1)

$$\begin{aligned}\dot{\mathbf{V}}_{\tilde{\mathbf{x}}}(t) = {} & [\mathbf{F}(t) - \mathbf{K}(t)\mathbf{H}(t)]\mathbf{V}_{\tilde{\mathbf{x}}}(t) - \mathbf{V}_{\tilde{\mathbf{x}}}(t)[\mathbf{F}(t) - \mathbf{K}(t)\mathbf{H}(t)]^T \\ & + \mathbf{G}(t)\boldsymbol{\psi}_{\mathbf{w}}(t)\mathbf{G}^T(t) + \mathbf{K}(t)\boldsymbol{\psi}_{\mathbf{v}}(t)\mathbf{K}^T(t)\end{aligned}$$

This relation is important in its own right, since it can be used to conduct studies of the effect of suboptimal gains, $\mathbf{K}(t)$, upon the filter error variance. To solve this equation, an initial condition is needed. Since $\hat{\mathbf{x}}(t_0)$ is a deterministic quantity, we have

$$\mathbf{V}_{\tilde{\mathbf{x}}}(t_0) = \text{var}\{\mathbf{x}(t_0) - \hat{\mathbf{x}}(t_0)\} = \text{var}\{\mathbf{x}(t_0)\} = \mathbf{V}_{\mathbf{x}}(t_0)$$

as the needed initial condition.

Our basic objective is to minimize

$$J(t) = \text{tr}[\mathbf{V}_{\tilde{\mathbf{x}}}(t)]$$

at each instant of time by proper choice of $\mathbf{K}(t)$. This will be accomplished if we minimize

$$\begin{aligned}\frac{dJ(t)}{dt} &= \text{tr}[\dot{\mathbf{V}}_{\tilde{\mathbf{x}}}(t)] \\ &= \text{tr}[(\mathbf{F} - \mathbf{K}\mathbf{H})\mathbf{V}_{\tilde{\mathbf{x}}} + \mathbf{V}_{\tilde{\mathbf{x}}}(\mathbf{F} - \mathbf{K}\mathbf{H})^T + \mathbf{G}\boldsymbol{\psi}_{\mathbf{w}}\mathbf{G}^T + \mathbf{K}\boldsymbol{\psi}_{\mathbf{v}}\mathbf{K}^T]\end{aligned}$$

where the (t) arguments have been dropped for simplicity. To minimize this scalar error quantity $dJ(t)/dt$ by choice of the matrix $\mathbf{K}(t)$, it is necessary to use concepts from matrix calculus. The reader can easily visualize the needed steps by analogy to the case of scalar $\mathbf{K}$ and $\mathbf{V}_{\tilde{\mathbf{x}}}$. We have[1]

[1] See Appendix B.

$$\frac{d\,\mathrm{tr}([\mathbf{F}-\mathbf{K}\mathbf{H})\mathbf{V}_{\tilde{\mathbf{x}}}]}{d\mathbf{K}} = -\mathbf{V}_{\tilde{\mathbf{x}}}\mathbf{H}^T$$

$$\frac{d\,\mathrm{tr}[\mathbf{V}_{\tilde{\mathbf{x}}}(\mathbf{F}-\mathbf{K}\mathbf{H})^T]}{d\mathbf{K}} = -\mathbf{V}_{\tilde{\mathbf{x}}}\mathbf{H}^T$$

$$\frac{d\,\mathrm{tr}[\mathbf{K}\boldsymbol{\psi}_{\mathbf{v}}\mathbf{K}^T]}{d\mathbf{K}} = 2\mathbf{K}\boldsymbol{\psi}_{\mathbf{v}}$$

such that the requirement $d\dot{J}(t)/d\mathbf{K} = \mathbf{0}$ leads to

$$-\mathbf{V}_{\tilde{\mathbf{x}}}\mathbf{H}^T + \mathbf{K}\boldsymbol{\psi}_{\mathbf{v}} = \mathbf{0}$$

Thus we have finally

$$\mathbf{K}(t) = \mathbf{V}_{\tilde{\mathbf{x}}}(t)\mathbf{H}^T(t)\boldsymbol{\psi}_{\mathbf{v}}^{-1}(t)$$

and

$$\dot{\mathbf{V}}_{\tilde{\mathbf{x}}} = \mathbf{F}\mathbf{V}_{\tilde{\mathbf{x}}} + \mathbf{V}_{\tilde{\mathbf{x}}}\mathbf{F}^T + \mathbf{G}\boldsymbol{\psi}_{\mathbf{w}}\mathbf{G}^T - \mathbf{K}\boldsymbol{\psi}_{\mathbf{v}}\mathbf{K}^T$$

as the optimum $\mathbf{K}(t)$ and $\mathbf{V}_{\tilde{\mathbf{x}}}$. This completes the derivation of the optimum unbiased minimum error variance linear filter. Figure 6.2-6 illustrates these algorithms in summary form. They are popularly

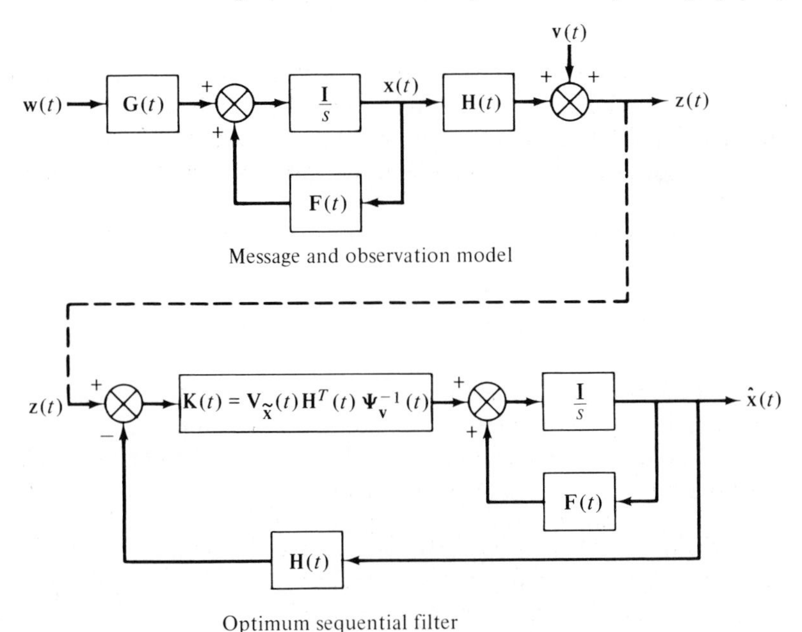

FIG. 6.2-6. Optimum linear sequential filter.

called the Kalman–Bucy, or just Kalman, filter algorithms. These algorithms were used in Chapter 1 to diagram the structure of an optimum estimator for position and velocity. They have received numerous applications to areas in which real-time sequential estimation is a requirement.

A important subclass of the foregoing development is the stationary (at least wide-sense-stationary) problem, in which the means are constant and the variance is only a function of the age variable $\gamma = t_1 - t_2$. The system coefficient matrix $\mathbf{F}(t)$ and distribution matrix $\mathbf{G}(t)$ are constant, so Eq. (6.2-1) becomes

$$\dot{\mathbf{x}}(t) = \mathbf{F}\mathbf{x}(t) + \mathbf{G}\mathbf{u}(t) \tag{6.2-36}$$

where $\mathbf{F}$ and $\mathbf{G}$ are *constant matrices.* Since the means are constant, Eq. (6.2-7) becomes simply

$$\dot{\boldsymbol{\mu}}_x = \mathbf{0} = \mathbf{F}\boldsymbol{\mu}_x + \mathbf{G}\boldsymbol{\mu}_u$$

or

$$\boldsymbol{\mu}_x = -\mathbf{F}^{-1}\mathbf{G}\boldsymbol{\mu}_u \tag{6.2-37}$$

If $\mathbf{F}$ is singular, then, in general, there will be no stationary value for $\boldsymbol{\mu}_x$ since it will be necessary for $\boldsymbol{\mu}_x$ to become unbounded. In simple physical terms, a singular $\mathbf{F}$ implies that the system has integrating capability (one or more poles at the origin), and a constant input will produce an unbounded output.

If the output process is stationary in the variance, setting $\dot{\mathbf{V}}_{\tilde{x}}(t) = \mathbf{0}$ in Eq. (6.2-34) results in

$$\mathbf{0} = \mathbf{F}\mathbf{V}_{\tilde{x}} + \mathbf{V}_{\tilde{x}}\mathbf{F}^T + \mathbf{G}\boldsymbol{\psi}_u\mathbf{G}^T \tag{6.2-38}$$

where of necessity $\mathbf{F}$, $\mathbf{G}$, and $\boldsymbol{\psi}_u$ must not be time functions. Unfortunately, this relation may not be simply solved for $\mathbf{V}_{\tilde{x}}$. Three approaches are possible. The variance equation (6.2-34) may be solved (on a computer) as a differential equation. Equation (6.2-38) may be solved (on a computer) as a linear matrix algebraic equation. Alternatively, a spectral-density approach may be used and advantage taken of the properties of the Laplace and Fourier transform. We shall accomplish this in the next section for single input–single output systems, and shall then briefly discuss the multiple input–output case.

EXERCISES 6.2

6.2-1. Consider the system

$$\dot{x}(t) = atx(t) + (\sin t)\, u(t)$$

where $u(t)$ is a white noise with zero mean and $V_u(t, \tau) = \psi_u \delta_D(t - \tau)$. Set up the equations for the mean $\mu_x(t)$ and the variance $V_x(t)$. Do not solve the equations.

Answers: $\dot{\mu}_x(t) = at\mu_x(t)$, $\dot{V}_x(t) = 2atV_x(t) + \psi_u \sin^2 t$.

6.2-2. Given the system

$$\dot{x}(t) = tx(t)$$

with $\mu_x(0) = 1$ and $V_x(0) = 10$, find (a) $\mu_x(t)$, (b) $V_x(t)$, and (c) $V_x(t, \tau)$ for $t > \tau > 0$.

Answers: (a) $\mu_x(t) = e^{t^2/2}$, (b) $V_x(t) = 10e^{t^2}$, (c) $V_x(t, \tau) = 10e^{(t^2+\tau^2)/2}$.

6.2-3. Find $\mu_x(t)$, $V_{xu}(t)$, $V_x(t)$, and $V_x(t, \tau)$ for t and $\tau > 0$ if

$$\dot{x}(t) = -x(t) + u(t)$$

and $x(0) = 1$. Here $u(t)$ is a white-noise process with $\mu_u(t) = 2$ and $V_u(t, \tau) = e^{-2t}\delta_D(t - \tau)$.

Answers:

$$\mu_x(t) = 2 - e^{-t}$$

$$V_{xu}(t) = \tfrac{1}{2}e^{-2t}$$

$$V_x(t) = te^{-2t}$$

$$V_x(t, \tau) = \begin{cases} e^{-(t-\tau)}e^{-2\tau} & t > \tau \\ e^{-(\tau-t)}te^{-2t} & t < \tau \end{cases}$$

6.2-4. If $x(t)$ satisfies

$$\dot{x}(t) = -2x(t) + u(t)$$

with $x(0) = 0$, where u is a random variable with zero mean and unit variance, find $\mu_x(t)$ and $V_x(t, \tau)$.

Answers: $\mu_x(t) = 0$, $V_x(t, \tau) = \frac{1}{4}(1 - e^{-2t})(1 - e^{-2\tau})$.

6.2-5. The impulse response of a linear stationary system is

$$h(\tau) = \begin{cases} 1 & 0 \leq \tau \leq 1 \\ 0 & \text{otherwise} \end{cases}$$

and the input is

$$u(t) = 1 + n(t)$$

where $n(t)$ is a zero-mean white noise with spectral density of 4. Find the mean and variance of the output $y(t)$ for $t \geq 0$ if $y(0) = 0$.

Answers:

$$\mu_y(t) = \begin{cases} t & 0 \leq t < 1 \\ 1 & t \geq 1 \end{cases}$$

$$V_y(t) = \begin{cases} 4t & 0 \leq t \leq 1 \\ 4 & t \geq 1 \end{cases}$$

6.3. *Frequency-domain approach to stationary continuous systems*

When the input random process to a constant coefficient linear system is stationary, and steady-state behavior of the moments of the output process is desired, there is considerable merit inherent in the frequency-domain approach to the problem. We consider the system shown in Fig. 6.3-1, which illustrates a block diagram of a single input–single output message- and noise-generating process.

We are considering only linear constant coefficient systems and are therefore able to use transform techniques to advantage. The state vector $\mathbf{x}(t)$ and scalar output $z(t)$ evolve from the scalar input $u(t)$ according to the relations

$$\dot{\mathbf{x}}(t) = \mathbf{A}\mathbf{x}(t) + \mathbf{b}u(t) \tag{6.3-1}$$

$$z(t) = \mathbf{c}^T\mathbf{x}(t) + eu(t) \tag{6.3-2}$$

where

$$\mathbf{A} = \begin{bmatrix} a_{11} & a_{12} & \cdots & a_{1n} \\ a_{21} & a_{22} & \cdots & a_{2n} \\ \cdot & \cdot & & \cdot \\ \cdot & \cdot & & \cdot \\ \cdot & \cdot & & \cdot \\ a_{n1} & a_{n2} & \cdots & a_{nn} \end{bmatrix} \qquad \mathbf{b} = \begin{bmatrix} b_1 \\ b_2 \\ \cdot \\ \cdot \\ \cdot \\ b_n \end{bmatrix} \qquad \mathbf{c} = \begin{bmatrix} c_1 \\ c_2 \\ \cdot \\ \cdot \\ \cdot \\ c_n \end{bmatrix}$$

We wish to obtain expressions relating the output $z(t)$ to the input $u(t)$. The frequency-domain approach will be considered first, and then we shall obtain equivalent results using the time-domain approach. The Laplace transform of Eqs. (6.3-1) and (6.3-2) is

$$s\mathbf{X}(s) - \mathbf{x}(t = 0^+) = \mathbf{A}\mathbf{X}(s) + \mathbf{b}U(s) \tag{6.3-3}$$

$$Z(s) = \mathbf{c}^T\mathbf{X}(s) + eU(s) \tag{6.3-4}$$

We are interested only in the steady-state response of the system, so we may assume that the system initial condition $\mathbf{x}(t = 0^+)$ is zero and thus have

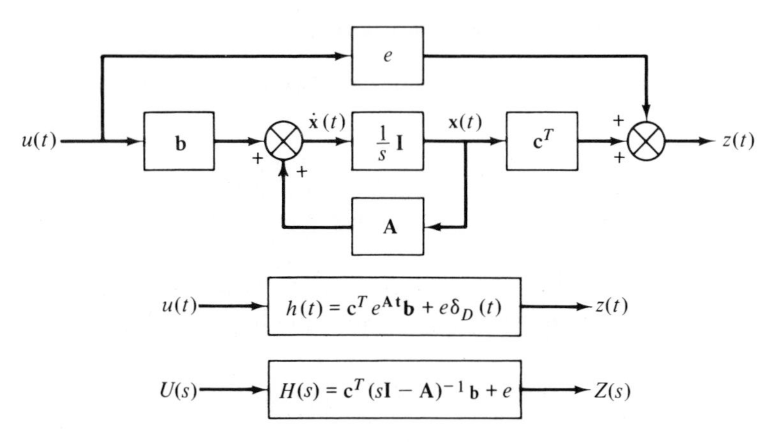

FIG. 6.3-1. Equivalent forms for single input–single output linear systems.

$$\mathbf{X}(s) = (s\mathbf{I} - \mathbf{A})^{-1}\mathbf{b}U(s) \tag{6.3-5}$$

$$Z(s) = [\mathbf{c}^T(s\mathbf{I} - \mathbf{A})^{-1}\mathbf{b} + e]U(s) = H(s)U(s) \tag{6.3-6}$$

The quantity

$$H(s) \triangleq \mathbf{c}^T(s\mathbf{I} - \mathbf{A})^{-1}\mathbf{b} + e \tag{6.3-7}$$

is known as the system transfer function. The inverse Laplace transform of Eq. (6.3-7) yields the impulse response of the time-invariant linear system

$$h(t) = \mathcal{L}^{-1}\{H(s)\} = \frac{1}{2\pi j}\int_{-j\infty}^{j\infty} H(s)e^{-st}\,ds \tag{6.3-8}$$

to a unit impulse occurring at $t = 0$.

As an alternative approach, we may write the time-domain response, for zero initial condition $\mathbf{x}(t = -\infty)$, of the system of Eq. (6.3-1) as

$$\mathbf{x}(t) = \int_{-\infty}^{t} \mathbf{\Phi}(t, \tau)\mathbf{b}u(\tau)\,d\tau \tag{6.3-9}$$

where

$$\frac{\partial \mathbf{\Phi}(t, \tau)}{\partial t} = \mathbf{A\Phi}(t, \tau) \qquad \mathbf{\Phi}(\tau, \tau) = \mathbf{I} \tag{6.3-10}$$

The solution to Eq. (6.3-10) may easily be obtained as

$$\mathbf{\Phi}(t, \tau) = \begin{cases} e^{\mathbf{A}(t-\tau)} & t \geq \tau \\ \mathbf{0} & t < \tau \end{cases} \tag{6.3-11}$$

For $\tau = 0$, it is a simple matter to verify that

$$\mathcal{L}\{\mathbf{\Phi}(t, 0)\} = (s\mathbf{I} - \mathbf{A})^{-1} \tag{6.3-12}$$

The scalar output of the system is

$$z(t) = \mathbf{c}^T \int_{-\infty}^{t} \mathbf{\Phi}(t, \tau)\mathbf{b}u(\tau)\, d\tau + eu(t) \tag{6.3-13}$$

We know that we can write the response of a linear constant coefficient system as a convolution of the system input and impulse response

$$z(t) = \int_{-\infty}^{t} h(t - \tau)u(\tau)\, d\tau \tag{6.3-14}$$

Thus, upon comparing the foregoing two equations, we see that the system impulse response is

$$\begin{aligned} h(t - \tau) &= \mathbf{c}^T\mathbf{\Phi}(t, \tau)\mathbf{b} + e\delta_D(t - \tau) \\ &= \begin{cases} \mathbf{c}^T e^{\mathbf{A}(t-\tau)}\mathbf{b} + e\delta_D(t - \tau) & t \geq \tau \\ 0 & t < \tau \end{cases} \end{aligned}$$

We may find the system transfer function by taking the Laplace transform of $h(t - \tau)$ with $\tau = 0$. This yields Eq. (6.3-7), as expected.

For most of the work which follows, the convolution expression of Eq. (6.3-14) is a bit inconvenient. A more convenient form results if we make the change of variable $\eta = t - \tau$ such that we have

$$z(t) = \int_{0}^{\infty} h(\eta)u(t - \eta)\, d\eta \tag{6.3-15}$$

The work that follows will be based upon the single input–single output block diagram of Fig. 6.3-1 and the relations (6.3-6) and (6.3-15), except for the concluding portion of this section, where we shall give brief mention to multiple input–multiple output linear systems.

We shall now develop relations necessary to obtain the mean value of the output of a linear system with a stationary random input. From Eq. (6.3-15) we have, upon taking the expected value,

$$\mu_z(t) = \int_{0}^{\infty} h(\eta)\mu_u\, d\eta = \mu_u \int_{0}^{\infty} h(\eta)\, d\eta \tag{6.3-16}$$

The cross correlation between input and output is stationary and is defined as

$$C_{zu}(\tau) = \mathcal{E}\{z(t + \tau)u(t)\} \tag{6.3-17}$$

From Eq. (6.3-15) we have

$$C_{zu}(\tau) = \mathcal{E}\left\{\int_0^\infty h(\eta)u(t+\tau-\eta)u(t)\,d\eta\right\} \tag{6.3-18}$$

By interchanging the operations of expectation and integration, we have

$$C_{zu}(\tau) = \int_0^\infty h(\eta)C_u(\tau-\eta)\,d\eta \tag{6.3-19}$$

In a similar way,

$$C_{uz}(\tau) = \mathcal{E}\{u(t+\tau)z(t)\} = \int_0^\infty h(\eta)C_u(\tau+\eta)\,d\tau \tag{6.3-20}$$

The cross-spectral-density expressions are also easily formulated for input–output relations in stationary random processes. The cross spectral density $C_{ZU}(s)$ is obtained directly from Eq. (6.3-19) and the definition of the bilateral Laplace transform[1] as

$$C_{ZU}(s) = \int_{-\infty}^\infty \left[\int_0^\infty h(\eta)C_u(\tau-\eta)\,d\eta\right]e^{-s\tau}\,d\tau$$

By interchanging the order of integration in the foregoing, we have

$$C_{ZU}(s) = \int_0^\infty h(\eta)\left[\int_{-\infty}^\infty C_u(\tau-\eta)e^{-s\tau}\,d\tau\right]d\eta$$

Now we insert an equivalent of unity

$$e^{-s\eta}e^{s\eta} = 1$$

into the integral in such a way that we have

$$C_{ZU}(s) = \int_0^\infty h(\eta)e^{-s\eta}\left[\int_{-\infty}^\infty C_u(\tau-\eta)e^{-s(\tau-\eta)}\,d\tau\right]d\eta$$

We change the variable of integration of the inner integral by letting $\lambda = \tau - \eta$ such that the inner integral becomes

$$\int_{-\infty}^\infty C_u(\tau-\eta)e^{-s(\tau-\eta)}\,d\tau = \int_{-\infty}^\infty C_u(\lambda)e^{-s\lambda}\,d\lambda = C_U(s)$$

which is *not* a function of η. Thus the expression for the cross spectral

[1] For notational convenience, we shall use the bilateral Laplace transform rather than the Fourier transform to express spectral density in this chapter.

density $C_{ZU}(s)$ becomes

$$C_{ZU}(s) = \int_0^\infty h(\eta)e^{-s\eta}C_U(s)\,d\eta$$

Thus we have finally

$$C_{ZU}(s) = H(s)C_U(s) \tag{6.3-21}$$

as the desired relation. In a similar way, we may show that

$$C_{UZ}(s) = H(-s)C_U(s) \tag{6.3-22}$$

This relation also follows, since $C_{UZ}(s) = C_{ZU}(-s)$, and $C_U(s)$ is an even function of s.

The output correlation function is obtained in a manner similar to the input–output correlation function. From Eq. (6.3-15) and the definition of $C_z(\tau)$ we have

$$\begin{aligned} C_z(\tau) &= \mathcal{E}\{z(t+\tau)z(t)\} \\ &= \mathcal{E}\left\{\int_0^\infty \int_0^\infty h(\eta_1)h(\eta_2)u(t+\tau-\eta_1)u(t-\eta_2)\,d\eta_1\,d\eta_2\right\} \end{aligned} \tag{6.3-23}$$

Interchanging the operations of expectation and integration, we have

$$C_z(\tau) = \int_0^\infty \int_0^\infty h(\eta_1)h(\eta_2)C_u(\tau-\eta_1+\eta_2)\,d\eta_1\,d\eta_2 \tag{6.3-24}$$

Analytical evaluation of this double integral expression is most difficult. It is usually far easier to obtain the output spectral density and take the inverse bilateral Laplace (or Fourier) transform. We have from Eq. (6.3-24), upon using the definition of spectral density and interchanging order of integration,

$$\begin{aligned} C_Z(s) &= \int_{-\infty}^\infty C_z(\tau)e^{-s\tau}\,d\tau \\ &= \int_0^\infty h(\eta_1)\,d\eta_1\left\{\int_0^\infty h(\eta_2)\,d\eta_2\left[\int_{-\infty}^\infty C_u(\tau-\eta_1+\eta_2)e^{-s\tau}\,d\tau\right]\right\} \end{aligned}$$

Inserting the relation

$$1 = e^{-s(\eta_1-\eta_2)}e^{s\eta_1}e^{-s\eta_2}$$

in the foregoing we have

$$\begin{aligned} C_Z(s) \\ = \int_0^\infty h(\eta_1)e^{-s\eta_1}\,d\eta_1\left\{\int_0^\infty h(\eta_2)e^{s\eta_2}\,d\eta_2\left[\int_{-\infty}^\infty C_u(\tau-\eta_1+\eta_2)e^{-s(\tau-\eta_1+\eta_2)}\,d\tau\right]\right\} \end{aligned}$$

By changing variables such that $\lambda = \tau - \eta_1 + \eta_2$, we recognize that the innermost integral in the foregoing is precisely $C_U(s)$ and is not a function of η_1 or η_2. Thus we have

$$C_Z(s) = H(s)H(-s)C_U(s) \tag{6.3-25}$$

as the desired relation for the output spectral density.

By taking the inverse transform of Eq. (6.3-25), we obtain the output correlation function

$$C_z(\tau) = \mathfrak{L}^{-1}\{C_Z(s)\} = \frac{1}{2\pi j}\int_{-j\infty}^{j\infty} C_Z(s)e^{s\tau}\, ds \tag{6.3-26}$$

Often we desire to obtain only the mean-square value of $z(t)$. A possible approach is to obtain $C_z(\tau)$ and let $\tau = 0$. Alternatively, we note, from the inverse transform relationship, that

$$P_z = C_z(0) = \frac{1}{2\pi j}\int_{-j\infty}^{j\infty} C_Z(s)\, ds \tag{6.3-27}$$

Thus if $C_Z(s)$ is the ratio of polynomials in s,

$$C_Z(s) = \frac{c(s)\, c(-s)}{d(s)\, d(-s)}$$

we may use the integral table of Appendix A to obtain the desired mean-square value.

We have just derived several most useful expressions for the output-correlation and spectral-density relations for linear constant coefficient systems with stationary random inputs. Table 6.3-1 presents these relations in summary form. We shall now present several examples that illustrate the usefulness of these relations.

Example *6.3-1*. To consider actual evaluation of the results of this section, let us consider a simple *RC* circuit with impulse response and transfer function

$$h(t) = ae^{-at} \qquad t > 0 \qquad H(s) = \frac{a}{s + a}$$

Table 6.3-1 may be used as a reference for all the desired input–output relations.

First we assume that the input noise is white with spectral density and correlation function

$$C_U(s) = \psi_u \qquad C_u(\tau) = \psi_u \delta_D(\tau)$$

TABLE 6.3-1. Input–Output Relations for Linear Constant Coefficient Systems with Stationary Random Inputs

Impulse response relation	$z(t) = \int_0^\infty h(\eta)u(\tau - \eta)\, d\tau$
Transfer function relation	$Z(s) = H(s)U(s)$
Input–output correlation	$C_{zu}(\tau) = \int_0^\infty h(\eta)C_u(\tau - \eta)\, d\eta$ $C_{uz}(\tau) = \int_0^\infty h(\eta)C_u(\tau + \eta)\, d\eta$
Input–output spectral densities	$C_{ZU}(s) = H(s)C_U(s)$ $C_{UZ}(s) = H(-s)C_U(s)$
Output correlation function	$C_z(\tau) = \int_0^\infty \int_0^\infty h(\eta_1)h(\eta_2)C_u(\tau - \eta_1 + \eta_2)\, d\eta_1\, d\eta_2$
Output spectral density	$C_Z(s) = H(s)H(-s)C_U(s)$

The output–input cross correlation is given by

$$C_{zu}(\tau) = \int_0^\infty h(\eta)C_u(\tau - \eta)\, d\eta$$

$$= \int_0^\infty ae^{-a\eta}\psi_u\delta_D(\tau - \eta)\, d\eta = \begin{cases} a\psi_u e^{a\tau} & \tau \geq 0 \\ 0 & \tau < 0 \end{cases}$$

The cross spectral density is given by the bilateral Laplace transform of the foregoing, or

$$C_{ZU}(s) = H(s)C_U(s) = \frac{a}{a + s}\psi_u$$

The output-correlation-function integral is determined as

$$C_z(\tau) = \int_0^\infty \int_0^\infty h(\eta_1)h(\eta_2)C_u(\tau - \eta_1 + \eta_2)\, d\eta_1\, d\eta_2$$

$$= \int_0^\infty \int_0^\infty a^2\psi_u e^{-a\eta_1}e^{-a\eta_2}\delta_D(\tau - \eta_1 + \eta_2)\, d\eta_1\, d\eta_2$$

We integrate first over η_1 and obtain

$$C_z(\tau) = \int_0^\infty a^2\psi_u e^{-a\eta_2}e^{-a(\tau+\eta_2)}\, d\eta_2$$

for $0 \leq \tau + \eta_2 \leq \infty$. If $\tau + \eta_2$ is negative, the integral is zero, since the impulse is not contained within the range of integration. Thus we see that there are two restrictions on η_2 in the foregoing integral. They are

$$\infty \geq \eta_2 \geq 0 \qquad \infty \geq \eta_2 \geq -\tau$$

The strongest inequality depends upon the value of τ. If τ is positive, $\eta_2 \geq 0$ is more restrictive than $\eta_2 \geq -\tau$. However, if τ is negative, $\eta_2 \geq -\tau$ is more restrictive than $\eta_2 \geq 0$. Thus the foregoing integral may be written as two integrals:

$$C_z(\tau) = \int_{-\tau}^{\infty} a^2 \psi_u e^{-a\eta_2} e^{-a(\tau+\eta_2)}\, d\eta_2 = \frac{a\psi_u}{2} e^{+a\tau} \qquad \tau \leq 0$$

$$C_z(\tau) = \int_{0}^{\infty} a^2 \psi_u e^{-a\eta_2} e^{-a(\tau+\eta_2)}\, d\eta_2 = \frac{a\psi_u}{2} e^{-a\tau} \qquad \tau \geq 0$$

the result of which may be written as

$$C_z(\tau) = \frac{a\psi_u}{2} e^{-a|\tau|}$$

This may also be determined more simply from the relation

$$C_Z(s) = H(s)H(-s)C_u(s) = \frac{a^2}{(a+s)(a-s)}\psi_u = \frac{a\psi_u/2}{a+s} + \frac{a\psi_u/2}{a-s}$$

such that inverse bilateral Laplace or Fourier transformation gives the autocorrelation function. The transfer-function approach is always recommended in determining input–output relations for constant coefficient linear systems as opposed to using the integrals of the correlation-function approach. To sense the enormous integration effort that may be involved, the reader is urged to solve the second part of this example using the correlation-function approach.

Now we consider that the input correlation and spectral density are

$$C_u(\tau) = V_u e^{-b|\tau|} \qquad C_U(s) = \frac{2bV_u}{b^2 - s^2}$$

The input–output cross spectral density is

$$C_{UZ}(s) = H(-s)C_U(s) = \frac{a}{a-s}\frac{2bV_u}{b^2 - s^2}$$

The input–output cross-correlation function is most easily obtained by taking the inverse transform of the cross spectral density to obtain

$$C_{UZ}(s) = \frac{2abV_u/(b^2 - a^2)}{a - s} + \frac{aV_u/(a + b)}{b + s} + \frac{aV_u/(a - b)}{b - s}$$

$$C_{uz}(\tau) = \begin{cases} \dfrac{aV_u}{a + b}e^{-b\tau} & \tau \geq 0 \\ \dfrac{2abV_u}{b^2 - a^2}e^{a\tau} + \dfrac{aV_u}{a - b}e^{b\tau} & \tau \leq 0 \end{cases}$$

The output spectral density is

$$C_Z(s) = H(s)H(-s)C_U(s) = \frac{a^2}{a^2 - s^2}\frac{2\psi_u b}{b^2 - s^2}$$

The output correlation function is obtained by taking the inverse transform of the foregoing expression. We have by partial fraction expansion

$$C_Z(s) = \frac{a\psi_u b/(b^2 - a^2)}{a + s} + \frac{a\psi_u b/(b^2 - a^2)}{a - s} + \frac{a^2\psi_u/(a^2 - b^2)}{b + s} + \frac{a^2\psi_u/(a^2 - b^2)}{b - s}$$

The output correlation function is therefore

$$C_z(\tau) = \frac{a\psi_u}{b^2 - a^2}[be^{-a|\tau|} - ae^{-b|\tau|}]$$

The mean-square value of the output is

$$P_z = C_z(0) = \frac{a\psi_u}{b + a}$$

If only this mean-square value is desired, it may be obtained more easily from the integral table of Appendix A than from the method used here, which involved partial fraction expansion and inverse transformation.

Example *6.3-2.* The techniques developed in this section may be used for fixed configuration optimization. Let us suppose that the input to a system is composed of a scalar signal $y(t)$ and noise $w(t)$ as shown in Fig. 6.3-2. The signal and noise are uncorrelated, and it is desired to

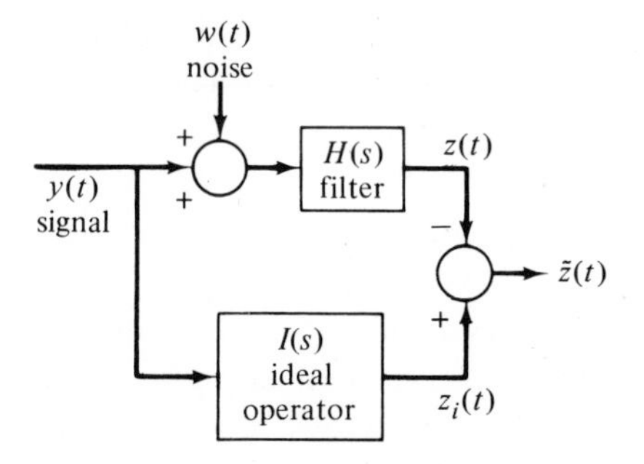

FIG. 6.3-2. Model for optimal filter.

obtain the "best" filter, $H(s)$, such as to minimize the mean-square error in estimating some linear operation on the input signal.

The ideal output from the filter should be

$$z_i(t) = \int_0^\infty i(\eta)y(t-\eta)\,d\eta$$
$$Z_i(s) = I(s)Y(s)$$

where $i(\eta)$ is the impulse response corresponding to the ideal operation on the input signal. The actual output is given by

$$z(t) = \int_0^\infty h(\eta)[y(t-\eta) + w(t-\eta)]\,d\eta$$
$$Z(s) = H(s)[Y(s) + W(s)]$$

such that the error in estimation is

$$\begin{aligned}\tilde{z}(t) &= z_i(t) - z(t)\\ &= \int_0^\infty \{[i(\eta) - h(\eta)]y(t-\eta) - h(\eta)w(t-\eta)\}\,d\eta\\ \tilde{Z}(s) &= Z_i(s) - Z(s)\\ &= [I(s) - H(s)]Y(s) - H(s)W(s)\end{aligned}$$

From Table 6.3-1 and the fact that $y(t)$ and $w(t)$ are uncorrelated, the correlation function and spectral density of the estimation error are given by

$$C_z(\tau) = \int_0^\infty \int_0^\infty \{[i(\eta_1) - h(\eta_1)][i(\eta_2) - h(\eta_2)]C_y(\tau - \eta_1 + \eta_2) + h(\eta_1)h(\eta_2)C_w(\tau - \eta_1 + \eta_2)\}\, d\eta_1\, d\eta_2$$

so that

$$C_Z(s) = [I(s) - H(s)][I(-s) - H(-s)]C_Y(s) + H(s)H(-s)C_W(s)$$

The relative simplicity of the frequency-domain expression is apparent. We are often interested in adjusting $h(t)$ or $H(s)$ such as to minimize the mean-square error. This is obtained from the frequency domain error spectral density as

$$P_z = C_z(\tau = 0) = \frac{1}{2\pi j}\int_{-j\infty}^{j\infty} C_Z(s)\, ds$$

If the system transfer function, $H(s)$, is specified except for a finite number of adjustable parameters, these parameters may be adjusted such as to minimize the mean-square estimation error.

For a specific example, let us suppose that the signal spectral density and autocorrelation function are

$$C_Y(s) = \frac{2aV_s}{a^2 - s^2} \qquad C_y(\tau) = V_y e^{-a|\tau|}$$

and that the noise is white with correlation function and spectral density

$$C_W(s) = \psi_w \qquad C_w(\tau) = \psi_w \delta_D(\tau)$$

We shall assume initially that the system transfer function is of the form

$$H(s) = \frac{b_0}{b_1 + s}$$

and that the ideal operation $I(s) = 1$ such that the desired output is the signal $s(t)$. The error spectral density is easily determined as

$$\begin{aligned} C_Z(s) &= [I(s) - H(s)][I(-s) - H(-s)]C_Y(s) + H(s)H(-s)C_W(s) \\ &= \left[1 - \frac{b_0}{b_1 + s}\right]\left[1 - \frac{b_0}{b_1 - s}\right]\frac{2aV_s}{a^2 - s^2} + \frac{b_0}{b_1 + s}\frac{b_0}{b_1 - s}\psi_w \\ &= \frac{(b_1 - b_0 + s)\sqrt{2aV_s}}{ab_1 + (a + b_1)s + s^2}\frac{(b_1 - b_0 - s)\sqrt{2aV_s}}{ab_1 - (a + b_1)s + s^2} \\ &\quad + \frac{b_0\sqrt{\psi_w}}{b_1 + s}\frac{b_0\sqrt{\psi_w}}{b_1 - s} \end{aligned}$$

We may evaluate the two integrals for determination of the mean-square value directly from Appendix A. The first integral is of the second order in s, and we have

$$\begin{aligned} c_0 &= (b_1 - b_0)\sqrt{2aV_s} \\ c_1 &= \sqrt{2aV_s} \\ d_0 &= ab_1 \\ d_1 &= a + b_1 \\ d_2 &= 1 \end{aligned}$$

Therefore

$$P_{zy} = \frac{c_0^2 d_2 + c_1^2 d_0}{2d_0 d_1 d_2} = \frac{2aV_s(b_1^2 - 2b_0 b_1 + b_0^2 + ab_1)}{2ab_1(a + b_1)}$$

The second integral is of the first order, and we have

$$P_{zw} = \frac{c_0^2}{2d_0 d_1} = \frac{b_0^2 \psi_w}{2b_1}$$

The total mean-square error is the sum of these two terms, or

$$P_z = \frac{2aV_y(b_1^2 - 2b_0 b_1 + b_0^2 + ab_1) + b_0^2 a\psi_w(a + b_1)}{2ab_1(a + b_1)}$$

We adjust b_0 and b_1 such as to minimize P_z. Thus we set $\partial P_z/\partial b_0 = \partial P_z/\partial b_1 = 0$ and solve the resulting expressions to obtain, after a modest amount of algebraic manipulation,

$$\hat{b}_1 = \left[\frac{2aV_y + a^2\psi_w}{\psi_w}\right]^{1/2}$$

$$\hat{b}_0 = \frac{2aV_s}{\psi_w[\hat{b}_1 + a]}$$

For specified values of V_y, ψ_w, and a, we may determine the specific values of $\hat{b}_0$ and $\hat{b}_1$ to use in the optimum filter and evaluate the resulting mean-square error.

It would be desirable to relax the restriction that $H(s)$ be known in form and find the optimum $H(s)$ such as to minimize the mean-square error. This leads us to the subject of the Wiener filter formulation in the frequency domain. Unfortunately, this topic is beyond the scope of the present work; the interested reader is referred to texts listed at the end of this chapter for material on this subject.

In concluding this section, we shall briefly mention the use of spectral-density concepts for multiple input–output constant coefficient linear systems with stationary random inputs. Two approaches are possible and we shall discuss both of them. The obvious generalization of Eqs. (6.3-1), (6.3-3), and (6.3-5) to the vector case is

$$\dot{\mathbf{x}}(t) = \mathbf{A}(t)\mathbf{x}(t) + \mathbf{B}(t)\mathbf{u}(t) \tag{6.3-28}$$

$$s\mathbf{X}(s) - \mathbf{x}(t = 0^+) = \mathbf{A}\mathbf{X}(s) + \mathbf{B}\mathbf{U}(s) \tag{6.3-29}$$

$$\mathbf{X}(s) = \mathbf{H}(s)\mathbf{U}(s) = (s\mathbf{I} - \mathbf{A})^{-1}\mathbf{B}\mathbf{U}(s) \tag{6.3-30}$$

Equation (6.3-9) becomes

$$\mathbf{x}(t) = \int_{-\infty}^{t} \mathbf{\Phi}(t, \tau)\mathbf{B}\mathbf{u}(\tau)\, d\tau \tag{6.3-31}$$

$$= \int_{0}^{\infty} \mathbf{H}(\eta)\mathbf{u}(t - \eta)\, d\eta \tag{6.3-32}$$

where

$$\mathbf{H}(\eta) = \mathbf{\Phi}(t, t - \eta)\mathbf{B} \tag{6.3-33}$$

and $\mathbf{H}(s)$ in Eq. (6.3-30) is the bilateral Laplace transform of $\mathbf{H}(\eta)$. The direct extension of the results of Table 6.3-1 to the vector case may now be made, and this leads immediately to the desired integral expressions for the correlation functions and transform multiplication results for the spectral densities.

For the stationary case, the expression for the input–output cross correlation, Eq. (6.2-25), becomes

$$\mathbf{C}_{\mathbf{xu}}(\gamma) \triangleq \mathbf{P}_{\mathbf{xu}}(t + \gamma, t) = \mathbf{\Phi}(t + \gamma, t_0)\mathbf{C}_{\mathbf{xy}}(t_0 - t) + \int_{t_0}^{t+\gamma} \mathbf{\Phi}(t + \gamma, \tau)\mathbf{B}\mathbf{C}_{\mathbf{u}}(\tau - t)\, d\tau$$

Since the dynamic system is stationary, the transition matrix becomes a function of the difference of its time arguments; that is,

$$\mathbf{\Phi}(t_1, t_2) = \mathbf{\Omega}(t_1 - t_2) \tag{6.3-34}$$

so that $\mathbf{C}_{\mathbf{xu}}(\gamma)$ is given by

$$\mathbf{C}_{\mathbf{xu}}(\gamma) = \mathbf{\Omega}(t + \gamma - t_0)\mathbf{C}_{\mathbf{xu}}(t_0 - t) + \int_{t_0}^{t+\gamma} \mathbf{\Omega}(t + \gamma - \tau)\mathbf{B}\mathbf{C}_{\mathbf{u}}(\tau - t)\, d\tau \tag{6.3-35}$$

To ensure that the correct stationary solution is obtained, it is necessary to set the initial condition $\mathbf{C}_{\mathbf{xu}}(t_0 - t)$ correctly. This is usually accomplished by letting $t_0 = -\infty$ and setting $\mathbf{C}_{\mathbf{xu}}(-\infty) = \mathbf{0}$ so that Eq. (6.3-35) becomes simply

$$\mathbf{C}_{\mathbf{xu}}(\gamma) = \int_{-\infty}^{t+\gamma} \mathbf{\Omega}(t + \gamma - \tau)\mathbf{B}\mathbf{C}_{\mathbf{u}}(\tau - t)\, d\tau$$

Although it would appear that this result is in fact a function of t, implying that the solution is not stationary, this conclusion is not correct. The explicit appearance of t can be removed by making the substitution $\lambda = \tau - t$ to obtain

$$\mathbf{C}_{\mathbf{xu}}(\gamma) = \int_{-\infty}^{\gamma} \mathbf{\Omega}(\gamma - \lambda)\mathbf{B}\mathbf{C}_{\mathbf{u}}(\lambda)\, d\lambda \tag{6.3-36}$$

Owing to the causal assumption about our systems, it is easy for us to show that $\mathbf{\Omega}(\gamma - \lambda)$ must be zero if $\gamma < \lambda$. Hence there is no change in the result if the upper limit of integration is increased from γ to ∞. Equation (6.3-36) may be written in an alternative form by letting $\xi = \gamma - \lambda$ such that we have

$$\mathbf{C}_{\mathbf{xu}}(\gamma) = \int_{-\infty}^{\infty} \mathbf{\Omega}(\xi)\mathbf{B}\mathbf{C}_{\mathbf{u}}(\gamma + \xi)\, d\xi \tag{6.3-37}$$

Because of the similarity of the foregoing to the standard convolution integral for deterministic systems, it is reasonable to consider the bilateral Laplace transform of Eq. (6.3-37), which will give the expression for the cross spectral density $\mathbf{C}_{\mathbf{XU}}(s)$ as

$$\mathbf{C}_{\mathbf{XU}}(s) = \mathbf{\Omega}(s)\mathbf{B}\mathbf{C}_{\mathbf{U}}(s) \tag{6.3-38}$$

Clearly, the cross spectral density $\mathbf{C}_{\mathbf{UX}}(s)$ is simply

$$\mathbf{C}_{\mathbf{UX}}(s) = \mathbf{C}_{\mathbf{XU}}^T(-s) = \mathbf{C}_{\mathbf{U}}(s)\mathbf{B}^T\mathbf{\Omega}^T(-s) \tag{6.3-39}$$

The reader will remember from basic state-variable theory[1] that the resolvent or transition matrix $\mathbf{\Omega}(s)$ is given by

$$\mathbf{\Omega}(s) = (s\mathbf{I} - \mathbf{A})^{-1}$$

Equations (6.3-38) and (6.3-39) are very easy to use, since they involve only algebraic manipulations and are obvious extensions of the scalar result.

[1] See Appendix B.

The output variance expression, Eq. (6.2-27), takes the following form in the stationary case:

$$\mathbf{C_x}(\gamma) \triangleq \mathbf{P_x}(t+\gamma, t) = \mathbf{\Omega}(t+\gamma-t_0)\mathbf{P_x}(t_0)\mathbf{\Omega}^T(t-t_0)$$
$$+ \int_{t_0}^{t+\gamma} d\tau_1 \int_{t_0}^{t} d\tau_2 \mathbf{\Omega}(t+\gamma-\tau_1)\mathbf{B}\mathbf{C_u}(\tau_1-\tau_2)\mathbf{B}^T\mathbf{\Omega}^T(t-\tau_2)$$

If we again let $t_0 = -\infty$ and $\mathbf{P_x}(t_0) = \mathbf{0}$, we have

$$\mathbf{C_x}(\gamma) = \int_{-\infty}^{t+\gamma} d\tau_1 \int_{-\infty}^{t} d\tau_2 \mathbf{\Omega}(t+\gamma-\tau_1)\mathbf{B}\mathbf{C_u}(\tau_1-\tau_2)\mathbf{B}^T\mathbf{\Omega}^T(t-\tau_2) \tag{6.3-40}$$

Now let us make the substitutions $\lambda_1 = t+\gamma-\tau_1$ and $\lambda_2 = t-\tau_2$. Equation (6.3-40) becomes

$$\mathbf{C_x}(\gamma) = \int_0^\infty d\lambda_1 \int_0^\infty d\lambda_2 \mathbf{\Omega}(\lambda_1)\mathbf{B}\mathbf{C_u}(\lambda_2+\gamma-\lambda_1)\mathbf{B}^T\mathbf{\Omega}^T(\lambda_2)$$

or

$$\mathbf{C_x}(\gamma) = \int_0^\infty \left[\int_0^\infty \mathbf{\Omega}(\lambda_1)\mathbf{B}\mathbf{C_u}(\lambda_2+\gamma-\lambda_1)\, d\lambda_1\right]\mathbf{B}^T\mathbf{\Omega}^T(\lambda_2)\, d\lambda_2$$

Here the lower limits on both integrals may be set to $-\infty$ without changing the results, since $\mathbf{\Omega}(t) = \mathbf{0}$ for $t < 0$. If this change is made, $\mathbf{C_x}(\gamma)$ becomes

$$\mathbf{C_x}(\gamma) = \int_{-\infty}^\infty \int_{-\infty}^\infty \mathbf{\Omega}(\lambda_1)\mathbf{B}\mathbf{C_u}(\lambda_2+\gamma-\lambda_1)\mathbf{B}^T\mathbf{\Omega}^T(\lambda_2)\, d\lambda_1\, d\lambda_2 \tag{6.3-41}$$

The expression for the spectral density $\mathbf{C_X}(s)$ can be obtained by taking the Fourier transform or Eq. (6.3-41) and following a procedure similar to that used to derive Eq. (6.3-25). The result is given by

$$\mathbf{C_X}(s) = \mathbf{\Omega}(s)\mathbf{B}\mathbf{C_U}(s)\mathbf{B}^T\mathbf{\Omega}^T(-s) \tag{6.3-42}$$

The spectral density relations of Eqs. (6.3-38) and (6.3-42), as well as those for the scalar case, can also be derived by a simple procedure. Let us think of $\mathbf{u}(t)$ and $\mathbf{x}(t)$ as deterministic stationary signals with the usual bilateral Laplace or Fourier transforms $\mathbf{U}(s)$ and $\mathbf{X}(s)$.[1] The spectral densities

[1] To be more rigorous, we may consider the Fourier transform of a single record and, if necessary, use the limiting arguments of Section 5.4 as applied to the truncated signals.

$\mathbf{C_{XU}}(s)$ and $\mathbf{C_X}(s)$ may then be written as

$$\mathbf{C_{XU}}(s) \triangleq \mathbf{X}(s)\mathbf{U}^T(-s) \tag{6.3-43}$$

and

$$\mathbf{C_X}(s) \triangleq \mathbf{X}(s)\mathbf{X}^T(-s) \tag{6.3-44}$$

The expression for $\mathbf{X}(s)$ may be obtained by taking the Laplace transform of Eq. (6.3-28) as

$$s\mathbf{X}(s) = \mathbf{AX}(s) + \mathbf{BU}(s)$$

or

$$\mathbf{X}(s) = (s\mathbf{I} - \mathbf{A})^{-1}\mathbf{BU}(s) = \mathbf{\Omega}(s)\mathbf{BU}(s)$$

Therefore $\mathbf{C_{XU}}(s)$ as given by Eq. (6.2-43) becomes

$$\mathbf{C_{XU}}(s) = \mathbf{\Omega}(s)\mathbf{BU}(s)\mathbf{U}^T(-s) = \mathbf{\Omega}(s)\mathbf{BC_U}(s) = \mathbf{C}^T_{\mathbf{UX}}(-s)$$

and

$$\begin{aligned}\mathbf{C_X}(s) &= \mathbf{\Omega}(s)\mathbf{BU}(s)[\mathbf{\Omega}(-s)\mathbf{BU}(-s)]^T = \mathbf{\Omega}(s)\mathbf{BU}(s)\mathbf{U}^T(-s)\mathbf{B}^T\mathbf{\Omega}^T(-s)\\ &= \mathbf{\Omega}(s)\mathbf{BC_U}(s)\mathbf{B}^T\mathbf{\Omega}^T(-s)\end{aligned}$$

We see that the previous rigorous results have been reproduced with relative ease. Although the above procedure is formal and one should hence be careful in its application to nonstandard situations, the procedure is nonetheless a useful one. The formal procedure may be used to develop results for complex, but otherwise standard, input–output relations and may also be used to develop plausible solutions to nonstandard problems.

Example *6.3-3.* Let us use spectral-density input–output relations, Eqs. (6.3-39) and (6.3-42), to find the spectral densities $C_{YU}(s)$ and $C_Y(s)$ for the simple stationary single input–single output system given by

$$\begin{aligned}\dot{\mathbf{x}}(t) &= \mathbf{Ax}(t) + \mathbf{b}u(t)\\ y(t) &= \mathbf{c}^T\mathbf{x}(t)\end{aligned}$$

We shall assume that the spectral density $C_U(s)$ of $u(t)$ is known.

Equations (6.3-39) and (6.3-42) may be used directly to obtain the spectral densities $\mathbf{C}_{\mathbf{X}U}(s)$ and $\mathbf{C_X}(s)$ as

$$\mathbf{C}_{\mathbf{X}U}(s) = \mathbf{\Omega}(s)\mathbf{b}C_U(s)$$

and

$$\mathbf{C_X}(s) = \mathbf{\Omega}(s)\mathbf{b}C_U(s)\mathbf{b}^T\mathbf{\Omega}(-s)$$

where $\mathbf{\Omega}(s) = (s\mathbf{I} - \mathbf{A})^{-1}$. The desired spectral density $C_{YU}(s)$ can be obtained from $\mathbf{C_{X}}_U(s)$ by the use of the formal procedure discussed before. However, we recognize $\mathbf{c}^T\mathbf{\Omega}(s)\mathbf{b}$ as the transfer function relating y and u, or

$$\frac{Y(s)}{U(s)} = H(s) = \mathbf{c}^T\mathbf{\Omega}(s)\mathbf{b}$$

Therefore, $C_{YU}(s)$ becomes simply

$$C_{YU}(s) = H(s)C_U(s)$$

In a completely similar fashion, it is easy to show that $C_Y(s)$ is given by

$$C_Y(s) = H(s)H(-s)C_U(s)$$

This completes our study of input–output relations for constant coefficient linear systems with stationary random inputs. Although the results are not as general as those in the preceding section, there are numerous applicational areas in which they are useful.

EXERCISES 6.3

6.3-1. For the system

$$\dot{\mathbf{x}}(t) = \begin{bmatrix} 0 & 1 \\ -2 & -2 \end{bmatrix}\mathbf{x}(t) + \begin{bmatrix} 0 \\ 1 \end{bmatrix}u(t)$$

$$z(t) = x_1(t)$$

find the spectral densities $\mathbf{C_{X}}_U(s)$, $\mathbf{C_X}(s)$, $C_{ZU}(s)$, and $C_Z(s)$ if

$$C_U(s) = \frac{1}{1 - s^2}$$

Answers:

$$\mathbf{C_{X}}_U(s) = \frac{1}{(s^2 + 2s + 2)(1 - s^2)}\begin{bmatrix} 1 \\ s \end{bmatrix}$$

$$\mathbf{C_X}(s) = \frac{1}{(s^2 + 2s + 2)(s^2 - 2s + 2)(1 - s^2)}\begin{bmatrix} 1 & -s \\ s & -s \end{bmatrix}$$

6.3-2. Find the transfer function of a linear stationary system such that if the spectral density of the input is $C_U(s) = 100/(1 - s^2)$, then the output–input cross spectral density is $C_{XU}(s) = 10/[(1 - s^2)(2 + s)]$.

Answer: $H(s) = 0.1/(s + 2)$.

6.3-3. Given the system

$$\dot{x}_1(t) = -2x_1(t) + x_2(t)$$
$$\dot{x}_2(t) = -x_2(t) + u(t)$$

with $C_U(s) = 10$, find $\mathbf{C}_{\mathbf{X}U}(s)$ and $\mathbf{C}_{\mathbf{X}}(s)$.

Answer:

$$\mathbf{C}_{\mathbf{X}U}(s) = \begin{bmatrix} \dfrac{10}{(s+2)(s+1)} \\ \dfrac{10}{s+1} \end{bmatrix}$$

6.4. *Discrete systems*

In many studies involving random processes it is necessary or at least highly desirable to obtain algorithms that may be conveniently processed on a digital computer. It is, of course, possible to discretize the continuous results of the previous two sections. Alternatively, we may find a discrete model for the random processes and system and find relations for the propagation of the mean, variance, and covariance function of the discrete random processes of interest. It is reasonable that there would be only slight differences in the results obtained by the two procedures. After presenting a discussion of the response of linear discrete systems to random inputs, we shall present an example that illustrates some of the differences which exist for these two approaches.

The treatment of discrete time systems in this section will parallel very closely the development for continuous systems presented in Section 6.2. For this reason, most of the proofs will be shortened considerably.

The transition mechanism of interest for the study of discrete systems will be described in state-variable form by the vector difference equation

$$\mathbf{x}(\overline{k+1}T) = \mathbf{\Phi}(\overline{k+1}T, kT)\mathbf{x}(kT) + \mathbf{\Gamma}(kT)\mathbf{u}(kT) \tag{6.4-1}$$

where T is the sampling period. Except where there is risk of confusion or where the sampling interval is of importance, as for example in arguments where the sampling interval is varied, we shall omit the explicit inclusion of T and write for Eq. (6.4-1)

$$\mathbf{x}(k+1) = \mathbf{\Phi}(k+1, k)\mathbf{x}(k) + \mathbf{\Gamma}(k)\mathbf{u}(k) \tag{6.4-2}$$

The mean and variance of $\mathbf{u}(k)$ will initially be assumed given by

$$\begin{aligned} \boldsymbol{\mu}_{\mathbf{u}}(k) &\triangleq \mathcal{E}\{\mathbf{u}(k)\} \\ \mathbf{V}_{\mathbf{u}}(k, j) &\triangleq \operatorname{cov}\{\mathbf{u}(k), \mathbf{u}(j)\} \end{aligned} \tag{6.4-3}$$

Also needed are the initial mean and variance of the state vector

$$\begin{aligned} \boldsymbol{\mu}_{\mathbf{x}}(0) &= \mathcal{E}\{\mathbf{x}(0)\} = \boldsymbol{\mu}_{\mathbf{x}_0} \\ \mathbf{V}_{\mathbf{x}}(0) &= \operatorname{var}\{\mathbf{x}(0)\} = \mathbf{V}_{\mathbf{x}_0} \end{aligned} \tag{6.4-4}$$

We wish to derive both summation and difference equations for the mean and variance of $\mathbf{x}(k)$ and the covariance of $\mathbf{x}(k)$ and $\mathbf{u}(j)$. It is easy to verify (see Appendix B) that the solution to Eq. (6.4-2) is given by

$$\mathbf{x}(j) = \mathbf{\Phi}(i, 0)\mathbf{x}(0) + \sum_{i=0}^{j-1} \mathbf{\Phi}(j, i+1)\mathbf{\Gamma}(i)\mathbf{u}(i) \tag{6.4-5}$$

where we have assumed for simplicity that the time interval of interest starts at stage 0.

If we take the expected value of both sides of this equation, we obtain

$$\boldsymbol{\mu}_{\mathbf{x}}(j) = \mathbf{\Phi}(j, 0)\boldsymbol{\mu}_{\mathbf{x}}(0) + \sum_{i=0}^{j-1} \mathbf{\Phi}(j, i+1)\mathbf{\Gamma}(i)\boldsymbol{\mu}_{\mathbf{u}}(i) \tag{6.4-6}$$

The difference equation expression for $\boldsymbol{\mu}_{\mathbf{x}}$ can be obtained most easily by directly taking the expectation of both sides of Eq. (6.4-2) to obtain

$$\boldsymbol{\mu}_{\mathbf{x}}(j+1) = \mathbf{\Phi}(j+1, j)\boldsymbol{\mu}_{\mathbf{x}}(j) + \mathbf{\Gamma}(j)\boldsymbol{\mu}_{\mathbf{u}}(j) \tag{6.4-7}$$

Alternatively, we may easily convert Eq. (6.4-5) into a difference equation.

The development of summation equations for $\mathbf{V}_{\mathbf{xu}}(j, k)$ and $\mathbf{V}_{\mathbf{x}}(j, k)$ follows directly from the definitions and the use of Eq. (6.4-5). The resulting expressions are easily obtained as

$$\mathbf{V}_{\mathbf{xu}}(j, k) = \mathbf{\Phi}(j, 0)\mathbf{V}_{\mathbf{xu}}(0, k) + \sum_{i=0}^{j-1} \mathbf{\Phi}(j, i+1)\mathbf{\Gamma}(i)\mathbf{V}_{\mathbf{u}}(i, k) \tag{6.4-8}$$

and

$$\begin{aligned} \mathbf{V}_{\mathbf{x}}(j, k) &= \mathbf{\Phi}(j, 0)\mathbf{V}_{\mathbf{x}}(0)\mathbf{\Phi}^T(k, 0) \\ &\quad + \sum_{i=0}^{j-1} \sum_{p=0}^{k-1} \mathbf{\Phi}(j, i+1)\mathbf{\Gamma}(i)\mathbf{V}_{\mathbf{u}}(i, p)\mathbf{\Gamma}^T(p)\mathbf{\Phi}^T(k, p+1) \end{aligned} \tag{6.4-9}$$

where $\mathbf{V}_{\mathbf{x}}(0) \triangleq \mathbf{V}_{\mathbf{x}}(0, 0) \triangleq \operatorname{cov}\{\mathbf{x}(0), \mathbf{x}(0)\}$.

If the stochastic process $\mathbf{u}(k)$ has uncorrelated sample values or is "white," so that

$$\mathbf{V}_{\mathbf{u}}(j, k) = \mathbf{V}_{\mathbf{u}}(j)\delta_K(j-k) \tag{6.4-10}$$

then the expression for $\mathbf{V}_{\mathbf{xu}}(j, k)$ becomes

$$\mathbf{V}_{\mathbf{xu}}(j, k) = \mathbf{\Phi}(j, 0)\mathbf{V}_{\mathbf{xu}}(0, k) + \sum_{i=0}^{j-1} \mathbf{\Phi}(j, i+1)\mathbf{\Gamma}(i)\mathbf{V}_{\mathbf{u}}(i)\delta_K(i-k)$$

This expression may be simplified by carrying out the summation such that we obtain

$$\mathbf{V}_{\mathbf{xu}}(j, k) = \mathbf{\Phi}(j, 0)\mathbf{V}_{\mathbf{xu}}(0, k) + \begin{cases} \mathbf{0} & \text{if } k > j - 1 \\ \mathbf{\Phi}(j, k+1)\mathbf{\Gamma}(k)\mathbf{V}_{\mathbf{u}}(k) & \text{if } k \leq j - 1 \end{cases} \tag{6.4-11}$$

For the case when $\mathbf{u}(k)$ is white, the expression for $\mathbf{V}_{\mathbf{x}}(j, k)$ in Eq. (6.4-9) becomes

$$\mathbf{V}_{\mathbf{x}}(j, k) = \mathbf{\Phi}(j, 0)\mathbf{V}_{\mathbf{x}}(0)\mathbf{\Phi}^T(k, 0) + \sum_{i=0}^{j-1}\sum_{p=0}^{k-1} \mathbf{\Phi}(j, k+1)\mathbf{\Gamma}(i)\mathbf{V}_{\mathbf{u}}(i)\delta_K(i-p)\mathbf{\Gamma}^T(p)\mathbf{\Phi}^T(k, p+1)$$

If the summation is first carried out over the larger of the two ranges, that is, the max $\{j, k\}$, we are assured that the Kronecker delta will occur during the first summation. The expression for $\mathbf{V}_{\mathbf{x}}(j, k)$ then becomes

$$\mathbf{V}_{\mathbf{x}}(j, k) = \mathbf{\Phi}(j, 0)\mathbf{V}_{\mathbf{x}}(0)\mathbf{\Phi}^T(k, 0) + \sum_{i=0}^{\min\{j-1, k-1\}} \mathbf{\Phi}(j, i+1)\mathbf{\Gamma}(i)\mathbf{V}_{\mathbf{u}}(i)\mathbf{\Gamma}^T(i)\mathbf{\Phi}^T(k, i+1) \tag{6.4-12}$$

To determine a difference equation for $\mathbf{V}_{\mathbf{x}}(j) \triangleq \mathbf{V}_{\mathbf{u}}(j, j)$, we use the definition of $\mathbf{V}_{\mathbf{x}}(j)$ and Eq. (6.4-2) to obtain

$$\begin{aligned}\mathbf{V}_{\mathbf{x}}(j+1) &= \text{cov}\{\mathbf{x}(j+1), \mathbf{x}(j+1)\} \\ &= \text{cov}\{[\mathbf{\Phi}(j+1, j), \mathbf{x}(j) + \mathbf{\Gamma}(j)\mathbf{u}(j)], [\mathbf{\Phi}(j+1, j)\mathbf{x}(j) + \mathbf{\Gamma}(j)\mathbf{u}(j)\}\end{aligned}$$

which becomes, after some algebraic manipulations,

$$\begin{aligned}\mathbf{V}_{\mathbf{x}}(j+1) = \mathbf{\Phi}(j+1, j)&\mathbf{V}_{\mathbf{x}}(j)\mathbf{\Phi}^T(j+1, j) \\ &+ \mathbf{\Gamma}(j)\text{cov}\{\mathbf{u}(j), \mathbf{x}(j)\}\mathbf{\Phi}^T(j+1, j) \\ &+ \mathbf{\Phi}(j+1, j)\text{cov}\{\mathbf{x}(j), \mathbf{u}(j)\}\mathbf{\Gamma}(j) \\ &+ \mathbf{\Gamma}(j)\mathbf{V}_{\mathbf{u}}(j)\mathbf{\Gamma}^T(j)\end{aligned}$$

From Eq. (6.4-11), or from physical reasoning, it is clear that $\mathbf{V}_{\mathbf{xu}}(j) = \mathbf{V}_{\mathbf{uv}}(j) = \mathbf{0}$, since $\mathbf{x}(j)$ depends only on $\mathbf{u}(i)$, $i < j$, and $\mathbf{u}(j)$ is uncorrelated with $\mathbf{u}(i)$, $i \neq j$. Therefore, the foregoing relation becomes

$$\mathbf{V}_{\mathbf{x}}(j+1) = \mathbf{\Phi}(j+1, j)\mathbf{V}_{\mathbf{x}}(j)\mathbf{\Phi}^T(j+1, j) + \mathbf{\Gamma}(j)\mathbf{V}_{\mathbf{u}}(j)\mathbf{\Gamma}^T(j) \tag{6.4-13}$$

It is a straightforward task to obtain the expression $\mathbf{V}_{\mathbf{x}}(j, k)$ by utilizing the definition of the covariance function and Eq. (6.4-5), for $j \geq k$:

$$\begin{aligned} \mathbf{V_x}(j, k) &= \text{cov}\{\mathbf{x}(j), \mathbf{x}(k)\} \\ &= \text{cov}\{\boldsymbol{\Phi}(j, k)\mathbf{x}(k) + \sum_{i=k}^{j-1} \boldsymbol{\Phi}(j, i+1)\boldsymbol{\Gamma}(i)\mathbf{u}(i), \mathbf{x}(k)\} \qquad (6.4\text{-}14) \\ &= \boldsymbol{\Phi}(j, k)\mathbf{V_x}(k) \qquad j \geq k \end{aligned}$$

By similar reasoning it follows that

$$\mathbf{V_x}(j, k) = \mathbf{V_x}(j)\boldsymbol{\Phi}^T(k, j) \qquad k \geq j \tag{6.4-15}$$

The reader is reminded that Eqs. (6.4-11), (6.4-13), (6.4-15), and (6.4-16) are only valid if $\mathbf{u}(k)$ has uncorrelated sample values. The case where $\mathbf{u}(k)$ is nonwhite may be treated by adjoining the state vector, as in the continuous case. Table 6.4-1 summarizes the mean-variance, covariance, and mean-square-value-propagation algorithms for discrete linear systems.

TABLE 6.4-1. State-Variable Relations for Discrete Linear System Response to White Random Inputs

State equation	$\mathbf{x}(k+1) = \boldsymbol{\Phi}(k+1, k)\mathbf{x}(k) + \boldsymbol{\Gamma}(k)\mathbf{u}(k)$
Moments	$\boldsymbol{\mu}_\mathbf{u}(k) = \mathcal{E}\{\mathbf{u}(k)\} \qquad \mathbf{V_u}(k) = \text{var}\{\mathbf{u}(k)\}$ $\boldsymbol{\mu}_\mathbf{x}(0) = \mathcal{E}\{\mathbf{x}(0)\} = \boldsymbol{\mu}_{\mathbf{x}_0} \qquad \mathbf{V_x}(0) = \text{var}\{\mathbf{x}(0)\} = \mathbf{V}_{\mathbf{x}_0}$
Mean-value algorithm	$\boldsymbol{\mu}_\mathbf{x}(k+1) = \boldsymbol{\Phi}(k+1, k)\boldsymbol{\mu}_\mathbf{x}(k) + \boldsymbol{\Gamma}(k)\boldsymbol{\mu}_\mathbf{u}(k)$ $\boldsymbol{\mu}_\mathbf{x}(0) = \boldsymbol{\mu}_{\mathbf{x}_0}$
Variance algorithm	$\mathbf{V_x}(k+1) = \boldsymbol{\Phi}(k+1, k)\mathbf{V_x}(k)\boldsymbol{\Phi}^T(k+1, k) + \boldsymbol{\Gamma}(k)\mathbf{V_u}(k)\boldsymbol{\Gamma}^T(k)$ $\mathbf{V_x}(0) = \mathbf{V}_{\mathbf{x}_0}$
Mean-square algorithm	$\mathbf{P_x}(k) = \mathbf{V_x}(k) + \boldsymbol{\mu}_\mathbf{x}(k)\boldsymbol{\mu}_\mathbf{x}^T(k)$
Alternative mean-square algorithm	$\mathbf{P_x}(k+1) = \boldsymbol{\Phi}(k+1, k)\mathbf{P_x}(k)\boldsymbol{\Phi}^T(k+1, k) + \boldsymbol{\Gamma}(k)\mathbf{P_u}(k)\boldsymbol{\Gamma}^T(k)$ $+ \boldsymbol{\Phi}(k+1, k)\boldsymbol{\mu}_\mathbf{x}(k)\boldsymbol{\mu}_\mathbf{u}^T(k)\boldsymbol{\Gamma}^T(k)$ $+ \boldsymbol{\Gamma}(k)\boldsymbol{\mu}_\mathbf{u}(k)\boldsymbol{\mu}_\mathbf{x}^T(k)\boldsymbol{\Phi}^T(k+1, k)$ $\mathbf{P_x}(0) = \mathbf{P}_{\mathbf{x}_0} = \mathbf{V}_{\mathbf{x}_0} + \boldsymbol{\mu}_{\mathbf{x}_0}\boldsymbol{\mu}_{\mathbf{x}_0}^T$
Covariance algorithm	$\mathbf{V_x}(j, k) = \begin{cases} \boldsymbol{\Phi}(j, k)\mathbf{V_x}(k) & j \geq k \\ \mathbf{V_x}(j)\boldsymbol{\Phi}^T(k, j) & k \geq j \end{cases}$
Correlation algorithm	$\mathbf{P_x}(j, k) = \mathbf{V_x}(j, k) + \boldsymbol{\mu}_\mathbf{x}(j)\boldsymbol{\mu}_\mathbf{x}^T(k)$

Example ***6.4-1.*** Let us consider a simple scalar example

$$x(k+1) = \alpha x(k) + u(k)$$

where $u(k)$ is a discrete random process with zero mean. It is nonwhite and evolves from the scalar difference equation

$$u(k+1) = \beta u(k) + w(k)$$

where $w(k)$ is white, with moments

$$\mu_w(k) = 0 \qquad \text{var}\{w(k)\} = V_w(k) = 1$$

The initial statistics of $x(k)$ and $u(k)$ are

$$\mu_x(0) = \mu_u(0) = 0 = V_x(0) = V_u(0) = 0$$

We can easily determine the moments of $x(k)$ by adjoining $u(k)$ to the state vector. We define

$$\mathbf{z}(k) = \begin{bmatrix} x(k) \\ \hdashline u(k) \end{bmatrix} \qquad \mathbf{\Phi} = \begin{bmatrix} \alpha & 1 \\ 0 & \beta \end{bmatrix} \qquad \mathbf{\Gamma} = \begin{bmatrix} 0 \\ 1 \end{bmatrix}$$

such that the adjoined message model becomes

$$\mathbf{z}(k+1) = \mathbf{\Phi}\mathbf{z}(k) + \mathbf{\Gamma} w(k)$$

with prior statistics

$$\boldsymbol{\mu}_{\mathbf{z}}(0) = \mathbf{0} \qquad \mathbf{V}_{\mathbf{z}}(0) = \mathbf{0}$$

The mean of $x(k)$ is always zero since $\boldsymbol{\mu}_{\mathbf{z}}(k)$ is determined by solution of

$$\boldsymbol{\mu}_{\mathbf{z}}(k+1) = \mathbf{\Phi}\boldsymbol{\mu}_{\mathbf{z}}(k) + \mathbf{\Gamma}\mu_w(k) \qquad \boldsymbol{\mu}_{\mathbf{z}}(0) = \mathbf{0}$$

and $\boldsymbol{\mu}_x(k)$ is the first component of the vector $\boldsymbol{\mu}_{\mathbf{z}}(k)$.

The variance of $\mathbf{z}(k)$ evolves from

$$\mathbf{V}_{\mathbf{z}}(k+1) = \mathbf{\Phi}\mathbf{V}_{\mathbf{z}}(k)\mathbf{\Phi}^T + \mathbf{\Gamma} V_w(k)\mathbf{\Gamma}^T$$
$$\mathbf{V}_{\mathbf{z}}(0) = \mathbf{0}$$

which in expanded form is

$$\begin{bmatrix} V_{z11}(k+1) & V_{z12}(k+1) \\ V_{z12}(k+1) & V_{z22}(k+1) \end{bmatrix} = \begin{bmatrix} \alpha & 1 \\ 0 & \beta \end{bmatrix} \begin{bmatrix} V_{z11}(k) & V_{z12}(k) \\ V_{z12}(k) & V_{z22}(k) \end{bmatrix} \begin{bmatrix} \alpha & 1 \\ 1 & \beta \end{bmatrix} + \begin{bmatrix} 0 \\ 1 \end{bmatrix} 1 [0 \quad 1]$$

with zero initial condition.

Normally, an equation of this sort is solved on a digital computer and the matrix representation of the algorithm for $\mathbf{V}_z$ is particularly appropriate. We may also determine the solution by hand for a few stages. In component form the foregoing equation becomes

$$V_{z_{11}}(k+1) = \alpha^2 V_{z_{11}}(k) + 2\alpha V_{z_{12}}(k) + V_{z_{22}}(k)$$
$$V_{z_{12}}(k+1) = \alpha\beta V_{z_{12}}(k) + \beta V_{z_{22}}(k)$$
$$V_{z_{22}}(k+1) = \beta V_{z_{22}}(k) + 1$$

Although it is a tedious process, these equations may be solved step by step with the initial conditions $\mathbf{V}_z(0) = \mathbf{0}$.

Of interest also is the steady-state solution of these equations. In steady state, if one exists, $\mathbf{V}_z(k+1) = \mathbf{V}_z(k)$. With this restriction we can easily solve the foregoing equations to obtain

$$V_{z_{22}}(\infty) = \frac{1}{1-\beta^2} = V_u(\infty)$$

$$V_{z_{12}}(\infty) = \frac{\beta}{(1-\beta^2)(1-\alpha\beta)} = V_{xu}(\infty)$$

$$V_{z_{11}}(\infty) = \frac{1+\alpha\beta}{(1-\beta^2)(1-\alpha\beta)(1-\alpha^2)} = V_x(\infty)$$

For the variance terms V_u and V_x to be finite and positive, we must require that

$$|\beta| < 1 \qquad |\alpha| < 1$$

It turns out that this is a stability requirement on the message-model equations.

As the next effort in the development of input–output relations for dynamic transformations, we consider the time-invariant case for discrete systems. The time-invariant form of Eq. (6.4-2) is given by

$$\mathbf{x}(k+1) = \mathbf{\Omega}\mathbf{x}(k) + \mathbf{\Gamma}\mathbf{u}(k) \tag{6.4-16}$$

where $\mathbf{\Omega} = \mathbf{\Omega}(T) = \mathbf{\Phi}(k+1, k) = \mathbf{\Phi}(\overline{k+1}T, kT)$. If the random process $\mathbf{u}(k)$ is stationary, the mean $\boldsymbol{\mu}_\mathbf{u}$ is constant and the variance $\mathbf{V}_\mathbf{u}(j, k)$ depends only on the age or separation interval $i = j - k$; that is,

$$\mathbf{R}_\mathbf{u}(i) \triangleq \mathbf{V}_\mathbf{u}(k+i, k) \triangleq \mathbf{V}_\mathbf{u}(k, k+i)$$

for all k.

The mean of the output can be determined easily by taking the expectation of both sides of Eq. (6.4-16) to obtain

$$\boldsymbol{\mu}_{\mathbf{x}}(k+1) = \boldsymbol{\Omega}\boldsymbol{\mu}_{\mathbf{x}}(k) + \boldsymbol{\Gamma}\boldsymbol{\mu}_{\mathbf{u}}$$

Since the input process is stationary and the system is time invariant, the output process is also stationary in the steady state, and $\boldsymbol{\mu}_{\mathbf{x}}(k+1) = \boldsymbol{\mu}_{\mathbf{x}}(k) = \boldsymbol{\mu}_{\mathbf{x}}$. Thus we have

$$\begin{aligned} \boldsymbol{\mu}_{\mathbf{x}} &= \boldsymbol{\Omega}\boldsymbol{\mu}_{\mathbf{x}} + \boldsymbol{\Gamma}\boldsymbol{\mu}_{\mathbf{x}} \\ \boldsymbol{\mu}_{\mathbf{x}} &= (\mathbf{I} - \boldsymbol{\Omega})^{-1}\boldsymbol{\Gamma}\boldsymbol{\mu}_{\mathbf{u}} \end{aligned} \tag{6.4-17}$$

In the discussion of continuous systems we developed the expressions for $\mathbf{C}_{\mathbf{xu}}$ and $\mathbf{C}_{\mathbf{x}}$ by the application of the general nonstationary results to the stationary case. As a variation, we shall proceed in the discrete case by making a direct application of fundamental definitions to the stationary equations.

The solution for Eq. (6.4-16) is given by

$$\mathbf{x}(j) = \sum_{k=-\infty}^{j-1} \boldsymbol{\Omega}(j-1-k)\boldsymbol{\Gamma}\mathbf{u}(k)$$

where, for simplicity, we have assumed that the system begins at rest, $\mathbf{x} = \mathbf{0}$, at $k = -\infty$. The covariance $\mathbf{R}_{\mathbf{xu}}(i)$ is therefore

$$\begin{aligned} \mathbf{R}_{\mathbf{xu}}(i) &\triangleq \operatorname{cov}\{\mathbf{x}(j+i), \mathbf{u}(j)\} \\ &= \operatorname{cov}\left\{\sum_{k=-\infty}^{j+i-1} \boldsymbol{\Omega}(j+i-1-k)\boldsymbol{\Gamma}\mathbf{u}(k), \mathbf{u}(j)\right\} \\ &= \sum_{k=-\infty}^{j+i-1} \boldsymbol{\Omega}(j+i-1-k)\boldsymbol{\Gamma}\mathbf{R}_{\mathbf{u}}(k-j) \end{aligned}$$

To remove the apparent dependence on j, let $m = k - j$ so that we have

$$\mathbf{R}_{\mathbf{xu}}(i) = \sum_{m=-\infty}^{i+1} \boldsymbol{\Omega}(i-1-m)\boldsymbol{\Gamma}\mathbf{R}_{\mathbf{u}}(m) \tag{6.4-18}$$

Because of the causal nature of our systems, $\boldsymbol{\Omega}(k)$ must be zero for $k < 0$, and hence the upper limit on the summation may be increased to infinity without changing the result. Hence we have

$$\mathbf{R}_{\mathbf{xu}}(i) = \sum_{m=-\infty}^{\infty} \boldsymbol{\Omega}(i-1-m)\boldsymbol{\Gamma}\mathbf{R}_{\mathbf{u}}(m) \tag{6.4-19}$$

Also by letting $i - 1 - m = \eta$, we obtain

$$\mathbf{R}_{\mathbf{xu}}(i) = \sum_{\eta=-\infty}^{\infty} \boldsymbol{\Omega}(\eta)\boldsymbol{\Gamma}\mathbf{R}_{\mathbf{u}}(i-1-\eta) \tag{6.4-20}$$

and we may, if we like, start the summation at $\eta = 0$ rather than $\eta = -\infty$ because of the causal nature of $\mathbf{\Omega}(k)$.

The discrete spectral-density expression may now be obtained by $\mathcal{Z}$ transforming both sides of Eq. (6.4-19). The resulting spectral-density expressions are quite useful for work in discrete data-control systems with random inputs, and their development may be found in the references at the end of this chapter.

Some interesting relations between discrete and continuous white noise were discovered in Example 5.8-3. In that example we demonstrated that as the samples became dense, the zero-mean discrete white noise $\mathbf{u}(kT)$ with associated variance

$$\mathbf{V}_{\mathbf{u}}(kT, jT) = \operatorname{cov}\{\mathbf{u}(kT), \mathbf{u}(jT)\} = \mathbf{V}_{\mathbf{u}}(k)\delta_K(k-j) = \frac{\mathbf{\Psi}_{\mathbf{u}}(kT)}{T}\delta_K(k-j) \tag{6.4-21}$$

became the continuous white noise $\mathbf{u}(t)$ with variance

$$\begin{aligned} \mathbf{V}_{\mathbf{u}}(t, \tau) = \operatorname{var}\{\mathbf{u}(t), \mathbf{u}(\tau)\} &= \lim_{\substack{k,\, j \to \infty \\ kT \to t \\ jT \to \tau \\ T \to 0}} \operatorname{var}\{\mathbf{u}(kT), \mathbf{u}(jT)\} \\ &= \mathbf{\Psi}_{\mathbf{u}}(t)\delta_D(t-\tau) \end{aligned} \tag{6.4-22}$$

Armed with this result we are prepared to show that the discrete variance equation (6.2-13) becomes the continuous equation (6.4-13). We first note that the system equation

$$\mathbf{x}(\overline{k+1}T) = \mathbf{\Phi}(\overline{k+1}T, kT)\mathbf{x}(kT) + \mathbf{\Gamma}(kT)\mathbf{x}(kT) \tag{6.4-23}$$

becomes the continuous equation

$$\dot{\mathbf{x}} = \mathbf{F}(t)\mathbf{x}(t) + \mathbf{G}(t)\mathbf{u}(t) \tag{6.4-24}$$

in the limit as the samples become dense. The matrices in the foregoing two equations are related by the definitions

$$\mathbf{F}(t) = \lim_{\substack{T \to 0 \\ kT \to t}} \frac{\mathbf{I} - \mathbf{\Phi}(\overline{k+1}T, kT)}{T} \tag{6.4-25}$$

$$\mathbf{G}(t) = \lim_{\substack{T \to 0 \\ kT \to t}} \frac{\mathbf{\Gamma}(kT)}{T} \tag{6.4-26}$$

It is equally simple to show that this limiting process changes the discrete variance equation

$$\mathbf{V}_{\mathbf{x}}(k+1) = \mathbf{\Omega}(k+1, k)\mathbf{V}_{\mathbf{x}}(k)\mathbf{\Phi}^{T}(k+1, k) + \mathbf{\Gamma}(k)\mathbf{V}_{\mathbf{u}}(k)\mathbf{\Gamma}^{T}(k) \tag{6.4-27}$$

to the continuous variance equation

$$\dot{\mathbf{V}}_{\mathbf{x}}(t) = \mathbf{F}(t)\mathbf{V}_{\mathbf{x}}(t) + \mathbf{V}_{\mathbf{x}}(t)\mathbf{F}^T(t) + \mathbf{G}(t)\boldsymbol{\psi}_{\mathbf{u}}(t)\mathbf{G}^T(t) \tag{6.4-28}$$

as we easily see by use of Eqs. (6.4-25) and (6.4-26) in Eq. (6.4-27)

This completes the study of the input–output expressions for the mean and variance propagation in discrete linear systems with random inputs. In any further study of stochastic processes in the information and control sciences, we shall have repeated need for the results of this section and this chapter.

Example ***6.4-2.*** We have indicated two possible approaches to the discrete modeling of stochastic processes. We may discretize a continuous model representing the process and then find the mean and variance equations corresponding to this discrete model. Alternatively, we may find the continuous differential equations for the evolution of the mean and variance of the continuous system and discretize them. Let us indicate the similarities and differences of the two approaches.

A continuous zero-mean stochastic process $x(t)$ is formed from zero-mean white noise $u(t)$ with covariance coefficient ψ_u by means of the dynamic transformation

$$\dot{x}(t) = -ax(t) + u(t)$$

The continuous variance equation is easily determined from Table 6.2-1 and is

$$\dot{V}_x(t) = -2aV_x(t) + \psi_{\mathbf{u}} \qquad V_x(0) = V_{x_0}$$

The true solution of this equation is obviously

$$V_x(t) = V_{x_0}e^{-2at} + \frac{\psi_u}{2a}(1 - e^{-2at})$$

and the steady-state value of the true variance $V_x(t)$ is

$$V_x(t \longrightarrow \infty) = \frac{\psi_u}{2a}$$

Many approaches may be used to discretize the continuous message model. The simplest approach is to use the first forward difference approximation to the derivative such that we have

$$\frac{x(\overline{k+1}T) - x(kT)}{T} = -ax(kT) + u(kT)$$

or

$$x(\overline{k+1}T) = (1 - aT)x(kT) + Tu(kT)$$

where $u(kT)$ is discrete white noise with zero mean and variance

$$V_u(kT) = \frac{\psi_u}{T}$$

The discrete variance equation of Table 6.4-1 yields for this method of discretization

$$V_x(\overline{k+1}T) = (1 - aT)^2 V_x(kT) + T\psi_u$$

The true solution of this discrete variance equation is

$$V_x(kT) = (1 - aT)^{2k} V_x(0) + T \sum_{i-1}^{k-1} (1 - aT)^{2(k-i+1)} \psi_u$$

and the true steady-state solution is

$$V_x(kT \longrightarrow \infty) = \frac{\psi_u}{2a - a^2 T}$$

It is straightforward to show that these solutions of this discrete variance equation become precisely the solutions of the continuous variance equation in the limit as $T \longrightarrow 0$.

An alternative method of discretizing the message model is to first obtain the solution to its differential equation, which is

$$x(t) = e^{-a(t-\eta)} x(\eta) + \int_{\eta}^{t} e^{-a(\tau-t)} u(\tau)\, d\tau$$

and then let $t = \overline{k+1}T$ and $\eta = kT$ to obtain

$$x(\overline{k+1}T) = e^{-aT} x(kT) + \int_{kT}^{\overline{k+1}T} e^{-a(\tau-\overline{k+1}T)} u(\tau)\, d\tau$$

It is now assumed that

$$u(\tau) \approx u(kT) \qquad \text{for } kT \leq \tau \leq \overline{k+1}T$$

and then noted that

$$\int_{kT}^{\overline{k+1}T} e^{-a(\tau-\overline{k+1}T)}\, d\tau = \frac{e^{aT} - 1}{a}$$

such that the (approximate) discrete model is now

$$x(\overline{k+1}T) = e^{-aT} x(kT) + \frac{e^{aT} - 1}{a} u(kT)$$

where $u(kT)$ is zero mean and white with variance ψ_u/T. Because of the method used in constructing the discrete model, we would expect it to be a better model of the continuous process than the first difference model. The variance equation corresponding to this model is

$$V_x(\overline{k+1}T) = e^{-2aT}V_x(kT) + \left[\frac{e^{aT}-1}{a}\right]^2 \frac{\psi_u}{T}$$

The steady-state solution for this variance equation is

$$\begin{aligned} V_x(kT \longrightarrow \infty) &= \left[\frac{e^{aT}-1}{a}\right]^2 \frac{\psi_u}{T(1-e^{-2aT})} \\ &= \frac{\psi_u(1-2e^{aT}+e^{2aT})}{a^2T(1-e^{-2aT})} \end{aligned}$$

and, again, this is the correct solution to the continuous problem for $T \longrightarrow 0$.

The alternative approach is to discretize the continuous variance equation such that it can be solved on a digital computer. The first difference approach to discretizing the continuous variance equation of this example leads to

$$\frac{V_x(\overline{k+1}T) - V_x(kT)}{T} = -2aV_x(kT) + \psi_u$$

or

$$V_x(\overline{k+1}T) = (1-2aT)V_x(kT) + T\psi_u$$

The steady-state solution of this equation is

$$V_x(kT \longrightarrow \infty) = \frac{\psi_u}{2a}$$

Alternatively, we may write the solution to the continuous variance equation as

$$V_x(t) = e^{-2a(t-\eta)}V_x(\eta) + \frac{\psi_u}{2a}[1-e^{-2a(t-\eta)}]$$

and then let $t = \overline{k+1}T$ and $\eta = kT$ to obtain

$$V_x(\overline{k+1}T) = e^{-2aT}V_x(kT) + \frac{\psi_u}{2a}[1-e^{-2aT}]$$

Obviously, this last difference equation has a solution that is the exact solution to the continuous variance equation. Unfortunately, it is in general quite impossible to obtain an exact difference equation representation of a continuous model.

The result of this discussion might lead us to believe that it would be better to obtain the continuous variance equation and discretize it, since the resulting equation has no steady-state error and, for this simple problem, it is possible to obtain an exact discrete representation of the continuous variance equation. However, the errors resulting from using the variance equation for the discrete message model are quite small. For aT as large as 0.1, the steady-state error for the variance equation corresponding to the first difference approximation discrete message model is 5 per cent, whereas for the variance equation corresponding to the more accurate discrete message model a value of $aT = 0.1$ gives rise to a steady-state error of an even lower value.

A considerable advantage may in fact accrue with the use of a discrete variance equation corresponding to the discrete message model. The discrete message model is not an exact representation of the continuous message model, but a discrete model must be used if the random process is to be generated on a digital computer. So we must of necessity in many situations use a discrete message-generating process. The difference equation representing the variance of this process is exact in the sense that it represents the variance of the discrete process. The discretized version of the continuous variance equation yields a solution that does not represent either the variance of the discrete or the continuous message model. (In this very simple example we were able to obtain an exact discretization of the continuous variance equation. In general this cannot be done and, in any case, the solution of the discretized continuous variance equation cannot represent the variance of the discrete message.) The advantage of using the discrete variance equation is that any errors are errors in modeling the continuous message process. In using a discretized version of the continuous variance equation there are two sources of modeling errors—one due to modeling the continuous stochastic process and one due to modeling the continuous variance equation. For a variety of problems it is preferable that the only source of modeling error be that due to modeling the stochastic process itself. Thus the use of the discrete message model and discrete variance equation corresponding to this discrete message model may well be preferable to use of the discrete message model and a discrete version of the continuous variance equation corresponding to the continuous message model.

EXERCISES 6.4

6.4-1. If $x(k+1) = k^2x(k) + u(k)$, $\mu_u(k) = 0$, and $V_u(k, j) = \delta_K(k-j)$, and $x(0) = 1$, find (a) $\mu_x(3)$, (b) $V_x(2, 1)$, and (c) $V_{xu}(3, 2)$.

Answers: (a) 0, (b) 1, (c) 1.

6.4-2. Find (a) $\mu_x(3)$, (b) $\mu_z(3)$, (c) $V_x(3)$, (d) $V_{xz}(3, 2)$, and (e) $V_z(1, 2)$ if

$$x(k+1) = 2x(k) + u(k)$$
$$z(k) = 3x(k) + v(k)$$

Here $\mu_u(k) = 2$, $\mu_v(k) = 0$, $V_{uv}(k, j) = \delta_K(k-j)$, $V_u(k, j) = 4\delta_K(k-j) = V_v(k, j)$, and $x(0) = 1$.

Answers: (a) 22, (b) 66, (c) 84, (d) 121, (e) 75.

6.5. *Conclusions*

The subject discussed in this chapter is an important one, and much of the advanced work in application areas for stochastic processes relies heavily on a good understanding of the response of linear systems to random inputs. For further study in these areas the references are recommended. References [1], [5], [7], [8], and [9] present the state-variable approach to the response of linear systems and applications, whereas references [2], [3], [4], [6], [7], and [10] discuss frequency-domain applications.

6.6. *References*

(1) BRYSON, A. E., and Y. C. HO, *Applied Optimal Control*, Ginn/Blaisdell, Waltham, Mass., 1969.

(2) DAVENPORT, W. B., JR., and W. L. ROOT, *Random Signals and Noise*, McGraw-Hill, New York, 1958.

(3) LANING, J. H., and R. H. BATTIN, *Random Processes in Automatic Control*, McGraw-Hill, New York, 1956.

(4) LEE, Y. W., *Statistical Theory of Communication*, Wiley, New York, 1960.

(5) MEDITCH, J. S., *Stochastic Optimal Linear Estimation and Control*, McGraw-Hill, New York, 1969.

(6) NEWTON, G. C., et al., *Analytical Design of Linear Feedback Controls*, Wiley, New York, 1957.

(7) SAGE, A. P., *Optimum Systems Control*, Prentice-Hall, Englewood Cliffs, N.J., 1968.

(8) SAGE, A. P., and J. L. MELSA, *Estimation Theory with Applications to Communications and Control*, McGraw-Hill, New York, 1971.

(9) SAGE, A. P., and J. L. MELSA, *System Identification*, Academic Press, New York, 1971.

(10) THOMAS, J. B., *An Introduction to Statistical Communication Theory*, Wiley, New York, 1969.

6.7. Problems

6.7-1. Consider the continuous time stochastic process that evolves from

$$\dot{x}(t) = \frac{-1}{1-t}x(t)$$

where $x(0)$ is a random initial condition with mean $\mu_x(0)$ and variance $V_x(0)$. Find
(a) the mean $\mu_x(t)$.
(b) the variance $V_x(t)$.
(c) the covariance $V_x(t_1, t_2)$.
(d) the correlation $P_x(t_1, t_2)$.

6.7-2. State the covariance function

$$V_z(t, \tau) = \text{cov}\{z(t), z(\tau)\}$$

where

$$z(t) = x(t)y(t)$$

and $x(t)$ and $y(t)$ are zero-mean independent random processes with

$$V_x(t, \tau) = e^{-a|t-\tau|}$$

$$V_y(t, \tau) = e^{-b|t-\tau|}$$

6.7-3. The output of a system is

$$z = x_1 + x_2 + x_3$$

where the gaussian vector $\mathbf{x} = [x_1 \quad x_2 \quad x_3]^T$ has the moments

$$\boldsymbol{\mu}_\mathbf{x} = \begin{bmatrix} \mu_1 \\ \mu_2 \\ \mu_3 \end{bmatrix} \qquad \mathbf{V}_\mathbf{x} = \begin{bmatrix} V_{11} & V_{12} & V_{13} \\ V_{12} & V_{22} & V_{23} \\ V_{13} & V_{23} & V_{33} \end{bmatrix}$$

What is the mean and variance of z?

6.7-4 Someone suggests that the scheme illustrated is a method for measuring the impulse response of a linear system. The noise $w(t)$ is zero mean, white, and ergodic with covariance

$$\text{cov}\{w(t), w(\tau)\} = \delta_D(t - \lambda)$$

Show that if the integration time T is sufficiently long,

$$z(T) = h(\tau)$$

which is the impulse of the unknown linear system at time τ.

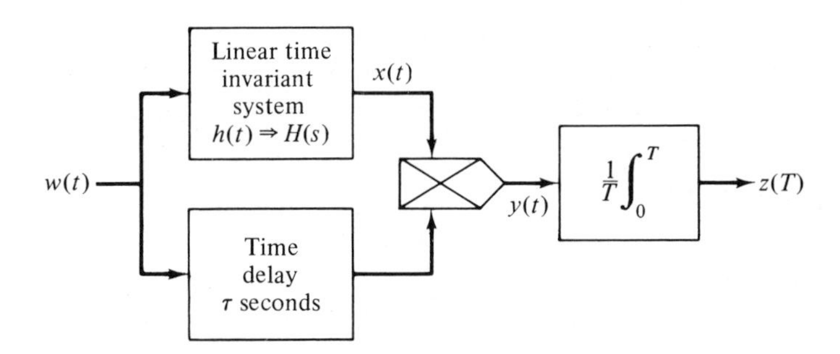

PROB. 6.7-4

6.7-5. Let n_T be defined by

$$n_T = \frac{1}{T}\int_0^T x(t)\,dt$$

where $x(t) = a + n(t)$. Here a is an unknown constant and $n(t)$ is a zero-mean white-noise process with $\text{cov}\{n(t), n(\tau)\} = \psi_n \delta(t - \tau)$. Find the mean and variance of n_T in terms of the integration time T.

6.7-6. For linear system

$$\dot{x} = a(t)x(t) + b(t)w(t)$$
$$w(t) = c(t)v(t) + \gamma(t)$$
$$\dot{v} = d(t)v(t) + u(t)$$

where $\gamma(t)$ and $u(t)$ are independent, zero mean, and white with

$$\text{cov}\{\gamma(t), \gamma(\tau)\} = \psi_\gamma \delta_D(t - \tau)$$
$$\text{cov}\{u(t), u(\tau)\} = \psi_u \delta_D(t - \tau)$$
$$\text{cov}\{\gamma(t), u(\tau)\} = 0$$

what are the explicit differential equations for the mean and variance $\mu_x(t)$ and $V_x(t)$?

6.7-7. A stochastic process $\mathbf{x}(t)$ is generated from the message model

$$\dot{x}_1(t) = x_2(t)$$
$$\dot{x}_2(t) = 0$$

where $x_1(0)$ and $x_2(0)$ are random with

$$\boldsymbol{\mu}_\mathbf{x}(0) = \mathcal{E}\left\{\begin{bmatrix} x_1(0) \\ x_2(0) \end{bmatrix}\right\} \qquad \mathbf{V}_\mathbf{x}(0) = \text{var}\left\{\begin{bmatrix} x_1(0) \\ x_2(0) \end{bmatrix}\right\}$$

A noisy observation of $x_1(t)$,

$$z(t) = x_1(t) + v(t)$$

is available, where $v(t)$ is zero-mean white noise with constant variance coefficient ψ_v.

Find the explicit equations for the optimum sequential linear (Kalman) filter of Example 6.2-6. Can you obtain an analytical solution to the error variance equation for this problem? What is it? What is the steady-state error variance? Why is it this value (physically)?

6.7-8. Someone has claimed that

$$\operatorname{cov}\{\tilde{\mathbf{x}}(t), \hat{\mathbf{x}}(t)\} = \mathbf{0}$$

where $\hat{\mathbf{x}}(t)$ is the Kalman filter estimate and $\tilde{\mathbf{x}}(t)$ the associated estimation error. Prove this result using the methods of Section 6.2.

6.7-9. The autocorrelation function of the input to a linear system is $C_u(\tau) = \mathcal{E}\{u(t+\tau)u(t)\} = e^{-|\tau|}$. The system transfer function is $H(s) = 2/(s+2)$. What is the autocorrelation function of the system output?

6.7-10. The system illustrated is driven by white noise with unity spectral density. What is the cross correlation $C_{x_2x_1}(\tau)$ and cross spectral density $C_{x_2x_1}(s)$ in the steady state, where as usual (for zero-mean random processes)

$$C_{x_2x_1}(\tau) = \mathcal{E}\{x_2(t+\tau)x_1(t)\}$$

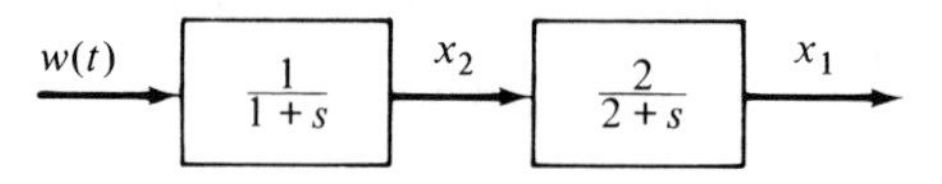

PROB. 6.7-10

6.7-11. Given the system illustrated and relations

$$C_{u_1u_1}(\tau) = \delta_D(\tau) \qquad C_{u_2u_2}(\tau) = \delta_D(\tau) \qquad C_{u_1u_2}(\tau) = 0$$

Find (a) $C_{x_2x_2}(\tau)$, $C_{X_2X_2}(s)$; (b) $C_{x_1x_2}(\tau)$, $C_{X_1X_2}(s)$.

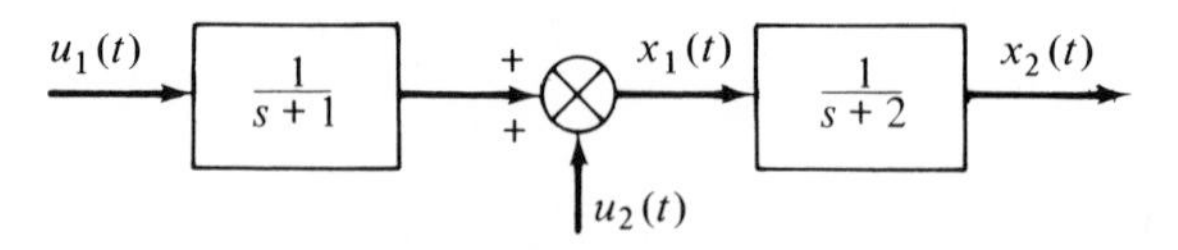

PROB. 6.7-11

6.7-12. For the given system, where $u(t)$ is a stationary random process, find (a) $C_X(s)$, (b) $C_Y(s)$, (c) $C_{YX}(s)$, and (d) $C_{YZ}(s)$. What do these spectral densities and the associated correlation functions become if

$$C_U(s) = \psi_u \qquad H_1(s) = \frac{1}{s+1} \qquad H_2(s) = \frac{2}{s+2} \qquad H_3(s) = \frac{3}{s+3}$$

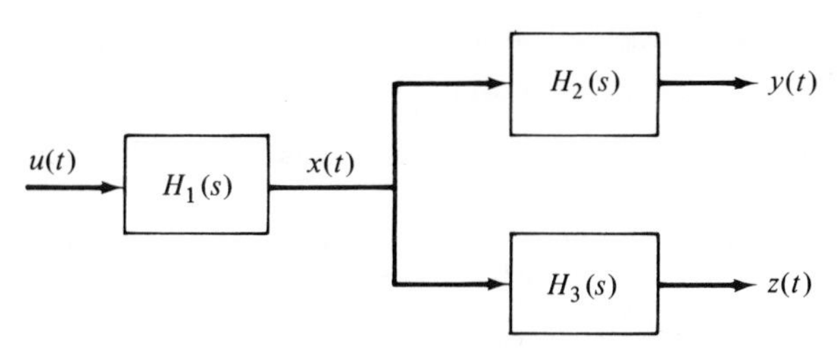

PROB. 6.7-12

6.7-13. For the systems illustrated, appropriate signal and noise spectral densities are

$$C_S(s) = \frac{-1}{s^2} \qquad C_N(s) = 1$$

and there is no correlation between signal and noise. Find the optimum linear filter gain K that gives minimum mean-square error. How does this error vary as a function of T? The error is defined as $c(t) - s(t)$.

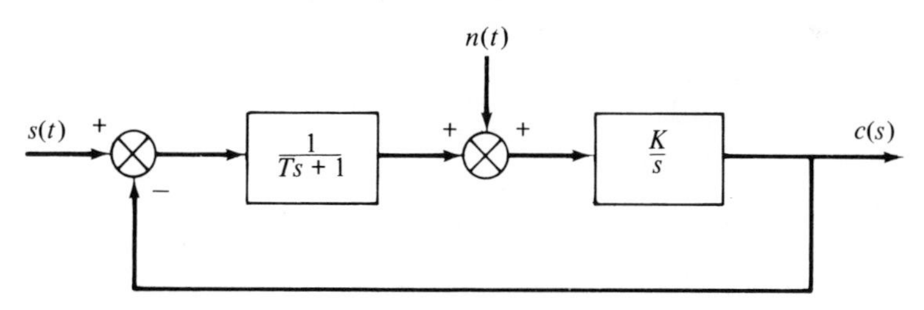

PROB. 6.7-13

6.7-14. If we have

$$\mathbf{X}(s) = \begin{bmatrix} \frac{1}{s+a} & 0 \\ \frac{1}{s} & \frac{1}{s+b} \end{bmatrix} \mathbf{U}(s)$$

find $\mathbf{C}_{\mathbf{XU}}(s)$ and $\mathbf{C}_{\mathbf{X}}(s)$.

6.7-15. A discrete system is characterized by the relationship

$$x(t_{k+2}) = x(t_k) + w(t_k)$$

where $w(t_k)$ is zero-mean white noise with variance $V_w(t_k)$. Derive a difference equation for $V_x(t_k)$, that is, the difference equation for the propagation of the variance of x.

6.7-16. We are given the continuous scalar random process

$$\dot{x} = ax(t) + bw(t)$$

where $w(t)$ is zero-mean white noise with

$$\text{cov}\{w(t), w(\tau)\} = \psi_w \delta_D(t - \tau)$$

A friend postulates that a discrete model for this random process is

$$x(k+1) = (1 + aT)x(k) + bTu(k) = \Phi x(k) + \Gamma w(k)$$

where k represents the state and T represents the sample period. He suggests that $w(k)$ should be discrete white noise with zero mean and covariance

$$\text{cov}\{w(k), w(j)\} = \frac{\Psi_w}{T}\delta_K(k-j) = V_w\delta_K(k-j)$$

We know that the state variance of the discrete model is

$$V_x(k+1) = \Phi^2 V_x(k) + \Gamma^2 V_w$$

Under what conditions is this a correct equation for the variance of the continuous system? Discuss the errors in the discretization process and suggest alternative means of discretization.

6.7-17. Find the mean and variance of the state vector for the system described by

$$\mathbf{x}(k+1) = \mathbf{\Phi}\mathbf{x}(k) + \mathbf{\Gamma}(k)w(k)$$

where

$$\mathbf{\Phi} = \begin{bmatrix} -0.5 & -0.5 \\ 1 & 0 \end{bmatrix} \qquad \mathbf{\Gamma} = \begin{bmatrix} 1 \\ 0 \end{bmatrix}$$

$$\mu_w(k) = 1 \qquad V_w(k, j) = \delta_K(k = j)$$

$$\boldsymbol{\mu}_{\mathbf{x}}(0) = \mathbf{0} \qquad \mathbf{V}_{\mathbf{x}}(0) = \mathbf{0}$$

6.7-18. Repeat Problem 6.7-17 for the case where $\mathbf{\Gamma}$ is changed to

$$\mathbf{\Gamma} = \begin{bmatrix} 0 \\ 1 \end{bmatrix}$$

6.7-19. A discrete linear system message model is

$$\mathbf{x}(k+1) = \mathbf{\Phi}\mathbf{x}(k) + \mathbf{\Gamma}\mathbf{w}(k)$$

and the observation model is

$$\mathbf{z}(k) = \mathbf{H}\mathbf{x}(k) + \mathbf{v}(k)$$

Necessary prior statistics are

$$\boldsymbol{\mu}_{\mathbf{x}}(0) = \boldsymbol{\mu}_{\mathbf{x}_0} \qquad \text{var}\{\mathbf{x}(0)\} = \mathbf{V}_{\mathbf{x}_0}$$

$$\boldsymbol{\mu}_{\mathbf{w}}(k) = \mathbf{0} \qquad \text{cov}\{\mathbf{w}(k), \mathbf{w}(j)\} = \mathbf{V}_{\mathbf{w}}(k)\delta_K(k-j)$$

$$\boldsymbol{\mu}_{\mathbf{v}}(k) = \mathbf{0} \qquad \text{cov}\{\mathbf{v}(k), \mathbf{v}(j)\} = \mathbf{V}_{\mathbf{v}}(k)\delta_K(k-j)$$

$$\text{cov}\{\mathbf{v}(k), \mathbf{w}(j)\} = \mathbf{0}$$

It is desired to estimate the state vector $\mathbf{x}(k)$ based upon observation $\mathbf{z}(j), 0 \le j \le k-1$. The filter

$$\hat{\mathbf{x}}(k+1) = \mathbf{\Phi}\hat{\mathbf{x}}(k) + \mathbf{K}(k)[\mathbf{z}(k) - \mathbf{H}\hat{\mathbf{x}}(k)]$$

$$\hat{\mathbf{x}}(0) = \boldsymbol{\mu}_{\mathbf{x}_0}$$

is postulated. (a) Show that the filter is unbiased. (b) Show that the error variance in filtering defined by $\mathbf{V}_{\tilde{\mathbf{x}}}(k) = \text{var}\{\tilde{\mathbf{x}}(k)\} = \text{var}\{\mathbf{x}(k) - \hat{\mathbf{x}}(k)\}$

evolves from the difference equation

$$\mathbf{V}_{\tilde{\mathbf{x}}}(k+1) = [\mathbf{\Phi} - \mathbf{K}(k)\mathbf{H}]\mathbf{V}_{\tilde{\mathbf{x}}}(k)[\mathbf{\Phi} - \mathbf{K}(k)\mathbf{H}]^T + \mathbf{\Gamma}\mathbf{V}_{\mathbf{w}}(k)\mathbf{\Gamma}^T - \mathbf{K}(k)\mathbf{V}_{\mathbf{v}}(k)\mathbf{K}^T(k)$$

with initial condition

$$\mathbf{V}_{\tilde{\mathbf{x}}}(0) = \mathbf{V}_{\mathbf{x}_0}$$

(c) Show that the optimum gain $\mathbf{K}(k)$ to minimize $J(k) = \mathrm{tr}[\mathbf{V}_{\tilde{\mathbf{x}}}(k)]$ is given by

$$\mathbf{K}(k) = \mathbf{\Phi}\mathbf{V}_{\tilde{\mathbf{x}}}(k)\mathbf{H}^T[\mathbf{H}^T\mathbf{V}_{\tilde{\mathbf{x}}}(k)\mathbf{H}^T + \mathbf{V}_{\mathbf{v}}(k)]^{-1}$$

The solution you have just obtained is called the one-stage prediction-solution discrete Kalman filter and is an important result in modern estimation theory.

7. Gaussian Processes

7.1. Introduction

The gaussian process has been discussed in several places throughout this book. Because of the importance of gaussian processes in the application of probability and stochastic processes, we wish to examine gaussian processes in some detail in this chapter. Although there are many reasons for the importance of gaussian processes, there are two principal reasons. First, the gaussian process has many convenient mathematical properties, including the important fact that linear operations on gaussian processes yield gaussian processes. This property will be studied in detail in Section 7.4. The second reason is that they represent good approximations to a wide variety of physical situations. In particular, we shall show that whenever a given physical effect is comprised of the addition of a series of independent effects, the associated density is often gaussian. This fact is a consequence of the central limit theorem developed in Section 7.3.

In Section 7.2 we shall define the general gaussian process and develop several properties associated with it, including the general characteristic function. In addition, third- and fourth-product moment expressions are developed, which are useful in the study of many measurement procedures. The final topic of Section 7.2 is the derivation of the distribution for certain often-used functions of gaussian processes.

Section 7.4 presents a study of the combined behavior of two multivariable gaussian processes with special emphasis on conditional probabilities.

7.2. *Multivariate gaussian processes*

In Section 3.5 we defined the scalar gaussian (normal) distribution with parameters μ and σ^2; the density function for this distribution is given by

$$p_x(\alpha) = \frac{1}{\sqrt{2\pi}\sigma} \exp\left\{-\frac{(\alpha - \mu)^2}{2\sigma^2}\right\} \tag{7.2-1}$$

Later, in Section 4.4, we showed that the parameters μ and σ^2 were, respectively, the mean and variance of the distribution.

For a two-dimensional random variable $\mathbf{x}$, the density function for the gaussian distribution takes the following form:

$$p_{x_1x_2}(\alpha_1, \alpha_2) = \frac{1}{2\pi\sigma_1\sigma_2\sqrt{1-\rho^2}} \exp\left\{-\frac{1}{2(1-\rho^2)}\left[\frac{(\alpha_1 - \mu_1)^2}{\sigma_1^2} - \frac{2\rho(\alpha_1 - \mu_1)(\alpha_2 - \mu_2)}{\sigma_1\sigma_2} + \frac{(\alpha_2 - \mu_2)^2}{\sigma_2^2}\right]\right\} \tag{7.2-2}$$

It is a routine, although somewhat lengthy, algebraic operation to show that

$$\mathcal{E}\{\mathbf{x}\} = \boldsymbol{\mu} = \begin{bmatrix} \mu_1 \\ \mu_2 \end{bmatrix} \tag{7.2-3}$$

and that

$$\operatorname{var}\{\mathbf{x}\} = \mathcal{E}\{(\mathbf{x} - \boldsymbol{\mu})(\mathbf{x} - \boldsymbol{\mu})^T\} = \mathbf{V} = \begin{bmatrix} \sigma_1^2 & \rho\sigma_1\sigma_2 \\ \rho\sigma_1\sigma_2 & \sigma_2^2 \end{bmatrix} \tag{7.2-4}$$

Here the parameter ρ is the normalized correlation coefficient:

$$\rho = \frac{\mathcal{E}\{(x_1 - \mu_1)(x_2 - \mu_2)\}}{\sigma_1\sigma_2}$$

To express the general n-vector gaussian density in the form of Eq. (7.2-2) is very complex. However, it is possible to express this general case in a simple form by the use of vector-matrix notation. The density function for the n-dimensional gaussian distribution with an n-vector parameter $\boldsymbol{\mu}$ and an $n \times n$ matrix parameter $\mathbf{V}$ is given by

$$p_{\mathbf{x}}(\boldsymbol{\alpha}) = \frac{1}{[(2\pi)^n \det \mathbf{V}]^{1/2}} \exp\left[-\frac{1}{2}(\boldsymbol{\alpha} - \boldsymbol{\mu})^T\mathbf{V}^{-1}(\boldsymbol{\alpha} - \boldsymbol{\mu})\right] \tag{7.2-5}$$

It is reasonably easy to show that $\boldsymbol{\mu} = \mathcal{E}\{\mathbf{x}\}$ and $\mathbf{V} = \text{var}\{\mathbf{x}\}$ just as in the scalar case.

It is perhaps instructive to demonstrate that the general form of Eq. (7.2-5) reduces to Eq. (7.2-2) for the two-dimensional case. If $\mathbf{V}$ is given by Eq. (7.2-4), then det $\mathbf{V}$ is

$$\det \mathbf{V} = \sigma_1^2\sigma_2^2(1 - \rho^2) \tag{7.2-6}$$

and $\mathbf{V}^{-1}$ becomes

$$\mathbf{V}^{-1} = \frac{1}{\sigma_1^2\sigma_2^2(1-\rho^2)}\begin{bmatrix} \sigma_2^2 & -\rho\sigma_1\sigma_2 \\ -\rho\sigma_1\sigma_2 & \sigma_1^2 \end{bmatrix} = \frac{1}{(1-\rho^2)}\begin{bmatrix} \dfrac{1}{\sigma_1^2} & -\dfrac{\rho}{\sigma_1\sigma_2} \\ -\dfrac{\rho}{\sigma_1\sigma_2} & \dfrac{1}{\sigma_2^2} \end{bmatrix}$$

Therefore, $(\boldsymbol{\alpha} - \boldsymbol{\mu})^T\mathbf{V}^{-1}(\boldsymbol{\alpha} - \boldsymbol{\mu})$ is simply

$$\begin{aligned}(\boldsymbol{\alpha} - \boldsymbol{\mu})^T&\mathbf{V}^{-1}(\boldsymbol{\alpha} - \boldsymbol{\mu}) \\ &= \frac{1}{1-\rho^2}\left[\frac{(\alpha_1-\mu_1)^2}{\sigma_1^2} - \frac{2\rho(\alpha_1-\mu_1)(\alpha_2-\mu_2)}{\sigma_1\sigma_2} + \frac{(\alpha_2-\mu_2)^2}{\sigma_2^2}\right]\end{aligned} \tag{7.2-7}$$

and substituting Eqs. (7.2-6) and (7.2-7) into the general form of Eq. (7.2-5) gives the desired result of Eq. (7.2-2). In the scalar case, $n = 1$, Eq. (7.2-5) trivially reduces to Eq. (7.2-1) with $\sigma^2 = V$.

For the work in Section 7.3, as well as several other general uses, it is desirable to find the characteristic function for the multivariate gaussian distribution. By definition, the characteristic function is given by

$$\begin{aligned}\Phi_{\mathbf{x}}(\boldsymbol{\omega}) = \mathcal{E}\{e^{j\boldsymbol{\omega}^T\mathbf{x}}\} = \int_{-\infty}^{\infty} &\frac{1}{[(2\pi)^n \det \mathbf{V}]^{1/2}} \\ &\cdot \exp\left\{j\boldsymbol{\omega}^T\boldsymbol{\alpha} - \frac{1}{2}(\boldsymbol{\alpha} - \boldsymbol{\mu})^T\mathbf{V}^{-1}(\boldsymbol{\alpha} - \boldsymbol{\mu})\right\} d\boldsymbol{\alpha}\end{aligned} \tag{7.2-8}$$

If we make the substitution $\boldsymbol{\alpha} = \mathbf{V}^{1/2}\boldsymbol{\beta} + \boldsymbol{\mu}$ and remember the "change-of-variable" rule that $d\boldsymbol{\alpha} = \det(\mathbf{V}^{1/2})\, d\boldsymbol{\beta} = (\det \mathbf{V})^{1/2}\, d\boldsymbol{\beta}$, then Eq. (7.2-8) becomes

$$\Phi_{\mathbf{x}}(\boldsymbol{\omega}) = \frac{1}{(2\pi)^{n/2}}\int_{-\infty}^{\infty} \exp\left\{j\boldsymbol{\omega}^T(\mathbf{V}^{1/2}\boldsymbol{\beta} + \boldsymbol{\mu}) - \frac{1}{2}\boldsymbol{\beta}^T\boldsymbol{\beta}\right\} d\boldsymbol{\beta} \tag{7.2-9}$$

Here $\mathbf{V}^{1/2}$ is the positive definite symmetric square root of the matrix $\mathbf{V}$ such that $\mathbf{V}^{1/2}\mathbf{V}^{1/2} = \mathbf{V}$. To conveniently express Eq. (7.2-9), we add and subtract the quantity $\boldsymbol{\omega}^T\mathbf{V}\boldsymbol{\omega}$ inside the exponential so that we obtain

$$\Phi_{\mathbf{x}}(\boldsymbol{\omega}) = \exp\left\{j\boldsymbol{\omega}^T\boldsymbol{\mu} - \frac{1}{2}\boldsymbol{\omega}^T\mathbf{V}\boldsymbol{\omega}\right\}\int_{-\infty}^{\infty}\frac{1}{(2\pi)^{n/2}}$$

$$\cdot \exp\left\{-\frac{1}{2}(\boldsymbol{\beta} - j\mathbf{V}^{1/2}\boldsymbol{\omega})^T(\boldsymbol{\beta} - j\mathbf{V}^{1/2}\boldsymbol{\omega})\right\} d\boldsymbol{\beta} \qquad (7.2\text{-}10)$$

We recognize the integrand as a gaussian density with mean $j\mathbf{V}^{1/2}\boldsymbol{\omega}$ and variance $\mathbf{I}$ so that the integral is just unity, and we have the desired result:

$$\Phi_{\mathbf{x}}(\boldsymbol{\omega}) = \exp\{j\boldsymbol{\omega}^T\boldsymbol{\mu} - \tfrac{1}{2}\boldsymbol{\omega}^T\mathbf{V}\boldsymbol{\omega}\} \qquad (7.2\text{-}11)$$

Note that the characteristic function for the gaussian random process has the same form as the density function.

An important property of gaussian processes is that any nonsingular linear transformation (plus translation) of a gaussian process is still a gaussian process. To be more specific, if we let $\mathbf{y}$ be defined by

$$\mathbf{y} = \mathbf{A}\mathbf{x} + \mathbf{b} \qquad (7.2\text{-}12)$$

where $\mathbf{A}$ is an $n \times n$ nonsingular matrix and $\mathbf{x}$ is an n-dimensional gaussian process, then $\mathbf{y}$ is also an n-dimensional gaussian process.[1] This property is sometimes referred to as the reproducibility property in reference to the fact that if a gaussian process is passed through a linear system, the output is still gaussian. To establish this result, we make use of the general transformation result of Eq. (4.3-11) repeated here for easy reference:

$$p_{\mathbf{y}}(\boldsymbol{\beta}) = p_{\mathbf{x}}[\mathbf{f}^{-1}(\boldsymbol{\beta})]\,|J_I(\boldsymbol{\beta})| \qquad (4.3\text{-}11)$$

Here $\mathbf{f}(\mathbf{x})$ is given by

$$\mathbf{f}(\mathbf{x}) = \mathbf{A}\mathbf{x} + \mathbf{b}$$

so that $\mathbf{f}^{-1}(\mathbf{y})$ is

$$\mathbf{x} = \mathbf{A}^{-1}(\mathbf{y} - \mathbf{b})$$

Therefore, the jacobian of the inverse transformation becomes

$$J_I(\mathbf{y}) = \det\left[\frac{\partial \mathbf{f}^{-1}(\mathbf{y})}{\partial \mathbf{y}}\right] = \det \mathbf{A}^{-1} = (\det \mathbf{A})^{-1}$$

so that $p_{\mathbf{y}}(\boldsymbol{\beta})$ is

$$p_{\mathbf{y}}(\boldsymbol{\beta}) = \frac{1}{[(2\pi)^n \det \mathbf{V}]^{1/2}} \exp\left\{-\frac{1}{2}[\mathbf{A}^{-1}(\boldsymbol{\beta} - \mathbf{b}) - \boldsymbol{\mu}]^T\mathbf{V}^{-1}\right.$$

$$\left.\cdot[\mathbf{A}^{-1}(\boldsymbol{\beta} - \mathbf{b}) - \boldsymbol{\mu}]\right\} |(\det \mathbf{A})^{-1}|$$

[1] We examined the proof of this statement in Example 4.3-1 for $\mathbf{b} = \mathbf{0}$.

We may rewrite this result in the form

$$p_{\mathbf{y}}(\boldsymbol{\beta}) = \frac{1}{[(2\pi)^n \det(\mathbf{AVA}^T)]^{1/2}} \exp\left\{-\frac{1}{2}[\boldsymbol{\beta} - (\mathbf{b} + \mathbf{A}\boldsymbol{\mu})]^T \cdot(\mathbf{AVA}^T)^{-1}[\boldsymbol{\beta} - (\mathbf{b} + \mathbf{A}\boldsymbol{\mu})]\right\} \tag{7.2-13}$$

We recognize this result as an n-dimensional gaussian density with mean $\mathbf{b} + \mathbf{A}\boldsymbol{\mu}$ and variance $\mathbf{AVA}^T$. We shall demonstrate in Section 7.4 that this statement is true for all $\mathbf{A}$.

Once we know that $\mathbf{y}$ has a gaussian distribution, we can easily find its exact density without remembering the result of Eq. (7.2-13). To characterize a gaussian distribution, we need only determine the mean and variance. The mean of $\mathbf{y}$ is given by

$$\mathcal{E}\{\mathbf{y}\} = \mathcal{E}\{\mathbf{Ax} + \mathbf{b}\}$$

Since the expectation operator is a linear operator and $\mathbf{A}$ and $\mathbf{b}$ are nonrandom constants, we have

$$\mathcal{E}\{\mathbf{y}\} = \mathbf{A}\mathcal{E}\{\mathbf{x}\} + \mathbf{b} = \mathbf{A}\boldsymbol{\mu} + \mathbf{b}$$

The variance of $\mathbf{y}$ is therefore

$$\begin{aligned} \text{var}\{\mathbf{y}\} &= \mathcal{E}\{[\mathbf{y} - \mathcal{E}\{\mathbf{y}\}][\mathbf{y} - \mathcal{E}\{\mathbf{y}\}]^T\} \\ &= \mathcal{E}\{[\mathbf{Ax} + \mathbf{b} - \mathbf{A}\boldsymbol{\mu} - \mathbf{b}][\mathbf{Ax} + \mathbf{b} - \mathbf{A}\boldsymbol{\mu} - \mathbf{b}]^T\} \\ &= \mathcal{E}\{[\mathbf{A}(\mathbf{x} - \boldsymbol{\mu})][\mathbf{A}(\mathbf{x} - \boldsymbol{\mu})]^T\} \end{aligned}$$

And, once again, since $\mathbf{A}$ is a nonrandom constant, we have

$$\text{var}\{\mathbf{y}\} = \mathbf{A}\mathcal{E}\{(\mathbf{x} - \boldsymbol{\mu})(\mathbf{x} - \boldsymbol{\mu})^T\}\mathbf{A}^T = \mathbf{AVA}^T$$

Hence we see that $\mathbf{y}$ is gaussian with mean $\mathbf{A}\boldsymbol{\mu} + \mathbf{b}$ and variance $\mathbf{AVA}^T$, and its density is therefore easily found to be given by Eq. (7.2-13).

In much work concerned with error analysis of measurements of random processes and in nonlinear filtering, there is the need to know the fourth-product moment of a gaussian random variable. In other words, we wish to find an expression for $\mathcal{E}\{x_1x_2x_3x_4\}$, where the $\mathbf{x} = [x_1\ x_2\ x_3\ x_4]^T$ is a four-dimensional gaussian process. By using the approach developed in Section 4.5, it is quite easy to show that the fourth-product moment may be obtained from the characteristic function of $\mathbf{x}$ in the following manner:

$$\mathcal{E}\{x_1x_2x_3x_4\} = \left.\frac{\partial^4 \Phi_{\mathbf{x}}(\boldsymbol{\omega})}{\partial\omega_1\,\partial\omega_2\,\partial\omega_3\,\partial\omega_4}\right|_{\boldsymbol{\omega}=0} \tag{7.2-14}$$

The characteristic function for this case takes the form

$$\Phi_{\mathbf{x}}(\boldsymbol{\omega}) = \exp\left\{ j \sum_{i=1}^{4} \omega_i \mu_i - \tfrac{1}{2} \sum_{i=1}^{4} \sum_{j=1}^{4} \omega_i \omega_j V_{ij} \right\}$$

where $\mu_i = \mathcal{E}\{x_i\}$ and $V_{ij} = \mathrm{cov}\{x_i, x_j\}$. If one carries out the routine, although somewhat tedious, algebra indicated by Eq. (7.2-14) with this characteristic function, one finds that the fourth-product moment takes the form:

$$\begin{aligned} \mathcal{E}\{x_1 x_2 x_3 x_4\} = {} & V_{12}V_{34} + V_{12}\mu_3\mu_4 + V_{34}\mu_1\mu_2 + V_{13}V_{24} \\ & + V_{13}\mu_2\mu_4 + V_{24}\mu_1\mu_3 + V_{14}V_{23} + V_{14}\mu_2\mu_3 \\ & + V_{23}\mu_1\mu_4 + \mu_1\mu_2\mu_3\mu_4 \end{aligned} \tag{7.2-15}$$

For zero-mean processes, this expression reduces to the simple relation

$$\mathcal{E}\{x_1 x_2 x_3 x_4\} = V_{12}V_{34} + V_{13}V_{24} + V_{14}V_{23} \tag{7.2-16}$$

The result of Eq. (7.2-15) can be more conveniently represented in the form:

$$\begin{aligned} \mathcal{E}\{x_1 x_2 x_3 x_4\} & = \mathcal{E}\{x_1 x_2\}\mathcal{E}\{x_3 x_4\} + \mathcal{E}\{x_1 x_3\}\mathcal{E}\{x_2 x_4\} \\ & \quad + \mathcal{E}\{x_1 x_4\}\mathcal{E}\{x_2 x_3\} - 2\mathcal{E}\{x_1\}\mathcal{E}\{x_2\}\mathcal{E}\{x_3\}\mathcal{E}\{x_4\} \\ & = P_{12}P_{34} + P_{13}P_{24} + P_{14}P_{23} - 2\mu_1\mu_2\mu_3\mu_4 \end{aligned} \tag{7.2-17}$$

as can be easily verified by direct substitution.

Example *7.2-1.* Let us use the expressions for the fourth product moment to find the covariance of $z(t) = x(t)x(t + \tau)$, where $x(t)$ is a zero-mean gaussian random process with $\mathrm{var}\{x(t)\} = V_x(t)$ and $\mathrm{cov}\{x(t), x(\tau)\} = V_x(t, \tau)$, so that

$$\begin{aligned} V_z(s, t) & = \mathrm{cov}\{x(s)x(s + \tau), x(t)x(t + \tau)\} \\ & = \mathcal{E}\{x(s)x(s + \tau)x(t)x(t + \tau)\} - \mathcal{E}\{x(s)x(s + \tau)\} \\ & \quad \cdot \mathcal{E}\{x(t)x(t + \tau)\} \end{aligned}$$

The use of the fourth-product moment expression yields

$$\begin{aligned} V_z(s, t) & = V_x(s, s + \tau)V_x(t, t + \tau) + V_x(s, t)V_x(s + \tau, t + \tau) \\ & \quad + V_x(s, t + \tau)V_x(s + \tau, t) - V_x(s, s + \tau)V_x(t, t + \tau) \\ & = V_x(s, t)V_x(s + \tau, t + \tau) + V_x(s, t + \tau)V_x(s + \tau, t) \end{aligned}$$

The fourth-product-moment relation can also be used to find the third-product-moment expression. This can be accomplished by letting one of the random variables, say x_4, be a nonrandom constant of value 1. Then $\mathcal{E}\{x_1x_2x_3x\} = \mathcal{E}\{x_1x_2x_3\}\mathcal{E}\{x_4\} = \mathcal{E}\{x_1x_2x_3\}$. If x_4 is identically equal to unity, then $\mu_4 = 1$ and all the variance terms involving x_4 become zero; so Eq. (7.2-15) is now

$$\mathcal{E}\{x_1x_2x_3\} = \mu_3V_{12} + \mu_2V_{13} + \mu_1V_{23} + \mu_1\mu_2\mu_3$$

This is the desired third-product expression. Note that the third-product moment is zero if $\boldsymbol{\mu} = \mathbf{0}$. (In fact all odd-product moments of a gaussian process, or any process with a symmetric density function, are zero if $\boldsymbol{\mu} = \mathbf{0}$.)

Example *7.2-2.* This example presents a brief discussion of the errors involved in measuring cross-correlation functions and covariances of two scalar stationary zero-mean random processes, $x(t)$ and $y(t)$. The operation of a correlator involves, in part, the multiplication of $x(t)$ and $y(t + \tau)$ to obtain

$$z(t) = x(t)y(t + \tau)$$

The multiplier, with output $z(t)$, is normally followed by a finite time integrator such that the approximation to the cross-correlation function is obtained as

$$\mathcal{E}\{x(t)y(t + \tau)\} = \frac{1}{T}\int_0^T x(t)y(t + \tau)\, dt$$

If we take the expected value of this result, we have

$$\mathcal{E}\left\{\frac{1}{T}\int_0^T x(t)y(t + \tau)\, dt\right\} = \frac{1}{T}\int_0^T R_{xy}(\tau)\, dt = R_{xy}(\tau)$$

The cross-correlation output is said to be *unbiased*, since the expected value of the output is the quantity we are trying to measure. The variance of the correlator output is easily obtained from the fourth-product-moment expression:

$$\begin{aligned}\operatorname{cov}\{z(t), z(\lambda)\} &= \operatorname{cov}\{x(t)y(t + \tau), x(\lambda)y(\lambda + \tau)\} \\ &= \operatorname{cov}\{x(t), x(\lambda)\}\operatorname{cov}\{y(t + \tau), y(\lambda + \tau)\} \\ &\quad + \operatorname{cov}\{x(t), y(\lambda + \tau)\}\operatorname{cov}\{y(t + \tau), x(\lambda)\}\end{aligned}$$

so that

$$\operatorname{cov}\{z(t), z(\lambda)\} = R_x(\lambda - t)R_y(\lambda - t) + R_{xy}(\lambda + \tau - t)R_{yx}(\lambda - t - \tau)$$

The expression for the variance of the error in the estimate of the correlation function is the variance expression

$$V_e(\tau) = \text{var}\left\{\frac{1}{T}\int_0^T z(t)\,dt\right\} = \frac{1}{T^2}\int_0^T\int_0^T \text{cov}\{z(t), z(\lambda)\}\,dt\,d\lambda$$

Thus, for any given gaussian random processes $x(t)$ and $y(t)$ whose first two moments are known, we can determine the error variance associated with the correlation measurement as a function of the integration time.

For example, if $y(t) = x(t)$ and $R_x(\tau) = \exp\{-|\tau|\}$, then

$$\text{cov}\{z(t), z(\lambda)\} = e^{-2|t-\lambda|} + e^{-|t-\lambda-\tau|}e^{-|t+\tau-\lambda|}$$

and the error variance depends upon the age variable τ. The "worst case" occurs where $\tau = 0$, and for this value of τ,

$$\text{cov}\{z(t), z(\lambda)\} = 2e^{-2|t-\lambda|}$$

$$V_e(0) = \frac{1}{T^2}\int_0^T\int_0^T 2e^{-2|t-\lambda|}\,dt\,d\lambda = \frac{2}{T} - \frac{1 - e^{-2T}}{T^2}$$

Therefore, the variance of the error in estimation of the correlation function decreases approximately as the inverse of T, the integration time. Much more discussion and amplification of these results concerning measurement of random-process characteristics is presented in Bendat and Piersol [1].

Because of the wide use that is made of gaussian processes, one often encounters functions of gaussian processes. In Section 7.4 it will be shown that any linear function of a gaussian process is another gaussian process. In Section 4.3 we demonstrated that if x_1 and x_2 are zero-mean identically distributed gaussian processes, then y_1 given by

$$y_1 = \tan^{-1}\left(\frac{x_2}{x_1}\right)$$

has a uniform distribution on $[-\pi/2, \pi/2]$, whereas

$$y_2 = \sqrt{x_1^2 + x_2^2}$$

has a Rayleigh distribution with parameter σ^2.

Other functions often of interest are the sample mean

$$\bar{\mu} = \frac{1}{n}\sum_{i=1}^{n} x_i \tag{7.2-19}$$

and sample variance

$$\bar{\sigma}^2 = \frac{1}{n}\sum_{i=1}^{n}(x_i - \bar{\mu}^2) \tag{7.2-20}$$

and the norm

$$y = \sqrt{x_1^2 + x_2^2 + \cdots + x_n^2} \tag{7.2-21}$$

and the norm squared

$$z = y^2 = \sum_{i=1}^{n} x_i^2 \tag{7.2-22}$$

Here we assume that the x_i's, $i = 1, 2, \ldots, n$, are scalar, zero-mean, unit variance independent gaussian processes. If the x_i's do not have zero mean and unit variance, they may be put into this form by using the transformation

$$x_i' = \frac{x_i - \mu_i}{\sigma_i}$$

where μ_i is the mean and σ_i is the variance of x_i. We wish to find the density function for $\bar{\mu}$, $\bar{\sigma}^2$, y, and z.

By using the results of Section 7.4, we can conclude that the density of $\bar{\mu}$ is gaussian. Hence we need only find the mean and variance of $\bar{\mu}$ to determine its density function. The mean of $\bar{\mu}$ is given by

$$\mathcal{E}\{\bar{\mu}\} = \frac{1}{n}\sum_{i=1}^{n}\mathcal{E}\{x_i\} = 0$$

The variance is

$$\text{var}\{\bar{\mu}\} = \frac{1}{n^2}\sum_{i=1}^{n}\text{var}\{x_i\} = \frac{1}{n}$$

since the x_i's are independent. The density function of $\bar{\mu}$ is therefore

$$p_{\bar{\mu}}(\alpha) = \sqrt{\frac{n}{2\pi}}\exp\left\{\frac{-n\alpha^2}{2}\right\} \tag{7.2-23}$$

We ignore for the moment $\bar{\sigma}^2$ and consider next y given by Eq. (7.2-21). If we make use of Eq. (4.3-8), we may write, for $a > 0$,

$$\mathcal{P}\{a < y < a + da\} = p_y(a)\,da = \mathcal{P}\left\{a < \sqrt{\sum_{i=1}^{n} x_i^2} < a + da\right\}$$

The quantity on the extreme right of this expression is the probability that $x = [x_1\ x_2\ \cdots\ x_n]^T$ will lie in the hypershell of inner radius a and outer radius $a + da$. Normally, to find this probability it would be necessary to integrate $p_{\mathbf{x}}(\boldsymbol{\alpha})$ over the region of the hypershell. In this case, the procedure is considerably simpler because $p_{\mathbf{x}}(\boldsymbol{\alpha})$ is constant on any hypershell in the $\mathbf{x}$ space so that, since $y = \|\mathbf{x}\|$,

$$\mathcal{P}\{a < \|\mathbf{x}\| < a + da\} = \int_{a<\|\boldsymbol{\alpha}\|<a+da} p_{\mathbf{x}}(\boldsymbol{\alpha})\, \mathbf{d}\boldsymbol{\alpha} = p_{\mathbf{x}}(\boldsymbol{\alpha})\Big|_{\|\boldsymbol{\alpha}\|=a} (\text{Vol})$$

$$= \frac{\text{Vol}}{(2\pi)^{n/2}} e^{-a^2/2}$$

where Vol is the volume of the region $a < \|\mathbf{x}\| < a + da$. From basic concepts of geometry it can be shown that

$$\text{Vol} = 2\left(\frac{\pi}{2}\right)^{n/2}\left[\Gamma\left(\frac{n}{2}\right)\right]^{-1} a^{n-1}\, da$$

where $\Gamma(\alpha)$ is the gamma function, so that

$$p_y(a) = \frac{2}{2^{n/2}\Gamma(n/2)} a^{n-1} e^{-a^2/2} \mathcal{u}(a) \tag{7.2-24}$$

Since $z = y^2$, we can make use of the results of Example 4.3-5 to write the density of z as

$$p_z(\beta) = \frac{1}{2\sqrt{\beta}}[p_y(\sqrt{\beta}) + p_y(-\sqrt{\beta})]\mathcal{u}(\beta)$$

Since $p_y(a) = 0$ for $\alpha < 0$, we have

$$p_z(\beta) = \frac{1}{2\sqrt{\beta}} p_y(\sqrt{\beta})\mathcal{u}(\beta)$$

which becomes

$$p_z(\beta) = \frac{1}{2^{n/2}\Gamma(n/2)} \beta^{n/2-1} e^{-\beta/2} \mathcal{u}(\beta) \tag{7.2-25}$$

We can recognize[1] this distribution as a chi-square distribution with n degrees of freedom.

By a rather involved algebraic procedure, it is possible to show that the sample variance has an unnormalized chi-squared distribution with $n - 1$ degrees of freedom so that $p_{\bar{\sigma}^2}$ becomes

$$p_{\bar{\sigma}^2}(\alpha) = \frac{n^{n-1}}{2^{(n-1)/2}\Gamma\left(\frac{n-1}{2}\right)} \beta^{(n-3)/2} e^{-n\alpha/2} \mathcal{u}(\alpha) \tag{7.2-26}$$

[1] See Table 3.5-1.

EXERCISES 7.2

7.2-1. We are often interested in the probability that the gaussian vector $\mathbf{x}$ lies inside the N-dimensional ellipsoid:

$$(\mathbf{x} - \boldsymbol{\mu}_{\mathbf{x}})^T \mathbf{V}_{\mathbf{x}}^{-1} (\mathbf{x} - \boldsymbol{\mu}_{\mathbf{x}}) = r^2$$

for constant r. Show that a linear transform of variable

$$\mathbf{y} = \mathbf{A}\mathbf{x} + \mathbf{b}$$

transforms this to

$$\sum_{n=1}^{N} \frac{y_n^2}{\sigma_n^2} = r^2$$

and that the transformation $z_n = y_n/\sigma_n$ results in the N-dimensional sphere

$$\sum_{n=1}^{N} y_n^2 = r^2$$

By using these transformations, show that the probability is

(a) for $N = 1$, $-\dfrac{1}{\sqrt{2\pi}} + 2 \operatorname{erf} r$.

(b) for $N = 2$, $1 - \exp(-0.5r^2)$.

(c) for $N = 3$, $-\dfrac{1}{\sqrt{\pi}} + 2 \operatorname{erf} r - \sqrt{\dfrac{2}{\pi}}\, r \exp(-0.5r^2)$.

7.2-2. Derive the fourth-product-moment expression of Eq. (7.2-15).

7.2-3. The output of a system is

$$z = x_1 + x_2 + x_3$$

where the gaussian vector $\mathbf{x} = [x_1 \quad x_2 \quad x_3]^T$ has the moments

$$\boldsymbol{\mu}_{\mathbf{x}} = \begin{bmatrix} \mu_1 \\ \mu_2 \\ \mu_3 \end{bmatrix} \qquad \mathbf{V}_{\mathbf{x}} = \begin{bmatrix} V_{11} & V_{12} & V_{13} \\ V_{12} & V_{22} & V_{23} \\ V_{13} & V_{23} & V_{33} \end{bmatrix}$$

What is the mean and variance of z?

Answer:

$$\mu_z = \mu_1 + \mu_2 + \mu_3$$

$$V_z = \sum_{i=1}^{3} \sum_{j=1}^{3} V_{ij}$$

7.2-4. Show that the sample variance

$$\bar{\sigma}^2 = \frac{1}{n} \sum_{i=1}^{n} (x_i - \bar{\mu})^2$$

has an unnormalized chi-squared distribution with $n - 1$ degrees of freedom.

7.3. Central limit theorem

There are many limit theorems in probability and statistics. The *central limit theorem* is a generic name referring to many theorems having essentially the same result but slightly different initial hypotheses (Dubes [3]). We shall use only one of them here. Let $\mathbf{x}^i$ denote a member of a set of K identically distributed independent zero-mean N-vector random variables with finite variance $\mathbf{V}_\mathbf{x}$. If we define a sequence of random variables $\{\mathbf{z}\}_K$ by

$$\mathbf{z}_K = \frac{1}{\sqrt{K}} \sum_{i=1}^{K} \mathbf{x}^i \tag{7.3-1}$$

then the sequence $\{\mathbf{z}_K\}$ *converges in distribution*[1] to a gaussian distribution with zero mean and variance $\mathbf{V}_\mathbf{x}$. In other words, we have

$$\lim_{K\to\infty} F_{\mathbf{z}_K}(\boldsymbol{\alpha}) = \int_{-\infty}^{\boldsymbol{\alpha}} \frac{1}{[(2\pi)^N \det \mathbf{V}_\mathbf{x}]^{1/2}} \exp\left\{-\frac{1}{2}\boldsymbol{\beta}^T\mathbf{V}_\mathbf{x}^{-1}\boldsymbol{\beta}\right\} d\boldsymbol{\beta} \tag{7.3-2}$$

This result may or may not imply that

$$\lim_{K\to\infty} p_{\mathbf{z}_K}(\boldsymbol{\alpha}) = \frac{1}{[(2\pi)^N \det \mathbf{V}_\mathbf{x}]^{1/2}} \exp\left\{-\frac{1}{2}\boldsymbol{\alpha}^T\mathbf{V}_\mathbf{x}^{-1}\boldsymbol{\alpha}\right\}$$

If, for example, the $\mathbf{x}^i$'s are discrete random variables, then $\mathbf{z}_K$ for each K will also be discrete, and the density function of $\mathbf{z}_K$ will not approach the gaussian density function as K increases, although the distribution function will approach the normal distribution. Figure 7.3-1 indicates this statement in a graphical form.

To establish the central limit theorem, we shall use the fact, stated in Section 4.6, that convergence in distribution may be established by proving the convergence of the characteristic functions. Hence we wish to show that

$$\lim_{K\to\infty} \Phi_{\mathbf{z}_K}(\boldsymbol{\omega}) = \exp\{-\tfrac{1}{2}\boldsymbol{\omega}^T\mathbf{V}_\mathbf{x}\boldsymbol{\omega}\} \tag{7.3-3}$$

The characteristic function of $\mathbf{z}_K$ is given by

$$\Phi_{\mathbf{z}_K}(\boldsymbol{\omega}) = \mathcal{E}\{e^{j\boldsymbol{\omega}^T\mathbf{z}_K}\}$$

[1] See Section 4.6.

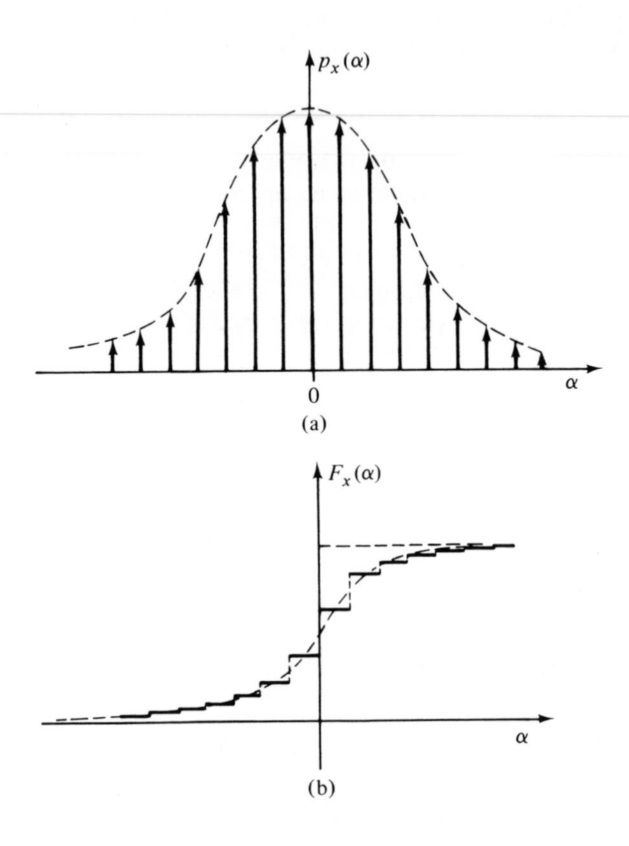

FIG. 7.3-1 Discrete density and distribution functions that limit in distribution to a gaussian distribution.

Upon substitution of Eq. (7.3-1), we have

$$\begin{aligned}\Phi_{\mathbf{z}_K}(\boldsymbol{\omega}) &= \mathcal{E}\left\{\exp\left[j\boldsymbol{\omega}^T \sum_{i=1}^{X} \frac{\mathbf{x}^i}{\sqrt{K}}\right]\right\} \\ &= \mathcal{E}\left\{\prod_{i=1}^{K} \exp \frac{j\boldsymbol{\omega} T \mathbf{x}^i}{\sqrt{K}}\right\} \end{aligned} \tag{7.3-4}$$

Because the $\mathbf{x}^i$'s are independent, the random variables $\exp(j\boldsymbol{\omega}^T\mathbf{x}^i/\sqrt{K})$ are also independent, and we may use the fact that the expectation of the product of independent random variables is the product of the expectations. Therefore, Eq. (7.3-4) may be written as

$$\Phi_{\mathbf{z}_K}(\boldsymbol{\omega}) = \prod_{i=1}^{K} \mathcal{E}\left\{\frac{j\boldsymbol{\omega}^T\mathbf{x}^i}{\sqrt{K}}\right\} \tag{7.3-5}$$

If we define the characteristic function of $\mathbf{x}^i$, $i = 1, 2, \ldots, K$, as $\Phi_{\mathbf{x}}(\boldsymbol{\omega})$, which is not a function of i since the $\mathbf{x}^i$ are identically distributed, we see that

$$\mathcal{E}\left\{\frac{j\boldsymbol{\omega}^T\mathbf{x}^i}{\sqrt{K}}\right\} = \Phi_{\mathbf{x}}\left(\frac{\boldsymbol{\omega}}{\sqrt{K}}\right)$$

Using this result, we may express Eq. (7.3-5) as

$$\Phi_{\mathbf{z}_K}(\omega) = \prod_{i=1}^{K} \Phi_{\mathbf{x}}\left(\frac{\boldsymbol{\omega}}{\sqrt{K}}\right) = \left[\Phi_{\mathbf{x}}\left(\frac{\omega}{\sqrt{K}}\right)\right]^K \tag{7.3-6}$$

Thus we could determine the characteristic function of $\mathbf{z}_K$ if we knew the characteristic function of $\mathbf{x}$. Here we do not know $\Phi_{\mathbf{x}}(\boldsymbol{\omega})$ and thus must obtain the limiting behavior of $\Phi_{\mathbf{z}_K}(\boldsymbol{\omega})$ without this knowledge.

Using the methods of Section 4.5, we may expand the characteristic function $\Phi_{\mathbf{x}}(\boldsymbol{\omega})$ in the following form:

$$\Phi_{\mathbf{x}}(\boldsymbol{\omega}) = 1 + j\boldsymbol{\omega}^T\boldsymbol{\mu}_{\mathbf{x}} - \frac{\boldsymbol{\omega}^T\mathbf{V}_{\mathbf{x}}\boldsymbol{\omega}}{2} + \xi \tag{7.3-7}$$

Here ξ is the summation of all the "higher power" terms involving third-order and higher moments and third and higher powers of $\boldsymbol{\omega}$. Since we have assumed that the $\mathbf{x}^i$'s have zero mean, $\boldsymbol{\mu}_{\mathbf{x}} = \mathbf{0}$, so that $\Phi_{\mathbf{x}}(\boldsymbol{\omega}/\sqrt{K})$ becomes

$$\Phi_{\mathbf{x}}\left(\frac{\boldsymbol{\omega}}{\sqrt{K}}\right) = 1 - \frac{\boldsymbol{\omega}^T\mathbf{V}_{\mathbf{x}}\boldsymbol{\omega}}{2K} + \xi' \tag{7.3-8}$$

where ξ' involves third or higher powers of $\boldsymbol{\omega}/\sqrt{K}$. If we take the logarithm of Eq. (7.3-6), we have

$$\ln \Phi_{\mathbf{z}_K}(\boldsymbol{\omega}) = K \ln\left[\Phi_{\mathbf{x}}\left(\frac{\boldsymbol{\omega}}{\sqrt{K}}\right)\right] = K \ln\left[1 - \frac{\boldsymbol{\omega}^T\mathbf{V}_{\mathbf{x}}\boldsymbol{\omega}}{2K} + \xi'\right] \tag{7.3-9}$$

If we make a Taylor series expansion of $\ln(1 - \eta)$ about $\eta = 0$, we obtain

$$\ln(1 - \eta) = -\eta + \frac{\eta^2}{2} - \frac{\eta^3}{3} + \cdots$$

Applying this expression to Eq. (7.3-9) with $\eta = \boldsymbol{\omega}^T\mathbf{V}_{\mathbf{x}}\boldsymbol{\omega}/2K - \xi'$, we obtain

$$\ln \Phi_{\mathbf{z}_K}(\boldsymbol{\omega}) = K\left\{\frac{-\boldsymbol{\omega}^T\mathbf{V}_{\mathbf{x}}\boldsymbol{\omega}}{2K} + \xi' + \frac{1}{2}\left[\frac{\boldsymbol{\omega}^T\mathbf{V}_{\mathbf{x}}\boldsymbol{\omega}}{2K} - \xi'\right]^2 + \text{higher order terms}\right\}$$

Now if we take the limit as K approaches infinity, we find that

$$\lim_{K\to\infty} \ln \Phi_{\mathbf{z}_K}(\boldsymbol{\omega}) = \frac{-\boldsymbol{\omega}^T \mathbf{V}_{\mathbf{x}} \boldsymbol{\omega}}{2}$$

since $\lim_{K\to\infty} K\xi' = 0$. By taking the inverse logarithm of this expression and interchanging the limit and inverse logarithm operation, we obtain the desired result:

$$\lim_{K\to\infty} \Phi_{\mathbf{z}_K}(\boldsymbol{\omega}) = \exp\left\{\frac{-\boldsymbol{\omega}^T \mathbf{V}_{\mathbf{x}} \boldsymbol{\omega}}{2}\right\}$$

It should be noted that we have placed no restriction on the distribution of the $\mathbf{x}^i$'s, except that $\mathbf{V}_{\mathbf{x}}$ be finite; in particular we emphasize that the distribution need *not* be gaussian. Thus we reach the important conclusion that almost any process obtained by adding a large number of independent zero-mean random (not necessarily gaussian) processes will be gaussian. If, in fact, the $\mathbf{x}^i$'s do have a gaussian distribution, then $\mathbf{z}_K$ will be exactly gaussian for all K.[1]

The central limit theorem is useful in estimating probabilities for the sums of nongaussian distributions. Let us consider the scalar case; then we may approximate (for finite K) the probability that $|z_K|$ exceeds ϵ as

$$\mathcal{P}\left\{|z_K| = \frac{1}{\sqrt{K}}\left|\sum_{i=1}^{K} x^i\right| \geq \epsilon\right\} \sim \int_{\epsilon}^{\infty} \frac{1}{\sqrt{2\pi V_x}} e^{-\alpha^2/2V_x}\, d\alpha = 1 - \operatorname{erf}\left(\frac{\epsilon}{V_x}\right) \tag{7.3-10}$$

This may be a good or a bad approximation. If x^i is gaussian, the approximation becomes an equality. If x^i is uniformly distributed from -1 to $+1$, then

$$\mathcal{P}\{|z_K| \geq \epsilon\} = 0$$

if $\epsilon > \sqrt{K}$, since the maximum value that $\sum_{i=1}^{K} x^i/\sqrt{K}$ can take is $\sqrt{K}$. Using Eq. (7.3-10) may lead to serious error, depending upon the use to which the resulting probability expression is put. For $\epsilon < \sqrt{K}$, the approximation is quite good, however, for modest values of K. It is often true that the approximation obtained from the central limit theorem does not work too well at the "tails" of the distribution but is excellent elsewhere. It should be obvious that the approximation of Eq. (7.3-10) would be more accurate as K increases.

Example ***7.3-1.*** We naturally wonder just how large K must be before the sum of nongaussian random variables approaches a gaussian ran-

[1] See Section 7.4.

dom variable. The answer obviously depends upon the actual probability density of the nongaussian random variable, but often K as small as 3 or 4 will result in an essentially gaussian distribution. To illustrate this, let us suppose that

$$p_x(\beta) = \begin{cases} 0.5 & -1 < \beta < 1 \\ 0 & \text{otherwise} \end{cases}$$

for which $\mu_x = 0$ and $V_x = \frac{1}{3}$. We form the random variable z_K as

$$z_K = \frac{1}{\sqrt{K}} \sum_{i=1}^{K} x^i$$

The actual characteristic function of z_K is easily determined from Eq. (7.3-6) and Table 4.5-1 as

$$\Phi_{z_K}(\omega) = \left[\frac{\sqrt{K}}{\omega} \sin \frac{\omega}{\sqrt{K}}\right]^K$$

The density function $p_{z_K}(\alpha)$ is determined by taking the inverse transform of the foregoing characteristic function so that

$$p_{z_K}(\alpha) = \frac{1}{2\pi} \int_{-\infty}^{\infty} \Phi_{z_K}(\omega) e^{-j\omega\alpha} \, d\omega$$

Figure 7.3-2 illustrates the density function $p_{z_K}(\alpha)$, for $K = 1, 2, 3,$ and ∞. As is apparent, the density function for $K = 3$ is quite close to the gaussian density function ($K = \infty$). Thus for K greater than 4 or 5, we would expect the approximation of Eq. (7.3-10) to be quite good, except for $\epsilon > \sqrt{K}$.

EXERCISES 7.3

7.3-1. Consider the density

$$p_x(\beta) = p^\beta q^{1-\beta} \qquad \beta = 0, 1$$

Show that the first two moments are $\mu_x = p$ and $\sigma_x^2 = pq$. From

$$z = \frac{1}{\sqrt{K}} \sum_{i=1}^{K} (x^i - \mu_x)$$

find μ_z and σ_z^2. Show that as K increases, the *distribution* function $F_z(\alpha)$ becomes gaussian. Is the density of z, $p_z(\alpha)$ gaussian? Illustrate your results graphically for several representative values of K.

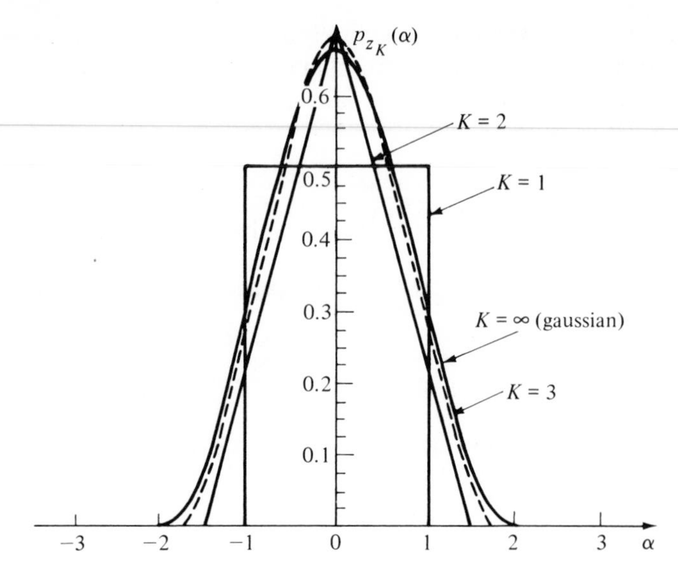

FIG. 7.3-2 Example 7.3-1, $p_{z_K}(\alpha)$ versus α for $K = 1, 2, 3, \infty$.

7.3-2. Show that the sum of two independent random variables with identical Cauchy distributions is a Cauchy distribution. Why doesn't this result obey the central limit theorem?

7.4. Conditional gaussian random processes

In the two preceding sections attention has been directed to the study of a single multivariate gaussian process. In this section we wish to examine the properties associated with sets (in particular, pairs) of multivariate gaussian processes. The major emphasis of this study will be on conditional densities, since they play such an important role in the fields of estimation and detection theory.

Let us suppose that we have an n-dimensional random-vector gaussian process $\mathbf{x}$ whose density function is given by

$$p_{\mathbf{x}}(\boldsymbol{\alpha}) = \frac{1}{[(2\pi)^n \det \mathbf{V}]^{1/2}} \exp\left\{-\frac{1}{2}(\boldsymbol{\alpha} - \boldsymbol{\mu})^T \mathbf{V}^{-1}(\boldsymbol{\alpha} - \boldsymbol{\mu})\right\} \qquad (7.4\text{-}1)$$

Now suppose that $\mathbf{x}$ is partitioned into two lower-dimensional vectors $\mathbf{x}_1$ and $\mathbf{x}_2$, where $\mathbf{x}_1$ is $(m < n)$-dimensional and $\mathbf{x}_2$ is $(n - m)$-dimensional so that

$$\mathbf{x} = \begin{bmatrix} \mathbf{x}_1 \\ \hdashline \mathbf{x}_2 \end{bmatrix} \tag{7.4-2}$$

In a similar manner, $\boldsymbol{\mu}$ and $\mathbf{V}$ may be partitioned as

$$\boldsymbol{\mu} = \begin{bmatrix} \boldsymbol{\mu}_1 \\ \hdashline \boldsymbol{\mu}_2 \end{bmatrix} \qquad \mathbf{V} = \begin{bmatrix} \mathbf{V}_{11} & \mathbf{V}_{12} \\ \mathbf{V}_{12}^T & \mathbf{V}_{22} \end{bmatrix} \tag{7.4-3}$$

where $\boldsymbol{\mu}_1 = \mathcal{E}\{\mathbf{x}_1\}$, $\boldsymbol{\mu}_2 = \mathcal{E}\{\mathbf{x}_2\}$, $\mathbf{V}_{11} = \text{var}\{\mathbf{x}_1\}$, $\mathbf{V}_{22} = \text{var}\{\mathbf{x}_2\}$, and $\mathbf{V}_{12} = \text{cov}\{\mathbf{x}_1, \mathbf{x}_2\}$. The vectors $\mathbf{x}_1$ and $\mathbf{x}_2$ are said to be *jointly gaussian*, since their joint density function given by Eq. (7.4-1) is gaussian. We wish to show that the two vectors are also *marginally* gaussian; that is, the marginal densities $p_{\mathbf{x}_1}(\boldsymbol{\alpha}_1)$ and $p_{\mathbf{x}_2}(\boldsymbol{\alpha}_2)$ are also gaussian. This fact, which may appear obvious, is quite easy to establish in a number of ways. We shall consider only one of the simpler approaches here.

The characteristic function of $\mathbf{x}$ is given by

$$\Phi_{\mathbf{x}}(\boldsymbol{\omega}) = \exp\{+j\boldsymbol{\omega}^T\boldsymbol{\mu} - \tfrac{1}{2}\boldsymbol{\omega}^T\mathbf{V}\boldsymbol{\omega}\}$$

In terms of the partitioned representation of $\mathbf{x}$, this becomes

$$\begin{aligned}\Phi_{\mathbf{x}_1\mathbf{x}_2}(\boldsymbol{\omega}_1, \boldsymbol{\omega}_2) = \exp\{&j\boldsymbol{\omega}_1^T\boldsymbol{\mu}_1 + j\boldsymbol{\omega}_2\boldsymbol{\mu}_2 - \tfrac{1}{2}\boldsymbol{\omega}_1^T\mathbf{V}_{11}\boldsymbol{\omega}_1 \\ &- \boldsymbol{\omega}_1^T\mathbf{V}_{12}\boldsymbol{\omega}_2 - \tfrac{1}{2}\boldsymbol{\omega}_2^T\mathbf{V}_{22}\boldsymbol{\omega}_2\}\end{aligned} \tag{7.4-4}$$

In Section 4.5 it was shown that a marginal characteristic function could be obtained from the joint characteristic function by setting the components of $\boldsymbol{\omega}$ associated with the unwanted variable equal to zero. In the case of interest here, we can obtain $\Phi_{\mathbf{x}_1}(\boldsymbol{\omega}_1)$ by letting $\boldsymbol{\omega}_2$ equal zero so that we have

$$\Phi_{\mathbf{x}_1}(\boldsymbol{\omega}_1) = \exp\{j\boldsymbol{\omega}_1^T\boldsymbol{\mu}_1 - \tfrac{1}{2}\boldsymbol{\omega}_1^T\mathbf{V}_{11}\boldsymbol{\omega}_1\} \tag{7.4-5}$$

This result reveals that $\mathbf{x}_1$ is a gaussian random variable with mean $\boldsymbol{\mu}_1$ and variance $\mathbf{V}_{11}$. In a similar manner, we could demonstrate that $\mathbf{x}_2$ is a gaussian process with mean $\boldsymbol{\mu}_2$ and variance $\mathbf{V}_{22}$ so that its density function is given by

$$p_{\mathbf{x}_2}(\boldsymbol{\alpha}_2) = \frac{1}{[(2\pi)^{n-m} \det \mathbf{V}_{22}]^{1/2}} \exp\left\{-\frac{1}{2}(\boldsymbol{\alpha}_2 - \boldsymbol{\mu}_2)^T\mathbf{V}_{22}^{-1}(\boldsymbol{\alpha} - \boldsymbol{\mu}_2)\right\} \tag{7.4-6}$$

Hence we may conclude that if $\mathbf{x}_1$ and $\mathbf{x}_2$ are jointly gaussian, they are also marginally gaussian. Since the partitioning of $\mathbf{x}$ was completely arbitrary, this result indicates that any subset of the components of $\mathbf{x}$ will be gaussian. The reader is cautioned *not* to conclude that the converse of this statement is true. It is quite possible for any or, in fact, all subsets of the components of $\mathbf{x}$ to be gaussian and for $\mathbf{x}$ to be nongaussian. In other words, in terms of the above partition, it is possible for $\mathbf{x}_1$ and $\mathbf{x}_2$ to be gaussian and $\mathbf{x}$ may still be nongaussian.

As the next step, we wish to show that the conditional density $p_{\mathbf{x}_1|\mathbf{x}_2}(\boldsymbol{\alpha}_1 | \boldsymbol{\alpha}_2)$ is also gaussian if $\mathbf{x}_1$ and $\mathbf{x}_2$ are jointly gaussian. To demonstrate this property, we write the conditional probability of $\mathbf{x}_1$ given $\mathbf{x}_2$ as

$$p_{\mathbf{x}_1|\mathbf{x}_2}(\boldsymbol{\alpha}_1 | \boldsymbol{\alpha}_2) = \frac{p_{\mathbf{x}}(\boldsymbol{\alpha})}{p_{\mathbf{x}_2}(\boldsymbol{\alpha}_2)} \tag{7.4-7}$$

If we substitute the expressions for $p_{\mathbf{x}}(\boldsymbol{\alpha})$ in partitioned form and for $p_{\mathbf{x}_2}(\boldsymbol{\alpha}_2)$ from Eq. (7.4-6), we obtain

$$\begin{aligned} p_{\mathbf{x}_1|\mathbf{x}_2}(\boldsymbol{\alpha}_1 | \boldsymbol{\alpha}_2) = \frac{p_{\mathbf{x}_1\mathbf{x}_2}(\boldsymbol{\alpha}_1, \boldsymbol{\alpha}_2)}{p_{\mathbf{x}_2}(\boldsymbol{\alpha}_2)} = K \exp\Big\{-\frac{1}{2}[(\boldsymbol{\alpha}_1 - \boldsymbol{\mu}_1)^T(\mathbf{V}^{-1})_{11}(\boldsymbol{\alpha}_1 - \boldsymbol{\mu}_1) \\ + 2(\boldsymbol{\alpha}_1 - \boldsymbol{\mu}_1)^T(\mathbf{V}^{-1})_{12}(\boldsymbol{\alpha}_2 - \boldsymbol{\mu}_2) \\ + (\boldsymbol{\alpha}_2 - \boldsymbol{\mu}_2)^T(\mathbf{V}^{-1})_{22}(\boldsymbol{\alpha}_2 - \boldsymbol{\mu}_2) \\ - (\boldsymbol{\alpha}_2 - \boldsymbol{\mu}_2)^T\mathbf{V}_{22}^{-1}(\boldsymbol{\alpha}_2 - \boldsymbol{\mu}_2)]\Big\} \end{aligned} \tag{7.4-8}$$

where we have partitioned $\mathbf{V}^{-1}$ as

$$\mathbf{V}^{-1} = \left[\begin{array}{c|c} (\mathbf{V}^{-1})_{11} & (\mathbf{V}^{-1})_{12} \\ \hline (\mathbf{V}^{-1})_{12}^T & (\mathbf{V}^{-1})_{22} \end{array}\right] \tag{7.4-9}$$

The normalizing constant K in Eq. (7.4-8) has not been expanded, because it simply introduces additional unnecessary complexity. It is quite easy to show by direct substitution that the partitioned elements of $\mathbf{V}^{-1}$ may be written in terms of the partitioned elements of $\mathbf{V}$ as

$$(\mathbf{V}^{-1})_{11} = (\mathbf{V}_{11} - \mathbf{V}_{12}\mathbf{V}_{22}^{-1}\mathbf{V}_{12}^T)^{-1} \tag{7.4-10}$$

$$(\mathbf{V}^{-1})_{12} = -(\mathbf{V}^{-1})_{11}\mathbf{V}_{12}\mathbf{V}_{22}^{-1} \tag{7.4-11}$$

$$(\mathbf{V}^{-1})_{22} = (\mathbf{V}_{22} - \mathbf{V}_{12}^T\mathbf{V}_{11}^{-1}\mathbf{V}_{12})^{-1} \tag{7.4-12}$$

By substituting these results in Eq. (7.4-8) and completing a modest amount of algebraic manipulation, we find that the conditional density may be written as

$$\begin{aligned} p_{\mathbf{x}_1|\mathbf{x}_2}(\boldsymbol{\alpha}_1 | \boldsymbol{\alpha}_2) = K \exp(-\tfrac{1}{2}\{\boldsymbol{\alpha}_1 - [\boldsymbol{\mu}_1 + \mathbf{V}_{12}\mathbf{V}_{22}^{-1}(\boldsymbol{\alpha}_2 - \boldsymbol{\mu}_2)]\}^T \\ \cdot(\mathbf{V}_{11} - \mathbf{V}_{12}\mathbf{V}_{22}^{-1}\mathbf{V}_{12}^T)^{-1}\{\boldsymbol{\alpha}_1 - [\boldsymbol{\mu}_1 + \mathbf{V}_{12}\mathbf{V}_{22}^{-1}(\boldsymbol{\alpha}_2 - \boldsymbol{\mu}_2)]\}) \end{aligned} \tag{7.4-13}$$

Hence we see that the conditional density of $\mathbf{x}_1$ given $\mathbf{x}_2$ is gaussian. The expected value of $\mathbf{x}_1$ given $\mathbf{x}_2$ is

$$\mathcal{E}\{\mathbf{x}_1 \mid \mathbf{x}_2\} = \boldsymbol{\mu}_1 + \mathbf{V}_{12}\mathbf{V}_{22}^{-1}(\mathbf{x}_2 - \boldsymbol{\mu}_2) \tag{7.4-14}$$

The conditional variance is

$$\operatorname{cov}\{\mathbf{x}_1 \mid \mathbf{x}_2\} = \mathbf{V}_{11} - \mathbf{V}_{12}\mathbf{V}_{22}^{-1}\mathbf{V}_{12}^T \tag{7.4-15}$$

We note that the expected value of $\mathcal{E}\{\mathbf{x}_1 \mid \mathbf{x}_2\}$ is just $\boldsymbol{\mu}_1$ since

$$\mathcal{E}\{\mathcal{E}\{\mathbf{x}_1 \mid \mathbf{x}_2\}\} = \mathcal{E}\{\mathbf{x}_1\} = \boldsymbol{\mu}_1 - \mathbf{V}_{12}\mathbf{V}_{22}^{-1}(\mathcal{E}\{\mathbf{x}_2\} - \boldsymbol{\mu}_2) = \boldsymbol{\mu}_1$$

The roles of $\mathbf{x}_1$ and $\mathbf{x}_2$ can obviously be reversed to demonstrate that the conditional density of $\mathbf{x}_2$ given $\mathbf{x}_1$ is also gaussian. The general conclusion that we may reach is therefore that any marginal or conditional densities derived from the components of a multivariable gaussian random process are always gaussian. It is also quite easy to reverse the above procedure and so conclude that, if $p_{\mathbf{x}_1|\mathbf{x}_2}$ and $p_{\mathbf{x}_2}$ are gaussian, $\mathbf{x}_1$ and $\mathbf{x}_2$ are jointly gaussian.

The conditional mean and variance expressions of Eqs. (7.4-14) and (7.4-15) are very important for many uses in the area of estimation theory [5]. To illustrate this fact, we consider a simple estimation problem in the following example.

Example *7.4-1.* We wish to estimate a gaussian process $\mathbf{x}$ based on an observation $\mathbf{z}$ that consists of a linear function of $\mathbf{x}$ plus an additive independent zero-mean gaussian noise $\mathbf{v}$. A model for this observation $\mathbf{z}$ can then be written as

$$\mathbf{z} = \mathbf{H}\mathbf{x} + \mathbf{v}$$

A "reasonable" estimate in this case is the expected value of $\mathbf{x}$ given $\mathbf{z}$. Here we assume that $\mathbf{x}$ has a mean $\boldsymbol{\mu}_\mathbf{x}$ and variance $\mathbf{V}_\mathbf{x}$, and $\mathbf{v}$ has a variance $\mathbf{V}_\mathbf{v}$.

It is quite easy to show that $\mathbf{x}$ and $\mathbf{z}$ are jointly gaussian. The conditional density of $\mathbf{z}$ given $\mathbf{x}$ is gaussian with mean $\mathbf{H}\mathbf{x}$ and variance $\mathbf{V}_\mathbf{v}$. Since $\mathbf{x}$ is also gaussian, then $p_{\mathbf{x}\mathbf{z}} = p_{\mathbf{z}|\mathbf{x}}p_\mathbf{x}$ is gaussian. Because $\mathbf{x}$ and $\mathbf{z}$ are jointly gaussian, Eq. (7.4-14) may be used to find $\mathcal{E}\{\mathbf{x} \mid \mathbf{z}\}$ as

$$\mathcal{E}\{\mathbf{x} \mid \mathbf{z}\} = \boldsymbol{\mu}_\mathbf{x} + \mathbf{V}_{\mathbf{x}\mathbf{z}}\mathbf{V}_\mathbf{z}^{-1}(\mathbf{z} - \boldsymbol{\mu}_\mathbf{z})$$

The quantities needed for this result can be expressed as

$$\mathbf{V}_{\mathbf{x}\mathbf{z}} = \operatorname{cov}\{\mathbf{x}, \mathbf{z}\} = \operatorname{cov}\{\mathbf{x}, \mathbf{H}\mathbf{x} + \mathbf{v}\} = \mathbf{V}_\mathbf{x}\mathbf{H}^T$$
$$\mathbf{V}_\mathbf{z} = \operatorname{var}\{\mathbf{z}\} = \operatorname{var}\{\mathbf{H}\mathbf{x} + \mathbf{v}\} = \mathbf{H}\mathbf{V}_\mathbf{x}\mathbf{H}^T + \mathbf{V}_\mathbf{v}$$

and

$$\boldsymbol{\mu}_z = \mathcal{E}\{\mathbf{z}\} = \mathcal{E}\{\mathbf{Hx} + \mathbf{v}\} = \mathbf{H}\boldsymbol{\mu}_x$$

so that $\mathcal{E}\{\mathbf{x} \mid \mathbf{z}\}$ becomes

$$\mathcal{E}\{\mathbf{x} \mid \mathbf{z}\} = \boldsymbol{\mu}_x + \mathbf{V}_x\mathbf{H}^T(\mathbf{H}\mathbf{V}_x\mathbf{H}^T + \mathbf{V}_v)^{-1}(\mathbf{z} - \mathbf{H}\boldsymbol{\mu}_x)$$

We note that the conditional expectation of $\mathbf{x}_1$ given $\mathbf{x}_2$ is a linear function of $\mathbf{x}_2$. It is of interest to inquire concerning the conditions that a probability density function must fulfill in order that the conditioned expectation be linear. It can be shown (Deutsch [2]) that if and only if

$$\left.\frac{\partial \Phi_{\mathbf{x}_1\mathbf{x}_2}(\boldsymbol{\omega}_1, \boldsymbol{\omega}_2)}{\partial \boldsymbol{\omega}_1}\right|_{\boldsymbol{\omega}_1 = 0} = \mathbf{a}\Phi_{\mathbf{x}_2}(\boldsymbol{\omega}_2) + \mathbf{A}\frac{\partial \Phi_{\mathbf{x}_2}(\boldsymbol{\omega}_2)}{\partial \boldsymbol{\omega}_2} \tag{7.4-16}$$

will the conditional expectation of two random vectors have the linear form

$$\mathcal{E}\{\mathbf{x}_1 \mid \mathbf{x}_2\} = \mathbf{a} + \mathbf{A}\mathbf{x}_2 \tag{7.4-17}$$

It is quite easy to show that the gaussian probability density does satisfy this requirement.

There are a multitude of interesting results that we can prove for conditional gaussian densities. For instance, we may show that the vector

$$\tilde{\mathbf{x}}_1 = \mathbf{x}_1 - \mathcal{E}\{\mathbf{x}_1 \mid \mathbf{x}_2\} \tag{7.4-18}$$

is independent of $\mathbf{x}_2$, the conditioning vector. We verify this result directly by showing that $\mathbf{x}_2$ and $\tilde{\mathbf{x}}_1$ are uncorrelated and using a result which we shall establish later in the section showing that if two gaussian processes are uncorrelated, they are independent. By the use of Eq. (7.4-14), we find that

$$\tilde{\mathbf{x}}_1 = \mathbf{x}_1 - \boldsymbol{\mu}_1 - \mathbf{V}_{12}\mathbf{V}_{22}^{-1}(\mathbf{x}_2 - \boldsymbol{\mu}_2) \tag{7.4-19}$$

and

$$\mathcal{E}\{\tilde{\mathbf{x}}_1\} = \mathcal{E}\{\mathbf{x}_1\} - \boldsymbol{\mu}_1 - \mathbf{V}_{12}\mathbf{V}_{22}^{-1}(\mathcal{E}\{\mathbf{x}_2\} - \boldsymbol{\mu}_2) = \mathbf{0} \tag{7.4-20}$$

so that $\text{cov}\{\tilde{\mathbf{x}}_1, \mathbf{x}_2\}$ becomes

$$\begin{aligned}
\text{cov}\{\tilde{\mathbf{x}}_1, \mathbf{x}_2\} &= \mathcal{E}\{[(\mathbf{x}_1 - \boldsymbol{\mu}_1) - \mathbf{V}_{12}\mathbf{V}_{22}^{-1}(\mathbf{x}_2 - \boldsymbol{\mu}_2)](\mathbf{x}_2 - \boldsymbol{\mu}_2)^T\} \\
&= \mathcal{E}\{(\mathbf{x}_1 - \boldsymbol{\mu}_1)(\mathbf{x}_2 - \boldsymbol{\mu}_2)^T\} - \mathbf{V}_{12}\mathbf{V}_{22}^{-1}\mathcal{E}\{(\mathbf{x}_2 - \boldsymbol{\mu}_2)(\mathbf{x}_2 - \boldsymbol{\mu}_2)^T\} \\
&= \mathbf{V}_{12} - \mathbf{V}_{12}\mathbf{V}_{22}^{-1}\mathbf{V}_{22} = \mathbf{0}
\end{aligned}$$

It is also possible to show that $\tilde{\mathbf{x}}_1$ and $\mathcal{E}\{\mathbf{x}_1 \mid \mathbf{x}_2\}$ are uncorrelated (and hence independent) in that

$$\text{cov}\{\tilde{\mathbf{x}}_1, \mathcal{E}\{\mathbf{x}_1 \,|\, \mathbf{x}_2\}\} = \mathbf{0} \tag{7.4-21}$$

Substituting from Eqs. (7.4-14) and (7.4-19), we have

$$\begin{aligned}
\text{cov}\{\tilde{\mathbf{x}}_1, \mathcal{E}\{\mathbf{x}_1 \,|\, \mathbf{x}_2\}\} &= \mathcal{E}\{[\mathbf{x}_1 - \boldsymbol{\mu}_1 - \mathbf{V}_{12}\mathbf{V}_{22}^{-1}(\mathbf{x}_2 - \boldsymbol{\mu}_2)][\mathbf{V}_{12}\mathbf{V}_{22}^{-1}(\mathbf{x}_2 - \boldsymbol{\mu}_2)]^T\} \\
&= \mathcal{E}\{(\mathbf{x}_1 - \boldsymbol{\mu}_1)(\mathbf{x}_1 - \boldsymbol{\mu}_2)^T\}\mathbf{V}_{22}^{-1}\mathbf{V}_{12}^T \\
&\quad - \mathbf{V}_{12}\mathbf{V}_{22}^{-1}\mathcal{E}\{(\mathbf{x}_2 - \boldsymbol{\mu}_2)(\mathbf{x}_2 - \boldsymbol{\mu}_2)^T\}\mathbf{V}_{22}^{-1}\mathbf{V}_{12}^T \\
&= \mathbf{V}_{12}\mathbf{V}_{22}^{-1}\mathbf{V}_{12}^T - \mathbf{V}_{12}\mathbf{V}_{22}^{-1}\mathbf{V}_{22}\mathbf{V}_{22}^{-1}\mathbf{V}_{12}^T = \mathbf{0}
\end{aligned}$$

Because $\tilde{\mathbf{x}}_1$ is a zero-mean process, we note that $\tilde{\mathbf{x}}_1$ and $\mathcal{E}\{\mathbf{x}_1 \,|\, \mathbf{x}_2\}$ are also orthogonal, since they are uncorrelated and one of the processes is zero mean. For this reason, Eq. (7.4-21) is often referred to as the *orthogonal projection lemma*, which is extremely valuable in the study of estimation theory [5].

In Section 7.2 we showed that if $\mathbf{x}$ is a gaussian process, then $\mathbf{y}$ given by

$$\mathbf{y} = \mathbf{A}\mathbf{x} + \mathbf{b} \tag{7.4-22}$$

is also gaussian if $\mathbf{A}$ is nonsingular. We wish to extend this result to include the cases when $\mathbf{A}$ is not square, or square and singular. Suppose first that the dimension of $\mathbf{y}$ is $m < n$ but that $\mathbf{A}$ is still full rank. Let us add $n - m$ rows to $\mathbf{A}$, which are linearly independent, so that a new $\mathbf{A}^*$ matrix is generated which is nonsingular. We also add $n - m$ zeros to $\mathbf{b}$, so Eq. (7.4-22) is now

$$\mathbf{y}^* = \mathbf{A}^*\mathbf{x} + \mathbf{b}^* \tag{7.4-23}$$

where the first m components of $\mathbf{y}^*$ are the components of $\mathbf{y}$. Using the result of Section 7.2, we know that $\mathbf{y}^*$ is gaussian; however, earlier in this section we established that any subset of the components of a gaussian random vector is also gaussian. Hence we can conclude that $\mathbf{y}$ is also gaussian.

Once we know that $\mathbf{y}$ is gaussian, it is quite easy to determine its density function, since we need only find the mean and variance. The mean of $\mathbf{y}$ is given by

$$\mathcal{E}\{\mathbf{y}\} = \mathbf{A}\mathcal{E}\{\mathbf{x}\} + \mathbf{b} = \mathbf{A}\boldsymbol{\mu}_{\mathbf{x}} + \mathbf{b} \tag{7.4-24}$$

The variance is given by

$$\text{var}\{\mathbf{y}\} = \text{var}\{\mathbf{A}\mathbf{x} + \mathbf{b}\} = \mathbf{A}\mathbf{V}_{\mathbf{x}}\mathbf{A}^T \tag{7.4-25}$$

This result is identical to that of Section 7.2, except that here $\mathbf{A}$ is not square.

Note, however, that since $\mathbf{A}$ is of full rank, $\mathbf{AV_xA}^T$ is nonsingular, so the probability density of $\mathbf{y}$ is

$$p_{\mathbf{y}}(\boldsymbol{\beta}) = \frac{1}{[(2\pi)^m \det(\mathbf{AV_xA}^T)]^{1/2}} \exp\left\{-\frac{1}{2}[\boldsymbol{\beta} - (\mathbf{A}\boldsymbol{\mu}_{\mathbf{x}} + \mathbf{b})]^T \cdot (\mathbf{AV_xA}^T)^{-1}[\boldsymbol{\beta} - (\mathbf{A}\boldsymbol{\mu}_{\mathbf{x}} + \mathbf{b})]\right\} \tag{7.4-26}$$

Whenever the dimension of $\mathbf{y}$ is greater than $\mathbf{x}$, or $\mathbf{A}$ does not have full rank, we have difficulty in applying the above result. Note, in particular, that the variance matrix $\mathbf{AV_xA}^T$ will not have an inverse. Even in this case, the density of $\mathbf{y}$ is gaussian if we expand the concept of a gaussian distribution slightly. Let us consider a density function equal to a Dirac delta function located at $\boldsymbol{\alpha} = \boldsymbol{\mu}$ so that $p_{\mathbf{x}}(\boldsymbol{\alpha})$ is

$$p_{\mathbf{x}}(\boldsymbol{\alpha}) = \delta_D(\boldsymbol{\alpha} - \boldsymbol{\mu}) \tag{7.4-27}$$

The characteristic function associated with this distribution is given by

$$\Phi_{\mathbf{x}}(\boldsymbol{\omega}) = \int_{-\infty}^{\infty} e^{j\boldsymbol{\omega}^T\boldsymbol{\alpha}}\delta_D(\boldsymbol{\alpha} - \boldsymbol{\mu})\, d\boldsymbol{\alpha} = e^{j\boldsymbol{\omega}^T\boldsymbol{\mu}} \tag{7.4-28}$$

If this result is compared with the characteristic function for a gaussian distribution, we see that $\mathbf{x}$ could be considered as a gaussian density with mean $\boldsymbol{\mu}$ and a variance of zero. Of course, $\mathbf{x}$ could also represent a constant of value $\boldsymbol{\mu}$. For our purposes here, it will be convenient to assume that the Dirac delta density of Eq. (7.4-27) is a degenerate form of a gaussian density function. With this convention, we can now demonstrate that any linear operation on a gaussian random variable will yield another gaussian random variable.

Suppose that $\mathbf{y}$ is given by

$$\mathbf{y} = \mathbf{Ax} + \mathbf{b} \tag{7.4-29}$$

where the dimension of $\mathbf{y}$ is not restricted to be less than the dimension of $\mathbf{x}$, and $\mathbf{A}$ may not be of full rank.

Let us find the largest set of rows of $\mathbf{A}$ such that the rows are linearly independent. Let $\mathbf{y}^*$ be the corresponding elements of $\mathbf{y}$; the dimension of $\mathbf{y}^*$ is $m \leq n$. The matrix formed from this set of linearly independent rows will be denoted by $\mathbf{A}^*$ and the corresponding elements of $\mathbf{b}$ by $\mathbf{b}^*$. By the argument presented above, $\mathbf{y}^*$ must be gaussian with mean $\mathbf{b}^*$ and variance $\mathbf{A}^*\mathbf{V_x}\mathbf{A}^{*T}$.

The remaining elements of $\mathbf{y}$, which we shall denote as $\mathbf{y}^{**}$, can be written in the form

$$\mathbf{y}^{**} = \mathbf{B}\mathbf{y}^{*} + \mathbf{c} \tag{7.4-30}$$

Now the conditional density of $\mathbf{y}^{**}$, given $\mathbf{y}^{*}$, is given by

$$p_{\mathbf{y}^{**}|\mathbf{y}^{*}}(\boldsymbol{\alpha}\,|\,\boldsymbol{\beta}) = \delta_D(\boldsymbol{\alpha} - \mathbf{B}\boldsymbol{\beta} - \mathbf{c})$$

But this is a degenerate form of gaussian density. Now since both $p_{\mathbf{y}^{*}}$ and $p_{\mathbf{y}^{**}|\mathbf{y}^{*}}$ are gaussian, $\mathbf{y}^{**}$ and $\mathbf{y}^{*}$ must be jointly gaussian, which means that $\mathbf{y}$ is gaussian. This is the desired result. Hence we may conclude that if $\mathbf{x}$ is gaussian, then $\mathbf{y}$ given by Eq. (7.4-29) is also gaussian.

In several places throughout this book we have made use of the fact that if two gaussian processes are uncorrelated, they are independent. This fact is both useful and quite easy to establish. Suppose that we partition $\mathbf{x}$ into two vectors $\mathbf{x}_1$ of dimension $m < n$ and $\mathbf{x}_2$ of dimension $n - m$, as we did earlier. In this partitioned form the joint density function of $\mathbf{x}_1$ and $\mathbf{x}_2$ may be written as

$$\begin{aligned} p_{\mathbf{x}_1|\mathbf{x}_2}(\boldsymbol{\alpha}_1, \boldsymbol{\alpha}_2) = \frac{1}{[(2\pi)^n \det \mathbf{V}]^{1/2}} \exp\Big\{-\frac{1}{2}[(\boldsymbol{\alpha}_1 - \boldsymbol{\mu}_1)^T(\mathbf{V}^{-1})_{11}(\boldsymbol{\alpha}_1 - \boldsymbol{\mu}_1) \\ + 2(\boldsymbol{\alpha}_1 - \boldsymbol{\mu}_1)^T(\mathbf{V}^{-1})_{12}(\boldsymbol{\alpha}_2 - \boldsymbol{\mu}_2) \\ + (\boldsymbol{\alpha}_2 - \boldsymbol{\mu}_2)^T(\mathbf{V}^{-1})_{22}(\boldsymbol{\alpha}_2 - \boldsymbol{\mu}_2)]\Big\} \end{aligned} \tag{7.4-31}$$

where $(\mathbf{V}^{-1})_{11}$, $(\mathbf{V}^{-1})_{12}$, and $(\mathbf{V}^{-1})_{22}$ are given by Eqs. (7.4-10) to (7.4-12). If $\mathbf{x}_1$ and $\mathbf{x}_2$ are uncorrelated, then $\mathbf{V}_{12} = \mathbf{0}$ and Eqs. (7.4-10) to (7.4-12) become

$$(\mathbf{V}^{-1})_{11} = \mathbf{V}_{11}^{-1} \qquad (\mathbf{V}^{-1})_{12} = \mathbf{0} \qquad (\mathbf{V}^{-1})_{22} = \mathbf{V}_{22}^{-1}$$

In addition, it is easy to show that

$$\det \mathbf{V} = \det \mathbf{V}_{11} \det \mathbf{V}_{22}$$

if $\mathbf{V}_{12} = \mathbf{0}$. Therefore, the joint density function of Eq. (7.4-26) may be written as

$$\begin{aligned} p_{\mathbf{x}_1\mathbf{x}_2}(\boldsymbol{\alpha}_1, \boldsymbol{\alpha}_2) &= \left(\frac{1}{[(2\pi)^m \det \mathbf{V}_{11}]^{1/2}} \exp\Big\{-\frac{1}{2}(\boldsymbol{\alpha}_1 - \boldsymbol{\mu}_1)^T\mathbf{V}_{11}^{-1}(\boldsymbol{\alpha}_1 - \boldsymbol{\mu}_1)\Big\}\right) \\ &\quad \cdot \left(\frac{1}{[(2\pi)^{n-m} \det \mathbf{V}_{22}]^{1/2}} \exp\Big\{-\frac{1}{2}(\boldsymbol{\alpha}_2 - \boldsymbol{\mu}_2)^T\mathbf{V}_{22}^{-1}(\boldsymbol{\alpha}_2 - \boldsymbol{\mu}_2)\Big\}\right) \\ &= p_{\mathbf{x}_1}(\boldsymbol{\alpha}_1)p_{\mathbf{x}_2}(\boldsymbol{\alpha}_2) \end{aligned}$$

and we have established that fact that $\mathbf{x}_1$ and $\mathbf{x}_2$ are independent.

EXERCISES 7.4

7.4-1. Show that two gaussian random variables $\mathbf{x}_1$ and $\mathbf{x}_2$ may be marginally gaussian but not jointly gaussian.

7.4-2. Derive Eqs. (7.4-10) to (7.4-12).

7.4-3. Show that if $\mathbf{x}_1$ and $\mathbf{x}_2$ are jointly gaussian, they satisfy Eq. (7.4-16). Derive Eq. (7.4-14) by the use of Eq. (7.4-17).

7.5. Conclusions

In this chapter we have studied several of the properties of multivariable gaussian processes. A complete study of gaussian processes would be long and extensive, and we have treated only a small part of the topic. Emphasis has been placed on those aspects of the field which are of most use in engineering applications.

Among the several results developed in this chapter, there are three of prime importance. The first result is the central limit theorem of Section 7.3, which indicates that many physical processes will have a gaussian distribution. Second is the fact that the marginal and conditional probability densities associated with two jointly gaussian processes are also gaussian. The final result of major importance is the fact that linear operations on gaussian processes yield other gaussian processes.

7.6. References

(1) Bendat, J. S. and A. G. Piersol, *Measurement and Analysis of Random Data*, John Wiley & Sons, Inc., New York, 1966.

(2) Deutsch, R. *Estimation Theory*, Prentice-Hall, Inc., Englewood Cliffs, N.J., 1965.

(3) Dubes, R. C., *The Theory of Applied Probability*, Prentice-Hall, Inc., Englewood Cliffs, N.J., 1968.

(4) Papoulis, A., *Probability, Random Variables, and Stochastic Processes*, McGraw-Hill, New York, 1965.

(5) Sage, A. P., and J. L. Melsa, *Estimation Theory with Applications to Communications and Control*, McGraw-Hill, New York, 1971.

7.7. Problems

7.7-1. State and prove the central limit theorem for the non-zero-mean case.

7.7-2. Show that all odd-product moments of the form $\mathcal{E}\{\prod_{i=1}^{2n-1} x_i\}$ are zero if $\mu_{x_i} = 0$ for all $i = 1, 2, \ldots, 2n - 1$. Here the x_i's are assumed to be identically distributed gaussian random variables.

7.7-3. The stochastic process $\mathbf{x}(t)$ satisfies the equation

$$\dot{\mathbf{x}}(t) = -\mathbf{x}(t) + \mathbf{u}(t)$$

with $\mathbf{x}(0) = \mathbf{0}$. Here $\mathbf{u}(t)$ is a gaussian, zero-mean, white-noise process with $\text{cov}\{\mathbf{u}(t), \mathbf{u}(\tau)\} = \mathbf{I}\delta_D(t - \tau)$. Find the probability density of $\mathbf{x}(t)$ for $t \geq 0$.

7.7-4. What is the probability density function for the process

$$y(t) = x(t) \sin (\omega t + \theta)$$

where $x(t)$ is zero mean and gaussian and θ is uniformly distributed in the interval 0 to 2π?

7.7-5. What is the correlation function associated with $y(t)$ in Problem *7.7-4.* if

$$C_x(\tau) = \mathcal{E}\{x(t)x(t + \tau)\} = e^{-|\tau|}?$$

7.7-6. What is the probability density function for

$$y(t) = x(t) + \sin(\omega t + \theta)$$

if $x(t)$ is zero mean and gaussian and θ is a uniformly distributed $[0, 2\pi]$ random variable?

7.7-7. What is the correlation function associated with $y(t)$ in Problem 7.7-6 if

$$C_x(\tau) = e^{-|\tau|}?$$

7.7-8. Show that the random process evolving from

$$x(k + 1) = \Phi x(k) + w(k)$$

is gaussian if $x(0)$ and $w(k)$ are gaussian. For convenience you may assume that $w(k)$ is also white.

7.7-9. What is

$$\hat{x}(k) = \mathcal{E}\{x(k) \,|\, Z(k)\}$$

where

$$Z(k) = \begin{bmatrix} z(0) \\ z(1) \\ \cdot \\ \cdot \\ \cdot \\ z(k) \end{bmatrix}$$

if $x(k)$ and $z(k)$ are gaussian and evolve from

$$x(k+1) = \Phi x(k) + \Gamma w(k)$$
$$z(k) = Hx(k) + v(k)$$

where $w(k)$ and $v(k)$ are zero mean and gaussian? Show that the $\hat{x}(k)$ is specified by the Kalman filter algorithms of Example 6.2-6.

8. *Markov Processes and Stochastic Differential Equations*

8.1. *Introduction*

When a stochastic process is Markov, many problems are considerably simplified. The Markov assumption states that a knowledge of the present separates the past from the future and, therefore, lends particular significance to conditional density functions, to which special emphasis will be given in the first part of this chapter. Some of the process models from Chapter 6 will be reexamined, and we shall show that most of the models of that chapter are, in fact, Markov models. The Wiener process, a particularly important Markov process, will be developed.

In this chapter random processes in nonlinear systems will also be studied. We shall find that ordinary rules of differential and integral calculus do not, unfortunately, always apply to stochastic processes. For instance, the integration of a product of two random variables is not defined by the rules of ordinary calculus. We shall develop the Ito stochastic calculus, which resolves many fundamental problems concerning random process in nonlinear systems. The Fokker–Planck partial differential equation, whose solution characterizes the evolution of the probability function of the state vector of a broad class of nonlinear systems with a white gaussian noise input, is also developed.

Finally, a discussion of mean and variance propagation for random processes in nonlinear systems will be presented.

8.2. Markov processes

In this section we shall study the concept of Markov processes. These concepts play an important role in the engineering application of stochastic processes because almost every stochastic process of interest is a Markov process. Use of Markov-process properties results in considerable simplification in many results or derivations. Consider, for example, the probabilistic description associated with K sample values $\mathbf{x}(t_i)$, $i = 1, 2, \ldots, K$, taken from a stochastic process $\mathbf{x}(t)$. Normally, it would be necessary to know the joint density function $p_{\mathbf{x}(t_1)\mathbf{x}(t_2)\cdots\mathbf{x}(t_K)}(\boldsymbol{\alpha}_1, \boldsymbol{\alpha}_2, \ldots, \boldsymbol{\alpha}_K)$, which may be difficult to obtain, in general, for large K. If the samples were jointly gaussian, the task would be somewhat simpler since we would only need to know the means $\boldsymbol{\mu}_i = \mathcal{E}\{\mathbf{x}(t_i)\}$ and the $\text{cov}\{\mathbf{x}(t_i), \mathbf{x}(t_j)\}$. The markovian property will make a similar reduction in effort possible.

Definition 8.2-1. ***Markov Process (Continuous Time):*** A stochastic process $\mathbf{x}(t)$, $t > t_0$, is said to be Markov if, for every $\tau \leq t$,

$$p_{\mathbf{x}(t) \mid X(\tau)}(\boldsymbol{\alpha} \mid B) = p_{\mathbf{x}(t) \mid \mathbf{x}(\tau)}(\boldsymbol{\alpha} \mid \boldsymbol{\beta}) \tag{8.2-1}$$

where $X(\tau) = \{\mathbf{x}(s) \colon t_0 \leq s \leq \tau \leq t\}$.

If $\mathbf{x}$ is a discrete-time stochastic process, then t will only take on the discrete time points; so the definition becomes

Definition 8.2-2. ***Markov Process (Discrete Time):*** A stochastic process $\mathbf{x}(\mathbf{t}_i)$, $t_i > t_0$, is said to be Markov if for every $t_j, j < i$,

$$p_{\mathbf{x}(t_i) \mid X(t_j)}(\boldsymbol{\alpha}_i \mid B_j) = p_{\mathbf{x}(t_i) \mid \mathbf{x}(t_j)}(\boldsymbol{\alpha}_i \mid \boldsymbol{\beta}_j) \tag{8.2-2}$$

where $X(t_j) = \{\mathbf{x}(t_k) \colon 0 \leq k \leq j \leq i\}$.

In simple terms, the markovian property means that the probability density of $\mathbf{x}(t)$ conditioned on a set of past values of $\mathbf{x}$ depends only on the most recent value from the set. We shall concentrate our attention on the discrete-time case initially; later the continuous-time case will be treated by letting the sample become dense. It should be noted that the samples need not be uniform; that is, $t_{i+1} - t_i$ is not necessarily equal to $t_{j+1} - t_j$, $j \neq i$, but ordering of the samples is necessary, and we assume that $t_i < t_j$ if $i < j$.

An elementary and prime characteristic of a Markov process is that joint probability densities may be expressed as the product of *transition probability densities* defined as

$$p_{\mathbf{x}(t_k)\mathbf{x}(t_{k-1})}(\boldsymbol{\alpha}, \boldsymbol{\beta}) = p_{\mathbf{x}(t_k)\,|\,\mathbf{x}(t_{k-1})}(\boldsymbol{\alpha}\,|\,\boldsymbol{\beta}) \tag{8.2-3}$$

The joint probability density of $X(t_i)$ may be written as

$$\begin{aligned} p_{X(t_i)}(B_i) &= p_{\mathbf{x}(t_i)\mathbf{x}(t_{i-1})\cdots\mathbf{x}(t_0)}(\boldsymbol{\beta}_i\, \boldsymbol{\beta}_{i-1}, \ldots, \boldsymbol{\beta}_0) \\ &= p_{\mathbf{x}(t_i)\,|\,X(t_{i-1})}(\boldsymbol{\beta}_i\,|\,B_{i-1})p_{X(t_{i-1})}(B_{i-1}) \end{aligned} \tag{8.2-4}$$

By using the markovian property we know that

$$p_{\mathbf{x}(t_i)\,|\,X(t_{i-1})} = p_{\mathbf{x}(t_i)\,|\,\mathbf{x}(t_{i-1})} = p_{\mathbf{x}(t_i)\mathbf{x}(t_{i-1})}$$

so that Eq. (8.2-4) becomes

$$p_{X(t_i)}(B_i) = p_{\mathbf{x}(t_i)\mathbf{x}(t_{i-1})}(\boldsymbol{\beta}_i, \boldsymbol{\beta}_{i-1})p_{X(t_{i-1})}(B_{i-1}) \tag{8.2-5}$$

By applying this procedure repeatedly, we can express the joint density $p_{X(t_i)}$ as

$$p_{X(t_i)}(B_i) = p_{\mathbf{x}(t_0)}(\boldsymbol{\beta}_0) \prod_{j=1}^{i} p_{x(t_j)x(t_{j-1})}(\boldsymbol{\beta}_j, \boldsymbol{\beta}_{j-1}) \tag{8.2-6}$$

If we know the initial probability density and the transition probability densities, we can completely determine the joint density $p_{X(t_i)}$.

A Markov process is said to be *homogeneous* or to have a stationary transition mechanism if $p_{\mathbf{x}(t_i)\mathbf{x}(t_{i-1})}$ depends only on the difference $t_i - t_{i-1}$ and not on either t_i or t_{i-1}. If a Markov process is homogeneous and the event points t_i, $i = 1, 2, \ldots$, are uniformly spaced, that is, $t_i - t_{i-1} = T$ for all i, then

$$p_{\mathbf{x}(t_i)\mathbf{x}(t_{i-1})} = p_{\mathbf{x}(t_j)\mathbf{x}(t_{j-1})} \qquad \text{for all } i \text{ and } j$$

We note that a homogeneous Markov process is not necessarily stationary, even though it has a stationary transition mechanism.

In addition to the one-stage transition probability density defined by Eq. (8.2-3), it is often desirable to work with the n-stage transition probability density defined as

$$p_{\mathbf{x}(t_i)\mathbf{x}(t_{i-n})}(\boldsymbol{\alpha}, \boldsymbol{\beta}) = p_{\mathbf{x}(t_i)\,|\,\mathbf{x}(t_{i-n})}(\boldsymbol{\alpha}\,|\,\boldsymbol{\beta}) \tag{8.2-7}$$

By the use of this transition probability density we can directly determine the probability density $p_{\mathbf{x}(t_j)}$ from the initial density $p_{\mathbf{x}(t_0)}$ as

$$p_{\mathbf{x}(t_j)}(\boldsymbol{\alpha}) = \int_{-\infty}^{\infty} p_{\mathbf{x}(t_j)\mathbf{x}(t_0)}(\boldsymbol{\alpha}, \boldsymbol{\beta})p_{\mathbf{x}(t_0)}(\boldsymbol{\beta})\, d\boldsymbol{\beta} \tag{8.2-8}$$

The similarity of Eq. (8.2-8) to the convolution integral for determining a state $\mathbf{x}(t_i)$ from $\mathbf{x}(t_0)$ should be noted. We see that the transition probability density plays a role parallel to that of the state transition matrix, except here probability density of the state rather than the state itself is being determined.

The n-stage transition probability density can be determined by combining the associated n one-stage transition probability densities. Let us consider the problem of computing a two-stage transition probability density. For any stochastic process, we know that

$$\int_{-\infty}^{\infty} p_{\mathbf{x}(t_i)\mathbf{x}(t_{i-1})\mathbf{x}(t_{i-2})}(\boldsymbol{\alpha}, \boldsymbol{\beta}, \boldsymbol{\gamma})\, d\boldsymbol{\beta} = p_{\mathbf{x}(t_i)\mathbf{x}(t_{i-2})}(\boldsymbol{\alpha}, \boldsymbol{\gamma}) \tag{8.2-9}$$

By the use of Eq. (8.2-5), we may express the joint density on the left-hand side for a Markov process as

$$p_{\mathbf{x}(t_i)\mathbf{x}(t_{i-1})\mathbf{x}(t_{i-2})}(\boldsymbol{\alpha}, \boldsymbol{\beta}, \boldsymbol{\gamma}) = \mathscr{p}_{\mathbf{x}(t_i)\mathbf{x}(t_{i-1})}(\boldsymbol{\alpha}, \boldsymbol{\beta})\mathscr{p}_{\mathbf{x}(t_{i-1})\mathbf{x}(t_{i-2})}(\boldsymbol{\beta}, \boldsymbol{\gamma})p_{\mathbf{x}(t_{i-2})}(\gamma) \tag{8.2-10}$$

The density of the right-hand side of Eq. (8.2-9) may be written as (whether it is Markov or not)

$$\begin{aligned} p_{\mathbf{x}(t_i)\mathbf{x}(t_{i-2})}(\boldsymbol{\alpha}, \boldsymbol{\gamma}) &= p_{\mathbf{x}(t_i)\,|\,\mathbf{x}(t_{i-2})}(\boldsymbol{\alpha}\,|\,\boldsymbol{\gamma})p_{\mathbf{x}(t_{i-2})}(\boldsymbol{\gamma}) \\ &= \mathscr{p}_{\mathbf{x}(t_i)\mathbf{x}(t_{i-2})}(\boldsymbol{\alpha}, \boldsymbol{\gamma})p_{\mathbf{x}(t_{i-2})}(\boldsymbol{\gamma}) \end{aligned} \tag{8.2-11}$$

If we substitute Eqs. (8.2-9) and (8.2-11) into Eq. (8.2-10) and cancel the common $p_{\mathbf{x}(t_{i-2})}$, we obtain

$$\int_{-\infty}^{\infty} \mathscr{p}_{\mathbf{x}(t_i)\mathbf{x}(t_{i-1})}(\boldsymbol{\alpha}, \boldsymbol{\beta})\mathscr{p}_{\mathbf{x}(t_{i-1})\mathbf{x}(t_{i-2})}(\boldsymbol{\beta}, \boldsymbol{\gamma})\, d\boldsymbol{\beta} = \mathscr{p}_{\mathbf{x}(t_i)\mathbf{x}(t_{i-2})}(\boldsymbol{\alpha}, \boldsymbol{\gamma}) \tag{8.2-12}$$

This result is known as the *Chapman–Kolmogorov* equation. By proceeding in this same manner, we can write the n-stage transition probability density as

$$\int_{-\infty}^{\infty} \cdots \int_{-\infty}^{\infty} \prod_{j=i-n+1}^{i} [\mathscr{p}_{\mathbf{x}(t_j)\mathbf{x}(t_{j-1})}(\boldsymbol{\alpha}_j, \boldsymbol{\alpha}_{j-1})\, d\boldsymbol{\alpha}_j] = \mathscr{p}_{\mathbf{x}(t_i)\mathbf{x}(t_{i-n})}(\boldsymbol{\alpha}_i, \boldsymbol{\alpha}_{i-n}) \tag{8.2-13}$$

Among the numerous properties of Markov processes, the four following properties are often of interest:

1. Any subsequence of a Markov process is also a Markov process.
2. A Markov process is Markov in reverse.
3. For a Markov process $\mathcal{E}\{\mathbf{x}(t_k)\,|\,X(t_j)\} = \mathcal{E}\{\mathbf{x}(t_k)\,|\,\mathbf{x}(t_j)\}$ for $j \leq k$.
4. If $\mathbf{x}(t_2) - \mathbf{x}(t_1)$ is independent of $\mathbf{x}(t)$ for every $t \leq t_1 \leq t_2$, the process $\mathbf{x}(t)$ is Markov.

The proof of these properties, although reasonably simple, is omitted for the sake of brevity.

Example ***8.2-1.*** To illustrate some of the concepts of Markov processes, let us consider a simple example of a Markov process. Suppose that $\{x(t_i)\}$ is a sequence of independent random variables with known probability densities $p_{x(t_i)}$. Now we define another sequence of random variables by the recursion relation

$$y(t_i) = y(t_{i-1}) + x(t_i)$$

with $y(t_0) = x(t_0)$. We wish to show that $\{y(t_i)\}$ is a Markov process. Using the independent property of $\{x(t_i)\}$, the joint density of $Y(t_i)$ is easily written as

$$p_{Y(t_n)}(B_n) = p_{x(t_0)}(\beta_0) \prod_{j=1}^{n} p_{x(t_j)}(\beta_j - \beta_{j-1})$$

Therefore, the conditional density of $y(t_i)$ given $Y(t_{i-1})$ is

$$\begin{aligned} p_{y(t_i) \mid Y(t_{i-1})}(\beta_i \mid B_{i-1}) &= \frac{p_{Y(t_i)}(B_i)}{p_{Y(t_{i-1})}(B_{i-1})} \\ &= p_{x(t_i)}(\beta_i - \beta_{i-1}) \end{aligned}$$

Since $p_{y(t_i) \mid Y(t_{i-1})}(\beta_i \mid B_{i-1})$ depends only on β_{i-1}, we see that

$$p_{y(t_i) \mid Y(t_{i-1})}(\beta_i \mid B_{i-1}) = p_{y(t_i) \mid y(t_{i-1})}(\beta_i \mid \beta_{i-1})$$

and the process $\{y(t_i)\}$ is Markov.

If we assume that $\mathcal{E}\{x(t_i)\} = 0$ for all i, then $\mathcal{E}\{y(t_i)\} = 0$ for all i. Now let us consider the conditional expected value of $y(t_i)$ given by

$$\begin{aligned} \mathcal{E}\{y(t_i) \mid y(t_{i-1})\} &= \mathcal{E}\{y(t_{i-1}) + x(t_i) \mid y(t_{i-1})\} \\ &= \mathcal{E}\{y(t_{i-1}) \mid y(t_{i-1})\} + \mathcal{E}\{x(t_i) \mid y(t_{i-1})\} \end{aligned}$$

Since $\{x(t_i)\}$ is a sequence of independent random variables, $x(t_i)$ is independent of $y(t_{i-1})$, which contains only information about $x(t_j)$, $j = 1, 2, \ldots, i-1$. Therefore, the expression above becomes

$$\mathcal{E}\{y(t_i) \mid y(t_{i-1})\} = y(t_{i-1}) + \mathcal{E}\{x(t_i)\} = y(t_{i-1})$$

Because $\{y(t_i)\}$ is Markov, we know from Property 3 that $\mathcal{E}\{y(t_i) \mid Y(t_{i-1})\} = \mathcal{E}\{y(t_i) \mid y(t_{i-1})\}$; so we have

$$\mathcal{E}\{y(t_i) \mid Y(t_{i-1})\} = y(t_{i-1})$$

A process that satisfies this property is known as a *martingale*. We will discuss this concept later in this chapter.

Although the general Chapman–Kolmogorov equation [Eq. (8.2-13)] can be used (with some difficulty) to determine the evolution of the probability density function of the state $\mathbf{x}(t_i)$ for a discrete-time Markov process, it cannot be used for the continuous-time case. In the continuous-time case, the number of stages, and hence the number of integrations, in Eq. (8.2-13) becomes uncountable. We shall develop a procedure for handling the continuous-time case in Section 8.4; first, however, we shall examine the concepts of white noise and the Wiener process in some detail.

EXERCISES 8.2

8.2-1. Show that any subsequence of a Markov process is also Markov.

8.2-2. Show that $\mathcal{E}\{\mathbf{x}(t_k)\,|\,\boldsymbol{X}(t_j)\} = \mathcal{E}\{\mathbf{x}(t_k)\,|\,\mathbf{x}(t_j)\}$ if $\mathbf{x}$ is Markov.

8.2-3. Show that the stochastic process defined by

$$x(k+1) = ax(k) + u(k)$$

is first-order Markov if and only if $u(k)$ is a discrete white noise such that $\operatorname{cov}\{u(k), u(j)\} = V_u(k)\delta_K(k-j)$.

8.2-4. Suppose that $\mathbf{x}(k) = [x_1(k)\,x_2(k)]^T$ is Markov. Show that $z(k)$ defined by

$$z(k) = x_1(k) + x_2(k)$$

is not (first-order) Markov.

8.2-5. Consider a Markov stochastic process $x(k)$ defined by

$$x(k+1) = x(k) + u(k)$$

where $x(0)$ and $u(k)$ for all k are gaussian with zero mean and unit variance, $\operatorname{cov}\{x(0), u(k)\} = 0$ for all $k \geq 0$, and $\operatorname{cov}\{u(k), u(j)\} = \delta_K(k-j)$. Find the probability density function for $x(1)$ and $x(2)$ by the use of Eqs. (8.2-8) and (8.2-12). Is there an easier way to obtain the answer?

8.3. Wiener process

In this and the next section we shall develop some rigorous mathematical methods for studying the behavior of a nonlinear dynamic system excited by white noise. The principal result of this study is an expression known as the Fokker–Planck equation, whose solution yields the propagation of the probability density of the system state. As we shall see, the fundamental difficulty is the desire to use the white-noise concept even though it involves many mathematical problems. In this section a simple example will be considered to illustrate some of these problems. Then we shall examine a rather unusual class of stochastic processes known as Wiener processes, which will give a firmer mathematical foundation to the white-noise concept.

Many random processes occurring in nature are approximately gaussian, approximately stationary, and have a power spectrum approximately flat up to frequencies far higher than the maximum frequency at which a system is capable of significant response. Thus it is convenient to use the white-noise process, even though such a process cannot have physical meaning since it requires infinite signal power. The fact that white noise is not physically realizable is no reason to disregard the white-noise concept any more than it would be reason to disregard the linear system impulse response concept because impulse inputs do not often exist in practice. Impulse or Dirac delta functions play vital roles as limiting functions, and so does the white-noise process.

We shall say that a continuous gaussian process $\mathbf{w}(t)$ is a zero-mean white-noise process if

$$\mathcal{E}\{\mathbf{w}(t)\} = \mathbf{0} \qquad \operatorname{cov}\{\mathbf{w}(t), \mathbf{w}(\tau)\} = \mathbf{\Psi}_{\mathbf{w}}(t)\delta_D(t - \tau) \tag{8.3-1}$$

Clearly, there is a lack of rigor in this definition, since the Dirac delta function can only be rigorously defined in terms of the integral

$$\int_{-\infty}^{\infty} f(\tau)\delta_D(\tau - \lambda)\, d\tau = f(\lambda) \tag{8.3-2}$$

To illustrate the problem that the use of the white-noise concept may cause, let us consider the variance equation for the linear system

$$\dot{\mathbf{x}}(t) = \mathbf{F}(t)\mathbf{x}(t) + \mathbf{G}(t)\mathbf{w}(t) \tag{8.3-3}$$

with $\mathcal{E}\{\mathbf{x}(t_0)\} = \mathbf{0}$ and $\mathbf{V}_{\mathbf{x}}(t_0) = \mathbf{V}_0$. In Section 6.2 the variance equation was developed by differentiating the expression

$$\mathbf{V}_{\mathbf{x}}(t) = \operatorname{var}\{\mathbf{x}(t)\} = \mathcal{E}\{\mathbf{x}(t)\mathbf{x}^T(t)\} \tag{8.3-4}$$

to obtain

$$\begin{aligned}\dot{\mathbf{V}}_{\mathbf{x}}(t) &= \mathcal{E}\{\dot{\mathbf{x}}(t)\mathbf{x}^T(t)\} + \mathcal{E}\{\mathbf{x}(t)\dot{\mathbf{x}}^T(t)\} \\ &= \mathbf{F}(t)\mathcal{E}\{\mathbf{x}(t)\mathbf{x}^T(t)\} + \mathbf{G}(t)\mathcal{E}\{\mathbf{w}(t)\mathbf{x}^T(t)\} \\ &\quad + \mathcal{E}\{\mathbf{x}(t)\mathbf{x}^T(t)\}\mathbf{F}^T(t) + \mathcal{E}\{\mathbf{x}(t)\mathbf{w}^T(t)\}\mathbf{G}^T(t)\end{aligned}$$

We recognize the definition of the variance expression and have

$$\dot{\mathbf{V}}_{\mathbf{x}}(t) = \mathbf{F}(t)\mathbf{V}_{\mathbf{x}}(t) + \mathbf{V}_{\mathbf{x}}(t)\mathbf{F}^T(t) + \mathbf{G}(t)\mathcal{E}\{\mathbf{w}(t)\mathbf{x}^T(t)\} + \mathcal{E}\{\mathbf{x}(t)\mathbf{w}^T(t)\}\mathbf{G}^T(t) \tag{8.3-5}$$

To evaluate this expression, we need to find $\mathcal{E}\{\mathbf{x}(t)\mathbf{w}^T(t)\}$. If we assume that $\mathcal{E}\{\mathbf{x}(t_0)\mathbf{w}^T(t)\} = \mathbf{0}$ for all $t \geq t_0$, we have

$$\begin{aligned}\mathcal{E}\{\mathbf{x}(t)\mathbf{w}^T(t)\} &= \int_{t_0}^{t} \mathbf{\Phi}(t, \tau)\mathbf{G}(\tau)\mathcal{E}\{\mathbf{w}(\tau)\mathbf{w}^T(t)\}\, d\tau \\ &= \int_{t_0}^{t} \mathbf{\Phi}(t, \tau)\mathbf{G}(\tau)\mathbf{\Psi}_{\mathbf{w}}(\tau)\delta_D(\tau - t)\, d\tau \end{aligned} \tag{8.3-6}$$

Because the impulse occurs at the end of the integration interval, the value of this integral depends on the type of delta function used. In Chapter 6 a symmetrical delta function was used with one half its (unit) area to the right of $\tau = t$ and half to the left. In this case, Eq. (8.3-6) becomes

$$\mathcal{E}\{\mathbf{x}(t)\mathbf{w}^T(t)\} = \tfrac{1}{2}\mathbf{G}(t)\mathbf{\Psi}_{\mathbf{w}}(t) \tag{8.3-7}$$

In a similar way, we have

$$\mathcal{E}\{\mathbf{w}(t)\mathbf{x}^T(t)\} = \tfrac{1}{2}\mathbf{\Psi}_{\mathbf{w}}(t)\mathbf{G}^T(t) \tag{8.3-8}$$

and the resulting variance equation is

$$\dot{\mathbf{V}}_{\mathbf{x}}(t) = \mathbf{F}(t)\mathbf{V}_{\mathbf{x}}(t) + \mathbf{V}_{\mathbf{x}}(t)\mathbf{F}^T(t) + \mathbf{G}(t)\mathbf{\Psi}_{\mathbf{w}}(t)\mathbf{G}^T(t) \tag{8.3-9}$$

If we were to use a nonsymmetrical impulse function with

$$\int_{0}^{\infty} \delta_D(t)\, dt = 1 - a = 1 - \int_{\infty}^{0} \delta_D(t)\, dt$$

where $a \leq 1$, we would find that Eqs. (8.3-7) and (8.3-8) become

$$\mathcal{E}\{\mathbf{w}(t)\mathbf{x}^T(t)\} = (1 - a)\mathbf{\Psi}_{\mathbf{w}}(t)\mathbf{G}^T(t) \tag{8.3-10}$$

$$\mathcal{E}\{\mathbf{x}(t)\mathbf{w}^T(t)\} = a\mathbf{G}(t)\mathbf{\Psi}_{\mathbf{w}}(t) \tag{8.3-11}$$

and we still obtain Eq. (8.3-9). However, we are now faced with a nonunique expression for $\mathcal{E}\{\mathbf{x}(t)\mathbf{w}^T(t)\}$.

We could rationalize the use of the symmetric delta function by recalling that it is part of a covariance function which must of necessity be symmetric. For linear systems, this approach, applied *very* carefully, can be used without difficulty. However, in the nonlinear case it is necessary to develop a new way of handling the problem.

It is of interest to consider a discrete approach and then let the samples become dense to see if the same uniqueness difficulty arises. We shall form a discrete time system that will approach the continuous time system of Eq. (8.3-3) as the samples become dense. We shall use the discrete system

$$\mathbf{x}(\overline{k+1T}) = \mathbf{\Phi}(\overline{k+1T}, kT)\mathbf{x}(kT) + \mathbf{\Gamma}(kT)\mathbf{w}(kT) \tag{8.3-12}$$

where

$$\mathbf{\Phi}(\overline{k+1T}, kT) = \mathbf{I} + T\mathbf{F}(kT) \tag{8.3-13}$$

$$\mathbf{\Gamma}(kT) = T\mathbf{G}(kT) \tag{8.3-14}$$

To demonstrate that this discrete model approaches the continuous model of Eq. (8.3-3), as T approaches zero and kT equals t, we use the definition of the time derivative to write

$$\dot{\mathbf{x}}(t) = \lim_{\substack{T\to 0\\ kT=t}} \frac{\mathbf{x}(\overline{k+1T}) - \mathbf{x}(kT)}{T} = \lim_{\substack{T\to 0\\ kT=t}} \{[\mathbf{\Phi}(\overline{k+1T}, kT) - \mathbf{I}]\mathbf{x}(kT) + \frac{\mathbf{\Gamma}(kT)\mathbf{w}(kT)\}}{T}$$

Now substituting for $\mathbf{\Phi}$ and $\mathbf{\Gamma}$ from Eqs. (8.3-13) and (8.3-14), we obtain

$$\dot{\mathbf{x}}(t) = \lim_{\substack{T\to 0\\ kT=t}} \{\mathbf{F}(kT)\mathbf{x}(kT) + \mathbf{G}(kT)\mathbf{w}(kT)\} = \mathbf{F}(t)\mathbf{x}(t) + \mathbf{G}(t)\mathbf{w}(t)$$

The discrete white noise $\mathbf{w}(kT)$ is defined as

$$\mathcal{E}\{\mathbf{w}(kT)\} = \mathbf{0}$$

$$\operatorname{cov}\{\mathbf{w}(kT), \mathbf{w}(jT)\} = \frac{\mathbf{\Psi}_{\mathbf{w}}(kT)}{T}\delta_K(k-j) \tag{8.3-15}$$

This discrete white noise will become continuous white noise as T approaches zero since

$$\lim_{\substack{T\to 0\\ kT=t\\ jT=\tau}} \mathbf{\Psi}_{\mathbf{w}}(kT)\frac{\delta_K(k-j)}{T} = \mathbf{\Psi}_{\mathbf{w}}(t)\delta_D(t-\tau)$$

The variance equation for this discrete system was developed in Section 6.4 as

$$\mathbf{V}_{\mathbf{x}}(\overline{k+1T}) = \mathbf{\Phi}(\overline{k+1T}, kT)\mathbf{V}_{\mathbf{x}}(kT)\mathbf{\Phi}^T(\overline{k+1T}, kT) + \mathbf{\Gamma}(kT)\mathbf{V}_{\mathbf{w}}(kT)\mathbf{\Gamma}^T(kT) \tag{8.3-16}$$

Now we let the samples become dense and have

$$\dot{\mathbf{V}}_{\mathbf{x}} = \frac{d\mathbf{V}_{\mathbf{x}}(t)}{dt} = \lim_{\substack{T\to 0\\ kT=t}} \frac{\mathbf{V}_{\mathbf{x}}(k+1T) - \mathbf{V}_{\mathbf{x}}(kT)}{T} = \mathbf{F}(t)\mathbf{V}_{\mathbf{x}}(t) + \mathbf{V}_{\mathbf{x}}(t)\mathbf{F}^T(t) + \mathbf{G}(t)\mathbf{\Psi}_{\mathbf{w}}(t)\mathbf{G}^T(t) \tag{8.3-17}$$

We recognize this equation as the variance equation for the continuous case. In the discrete case, note that $\mathcal{E}\{\mathbf{x}(kT)\mathbf{w}^T(kT)\} = \mathbf{0}$ for all k. Thus the use of a discrete time model, if carefully handled, provides another method of correctly dealing with the white-noise problem. This approach becomes somewhat unwieldy and full of possible pitfalls when nonlinear systems are involved. The stochastic calculus, discussed in Section 8.4, provides a rigorous method for handling the problem. Central to the study of stochastic calculus is the Wiener process.

The Wiener process was devised by Wiener as a simple model for brownian motion. Let $u(t)$ denote the displacement from the origin at time t of a particle. A "brownian" particle undergoes motion caused by force-field impacts upon the particle. The displacement of the particle over the time interval s to t, which is quite long compared to the time between impacts, is the sum of a large number of small disturbances and therefore subject to application of the central limit theorem. Thus $u(s) - u(t)$ has a gaussian density. The probability density of the displacement from time s to time t is the same as from time $s + \tau$ to time $t + \tau$, since the density of the particles displacement depends upon the length of the time interval and not upon the specific time of observation.

Definition 8.3-1. ***Wiener Process:*** The stochastic process $\mathbf{u}(t)$, $t \geq 0$, is a Wiener process if
1. $\mathbf{u}(t)$ has stationary independent increments.
2. $\mathbf{u}(t)$ is gaussianly distributed for all $t > 0$.
3. $\mathcal{E}\{\mathbf{u}(t)\} = \mathbf{0}$ for all $t > 0$.
4. $\mathbf{u}(0) = \mathbf{0}$.

A stochastic process $\mathbf{x}(t)$ is said to have *independent increments* if $\mathbf{x}(0) = \mathbf{0}$ and, for all $0 < t_1 < t_2 < \cdots < t_{n-1} < t_n$, the n random variables $\mathbf{x}(t_1) - \mathbf{x}(0)$, $\mathbf{x}(t_2) - \mathbf{x}(t_1)$, $\cdots$, $\mathbf{x}(t_n) - \mathbf{x}(t_{n-1})$ are independent. If $\mathbf{x}(t_i) - \mathbf{x}(t_{i-1})$ and $\mathbf{x}(t_i + h) - \mathbf{x}(t_{i-1} + h)$ have the same distribution for all $h > 0$, the increments are said to be *stationary*.

We wish to show that the process $\mathbf{u}(t)$ defined by

$$\mathbf{u}(t) = \int_0^t \mathbf{w}(\lambda)\, d\lambda \tag{8.3-18}$$

where $\mathbf{w}(\lambda)$ is a zero-mean, white, gaussian process with covariance parameter $\mathbf{\Psi}_\mathbf{w}$, is a Wiener process. Since $\mathbf{u}(0)$ is given by

$$\mathbf{u}(0) = \int_0^0 \mathbf{w}(\lambda)\, d\lambda$$

it is easy to conclude that $\mathbf{u}(0) = \mathbf{0}$ as required. For all $t > 0$, the $\mathcal{E}\{\mathbf{u}(t)\} = \mathbf{0}$, since

$$\mathcal{E}\{\mathbf{u}(t)\} = \mathcal{E}\left\{\int_0^t \mathbf{w}(\lambda)\, d\lambda\right\} = \int_0^t \mathcal{E}\{\mathbf{w}(\lambda)\}\, d\lambda = \mathbf{0}$$

Because $\mathbf{w}(t)$ is gaussian, $\mathbf{u}(t)$ is also gaussian for all $t > 0$ since it is obtained by a linear operation on a gaussian process. Hence, to demonstrate that $\mathbf{u}(t)$ is a Wiener process, it is only necessary to show that $\mathbf{u}(t)$ has stationary independent increments.

The increments $\mathbf{u}(\tau) - \mathbf{u}(s)$, $s < \tau$, are gaussian, because the process $\mathbf{u}(t)$ is gaussian. In addition, the increments have zero mean since $\mathbf{u}(t)$ has zero mean. Therefore, we can find the density of the increments by finding the variance of the increment $\mathbf{u}(\tau) - \mathbf{u}(s)$. This variance is given by

$$\begin{aligned}\text{var}\{\mathbf{u}(\tau) - \mathbf{u}(s)\} &= \text{var}\left\{\int_0^\tau \mathbf{w}(\lambda)\, d\lambda - \int_0^s \mathbf{w}(\lambda)\, d\lambda\right\} \\ &= \text{var}\left\{\int_s^\tau \mathbf{w}(\lambda)\, d\lambda\right\}\end{aligned}$$

Now, interchanging the orders of the integration and variance operators, we obtain

$$\text{var}\{\mathbf{u}(\tau) - \mathbf{u}(s)\} = \int_s^\tau \text{var}\{\mathbf{w}(\lambda)\}\, d\lambda = (\tau - s)\mathbf{\Psi}_\mathbf{w} \tag{8.3-19}$$

Hence we may conclude that the increments are stationary since they depend only on the time difference $(\tau - s)$.

To show that the increments are independent, let us consider the covariance of the increments $\mathbf{u}(\tau) - \mathbf{u}(s)$ and $\mathbf{u}(s) - \mathbf{u}(p)$, where $\tau > s > p$. Since we know that

$$\mathbf{u}(\tau) - \mathbf{u}(s) = \int_s^\tau \mathbf{w}(\lambda)\, d\lambda$$

and

$$\mathbf{u}(s) - \mathbf{u}(p) = \int_p^s \mathbf{w}(\gamma)\, d\gamma$$

we see that

$$\operatorname{cov}\{\mathbf{u}(\tau)-\mathbf{u}(s), \mathbf{u}(s)-\mathbf{u}(p)\} = \operatorname{cov}\left\{\int_s^\tau \mathbf{w}(\lambda)\,d\lambda, \int_p^s \mathbf{w}(\gamma)\,d\gamma\right\}$$

$$= \int_s^\tau d\lambda \int_p^s d\gamma \operatorname{cov}\{\mathbf{w}(\lambda), \mathbf{w}(\gamma)\} \quad (8.3\text{-}20)$$

Since $\mathbf{w}(t)$ is a white gaussian process with covariance $\operatorname{cov}\{\mathbf{w}(t), \mathbf{w}(\tau)\} = \mathbf{\Psi}_{\mathbf{w}}\delta_D(t-\tau)$, Eq. (8.3-20) becomes

$$\operatorname{cov}\{\mathbf{u}(\tau)-\mathbf{u}(s), \mathbf{u}(s)-\mathbf{u}(p)\} = \mathbf{\Psi}_{\mathbf{w}} \int_s^\tau d\lambda \int_p^s d\gamma\, \delta_D(\lambda - r)$$

Note that

$$\int_p^s \delta_D(\lambda-\gamma)\,d\gamma = \begin{cases} a & 0 \le a \le 1 \quad \text{if } \lambda = s \\ 0 & \text{otherwise} \end{cases}$$

so that the right side of the previous equation becomes

$$\mathbf{\Psi}_{\mathbf{w}} \int_s^\tau d\lambda \int_p^s d\gamma \delta_D(\lambda-\gamma) = 0$$

Therefore, we have established that the increments are independent in addition to being stationary. Hence we have satisfied all the requirements of the Wiener process, and we may conclude that $\mathbf{u}(t)$ defined by Eq. (8.3-18) is a Wiener process.

In Section 8.4 we shall show how the problem associated with the white-noise concept can be eliminated by the use of the Wiener process. Before leaving this section, let us note a few additional properties of the Wiener process.

1. The Wiener process is a Markov process in that

$$p_{\mathbf{u}(t_3)\mid\mathbf{u}(t_1),\mathbf{u}(t_2)}(\gamma_3 \mid \gamma_1, \gamma_2) = p_{\mathbf{u}(t_3)\mid\mathbf{u}(t_2)}(\gamma_3 \mid \gamma_2) \qquad t_1 \le t_2 \le t_3 \quad (8.3\text{-}21)$$

2. The Wiener process is a *martingale process* in that the conditional expectation of $\mathbf{u}(t_k)$, given the values $\mathbf{u}(t_0), \mathbf{u}(t_1), \ldots, \mathbf{u}(t_{k-1})$, is equal to the most recently observed value, $\mathbf{u}(t_{k-1})$, where $t_k > t_{k-1} > \cdots > t_0$. Thus

$$\mathcal{E}\{\mathbf{u}(t_k) \mid \mathbf{u}(t_{k-1}), \mathbf{u}(t_{k-2}), \ldots, \mathbf{u}(t_0)\} = \mathbf{u}(t_{k-1}) \quad (8.3\text{-}22)$$

3. The Wiener process possesses the *Lévy oscillation property* in that if $\tau = t_0, t_1, \ldots, t_K$ is a partition of the interval $[0, T]$, such that $t_0 = 0$, $t_K = T$, then (Doob [1])

$$\underset{\max(t_k - t_{k-1})\to 0}{\text{l. i. m.}} \sum_{k=1}^{K} [\mathbf{u}(t_k) - \mathbf{u}(t_{k-1})][\mathbf{u}(t_k) - \mathbf{u}(t_{k-1})]^T = T\mathbf{\Psi}_{\mathbf{w}} \quad (8.3\text{-}23)$$

This property is sometimes represented symbolically as $\mathbf{du}\,\mathbf{du}^T = [\mathbf{u}(t + dt) - \mathbf{u}(t)][\mathbf{u}(t + dt) - \mathbf{u}(t)]^T \sim \mathbf{\Psi}_w \, dt$. This property will play an important role in the developments of Section 8.4.

EXERCISES 8.3

8.3-1. Show that the covariance of the Wiener process $u(t)$ is given by

$$\text{cov}\{u(t), u(s)\} = \Psi_w \min\{t, s\}$$

8.3-2. Prove that the Wiener process is a Markov process.

8.3-3. Show that the Wiener process is a martingale process.

8.4. Stochastic differential equations

In the preceding section we considered the Wiener process. Doob [1] has shown that the sample functions of a Wiener process are continuous functions but do not have bounded variation and are almost nowhere differentiable. In a nonrigorous fashion, this last statement follows from the fact that the increment $\mathbf{\Delta u} = \mathbf{u}(t + \Delta t) - \mathbf{u}(t)$ is of the order $\sqrt{\Delta t}$ (Lévy oscillation property). Hence we see that the derivative $\mathbf{\Delta u}/\Delta t$ diverges as Δt approaches zero, since it is of the form $\sqrt{\Delta t}/\Delta t$.

Thus, if $\mathbf{u}(t)$ is a Wiener process, $\mathbf{du}/dt$ has no meaning, and we may question whether the Riemann integral

$$\mathbf{I}(t) = \int_0^t \mathbf{\Gamma}(t, \lambda) \frac{d\mathbf{u}(\lambda)}{d\lambda} \, d\lambda$$

has any meaning. We have worked with integrals of this type before when we wrote the response of a linear system excited by white noise,

$$\dot{\mathbf{x}} = \mathbf{F}(t)\mathbf{x}(t) + \mathbf{G}(t)\mathbf{w}(t) \qquad \mathbf{x}(0) = \mathbf{0}$$

as

$$\mathbf{x}(t) = \int_0^t \mathbf{\Phi}(t, \lambda)\mathbf{G}(\lambda)\mathbf{w}(\lambda) \, d\lambda = \int_0^t \mathbf{\Gamma}(t, \lambda) \frac{\mathbf{du}(\lambda)}{d\lambda} \, d\lambda$$

where $\mathbf{\Gamma}(t, \lambda) = \mathbf{\Phi}(t, \lambda)\mathbf{G}(\lambda)$. This result is meaningless since $\mathbf{du}(t)/dt$ is not defined.

We might attempt to use concepts from *Lebesgue–Stieltjes integration* and write

$$\mathbf{I}(t) = \int_0^t \mathbf{\Gamma}(t, \lambda)\frac{\mathbf{du}(\lambda)}{d\lambda}d\lambda = \int_0^t \mathbf{\Gamma}(t, \lambda)\,\mathbf{du}(\lambda)$$

However, this does not alleviate the difficulty, since $\mathbf{u}(t)$ is not a function of bounded variation and so the Lebesgue–Stieltjes integral is not defined.

The stochastic integral is a statistically meaningful generalization of the Lebesgue–Stieltjes integral that allows us to resolve this problem. Let $\xi(t)$ be a stochastic process and $\mathbf{\Gamma}(t, \lambda)$ be a piecewise continuous arbitrary matrix function, for all $\lambda \in [0, t]$, that depends at most on the present and past values of ξ, that is, $\xi(\tau)$, $0 \leq \tau \leq \lambda$. This requirement means that

$$\operatorname{cov}\{\mathbf{\Gamma}(t, \lambda_1), \xi(\lambda_2)\} = \mathbf{0} \qquad \text{for all } \lambda_2 > \lambda_1$$

Let $\mathbf{\Xi}$ denote the set of functions $\mathbf{\Gamma}(t, \lambda)$ that can be assigned a probability, and identify three special subclasses of the set $\mathbf{\Xi}$:

I. $\mathbf{\Xi}^0$ is the set of functions in $\mathbf{\Xi}$ that are piecewise constant for $\lambda \in [0, t]$.

II. $\mathbf{\Xi}^1$ is the set of functions in $\mathbf{\Xi}$ that are mean-square integrable, $\lambda \in [0, t]$.

III. $\mathbf{\Xi}^2$ is the set of functions in $\mathbf{\Xi}$ that are square integrable with probability 1, $\lambda \in [0, t]$.

If $\mathbf{\Gamma}(t, \lambda)$ is in the set $\mathbf{\Xi}^0$, there are points in time $\lambda = t_1, t_2, \ldots, t_i, \ldots,$ such that $\mathbf{\Gamma}(t, \lambda) = \mathbf{\Gamma}(t, t_i)$ for $t_i \leq \lambda < t_{i+1}$, and we may define stochastic or Ito integration by (where $⨍$ represents stochastic integration)

$$\mathbf{I}(t) = ⨍_0^t \mathbf{\Gamma}(t, \lambda)\, d\xi(\lambda) = \sum_i \mathbf{\Gamma}(t, t_i)[\xi(t_{i+1}) - \xi(t_i)] \tag{8.4-1}$$

where the $\{t_i\}$ are the jump points of $\mathbf{\Gamma}(t, \lambda)$. When $\mathbf{\Gamma}(t, \lambda) \in \mathbf{\Xi}$ but is not a member of $\mathbf{\Xi}^0$, then standard limit procedures (Doob [1]) are used to extend the definition of Eq. (8.4-1) to the form

$$\mathbf{I}(t) = ⨍_0^t \mathbf{\Gamma}(t, \lambda)\, d\xi(\lambda) = \underset{\substack{k \to \infty \\ \delta \to 0}}{\text{l.i.m.}} \sum_{i=1}^{k-1} \mathbf{\Gamma}(t, t_i)[\xi(t_{i+1}) - \xi(t_i)] \tag{8.4-2}$$

where l.i.m. indicates limit in the mean (as $k \longrightarrow \infty$ and $\delta \longrightarrow 0$), and $t = t_k > t_{k-1} > t_{k-2} > \cdots > t_1 = 0$, and $\delta = \max(t_{i+1} - t_i)$, for all i.

It should be noted that there are other ways in which stochastic integration could be defined, such as

$$\mathbf{I}^*(t) = \underset{\substack{k \to \infty \\ \delta \to 0}}{\text{l.i.m.}} \sum_{i=1}^{k-1} \mathbf{\Gamma}(t, t_i')[\xi(t_{i+1}) - \xi(t_i)] \tag{8.4-3}$$

where $t_{i-1} > t'_i > t_i$ or, for arbitrarily small positive ϵ,

$$\mathbf{I}^{**}(t) = \underset{\substack{k\to\infty \\ \delta\to 0}}{\text{l.i.m.}} \sum_{i=1}^{k-1} \left[\frac{\mathbf{\Gamma}(t, t_{i+1} - \epsilon) + \mathbf{\Gamma}(t, t_i)}{2}\right][\boldsymbol{\xi}(t_{i+1}) - \boldsymbol{\xi}(t_i)] \quad (8.4\text{-}4)$$

When $\mathbf{\Gamma}$ is in the set Ξ^0, it is not difficult to see that the three definitions of Eqs. (8.4-2) to (8.4-4), along with many other possible definitions, are all equivalent. However, if $\mathbf{\Gamma}$ is not a member of Ξ^0, the definitions are, in general, not equivalent because of the Lévy oscillation property. Although there are certain advantages associated with the definitions of Eqs. (8.4-3) and (8.4-4), we shall show later that the definition of Eq. (8.4-2) is normally the more convenient. The relationships among the various definitions of the stochastic integral and the ordinary calculus will be discussed later.

The integral defined in Eq. (8.4-2) is referred to as the *Ito integral* (Skorokhod [6]); it has several interesting properties. The Ito integral represents a linear operation on $\mathbf{\Gamma}$ in the sense that

$$\int_0^t [\alpha\mathbf{\Gamma}_1(t, \lambda) + \beta\mathbf{\Gamma}_2(t,\lambda)]\, d\boldsymbol{\xi}(\lambda) = \alpha \int_0^t \mathbf{\Gamma}_1(t, \lambda)\, d\boldsymbol{\xi}(\lambda) + \beta \int_0^t \mathbf{\Gamma}_2(t, \lambda)\, d\boldsymbol{\xi}(\lambda) \quad (8.4\text{-}5)$$

for every admissible $\mathbf{\Gamma}_1$ and $\mathbf{\Gamma}_2$ and real α and β. If $\mathbf{\Gamma}$ is in Ξ^1 and $\boldsymbol{\xi}(t)$ is a Wiener process $\mathbf{u}(t)$, the stochastic integral has the following two properties that are useful later:

$$\mathcal{E}\left\{\int_0^t \mathbf{\Gamma}(t, \lambda)\, \mathbf{du}(\lambda)\right\} = \mathbf{0} \quad (8.4\text{-}6)$$

and

$$\mathcal{E}\left\{\left[\int_0^t \mathbf{\Gamma}(t, \lambda)\, \mathbf{du}(\lambda)\right]\left[\int_0^t \mathbf{\Gamma}(t, \lambda)\, \mathbf{du}(\tau)\right]^T\right\} = \int_0^t \mathcal{E}\{\mathbf{\Gamma}(t, \lambda)\mathbf{\Psi}_w\mathbf{\Gamma}^T(t, \lambda)\}\, d\lambda \quad (8.4\text{-}7)$$

The proof of these properties is based on the fact that the increment $[\mathbf{u}(t_{i+1}) - \mathbf{u}(t_i)]$ is independent of $\mathbf{\Gamma}(t, t_j)$ by assumption, for $j \leq i$, and of $[\mathbf{u}(t_{j+1}) - \mathbf{u}(t_j)]$, $j \neq i$, owing to the independent increment property of the Wiener process.

For simplicity, we carry out the proof assuming that $\boldsymbol{\eta} \in \Xi^0$; the general case follows directly. By using Eq. (8.4-1), we may write Eq. (8.4-6) as

$$\begin{aligned}\mathcal{E}\left\{\int_0^t \mathbf{\Gamma}(t, \lambda)\, \mathbf{du}(\lambda)\right\} &\triangleq \mathcal{E}\{\sum_i \mathbf{\Gamma}(t, t_i)[\mathbf{u}(t_{i+1}) - \mathbf{u}(t_i)]\} \\ &= \sum_i \mathcal{E}\{\mathbf{\Gamma}(t, t_i)[\mathbf{u}(t_i + 1) - \mathbf{u}(t_i)]\}\end{aligned}$$

Using the fact that $\mathbf{\Gamma}(t, t_i)$ is independent of $[\mathbf{u}(t_{i+1}) - \mathbf{u}(t_i)]$, we can write the expected value of the product as the product of the expected values, or

$$\mathcal{E}\left\{\int_0^t \mathbf{\Gamma}(t, \lambda)\, \mathbf{du}(\lambda)\right\} = \sum_i \mathcal{E}\{\mathbf{\Gamma}(t, t_i)\}\mathcal{E}\{\mathbf{u}(t_{i+1}) - \mathbf{u}(t_i)\}$$

Since the Wiener process has zero-mean increments, the right side of this equation is zero and we have established the validity of Eq. (8.4-6). The proof of Eq. (8.4-7) follows a similar pattern. If $\boldsymbol{\eta}$ is not in $\mathbf{\Xi}^0$ and an integration rule different than Eq. (8.4-2) is used, then Eqs. (8.4-6) and (8.4-7) are, in general, not valid. This fact is one of the major reasons for selecting the definition of Eq. (8.4-2), since Eqs. (8.4-6) and (8.4-7) will be useful in later developments.

If the function $\mathbf{\Gamma}(t, \lambda)$ is not a function of t, the Ito integral, as defined by Eq. (8.4-2), is a martingale process so that

$$\mathcal{E}\{\mathbf{I}(t) \,|\, \mathbf{u}(\lambda), 0 \leq \tau \leq s < t\} = \mathbf{I}(s) \tag{8.4-8}$$

The proof of this property follows directly from Eq. (8.4-6) by writing $\mathbf{I}(t)$ as

$$\mathbf{I}(t) = \int_0^t \mathbf{\Gamma}(\lambda)\, \mathbf{du}(\lambda) = \int_0^s \mathbf{\Gamma}(\lambda)\, \mathbf{du}(\lambda) + \int_s^t \mathbf{\Gamma}(\lambda)\, \mathbf{du}(\lambda)$$

Therefore, the conditional mean of $\mathbf{I}(t)$ required in Eq. (8.4-8) is

$$\begin{aligned}\mathcal{E}\{\mathbf{I}(t) \,|\, \mathbf{u}(\tau), 0 \leq \tau \leq s < t\} &= \mathcal{E}\left\{\int_0^s \mathbf{\Gamma}(\lambda)\, \mathbf{du}(\lambda) \,|\, \mathbf{u}(\tau), 0 \leq \tau \leq s < t\right\} \\ &\quad + \mathcal{E}\left\{\int_s^t \mathbf{\Gamma}(\lambda)\, \mathbf{du}(\lambda) \,|\, \mathbf{u}(\tau), 0 \leq \tau \leq s < t\right\}\end{aligned}$$

In the first integral, $\mathbf{u}$ is completely known for the time range of integration, since it is given in the conditioning argument, and the integral becomes deterministic. In the second integral, on the other hand, the conditioning has no effect (owing to the independent increment nature of $\mathbf{u}$) and the expected value is zero by Eq. (8.4-6), so we have

$$\mathcal{E}\{\mathbf{I}(t) \,|\, \mathbf{u}(\tau), 0 \leq \tau \leq s < t\} = \int_0^s \mathbf{\Gamma}(\lambda)\, \mathbf{du}(\lambda) = \mathbf{I}(s)$$

In later developments we shall have need for the variance of the differential increment $\mathbf{du}$. From Eq. (8.3-19) we may write

$$\text{var}\{\mathbf{u}(s) - \mathbf{u}(t)\} = \mathbf{\Psi}_{\mathbf{w}} |s - t|$$

Now if we let $s = t + dt$, then $\mathbf{u}(s) - \mathbf{u}(t)$ becomes $\mathbf{du}(t)$, and we have, for $dt \geq 0$,

$$\text{var}\{\mathbf{du}(t)\} = \mathbf{\Psi}_{\mathbf{w}}\, dt$$

It is not difficult to show that the higher moments (higher than the first moment) of the quantity $[\mathbf{du}(t)\,\mathbf{du}^T(t)]$ are of order greater than dt. Hence $\mathcal{E}\{\mathbf{du}(t)\,\mathbf{du}^T(t)\}/dt$ is constant and equal to $\boldsymbol{\Psi}_w$ for vanishingly small dt, whereas $\mathcal{E}\{\mathbf{du}(t)\,\mathbf{du}^T(t)]^k\}/dt = \mathbf{0}$ for $k = 2, 3, \ldots$. Hence we conclude that the process $\mathbf{du}(t)\,\mathbf{du}^T(t)/dt$ is essentially deterministic and equal to $\boldsymbol{\Psi}_w$ for infinitesimally small dt, since the expected value or mean of the process is $\boldsymbol{\Psi}_w$, whereas all the higher-order moments are zero. Therefore, we have established that

$$\mathcal{E}\left\{\frac{\mathbf{du}(t)\,\mathbf{du}^T(t)}{dt}\right\} = \frac{\mathbf{du}(t)\,\mathbf{du}^T(t)}{dt} = \boldsymbol{\Psi}_w$$

which means that

$$\mathcal{E}\{\mathbf{du}(t)\,\mathbf{du}^T(t)\} = \mathbf{du}(t)\,\mathbf{du}^T(t) = \boldsymbol{\Psi}_w\,dt \tag{8.4-9}$$

We see that the Wiener process is indeed unusual since it is everywhere continuous but almost nowhere differentiable, and the variance of the infinitesimal increment is just the "square" of the increment. In a similar manner, one can show that

$$dt\,\mathbf{du}(t) = \mathbf{0} \tag{8.4-10}$$

We now wish to turn our attention to study of the response of the nonlinear system

$$\frac{d\mathbf{x}(t)}{dt} = \mathbf{f}[\mathbf{x}(t), t] + \mathbf{G}[\mathbf{x}(t), t]\mathbf{w}(t) \tag{8.4-11}$$

to a zero-mean, gaussian, white-noise input R vector $\mathbf{w}(t)$ with $\mathcal{E}\{\mathbf{w}(t)\} = \mathbf{0}$, $\operatorname{cov}\{\mathbf{w}(t), \mathbf{w}(\tau)\} = \boldsymbol{\Psi}_w(t)\delta_D(t-\tau)$, where $\mathbf{x}(t)$ is an N-dimensional random state vector, $\mathbf{f}[\mathbf{x}(t), t]$ is an N-dimensional nonlinear vector function of $\mathbf{x}(t)$ and t, and $\mathbf{G}[\mathbf{x}(t), t]$ is an $N \times R$ matrix function of $\mathbf{x}(t)$ and t.

We can (implicitly) find the response or state vector by formal integration of Eq. (8.4-11) to obtain

$$\mathbf{x}(t) - \mathbf{x}(t_0) = \int_{t_0}^{t} \mathbf{f}[\mathbf{x}(\lambda), \lambda]\,d\lambda + \int_{t_0}^{t} \mathbf{G}[\mathbf{x}(\lambda), \lambda]\,\mathbf{du}(\lambda) \tag{8.4-12}$$

where $\mathbf{du}(t)$ is a Wiener process, $\mathbf{du}(t) = \mathbf{dw}(t)\,dt$, that has the moments $\mathcal{E}\{\mathbf{du}(t)\} = \mathbf{0}$ and $\operatorname{var}\{\mathbf{du}(t)\} = \boldsymbol{\Psi}_w(t)\,dt$. We note that the first integral to the right of the equality in Eq. (8.4-12) is an ordinary integral, whereas the second integral is a stochastic integral in the general case in which $\mathbf{G}$ is a function of $\mathbf{x}(t)$. Equation (8.4-12) will be represented in the symbolic shorthand form as

$$\mathbf{dx}(t) = \mathbf{f}[\mathbf{x}(t), t]\,dt + \mathbf{G}[\mathbf{x}(t), t]\mathbf{du} \tag{8.4-13}$$

which is referred to as a stochastic differential equation. We shall say that $\mathbf{x}(t)$ is an *Ito process* relative to $\mathbf{f}$ and $\mathbf{G}$ if the stochastic motion of $\mathbf{x}(t)$ is given by Eq. (8.4-12) or symbolically by Eq. (8.4-13).

In carrying out operations on Ito processes, it is necessary to use rules which differ from those for ordinary stochastic processes. If $\psi[\mathbf{x}(t), t]$ is a scalar function of an Ito process $\mathbf{x}(t)$ and t with continuous second-order partial derivatives in both $\mathbf{x}$ and t, the *Ito differential rule* states that ψ is also an Ito process described by

$$d\psi = \left[\frac{\partial \psi}{\partial t} + \left(\frac{\partial \psi}{\partial \mathbf{x}}\right)^T \mathbf{f} + \frac{1}{2}\operatorname{tr}\left\{\frac{\partial^2 \psi}{\partial \mathbf{x}^2}\mathbf{G}\mathbf{\Psi}_w\mathbf{G}^T\right\}\right] dt + \left(\frac{\partial \psi}{\partial \mathbf{x}}\right)^T \mathbf{G}\,\mathbf{du} \tag{8.4-14}$$

The standard chain rule of differentiation would give

$$\begin{aligned} d\psi &= \left(\frac{\partial \psi}{\partial \mathbf{x}}\right)^T d\mathbf{x} + \frac{\partial \psi}{\partial t}\,dt \\ &= \left[\left(\frac{\partial \psi}{\partial \mathbf{x}}\right)^T \mathbf{f} + \frac{\partial \psi}{\partial t}\right] dt + \left(\frac{\partial \psi}{\partial \mathbf{x}}\right)^T \mathbf{G}\,\mathbf{du} \end{aligned}$$

Note that this expression does not agree with the Ito differential rule; the discrepancy is due to the Lévy oscillation property.

We may derive the Ito differential rule of Eq. (8.4-14) in a formal manner by expanding $\psi[\mathbf{x}(t) + \mathbf{dx}, t + dt]$ in a Taylor series about $\mathbf{x}(t)$ and t as

$$\begin{aligned} \psi[\mathbf{x}(t) + \mathbf{dx}, t + dt] = \psi[\mathbf{x}(t), t] &+ \left(\frac{\partial \psi[\mathbf{x}(t), t]}{\partial \mathbf{x}(t)}\right)^T \mathbf{dx} \\ &+ \frac{\partial \psi[\mathbf{x}(t), t]}{\partial t}\,dt + \frac{1}{2}\operatorname{tr}\left\{\frac{\partial^2 \psi[\mathbf{x}(t), t]}{\partial \mathbf{x}(t)^2}\,\mathbf{dx}\,\mathbf{dx}^T\right\} \\ &+ \frac{1}{2}\frac{\partial^2 \psi[\mathbf{x}(t), t]}{\partial t^2}\,dt^2 + \left(\frac{\partial^2 \psi[\mathbf{x}(t), t]}{\partial \mathbf{x}(t)\,\partial t}\right)^T \mathbf{dx}\,dt \\ &+ \text{higher-order terms} \end{aligned}$$

Here the higher-order terms are of third order in dt and/or $\mathbf{dx}$ and higher; we shall show later that these terms are all zero.

Let us consider first the second-order terms. Because dt is infinitesimal, dt^2 is zero. The term $\mathbf{dx}\,dt$ is given by

$$\mathbf{dx}dt = (\mathbf{f}\,dt + \mathbf{G}\,\mathbf{du})\,dt = \mathbf{f}\,dt^2 + \mathbf{G}\,\mathbf{du}\,dt$$

Here dt^2 is again zero and $\mathbf{du}\,dt$ is also zero from Eq. (8.4-10). Now let us consider $\mathbf{dx}\,\mathbf{dx}^T$ given by

$$\begin{aligned} \mathbf{dx}\,\mathbf{dx}^T &= (\mathbf{f}\,dt + G\,\mathbf{du})(\mathbf{f}\,dt + \mathbf{G}\,\mathbf{du})^T \\ &= \mathbf{ff}^T\,dt^2 + \mathbf{G}\,\mathbf{du}\,dt\,\mathbf{f}^T + \mathbf{f}\,dt\,\mathbf{du}^T\,\mathbf{G}^T + \mathbf{G}\,\mathbf{du}\,\mathbf{du}^T\,\mathbf{G}^T \end{aligned}$$

Here dt^2 and $\mathbf{du}\,dt$ equal zero, but $\mathbf{du}\,\mathbf{du}^T = \mathbf{\Psi}_w\,dt$ from Eq. (8.4-9), so that $\mathbf{dx}\,\mathbf{dx}^T$ becomes

$$\mathbf{dx}\,\mathbf{dx}^T = \mathbf{G\Psi}_w\mathbf{G}^T\,dt$$

Now we can show that all the higher-order terms in Eq. (8.4-15) are zero. Obviously, dt^3 and dt^2 terms are zero since dt^2 is zero. The $\mathbf{dx}\,\mathbf{dx}^T\,dt$ term is also zero since

$$\mathbf{dx}\,\mathbf{dx}^T\,dt = \mathbf{G\Psi}_w\mathbf{G}^T\,dt^2 = \mathbf{0}$$

The third-order term in $\mathbf{dx}$ will contain terms of the order dt^2 and $\mathbf{du}\,dt$, both of which are zero. Since the third-order terms are all zero, the fourth- and higher-order terms must also be zero; so Eq. (8.4-15) becomes

$$\psi[\mathbf{x}(t) + d\mathbf{x}, t + dt] = \psi[\mathbf{x}(t), t] + \left(\frac{\partial\psi[\mathbf{x}(t), t]}{\partial\mathbf{x}(t)}\right)^T \mathbf{dx} + \frac{\partial\psi[\mathbf{x}(t), t]}{\partial t}\,dt + \frac{1}{2}\operatorname{tr}\left\{\frac{\partial^2\psi[\mathbf{x}(t),t]}{\partial\mathbf{x}(t)^2}\mathbf{G\Psi}_w\mathbf{G}^T\right\}dt$$

If we substitute for $\mathbf{dx}$ from Eq. (8.4-13) and let $d\psi = \psi[\mathbf{x}(t) + \mathbf{dx}, t + dt] - \psi[\mathbf{x}(t), t]$, we have

$$d\psi = \left[\left(\frac{\partial\psi}{\partial\mathbf{x}}\right)^T\mathbf{f} + \frac{\partial\psi}{\partial t} + \frac{1}{2}\operatorname{tr}\left\{\frac{\partial^2\psi}{\partial\mathbf{x}^2}\mathbf{G\Psi}_w\mathbf{G}^T\right\}\right]dt + \left(\frac{\partial\psi}{\partial\mathbf{x}}\right)^T\mathbf{G}\,\mathbf{du}$$

which is the desired result. We note that the difference between Eq. (8.4-15) and the ordinary result occurs because $\mathbf{dx}\,\mathbf{dx}^T$ is not zero.

If we have a scalar function $\psi[\mathbf{x}(t), \mathbf{u}(t), t]$, we can treat this as a special case of the above result by adjoining new components $\mathbf{x}'$, which are given by

$$\mathbf{dx}' = \mathbf{du}$$

so that $\psi[\mathbf{x}(t), \mathbf{u}(t), t] = \psi[\mathbf{x}(t), \mathbf{x}'(t), t]$. Now $d\psi$ can be written as

$$\begin{aligned} d\psi = &\left[\frac{\partial\psi}{\partial t} + \left(\frac{\partial\psi}{\partial\mathbf{x}}\right)^T\mathbf{f} + \frac{1}{2}\operatorname{tr}\left\{\frac{\partial^2\psi}{\partial\mathbf{x}^2}\mathbf{G\Psi}_w\mathbf{G}^T\right\}\right]dt \\ &+ \left(\frac{\partial\psi}{\partial\mathbf{x}}\right)^T\mathbf{G}\,\mathbf{du} + \left(\frac{\partial\psi}{\partial\mathbf{u}}\right)^T\mathbf{du} + \frac{1}{2}\operatorname{tr}\left\{\frac{\partial^2\psi}{\partial\mathbf{u}^2}\mathbf{\Psi}_w\right\}dt \\ &+ \operatorname{tr}\left\{\frac{\partial^2\psi}{\partial\mathbf{x}\partial\mathbf{u}}\mathbf{G\Psi}_w\right\}dt \end{aligned} \tag{8.4-16}$$

Example *8.4-1*. Let us consider the solution of the nonlinear stochastic system

$$dx_1 = x_2(t)\,du$$
$$dx_2 = du$$

with $\mathbf{x}(0) = \mathbf{0}$. In terms of Eq. (8.4-13), we have

$$\mathbf{f} = \begin{bmatrix} 0 \\ 0 \end{bmatrix} \quad \text{and} \quad \mathbf{G} = \begin{bmatrix} x_2(t) \\ 1 \end{bmatrix}$$

The "solution" is easily found by the use of ordinary (Riemann) calculus to be

$$x_1(t) = \frac{u^2(t)}{2} = \frac{x_2^2(t)}{2}$$
$$x_2(t) = u(t)$$

To see if this is truly a solution to the stochastic differential equation, we may make use of the Ito differential rule. Let $\psi[x(t), t] = x_1(t) = x_2^2(t)/2$; then $d\psi$ becomes

$$d\psi = dx_1(t) = \tfrac{1}{2}\Psi_w\,dt + x_2(t)\,du$$

We see that this is not correct, since dx_1 is supposed to equal $x_2(t)\,du$. It is easy to show that

$$x_1(t) = \frac{x_2^2(t)}{2} - \frac{\Psi_w t}{2}$$

is the correct solution of the stochastic differential equation. This solution can also be obtained by the use of the Ito integral, since

$$x_1(t) = \int_0^t x_2(\tau)\,du(\tau) = \int_0^t u(\tau)\,du(\tau)$$

The use of any other stochastic integral definition will not yield the correct solution. In particular, if the definition of Eq. (8.4-4) is used, the incorrect solution given by ordinary integration rules is obtained.

Two interfaces between a physical problem and computational algorithms are of prime importance. One interface is the *modeling problem* of determining a stochastic differential equation that represents a physical process. The model that eventually evolves as representative of a process is generally a compromise between mathematical accuracy and computation convenience. The second interface consists of the *determination of computational algorithms*, which must of necessity take place after the mathematical model of the process has been discerned. Here we are concerned with some aspects of the second interface.

We have discovered that we may not, in all circumstances, treat a stochastic integral (an integral involving the product of two random processes) as an ordinary integral. The previous example illustrated this point. When a digital computer is used to simulate the computational algorithms, the proper interpretation of the stochastic integral is not trivial. Two approaches that may be used are approaches based on the stochastic calculus. Some rules for choice between the two approaches are (Gray and Caughey [2]):

1. If $\mathbf{G}[\mathbf{x}(t), t]$ is not a function of $\mathbf{x}$, the ordinary calculus and stochastic calculus lead to precisely the same results, and there is no need for special treatment of stochastic integrals. We have previously alluded to this fact. This simplifies considerably the treatment of linear estimation problems.
2. If the problem is strictly a mathematical one, the stochastic calculus must be used:

   $$\frac{\mathbf{x}(t_{i+1}) - \mathbf{x}(t_i)}{t_{i+1} - t_i} = \mathbf{f}[\mathbf{x}(t_i), t_i] + \mathbf{G}[\mathbf{x}(t_i), t_i]\mathbf{w}(t_i)$$

   As the samples become dense, the stochastic calculus must be used.
3. If the white-noise input is an approximation (or the limit) for a problem with a short correlation time, the ordinary calculus must be used.

The computational difference between the two approaches is relatively easy to determine (Wong and Zakai [8]). If we use ordinary (Riemann) calculus concepts, obtaining $\mathbf{x}(t)$ from Eq. (8.4-13), we actually obtain a solution for[1]

$$\mathbf{dx} = \mathbf{f}[\mathbf{x}(t),t]\,dt + \frac{1}{2}\left(\left\{\mathbf{G}^T[\mathbf{x}(t), t]\frac{\partial}{\partial \mathbf{x}}\right\}^T \mathbf{\Psi}_w \mathbf{G}^T[\mathbf{x}(t), t]\right)^T dt + \mathbf{G}[\mathbf{x}(t), t]\,\mathbf{du} \tag{8.4-17}$$

Note that when $\mathbf{G}$ is not a function of $\mathbf{x}$, as in the linear problem, the use of ordinary calculus rules will yield the correct solution to Eq. (8.4-13). We shall only establish Eq. (8.4-17) for the scalar case; the general case follows a similar pattern with a great deal of additional notational complexity. By the use of ordinary calculus, we would write the solution of Eq. (8.4-13) for the

[1] The notation $(\mathbf{A}\partial/\partial\mathbf{x})^T$ signifies a vector operation that operates on the quantities which follow.

scalar case as

$$x(t) = x(0) + \int_0^t f[x(t), t]\, dt + \int_0^t G[x(t), t]\, du \tag{8.4-18}$$

The problem centers on the second integral, since it should be written as an Ito integral. Let us examine this term more carefully. By the use of the ordinary definition, this integral can be written as

$$\int_0^t G[x(t), t]\, du = \underset{\substack{N\to\infty \\ \delta\to 0}}{\text{l.i.m.}} \sum_{i=0}^{N-1} \frac{G[x(t_{i+1}), t_{i+1}] + G[x(t_i), t_i]}{2}\, du \tag{8.4-19}$$

where $t_N = t$ and $\delta = \max\{t_{i+1} - t_i\}$. Now we use the Taylor series to write

$$\begin{aligned} G[x(t_{i+1}), t_{i+1}] &= G[x(t_i), t_i] + \frac{\partial G[x(t_i), t_i]}{\partial t_i}\, dt + \frac{\partial G[x(t_i), t_i]}{\partial x(t_i)}\, dx \\ &= G[x(t_i), t_i] + \frac{\partial G[x(t_i), t_i]}{\partial t_i}\, dt + \frac{\partial G[x(t_i), t_i]}{\partial x(t_i)} \\ &\quad \cdot \{f[x(t_i), t_i]\, dt + G[x(t_i), t_i]\, du\} \end{aligned}$$

Here we have omitted second- and higher-order terms because we are going to multiply the result by du. Equation (8.4-19) now becomes

$$\begin{aligned} \int_0^t G[x(t), t]\, du &= \underset{\substack{N\to\infty \\ \delta\to 0}}{\text{l.i.m.}} \sum_{i=0}^{N-1} G[x(t_i), t_i]\, du \\ &\quad + \underset{\substack{N\to\infty \\ \epsilon\to 0}}{\text{l.i.m.}} \frac{1}{2} \sum_{i=1}^{N-1} \frac{\partial G[x(t_i), t_i]}{\partial x(t_i)} G[x(t_i), t_i]\, du^2 \end{aligned} \tag{8.4-20}$$

since $du\, dt = 0$. Because $du^2 = \Psi_w dt$, Eq. (8.4-20) can be written as

$$\begin{aligned} \int_0^t G[x(t), t]\, du &= \underset{\substack{N\to\infty \\ \delta\to 0}}{\text{l.i.m.}} \sum_{i=0}^{N-1} G[x(t_i), t_i]\, du \\ &\quad + \underset{\substack{N\to\infty \\ \epsilon\to 0}}{\text{l.i.m.}} \frac{1}{2} \sum_{i=0}^{N-1} \frac{\partial G[x(t_i), t_i]}{\partial x(t_i)} G[x(t_i), t_i] \Psi_w\, dt \end{aligned} \tag{8.4-21}$$

by the use of Eq. (8.4-2), which becomes

$$\int_0^t G[x(t), t]\, du = \int_0^t G[x(t), t]\, du + \frac{1}{2} \int_0^t \frac{\partial G[x(t), t]}{\partial x(t)} \Psi_w G[x(t), t]\, dt \tag{8.4-22}$$

Therefore, Eq. (8.4-18) becomes

$$x(t) = x(0) + \int_0^t \left\{ f[x(t), t] + \frac{\partial G[x(t), t]}{\partial x(t)} \Psi_w G[x(t), t] \right\} dt + \int_0^t G[x(t), t]\, du \tag{8.4-23}$$

as desired.

In a similar manner, we can show that if we use ordinary calculus rules to solve the stochastic differential equation

$$\mathbf{dx} = \mathbf{f}[\mathbf{x}(t), t] - \frac{1}{2}\left(\left\{\mathbf{G}^T[\mathbf{x}(t), t]\frac{\partial}{\partial \mathbf{x}}\right\}^T \mathbf{\Psi}_w \mathbf{G}^T[\mathbf{x}(t), t]\right)^T dt + \mathbf{G}[\mathbf{x}(t), t]\,\mathbf{du} \tag{8.4-24}$$

we obtain the solution of Eq. (8.4-13). This result provides a convenient method for obtaining the solution of stochastic differential equations.

Example *8.4-2*. Let us use Eq. (8.4-24) to find the solution of the stochastic differential equation

$$dx = -\tfrac{1}{2}x\, dt + x\, du$$

with $x(0) = 1$ and $\Psi_w = 1$. We form the "equivalent" ordinary differential equation by the use of Eq. (8.4-24) as

$$dx = -\tfrac{1}{2}x\, dt - \tfrac{1}{2}x\, dt + x\, du = -x\, dt + x\, du$$

If we divide by x and integrate, we have

$$\ln x \Big|_0^{x(t)} = -t + u(t)$$

or

$$x(t) = \exp\{-t + u(t)\}$$

We can check this result by the use of the general form of the Ito differential rule in Eq. (8.4-16).

We are now ready to make use of the Ito differential rule to derive the *forward diffusion*, or *Fokker–Planck* equation, which describes the propagation of the probability density function of the state of the nonlinear stochastic system

$$\mathbf{dx} = \mathbf{f}[\mathbf{x}(t), t]\, dt + \mathbf{G}[\mathbf{x}(t), t]\,\mathbf{du}$$

as a function of time. Let us consider the scalar function

$$\psi[\mathbf{x}(t), t] = \exp\{j\boldsymbol{\omega}^T\mathbf{x}(t)\} \tag{8.4-25}$$

By the use of the Ito differential rule, $d\psi$ is given by

$$d\psi = \{j\mathbf{f}^T[\mathbf{x}(t), t]\boldsymbol{\omega} + \tfrac{1}{2}j^2\boldsymbol{\omega}^T\mathbf{G}[\mathbf{x}(t), t]\boldsymbol{\Psi}_w\mathbf{G}^T[\mathbf{x}(t), t]\boldsymbol{\omega}\}\psi\, dt + j\boldsymbol{\omega}^T\mathbf{G}[\mathbf{x}(t), t]\psi\, \mathbf{du}$$

If we integrate from t_0 to t, we obtain

$$\psi[\mathbf{x}(t), t] - \psi[\mathbf{x}(t_0), t_0] = \int_{t_0}^{t} [j\mathbf{f}^T\boldsymbol{\omega} + \tfrac{1}{2}j^2\boldsymbol{\omega}^T\mathbf{G}\boldsymbol{\Psi}_w\mathbf{G}^T\boldsymbol{\omega}]\psi\, dt + j\boldsymbol{\omega}^T \int_{t_0}^{t} \mathbf{G}\psi\, \mathbf{du} \tag{8.4-26}$$

We note that the expected value of $\psi[\mathbf{x}(t), t]$, given $\mathbf{x}(t_0) = \boldsymbol{\alpha}_0$, is simply the characteristic function of $\mathbf{x}(t)$, given $\mathbf{x}(t_0) = \boldsymbol{\alpha}_0$. If we take the expected value of both sides of Eq. (8.4-26), conditioned on $\mathbf{x}(t_0) = \boldsymbol{\alpha}_0$, we obtain

$$\begin{aligned}\Phi_{\mathbf{x}(t)\,|\,\mathbf{x}(t_0)=\boldsymbol{\alpha}_0}(\boldsymbol{\omega}) = e^{j\boldsymbol{\omega}^T\boldsymbol{\alpha}_0} &+ \int_{t_0}^{t} \mathcal{E}\{[j\mathbf{f}^T\boldsymbol{\omega} + \tfrac{1}{2}j^2\boldsymbol{\omega}^T\mathbf{G}\boldsymbol{\Psi}_w\mathbf{G}^T\boldsymbol{\omega}]\psi \,|\, \mathbf{x}(t_0) = \boldsymbol{\alpha}_0\}\, dt \\ &+ j\boldsymbol{\omega}^T\mathcal{E}\left\{\int_{t_0}^{t} \mathbf{G}\psi\, \mathbf{du} \,|\, \mathbf{x}(t_0) = \boldsymbol{\alpha}_0\right\}\end{aligned} \tag{8.4-27}$$

By the use of Eq. (8.4-6), the second integral in Eq. (8.4-27) is zero. If we take the partial derivative with respect to t on both sides of Eq. (8.4-27), we have[1]

$$\frac{\partial\Phi_{\mathbf{x}(t)\,|\,\mathbf{x}(t_0)=\boldsymbol{\alpha}_0}(\boldsymbol{\omega})}{\partial t} = \mathcal{E}\left\{\left[j\mathbf{f}^T\boldsymbol{\omega} + \frac{1}{2}j^2\boldsymbol{\omega}^T\mathbf{G}\boldsymbol{\Psi}_x\mathbf{G}^T\boldsymbol{\omega}\right]\psi \,|\, \mathbf{x}(t_0) = \boldsymbol{\alpha}_0\right\} \tag{8.4-28}$$

which we may write as

$$\frac{\partial\Phi_{\mathbf{x}(t)\,|\,\mathbf{x}(t_0)=\boldsymbol{\alpha}_0}(\boldsymbol{\omega})}{\partial t} = \int_{-\infty}^{\infty} \left[j\mathbf{f}^T\boldsymbol{\omega} + \frac{1}{2}j^2\boldsymbol{\omega}^T\mathbf{G}\boldsymbol{\Psi}_w\mathbf{G}^T\boldsymbol{\omega}\right] p_{\mathbf{x}(t)\,|\,\mathbf{x}(t_0)}(\boldsymbol{\alpha}\,|\,\boldsymbol{\alpha}_0) \cdot e^{j\boldsymbol{\omega}^T\mathbf{x}(t)}\, d\mathbf{x}(t) \tag{8.4-29}$$

If we take the inverse transform of both sides of this expression and remember that $-j\boldsymbol{\omega}$ denotes partial differentiation with respect to $\mathbf{x}(t)$, Eq. (8.4-29)

[1] Note that $\Phi_{\mathbf{x}(t)\,|\,\mathbf{x}(t_0)=\boldsymbol{\alpha}_0}(\boldsymbol{\omega})$ may, and in general will, have an explicit dependence on t, so partial differentiation with respect to t does make sense. Thus a better notational representation would be $\Phi_{\mathbf{x}(t)\,|\,\mathbf{x}(t_0)=\boldsymbol{\alpha}_0}(\boldsymbol{\omega}, t)$. This expanded notation will not be used, but it must be remembered that both $\Phi_{\mathbf{x}(t)\,|\,\mathbf{x}(t_0)=\boldsymbol{\alpha}_0}$ and $p_{\mathbf{x}(t)\,|\,\mathbf{x}(t_0)=\boldsymbol{\alpha}_0}$ may depend explicity on t.

becomes

$$\frac{\partial p_{\mathbf{x}(t)\,|\,\mathbf{x}(t_0)}(\boldsymbol{\alpha}\,|\,\boldsymbol{\alpha}_0)}{\partial t} = -\left(\frac{\partial}{\partial \boldsymbol{\alpha}}\right)^T \mathbf{f}[\boldsymbol{\alpha}, t]\, p_{\mathbf{x}(t)\,|\,\mathbf{x}(t_0)}(\boldsymbol{\alpha}\,|\,\boldsymbol{\alpha}_0) + \frac{1}{2}\left(\frac{\partial}{\partial \boldsymbol{\alpha}}\right)^T \left[\left(\frac{\partial}{\partial \boldsymbol{\alpha}}\right)^T \mathbf{G}(\boldsymbol{\alpha}, t)\boldsymbol{\Psi}_{\mathbf{w}}\mathbf{G}^T(\boldsymbol{\alpha}, t) \cdot p_{\mathbf{x}(t)\,|\,\mathbf{x}(t_0)}(\boldsymbol{\alpha}\,|\,\boldsymbol{\alpha}_0)\right]^T \tag{8.4-30}$$

which can be equivalently written as

$$\frac{\partial p_{\mathbf{x}(t)\,|\,\mathbf{x}(t_0)}(\boldsymbol{\alpha}\,|\,\boldsymbol{\alpha}_0)}{\partial t} = -\operatorname{tr}\left\{\frac{\partial \mathbf{f}(\boldsymbol{\alpha}, t) p_{\mathbf{x}(t)\,|\,\mathbf{x}(t_0)}(\boldsymbol{\alpha}\,|\,\boldsymbol{\alpha}_0)}{\partial \boldsymbol{\alpha}}\right\} + \frac{1}{2}\operatorname{tr}\left\{\frac{\partial}{\partial \boldsymbol{\alpha}}\left(\frac{\partial}{\partial \boldsymbol{\alpha}}\right)^T \mathbf{G}(\boldsymbol{\alpha}, t)\boldsymbol{\Psi}_{\mathbf{w}}\mathbf{G}^T(\boldsymbol{\alpha}, t) p_{\mathbf{x}(t)\,|\,\mathbf{x}(t_0)}(\boldsymbol{\alpha}\,|\,\boldsymbol{\alpha}_0)\right\} \tag{8.4-31}$$

The solution of this partial differential equation with the boundary condition

$$p_{\mathbf{x}(t_0)\,|\,\mathbf{x}(t_0)}(\boldsymbol{\alpha}\,|\,\boldsymbol{\alpha}_0) = \delta_D(\boldsymbol{\alpha} - \boldsymbol{\alpha}_0) \tag{8.4-32}$$

will yield the conditional (or transitional) probability density of $\mathbf{x}(t)$ given $\mathbf{x}(t_0) = \boldsymbol{\alpha}_0$. If we multiply both sides of Eq. (8.4-30) or (8.4-31) by $p_{\mathbf{x}(t_0)}(\boldsymbol{\alpha}_0)$, we obtain an equation, identical in form to Eq. (8.4-30) or (8.4-31), for the joint density $p_{\mathbf{x}(t)\mathbf{x}(t_0)}(\boldsymbol{\alpha}, \boldsymbol{\alpha}_0)$. By integrating over $\mathbf{x}(t_0)$, we obtain an equation for $p_{\mathbf{x}(t)}(\boldsymbol{\alpha})$ that once again has the same form as Eq. (8.4-30) or (8.4-31), or

$$\frac{\partial p_{\mathbf{x}(t)}(\boldsymbol{\alpha})}{\partial t} = -\operatorname{tr}\left\{\frac{\partial \mathbf{f}(\boldsymbol{\alpha}, t) p_{\mathbf{x}(t)}(\boldsymbol{\alpha})}{\partial \boldsymbol{\alpha}}\right\} + \frac{1}{2}\operatorname{tr}\left\{\frac{\partial}{\partial \boldsymbol{\alpha}}\left(\frac{\partial}{\partial \boldsymbol{\alpha}}\right)^T \mathbf{G}(\boldsymbol{\alpha}, t)\boldsymbol{\Psi}_{\mathbf{w}}\mathbf{G}^T(\boldsymbol{\alpha}, t) p_{\mathbf{x}(t)}(\boldsymbol{\alpha})\right\} \tag{8.4-33}$$

The boundary condition in this case is given by

$$p_{\mathbf{x}(t)}(\boldsymbol{\alpha})\big|_{t_0} = p_{\mathbf{x}(t_0)}(\boldsymbol{\alpha}) \tag{8.4-34}$$

To summarize: $p_{\mathbf{x}(t)}$, $p_{\mathbf{x}(t)\,|\,\mathbf{x}(t_0)}$, and $p_{\mathbf{x}(t)\mathbf{x}(t_0)}$ are obtained by solving the same partial differential equation with different boundary conditions.

Example 8.4-3 We shall consider the solution of the Fokker–Planck equation for $p_{x(t)}(\alpha)$ given by Eq. (8.4-33) for the simple scalar, linear system

$$dx = -ax(t)\,dt + du$$

with

$$p_{x(t_0)}(\alpha_0) = \frac{1}{\sqrt{2\pi V_0}} \exp\left\{-\frac{(\alpha_0 - \mu_0)^2}{2V_0}\right\}$$

The Fokker–Planck equation for this case is given by

$$\frac{\partial p_{x(t)}(\alpha)}{\partial t} = a\frac{\partial \alpha p_{x(t)}(\alpha)}{\partial \alpha} + \frac{1}{2}\Psi_w \frac{\partial^2 p_{x(t)}(\alpha)}{\partial \alpha^2}$$

Since $p_{x(t)}$ is gaussian at t_0, let us assume that $p_{x(t)}$ is gaussian for all t, so that

$$p_{x(t)}(\alpha) = \frac{1}{\sqrt{2\pi V(t)}} \exp\left\{-\frac{[\alpha - \mu(t)]^2}{2V(t)}\right\}$$

If we substitute this expression into the Fokker–Planck equation, we obtain

$$\frac{-1}{2[V(t)]^{3/2}}\dot{V}(t) + \frac{\alpha - \mu(t)}{[V(t)]^{3/2}}\dot{\mu}(t) + \frac{[\alpha - \mu(t)]^2}{2[V(t)]^{5/2}}\dot{V}(t)$$
$$= \frac{a}{[V(t)]^{1/2}} - \frac{a\alpha[\alpha - \mu(t)]}{[V(t)]^{3/2}} + \frac{1}{2}\Psi_w\left\{\frac{-1}{[V(t)]^{3/2}} + \frac{[\alpha - \mu(t)]^2}{[V(t)]^{5/2}}\right\}$$

where we have canceled the common exponential $\sqrt{2\pi}$ terms from both sides of the above equation. Let us separate the terms involving α and $\mu(t)$ from those not involving α and $\mu(t)$ to obtain

$$\frac{-V(t)}{2[V(t)]^{3/2}} = \frac{a}{[V(t)]^{1/2}} - \frac{\Psi_w}{2[V(t)]^{3/2}}$$

$$\frac{\alpha - \mu(t)}{[V(t)]^{3/2}}\dot{\mu}(t) + \frac{[\alpha - \mu(t)]^2}{2[V(t)]^{5/2}}\dot{V}(t) = \frac{a\alpha[\alpha - \mu(t)]}{[V(t)]^{3/2}} + \frac{1}{2}\frac{\Psi_w[\alpha - \mu(t)]^2}{[V(t)]^{5/2}}$$

or

$$\dot{V}(t) = -2aV(t) + \Psi_w$$
$$\dot{\mu}(t) = -\frac{[\alpha - \mu(t)]\dot{V}(t)}{2V(t)} - a\alpha + \frac{1}{2}\frac{\Psi_w[\alpha - \mu(t)]}{V(t)}$$

By substituting for $\dot{V}(t)$ in the second equation, we obtain

$$\dot{\mu}(t) = -a\mu(t)$$

If $\mu(t)$ and $V(t)$ satisfy these equations with $\mu(t_0) = \mu_0$ and $V(t_0) = V_0$, the assumed form for $p_{x(t)}$ will satisfy the Fokker–Planck equation.

We recognize the equations for $V(t)$ and $\mu(t)$ as the mean- and variance-propagation equations obtained in Section 6.2. A useful indirect result of this example is that we have offered an ancillary proof for the statement that the output of a linear system with a gaussian random input is itself gaussian.

Unfortunately, the Fokker–Planck equation is impossible to solve in closed form for most nontrivial problems. Nevertheless, it is very useful in studying the behavior of nonlinear or nongaussian systems. This is the topic of Section 8.5.

EXERCISES 8.4

8.4-1. Derive Eq. (8.4-7), assuming that $\boldsymbol{\eta} \in \boldsymbol{\Xi}^0$.

8.4-2. If $u(t)$ is a Wiener process, use the Ito integral definition to evaluate

$$\int_0^t [u(\lambda)]^N \, du(\lambda)$$

for $N = 1, 2, \ldots$.

8.4-3. Rework Exercise 8.4-2 by the use of the Ito differential rule.

8.4-4. What is the response of the system

$$dx(t) = 0.5x(t)\,dt + x(t)\,du(t)$$

if $u(t)$ is a Wiener process with $\Psi_w = 1$ and $x(0) = 1$.

Answer: $x(t) = \exp[-t + u(t)]$.

8.4-5. Rework Example 8.4-1 using Eq. (8.4-24).

8.5. Response of nonlinear systems

In Section 8.4 we derived the Fokker–Planck equation, which described the propagation of the probability density function for the general stochastic differential equation

$$\mathbf{dx} = \mathbf{f}[\mathbf{x}(t), t]\,dt + \mathbf{G}[\mathbf{x}(t), t]\,\mathbf{du} \tag{8.5-1}$$

Because the Fokker–Planck equation is a partial differential equation and processes no general solution except for linear systems and gaussian densities, it will generally not be possible to use it to determine the complete probability density for $\mathbf{x}(t)$. Fortunately, for many practical applications, a knowledge of the mean $\boldsymbol{\mu}_x(t)$ and variance $\mathbf{V}_x(t)$ are sufficient. We shall show how it is possible to find approximate equations for the propagation of the mean and variance for general nonlinear systems without actually solving the Fokker–Planck equation.

For reasons of notational simplicity, the developments of this section will be limited to the scalar version of Eq. (8.5-1), given by

$$dx = f[x(t), t] + g[x(t), t]\, du \tag{8.5-2}$$

For this system the Fokker–Planck equation (8.4-33) becomes

$$\frac{\partial p_{x(t)}(\alpha)}{\partial t} = -\frac{\partial f(\alpha, t)p_{x(t)}(\alpha)}{\partial \alpha} + \frac{1}{2}\frac{\partial^2 g^2(\alpha, t)\Psi_w p_{x(t)}(\alpha)}{\partial \alpha^2} \tag{8.5-3}$$

We shall give the results for the general case but leave the proof as an exercise for the reader.

We remember that the mean of $x(t)$ is given by

$$\mu_x(t) = \int_{-\infty}^{\infty} \alpha p_{x(t)}(\alpha)\, d\alpha \tag{8.5-4}$$

We can obtain an expression for $\mu_x(t)$ by multiplying both sides of Eq. (8.5-3) by α and integrating so that

$$\int_{-\infty}^{\infty} \alpha \frac{\partial p_{x(t)}(\alpha)}{\partial t}\, d\alpha = -\int_{-\infty}^{\infty} \alpha \frac{\partial f(\alpha, t)p_{x(t)}(\alpha)}{\partial \alpha}\, d\alpha + \frac{1}{2}\int_{-\infty}^{\infty} \alpha \frac{\partial^2 g^2(\alpha, t)\Psi_w p_{x(t)}(\alpha)}{\partial \alpha^2}\, d\alpha \tag{8.5-5}$$

We may interchange the partial derivative and the integral on the left-hand side of Eq. (8.5-5) so that we have

$$\int_{-\infty}^{\infty} \alpha \frac{\partial p_{x(t)}(\alpha)}{\partial t}\, d\alpha = \frac{\partial}{\partial t}\int_{-\infty}^{\infty} \alpha p_{x(t)}(\alpha)\, d\alpha = \frac{\partial}{\partial t}\mu_x(t) = \dot{\mu}_x(t) \tag{8.5-6}$$

Now let us use integration by parts on the right-hand side of Eq. (8.5-5) to obtain

$$-\int_{-\infty}^{\infty} \alpha \frac{\partial f(\alpha, t)p_{x(t)}(\alpha)}{\partial \alpha}\, d\alpha + \frac{1}{2}\int_{-\infty}^{\infty} \alpha \frac{\partial^2 g^2(\alpha, t)\Psi_w p_{x(t)}(\alpha)\, d\alpha}{\partial \alpha^2}$$

$$= -f(\alpha, t)p_{x(t)}(\alpha)\Big|_{\alpha=-\infty}^{\alpha=\infty} + \int_{-\infty}^{\infty} f(\alpha, t)p_{x(t)}(\alpha, t)\, d\alpha$$

$$+ \frac{1}{2}\alpha \frac{\partial g^2(\alpha, t)\Psi_w p_{x(t)}(\alpha)}{\partial \alpha}\Big|_{\alpha=-\infty}^{\alpha=\infty} - \frac{1}{2}g^2(\alpha, t)\Psi_w p_{x(t)}(\alpha)\Big|_{\alpha=-\infty}^{\alpha=\infty} \tag{8.5-7}$$

We assume the following boundary conditions:

$$-\alpha f(\alpha, t)p_{x(t)}(\alpha)\Big|_{\alpha=-\infty}^{\alpha=\infty} = 0 \tag{8.5-8}$$

$$\frac{1}{2}\alpha^2\frac{\partial g^2(\alpha, t)\Psi_w p_{x(t)}(\alpha)}{\partial\alpha}\Big|_{\alpha=-\infty}^{\alpha=\infty} = 0 \tag{8.5-9}$$

$$-\frac{1}{2}g^2(\alpha, t)\Psi_w p_{x(t)}(\alpha)\Big|_{\alpha=-\infty}^{\alpha=\infty} = 0 \tag{8.5-10}$$

So Eq. (8.5-9) becomes

$$-\int_{-\infty}^{\infty}\alpha\frac{\partial f(\alpha, t)p_{x(t)}(\alpha)}{\partial\alpha}\,d\alpha + \frac{1}{2}\int_{-\infty}^{\infty}\alpha\frac{\partial^2 g^2(\alpha, t)\Psi_w p_{x(t)}(\alpha)}{\partial\alpha^2}\,d\alpha$$

$$= \int_{-\infty}^{\infty} f(\alpha, t)p_{x(t)}(\alpha)\,d\alpha = \mathcal{E}\{f[x(t), t]\} \tag{8.5-11}$$

The combination of Eqs. (8.5-6) and (8.5-11) yields

$$\dot{\mu}_x(t) = \mathcal{E}\{f[x(t), t]\} \tag{8.5-12}$$

We note that this is an exact equation for the propagation of the mean. Unfortunately, we cannot use this result directly because we do not know $\mathcal{E}\{f[x(t), t]\}$. However, we obtain an approximate expression by expanding $f[x(t), t]$ in a Taylor series about $\mu_x(t)$. We shall retain terms up to the second order so that

$$f[x(t), t] \simeq f[\mu_x(t), t] + \frac{\partial f[\mu_x(t), t]}{\partial\mu_x(t)}[x(t) - \mu_x(t)] + \frac{1}{2}\frac{\partial^2 f[\mu_x(t), t]}{\partial\mu_x^2(t)}[x(t) - \mu_x(t)]^2 \tag{8.5-13}$$

We retain only the second-order terms because higher-order terms would require that we compute third- and higher-order moments. If we take the expected value of both sides of Eq. (8.5-13), we obtain

$$\mathcal{E}\{f[x(t), t]\} \simeq f[\mu_x(t), t] + \frac{1}{2}\frac{\partial^2 f[\mu_x(t), t]}{\partial\mu_x^2(t)}V_x(t) \tag{8.5-14}$$

By substituting this result into Eq. (8.5-12), we obtain the following second-order approximation to the equation for the propagation of the mean:

$$\dot{\mu}_x(t) = f[\mu_x(t), t] + \frac{1}{2}\frac{\partial^2 f[\mu_x(t), t]}{\partial\mu_x^2(t)}V_x(t) \tag{8.5-15}$$

We note that the equation for the mean requires knowledge of the variance. A first-order or linearized equation for $\mu_x(t)$ can be obtained by retaining only the first-order terms in Eq. (8.5-13). In this case, the equation for the propagation of the mean becomes

$$\dot{\mu}_x(t) = f[\mu_x(t), t] \tag{8.5-16}$$

We note that if the system is linear so that $f[x(t), t] = A(t)x(t)$, then Eq. (8.5-16) becomes exact and the computation of the mean is independent of the variance.

For the general vector case, the second-order equation may be written as

$$\dot{\boldsymbol{\mu}}_\mathbf{x}(t) = \mathbf{f}[\boldsymbol{\mu}_\mathbf{x}(t), t] + \frac{1}{2}\frac{\partial^2\mathbf{f}[\boldsymbol{\mu}_\mathbf{x}(t), t]}{\partial\boldsymbol{\mu}_\mathbf{x}^2(t)} : \mathbf{V}_\mathbf{x}(t) \tag{8.5-17}$$

where

$$\frac{\partial^2\mathbf{f}[\boldsymbol{\mu}_\mathbf{x}(t), t]}{\partial\boldsymbol{\mu}_\mathbf{x}^2(t)} : \mathbf{V}_\mathbf{x}(t) = \sum_{i=1}^{N}\sum_{j=1}^{N}[\mathbf{V}_\mathbf{x}(t)]_{ij}\frac{\partial^2\mathbf{f}[\boldsymbol{\mu}_\mathbf{x}(t), t]}{[\partial\boldsymbol{\mu}_\mathbf{x}(t)]_i[\partial\boldsymbol{\mu}_\mathbf{x}(t)]_j} \tag{8.5-18}$$

The last term can also be conveniently expressed by the alternative definition

$$\left[\frac{\partial^2\mathfrak{N}}{\partial\mathbf{x}^2} : \mathfrak{M}\right]_{ij} = \mathrm{tr}\left[\frac{\partial^2[\mathfrak{N}]_{ij}}{\partial\mathbf{x}^2}\mathfrak{M}\right] \tag{8.5-19}$$

which will henceforth be used for any $\mathfrak{N}$ matrix and square symmetric $\mathfrak{M}$ matrix. The corresponding first-order approximation is

$$\dot{\boldsymbol{\mu}}_\mathbf{x}(t) = \mathbf{f}[\mathbf{x}(t), t] \tag{8.5-20}$$

Let us turn attention to the computation of the variance. To determine an equation for V_x, we multiply both sides of Eq. (8.5-3) by $[\alpha - \mu_x(t)]^2$ and integrate so that we have

$$\int_{-\infty}^{\infty}[\alpha - \mu_x(t)]^2\frac{\partial p_{x(t)}(\alpha)}{\partial t}\,d\alpha = -\int_{-\infty}^{\infty}[\alpha - \mu_x(t)]^2\frac{\partial f(\alpha, t)p_{x(t)}(\alpha)}{\partial\alpha}\,d\alpha$$
$$+ \frac{1}{2}\int_{-\infty}^{\infty}[\alpha - \mu_x(t)]^2\frac{\partial^2 g^2(\alpha, t)\Psi_w p_{x(t)}(\alpha)}{\partial\alpha^2}\,\alpha \tag{8.5-21}$$

Now we interchange the order of partial differentiation and integration on the left-hand side of Eq. (8.5-21). On the right-hand side we use integration by parts, once on the first integral and twice on the second integral, so that Eq.

(8.5-21) can be written as

$$\frac{\partial}{\partial t}\int_{-\infty}^{\infty}[\alpha-\mu_x(t)]^2 p_{x(t)}(\alpha)\,d\alpha = -[\alpha-\mu_x(t)]^2 f(\alpha,t)p_{x(t)}(\alpha)\Bigg|_{\alpha=-\infty}^{\alpha=\infty}$$
$$+2\int_{-\infty}^{\infty}[\alpha-\mu_x(t)]f(\alpha,t)p_{x(t)}(\alpha)\,d\alpha$$
$$+\frac{1}{2}[\alpha-\mu_x(t)]^2\frac{\partial[g^2(\alpha,t)\Psi_w p_{x(t)}(\alpha)]}{\partial\alpha}\Bigg|_{\alpha=-\infty}^{\alpha=\infty}$$
$$-[\alpha-\mu_x(t)]g^2(\alpha,t)\Psi_w p_{x(t)}(\alpha)\Bigg|_{\alpha=-\infty}^{\alpha=\infty}$$
$$+\int_{-\infty}^{\infty}g^2(\alpha,t)\Psi_w p_{x(t)}(\alpha)\,d\alpha \qquad (8.5\text{-}22)$$

Again we assume that the evaluations of the various terms at $\alpha=\pm\infty$ are zero, so that Eq. (8.5-22) becomes

$$\frac{\partial}{\partial t}\int_{-\infty}^{\infty}[\alpha-\mu_x(t)]^2 p_{x(t)}(\alpha)\,d\alpha = 2\int_{-\infty}^{\infty}[\alpha-\mu_x(t)]f(\alpha,t)p_{x(t)}(\alpha)\,d\alpha$$
$$+\int_{-\infty}^{\infty}g^2(\alpha,t)\Psi_w p_{x(t)}(\alpha)\,d\alpha \qquad (8.5\text{-}23)$$

or, equivalently,

$$\dot{V}_x(t) = 2\mathcal{E}\{[x(t)-\mu_x(t)]f[x(t),t]\} + \mathcal{E}\{g^2[x(t),t]\Psi_w\} \qquad (8.5\text{-}24)$$

Once again this equation is exact but not too useful, since we do not know the values for the two terms on the right-hand side of Eq. (8.5-24). Again, we use the Taylor series expansion to find approximate expressions for these terms.

If we let $[x(t)-\mu_x(t)]f[x(t),t]$ be written as

$$[x(t)-\mu_x(t)]f[x(t),t] \simeq [x(t)-\mu_x(t)]f[\mu_x(t),t] + \frac{\partial f[\mu_x(t),t]}{\partial\mu_x(t)}[x(t)-\mu_x(t)]^2$$

the expected value becomes

$$\mathcal{E}\{[x(t)-\mu_x(t)]f[x(t),t]\} = \frac{\partial f[\mu_x(t),t]}{\partial\mu_x(t)}V_x(t) \qquad (8.5\text{-}25)$$

Here we have only retained the first order in the expansion of $f[x(t),t]$; this is necessary because higher-order terms would involve higher-order moments of $x(t)$. In a similar fashion, we may write

$$\mathcal{E}\{g^2[x(t), t]\Psi_w\} \simeq g^2[\mu_x(t), t]\Psi_w + \frac{1}{2}\frac{\partial^2 g^2[\mu_x(t), t]\Psi_w}{\partial \mu_x^2(t)} V_x(t) \quad (8.5\text{-}26)$$

If we substitute Eqs. (8.5-25) and (8.5-26) into Eq. (8.5-24), we obtain the second-order approximate expression for the variance as

$$\dot{V}_x(t) = 2\frac{\partial f[\mu_x(t), t]}{\partial \mu_x(t)} V_x(t) + g^2[\mu_x(t), t]\Psi_w + \frac{1}{2}\frac{\partial^2 g^2[\mu_x(t), t]}{\partial \mu_x^2(t)}\Psi_w V_x(t) \quad (8.5\text{-}27)$$

In this case, we note that the variance equation is coupled to the equation for the mean. If we wish the first-order, linearized equation, we drop the last term in Eq. (8.5-27) and obtain

$$\dot{V}_x(t) = 2\frac{\partial f[\mu_x(t), t]}{\partial \mu_x(t)} V_x(t) + g^2[\mu_x(t), t]\Psi_w \quad (8.5\text{-}28)$$

For the general vector case, the second-order approximate expression becomes

$$\dot{\mathbf{V}}_x(t) = \frac{\partial f[\mu_x(t), t]}{\partial \mu_x(t)}\mathbf{V}_x(t) + \mathbf{V}_x(t)\frac{\partial \mathbf{f}^T[\boldsymbol{\mu}_x(t), t]}{\partial \boldsymbol{\mu}_x(t)} + \mathbf{G}[\boldsymbol{\mu}_x(t), t]\boldsymbol{\Psi}_w\mathbf{G}^T[\boldsymbol{\mu}_x(t), t] + \frac{1}{2}\frac{\partial^2 \mathbf{G}[\boldsymbol{\mu}_x(t), t]\boldsymbol{\Psi}_w\mathbf{G}[\boldsymbol{\mu}_x(t), t]}{\partial \boldsymbol{\mu}_x^2(t)} : \mathbf{V}_x(t) \quad (8.5\text{-}29)$$

The corresponding first-order equation is obtained by deleting the last term in Eq. (8.5-29):

$$\dot{\mathbf{V}}_x(t) = \frac{\partial \mathbf{f}[\boldsymbol{\mu}_x(t), t]}{\partial \boldsymbol{\mu}_x(t)}\mathbf{V}_x(t) + \mathbf{V}_x(t)\frac{\partial \mathbf{f}^T[\boldsymbol{\mu}_x(t), t]}{\partial \boldsymbol{\mu}_x(t)} + \mathbf{G}[\mathbf{x}(t), t]\boldsymbol{\Psi}_w(t)\mathbf{G}^T[\boldsymbol{\mu}_x(t), t] \quad (8.5\text{-}30)$$

For easy reference, the first-order approximate expressions for computing the mean and variance are given in Table 8.5-1. The second order expressions are summarized in Table 8.5-2.

TABLE 8.5-1. First-Order Approximate Expressions for the Mean and Variance of Continuous Nonlinear Systems

Mean	$\dot{\boldsymbol{\mu}}_x(t) = \mathbf{f}[\boldsymbol{\mu}_x(t), t]$
Variance	$\dot{\mathbf{V}}_x(t) = \frac{\partial \mathbf{f}[\boldsymbol{\mu}_x(t), t]}{\partial \boldsymbol{\mu}_x(t)}\mathbf{V}_x(t) + \mathbf{V}_x(t)\frac{\partial \mathbf{f}^T[\boldsymbol{\mu}_x(t), t]}{\partial \boldsymbol{\mu}_x(t)} + \mathbf{G}[\boldsymbol{\mu}_x(t), t]\boldsymbol{\Psi}_w(t)\mathbf{G}^T[\boldsymbol{\mu}_x(t), t]$

TABLE 8.5-2. Second-Order Approximate Expressions for the Mean and Variance of Continuous Nonlinear Systems

Mean	$\dot{\boldsymbol{\mu}}_\mathbf{x}(t) = \mathbf{f}[\boldsymbol{\mu}_\mathbf{x}(t), t] + \frac{1}{2}\frac{\partial^2 \mathbf{f}[\boldsymbol{\mu}_\mathbf{x}(t), t]}{[\partial \boldsymbol{\mu}_\mathbf{x}(t)]^2} : \mathbf{V}_\mathbf{x}(t)$
Variance	$\dot{\mathbf{V}}_\mathbf{x}(t) = \frac{\partial \mathbf{f}[\boldsymbol{\mu}_\mathbf{x}(t), t]}{\partial \boldsymbol{\mu}_\mathbf{x}(t)}\mathbf{V}_\mathbf{x}(t) + \mathbf{V}_\mathbf{x}(t)\frac{\partial \mathbf{f}^T[\boldsymbol{\mu}_\mathbf{x}(t), t]}{\partial \boldsymbol{\mu}_\mathbf{x}(t)} + \mathbf{G}[\boldsymbol{\mu}_\mathbf{x}(t), t]\boldsymbol{\Psi}_\mathbf{w}(t)\mathbf{G}^T[\boldsymbol{\mu}_\mathbf{x}(t), t] + \frac{1}{2}\frac{\partial^2 \mathbf{G}[\boldsymbol{\mu}_\mathbf{x}(t), t]\boldsymbol{\Psi}_\mathbf{w}(t)\mathbf{G}^T[\boldsymbol{\mu}_\mathbf{x}(t), t]}{[\partial \boldsymbol{\mu}_\mathbf{x}(t)]^2} : \mathbf{V}_\mathbf{x}(t)$

To solve these sets of equations [Eqs. (8.5-20) and (8.5-30) or Eqs. (8.5-17) and (8.5-29)], we need initial conditions $\boldsymbol{\mu}_\mathbf{x}(t_0)$ and $\mathbf{V}_\mathbf{x}(t_0)$. These terms are the prior mean and variance of $\mathbf{x}$:

$$\boldsymbol{\mu}_\mathbf{x}(t_0) = \mathcal{E}\{\mathbf{x}(t_0)\} = \int_{-\infty}^{\infty} \boldsymbol{\beta} p_{\mathbf{x}(t_0)}(\boldsymbol{\beta})\, d\boldsymbol{\beta} \tag{8.5-31}$$

$$\mathbf{V}_\mathbf{x}(t_0) = \operatorname{var}\{\mathbf{x}(t_0)\} = \int_{-\infty}^{\infty} [\boldsymbol{\beta} - \boldsymbol{\mu}_\mathbf{x}(t_0)][\boldsymbol{\beta} - \boldsymbol{\mu}_\mathbf{x}(t_0)]^T p_{\mathbf{x}(t_0)}(\boldsymbol{\beta})\, d\boldsymbol{\beta} \tag{8.5-32}$$

It is quite possible to establish the first-order approximations of Eqs. (8.5-20) and (8.5-30) by linearizing the differential system of Eq. (8.5-1) by expanding certain terms about the mean of $\mathbf{x}(t)$ to obtain

$$\dot{\mathbf{x}}(t) = \mathbf{f}[\boldsymbol{\mu}_\mathbf{x}(t), t] + \frac{\partial \mathbf{f}[\boldsymbol{\mu}_\mathbf{x}(t), t]}{\partial \boldsymbol{\mu}_\mathbf{x}(t)}[\mathbf{x}(t) - \boldsymbol{\mu}_\mathbf{x}(t)] + \mathbf{G}[\boldsymbol{\mu}_\mathbf{x}(t), t]\mathbf{w}(t) \tag{8.5-33}$$

which may be written as

$$\frac{d\mathbf{x}(t)}{dt} = \mathbf{F}(t)\mathbf{x}(t) + \quad (t) \tag{8.5-34}$$

where the linearized coefficient matrix of the system is

$$\mathbf{F}(t) = \frac{\partial \mathbf{f}[\boldsymbol{\mu}_\mathbf{x}(t), t]}{\partial \boldsymbol{\mu}_\mathbf{x}(t)} \tag{8.5-35}$$

and the noise input is

$$(t) = \mathbf{f}[\boldsymbol{\mu}_\mathbf{x}(t), t] - \frac{\partial \mathbf{f}[\boldsymbol{\mu}_\mathbf{x}(t), t]}{\partial \boldsymbol{\mu}_\mathbf{x}(t)}\boldsymbol{\mu}_\mathbf{x}(t) + \mathbf{G}[\boldsymbol{\mu}_\mathbf{x}(t), t]\mathbf{w}(t) \tag{8.5-36}$$

which has the moments

$$\mathcal{E}\{\mathbf{w}(t)\} = \mathbf{f}[\boldsymbol{\mu}_{\mathbf{x}}(t), t] - \frac{\partial \mathbf{f}[\boldsymbol{\mu}_{\mathbf{x}}(t), t]}{\partial \boldsymbol{\mu}_{\mathbf{x}}(t)} \boldsymbol{\mu}_{\mathbf{x}}(t) = \boldsymbol{\mu}_{\mathbf{w}}(t) \tag{8.5-37}$$

$$\text{cov}\{\mathbf{w}(t), \mathbf{w}(\tau)\} = \mathbf{G}[\boldsymbol{\mu}_{\mathbf{x}}(t), t]\boldsymbol{\Psi}_{\mathbf{w}}(t)\mathbf{G}^{T}[\boldsymbol{\mu}_{\mathbf{x}}(t), t]\delta_{D}(t-\tau) = \boldsymbol{\Psi}_{\mathbf{w}}(t)\delta_{D}(t-\tau) \tag{8.5-38}$$

We now use previously developed relations for the mean and variance propagation in a linear system. Equations (6.2-10) and (6.2-23) are the directly applicable relations:

$$\dot{\boldsymbol{\mu}}_{\mathbf{x}}(t) = \mathbf{F}(t)\boldsymbol{\mu}_{\mathbf{x}}(t) + \boldsymbol{\mu}_{\mathbf{w}}(t) \tag{8.5-39}$$

$$\dot{\mathbf{V}}_{\mathbf{x}}(t) = \mathbf{F}(t)\mathbf{V}_{\mathbf{x}}(t) + \mathbf{V}_{\mathbf{x}}(t)\mathbf{F}^{T}(t) + \boldsymbol{\Psi}_{\mathbf{w}}(t) \tag{8.5-40}$$

Substitution of the definitions of $\mathbf{F}(t)$ from Eq. (8.5-35) and $\boldsymbol{\mu}_{\mathbf{w}}(t)$ and $\boldsymbol{\Psi}_{\mathbf{w}}(t)$ from Eqs. (8.5-37) and (8.5-38) into the foregoing two equations easily yields Eqs. (8.5-20) and (8.5-30).

It is, of course, not possible to establish the second-order equations (8.5-17) and (8.5-29) using the linear theory of Chapter 6. Often the linearized or first-order equations are quite sufficient in accuracy for many applications.

Example *8.5-1*. We desire to compute the time-varying mean and variance of $x(t)$, where $x(t)$ evolves from the system

$$\dot{x} = -x^3(t) + w(t)$$

where $w(t)$ is zero-mean white gaussian noise with $\Psi_w(t) = 1$.

The first-order approximations, Eqs. (8.5-20) and (8.5-29), yield for this example

$$\dot{\mu}_x = -\mu_x^3(t)$$

$$\dot{V}_x = -6\mu_x^2(t)V_x(t) + 1$$

The solution of these two equations is easily obtained for the case where $\mu_x(0) = V_x(0) = 0$ as

$$\mu_x(t) = 0 \qquad V_x(t) = t$$

which we recognize as the mean and variance of the output of a single integration linear system

$$\dot{y} = w(t) \qquad y(0) = 0 \qquad \text{var}\{y(0)\} = 0$$

to the white-noise input.

The second-order approximations, Eqs. (8.5-17) and (8.5-29), result in

$$\dot{\mu}_x = -\mu_x^3(t) - 3\mu_x(t)V_x(t)$$
$$\dot{V}_x = -6\mu_x^2(t)V_x(t) + 1$$

Again, if $\mu_x(0) = V_x(0) = 0$, we have $\mu_x(t) = 0$ and $V_x(t) = t$. For this particular example then, if $\mu_x(0) = V_x(0) = 0$, the approximations yield the same result as a linear system. This will not occur in general.

The Fokker–Planck equation is, we realize, formally inapplicable to discrete-time systems. Thus it is not possible for us to obtain exact relations of the form of Eqs. (8.5-12) and (8.5-24) for the mean and variance propagation in nonlinear discrete-time systems. Thus we shall consider an ancillary approach.

We desire to obtain the mean and variance propagation of the state vector $\mathbf{x}_k$ or $\mathbf{x}(k)$ or $\mathbf{x}(t_k)$, which evolves from the nonlinear Markov model[1]

$$\mathbf{x}(k+1) = \boldsymbol{\phi}[\mathbf{x}(k), k] + \boldsymbol{\Gamma}[\mathbf{x}(k), k]\mathbf{w}(k) \qquad (8.5\text{-}41)$$

where $\mathbf{x}(k)$ is zero-mean white gaussian noise with moments

$$\boldsymbol{\mu}_{\mathbf{w}}(k) = \mathcal{E}\{\mathbf{w}(k)\} = \mathbf{0} \qquad \mathbf{V}_{\mathbf{w}}(j, k) = \operatorname{cov}\{\mathbf{w}(j), \mathbf{w}(k)\} = \mathbf{V}_{\mathbf{w}}(k)\delta_K(k-j) \qquad (8.5\text{-}42)$$

We expand $\boldsymbol{\phi}$ and $\boldsymbol{\Gamma}\mathbf{V}_{\mathbf{w}}\boldsymbol{\Gamma}^T$ in a Taylor series about the assumed true mean $\boldsymbol{\mu}_{\mathbf{x}}(t)$, as above. We then take the expected value of Eq. (8.5-41) with the approximations used for $\boldsymbol{\phi}$ in this equation and obtain the mean-value propagation relation. In a similar way, we obtain var$\{\mathbf{x}(k)\}$ from the Markov model (8.5-41) with the series expansion terms for $\boldsymbol{\phi}$ and $\boldsymbol{\Gamma}\mathbf{V}_{\mathbf{w}}\boldsymbol{\Gamma}^T$ used instead of the true values. There results the variance equation $\mathbf{V}_{\mathbf{x}}(k)$. Since the steps involved are so similar to those used for the continuous case, we shall not repeat the details of the derivation here. We obtain the first-order (approximate) mean- and variance-propagation equation

$$\boldsymbol{\mu}_{\mathbf{x}}(k+1) = \boldsymbol{\phi}[\boldsymbol{\mu}_{\mathbf{x}}(k), k] \qquad (8.5\text{-}43)$$

$$\mathbf{V}_{\mathbf{x}}(k+1) = \frac{\partial \boldsymbol{\phi}[\boldsymbol{\mu}_{\mathbf{x}}(k), k]}{\partial \boldsymbol{\mu}_{\mathbf{x}}(k)}\mathbf{V}_{\mathbf{x}}(k)\frac{\partial \boldsymbol{\phi}^T[\boldsymbol{\mu}_{\mathbf{x}}(k), k]}{\partial \boldsymbol{\mu}_{\mathbf{x}}(k)} + \boldsymbol{\Gamma}[\boldsymbol{\mu}_{\mathbf{x}}(k), k]\mathbf{V}_{\mathbf{w}}(k)\boldsymbol{\Gamma}^T[\boldsymbol{\mu}_{\mathbf{x}}(k), k] \qquad (8.5\text{-}44)$$

[1] If the Markov model is linear, we know that $\boldsymbol{\Gamma}[\mathbf{x}(k), k]$ will not be a function of $\mathbf{x}(k)$ and that $\boldsymbol{\phi}[\mathbf{x}(k), k]$ will be of the form $\boldsymbol{\Phi}[k+1, k]\mathbf{x}(k)$.

and the second-order (approximate) mean- and variance-propagation equation

$$\boldsymbol{\mu}_{\mathbf{x}}(k+1) = \boldsymbol{\phi}[\boldsymbol{\mu}_{\mathbf{x}}(k), k] + \frac{1}{2}\frac{\partial^2\boldsymbol{\phi}[\boldsymbol{\mu}_{\mathbf{x}}(k), k]}{[\partial\boldsymbol{\mu}_{\mathbf{x}}(k)]^2} : \mathbf{V}_{\mathbf{x}}(k) \tag{8.5-45}$$

$$\begin{aligned}\mathbf{V}_{\mathbf{x}}(k+1) = {} & \frac{\partial\boldsymbol{\phi}[\boldsymbol{\mu}_{\mathbf{x}}(k), k]}{\partial\boldsymbol{\mu}_{\mathbf{x}}(k)}\mathbf{V}_{\mathbf{x}}(k)\frac{\partial\boldsymbol{\phi}^T[\boldsymbol{\mu}_{\mathbf{x}}(k), k]}{\partial\boldsymbol{\mu}_{\mathbf{x}}(k)} + \boldsymbol{\Gamma}[\boldsymbol{\mu}_{\mathbf{x}}(k), k]\mathbf{V}_{\mathbf{w}}(k)\boldsymbol{\Gamma}^T[\boldsymbol{\mu}_{\mathbf{x}}(k), k] \\ & + \frac{1}{2}\frac{\partial^2\boldsymbol{\Gamma}[\boldsymbol{\mu}_{\mathbf{x}}(k), k]\mathbf{V}_{\mathbf{w}}(k)\boldsymbol{\Gamma}^T[\boldsymbol{\mu}_{\mathbf{x}}(k), k]}{[\partial\boldsymbol{\mu}_{\mathbf{x}}(k)]^2} : \mathbf{V}_{\mathbf{x}}(k) + \boldsymbol{\Xi}(k)\end{aligned} \tag{8.5-46}$$

where

$$\begin{aligned}\boldsymbol{\Xi}(k) = {} & \frac{1}{4}\sum_{i=1}^{N}\sum_{j=1}^{N}\sum_{l=1}^{N}\sum_{m=1}^{N} + \{[\mathbf{V}_{\mathbf{x}}(k)]_{im}[\mathbf{V}_{\mathbf{x}}(k)]_{lj} + [\mathbf{V}_{\mathbf{x}}(k)]_{mj}[\mathbf{V}_{\mathbf{x}}(k)_{li}\} \\ & \cdot \frac{\partial^2\boldsymbol{\phi}[\boldsymbol{\mu}_{\mathbf{x}}(k), k]}{\partial[\boldsymbol{\mu}_{\mathbf{x}}(k)]_i\,\partial[\boldsymbol{\mu}_{\mathbf{x}}(k)]_j}\frac{\partial\boldsymbol{\phi}^T[\boldsymbol{\mu}_{\mathbf{x}}(k), k]}{\partial[\boldsymbol{\mu}_{\mathbf{x}}(k)]_l\,\partial[\boldsymbol{\mu}_{\mathbf{x}}(k)]_m}\end{aligned} \tag{8.5-47}$$

is obtained from the tensor-type term

$$\begin{aligned}\boldsymbol{\Xi}(k) = \boldsymbol{\mathcal{E}}\Big\{ & \Big[\frac{\partial^2\boldsymbol{\phi}[\boldsymbol{\mu}_{\mathbf{x}}(k), k]}{[\partial\boldsymbol{\mu}_{\mathbf{x}}(k)]^2} : [\mathbf{x}(k) - \boldsymbol{\mu}_{\mathbf{x}}(k)][\mathbf{x}(k) - \boldsymbol{\mu}_{\mathbf{x}}(k)]\Big] \\ & \cdot \Big[\frac{\partial^2\boldsymbol{\phi}[\boldsymbol{\mu}_{\mathbf{x}}(k), k]}{[\partial\boldsymbol{\mu}_{\mathbf{x}}(k)]^2} : [\mathbf{x}(k) - \boldsymbol{\mu}_{\mathbf{x}}(k)][\mathbf{x}(k) - \boldsymbol{\mu}_{\mathbf{x}}(k)]\Big]^T\Big\}\end{aligned} \tag{8.5-48}$$

by use of the fourth-product-moment expansion of a gaussian random variable. Also, we use the relation for gaussian random variables that third-order moments about the mean value are zero.

Often we can neglect the tensor algorithms $\boldsymbol{\Xi}(k)$ as involving products of small terms (that is, of the order $[\partial^2\boldsymbol{\phi}/\partial\boldsymbol{\mu}_{\mathbf{x}}^2]^2$). The initial conditions necessary to implement either the first- or second-order algorithms are the prior mean and variance

$$\boldsymbol{\mu}_{\mathbf{x}}(0) = \boldsymbol{\mathcal{E}}\{\mathbf{x}(0)\} \qquad \mathbf{V}_{\mathbf{x}}(0) = \operatorname{var}\{\mathbf{x}(0)\} \tag{8.5-49}$$

which are given as part of the problem specification.

It is of interest to let the samples become dense in these approximate discrete expressions. We assume the usual limiting relations between discrete and continuous white noise,

$$\mathbf{V}_{\mathbf{x}}(t) = \lim_{\substack{k\to\infty \\ kT\to t \\ T\to 0}} \mathbf{V}_{\mathbf{x}}(k)$$

$$\boldsymbol{\Psi}_{\mathbf{w}}(t) = \lim_{\substack{k\to\infty \\ kT\to t \\ T\to 0}} T\mathbf{V}_{\mathbf{w}}(k)$$

and the relations

$$\mathbf{f}[\mathbf{x}(t), t] = \lim_{\substack{k\to\infty \\ kT\to t \\ T\to 0}} \frac{1}{T}\{\boldsymbol{\phi}[\mathbf{x}(k), k] - \mathbf{x}(k)\}$$

$$\mathbf{G}[\mathbf{x}(t), t] = \lim_{\substack{k\to\infty \\ kT\to t \\ T\to 0}} \frac{1}{T}\{\boldsymbol{\Gamma}[\mathbf{x}(k), k]\}$$

such that we do in fact obtain the continuous first-order approximate algorithms of Eqs. (8.5-20) and (8.5-30) as the samples become dense in Eqs. (8.5-43) and (8.5-44). In a similar way, the second-order algorithms of Eqs. (8.5-45) and (8.5-46) become the continuous second-order algorithms of Eqs. (8.5-17) and (8.5-29). It is important to note that the tensor expression $\boldsymbol{\Xi}(k)$ becomes zero as the samples become dense.

EXERCISES 8.5

8.5-1. Derive Eqs. (8.5-17) and (8.5-29).

8.5-2. Find the first- and second-order approximate equations for the mean and variance propagation for the nonlinear system

$$dx = -x^2\, dt + x\, du$$

where $u(t)$ is a unit variance parameter Wiener process.

Answers: $\dot{\mu}_x = -\mu_x^2 - 2V_x$, $\dot{V}_x = -4\mu_x V_x + \mu_x^2 + 2V_x$.

8.6. Conclusions

In this chapter we studied two special types of stochastic processes: the Markov process and the Wiener process. The markovian property is extremely useful in simplifying many developments in the applications of stochastic processes. The Wiener process allows us to give a rigorous interpretation to the concept of continuous white noise.

To properly study nonlinear systems excited by Wiener-process input, it was necessary to introduce the Ito calculus, since the use of ordinary calculus rules may lead to errors. The major result of this development was the Fokker–Planck equation, which describes the propagation of the probability density function for the system state. The last section was directed to the derivation of equations for the mean and variance propagation in nonlinear systems. These relations are useful for a variety of advanced work in the information and control sciences. Reference [6] discusses some of this work and provides additional problems and references.

8.7. References

(1) DOOB, J. L., *Stochastic Processes*, Wiley, New York, 1953.

(2) GRAY, A. H., and T. K. CAUGHEY, "A Controversy in Problems Involving Random Parametric Excitation," *J. Math. Phys.*, vol. 44, pp. 288–296, September 1965.

(3) KAILATH, T., and P. FROST, "Mathematical Modeling for Stochastic Processes," *Stochastic Problems in Control* (Proc. Symp. AACC), published by ASME, Ann Arbor, Mich., June, 1968.

(4) PARZEN, E., *Stochastic Processes*, Holden-Day, San Francisco, 1962.

(5) SAGE, A. P., and J. L. MELSA, *Estimation Theory with Applications to Communications and Control*, McGraw-Hill, New York, 1971.

(6) SKOROKHOD, A. V., *Studies in the Theory of Random Processes*, Addison-Wesley, Reading, Mass., 1965.

(7) VITERBI, A. J., *Principles of Coherent Communication*, McGraw-Hill, New York, 1966.

(8) WONG E., and M. ZAKAI, "On the Relation between Ordinary and Stochastic Differential Equations," *Intern. J. Engr. Sci.*, vol. 3, pp. 213–229, 1965.

8.8. Problems

8.8-1. What is the Ito differential rule for

$$\phi[x(t), t] = \exp\{x(t)\}$$

for the system

$$dx(t) = f[x(t), t]\,dt + g[x(t), t]\,du(t)$$

8.8-2. The two systems $dx(t) = du(t)$ and $0.5\,dx^2(t) = x(t)\,du(t)$ are equivalent by ordinary calculus. Show that the results obtained by stochastic calculus are different.

8.8-3. Find the two approximate solutions for $\mu_x(t)$ and $V_x(t)$ for the system

$$\frac{dx(t)}{dt} = -\frac{x(t)}{1 + x^2(t)} + w(t)$$

where $\mathcal{E}\{w(t)\} = 0$ and $\text{var}\{w(t), w(\tau)\} = \Psi_v(t)\delta_D(t - \tau)$.

8.8-4. Find the two approximate solutions for $\mu_x(t)$ and $V_x(t)$ for the system

$$\frac{dx(t)}{dt} = -x^3(t) + x^2(t)w(t)$$

where $\mathcal{E}\{w(t)\} = 0$ and $\text{cov}\{w(t), w(\tau)\} = \Psi_w(t)\delta_D(t - \tau)$.

8.8-5. Find the two approximate solutions for $\mu_x(t)$ and $V_x(t)$ for the system

$$\frac{dx(t)}{dt} = -x^3(t) + w(t)$$

where $\mathcal{E}\{w(t)\{ = 1$ and $\text{cov}\{w(t), w(\tau)\} = e^{-|t-\tau|}$.

8.8-6. Show that only if the Ito definition of stochastic integration is used is the $x(t)$ given by

$$\frac{d\mathbf{x}}{dt} = \mathbf{f}[\mathbf{x}(t), t] + \mathbf{G}[\mathbf{x}(t), t]\mathbf{w}(t)$$

a Markov process. Assume that $\mathbf{w}(t)$ is a zero-mean, white gaussian process.

8.8-7. For the system of Problem 8.8-6, find the (a) unconditional Fokker–Planck equation; (b) Ito differential rule for $\psi(x, t)$.

Appendix A Operational Mathematics and the Laplace Transform

A.1. Introduction

The purpose of this appendix is to organize the large field of operational mathematics, and in particular that portion of the field which deals with the Laplace transform, into a form useful as background material for this book. It is assumed that the reader has previously studied this subject, and the main purpose here is to review this material in such a way as to optimize the study of the response of linear time-invariant systems to stationary stochastic-process inputs. In all cases we shall be concerned with characterizing the input–output relationship of the unilateral system shown in Fig. A.1-1.

For a great number of systems it is possible to break the system input into simpler inputs consisting of sums of impulses. Consider, for example, the unit pulse defined in Fig. A.1-2. The pulse is centered at time t^* and is $1/\Delta t$ $(\text{seconds})^{-1}$ high and Δt seconds in width, such that the pulse contains unit area. It is convenient to give this unit pulse the symbol $p(t - t^*, \Delta t)$. As $\Delta t \longrightarrow 0$, the resulting waveform will be called the unit impulse and given the symbol $\delta_D(t - t^*)$. Although this is not the only way of defining a unit impulse, it is sufficient for our purposes here.

Next we develop the response of a linear system with a single input and output to an arbitrary input $r(t)$. Suppose that the input $r(t)$ is as shown in Fig. A.1-3. By using the unit pulse we can approximate the system input

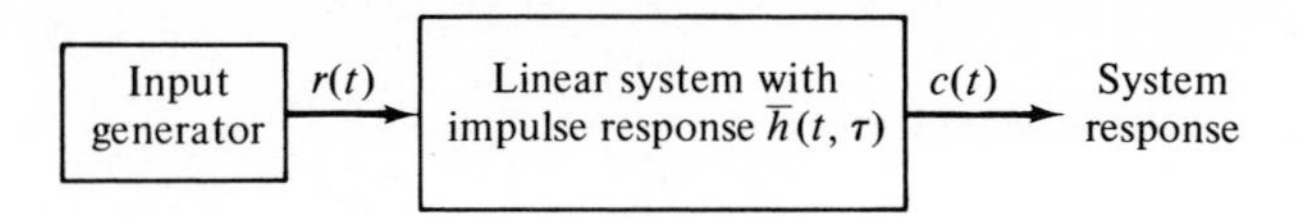

FIG. A.1-1 Unilateral linear system.

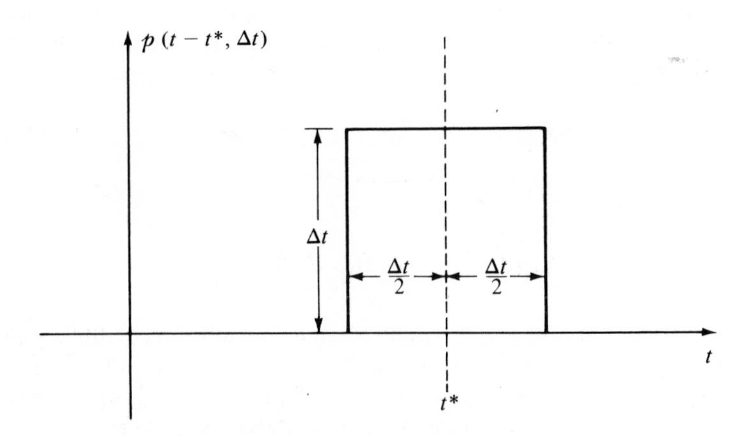

FIG. A.1-2 Definition of unit pulse.

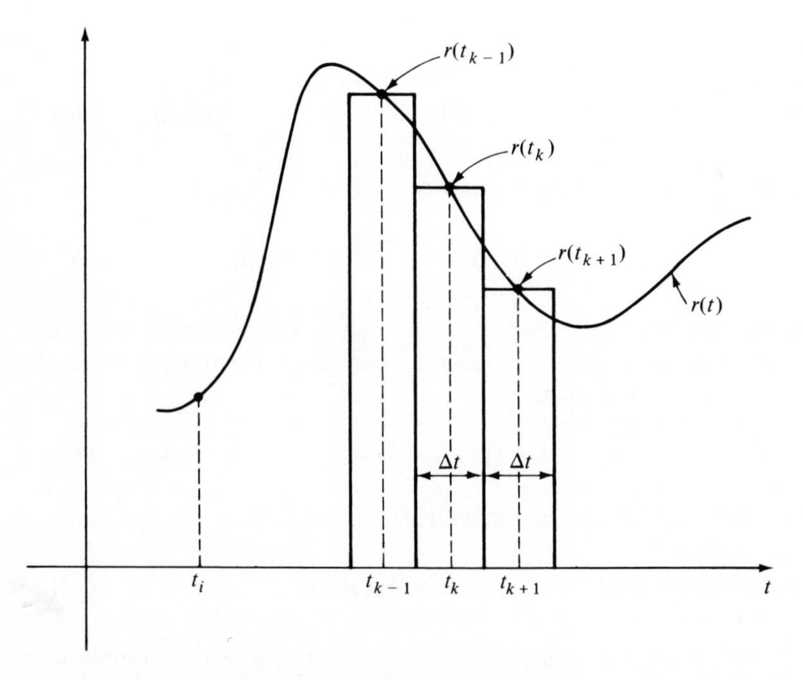

FIG. A.1-3 Arbitrary input $r(t)$.

$r(t)$ by the unit pulse relation

$$r(t) \cong \mathfrak{r}(t, t_k, \Delta t) = \sum_{k=-\infty}^{\infty} p(t - t_k, \Delta t) r(t_k)\, \Delta t \tag{A.1-1}$$

Let us now define the response of the initially unexcited system of Fig. A.1-1 to a unit pulse $p(t - t^*, \Delta t)$ as $\hbar(t, t^*, \Delta t)$, where t is the present time, t^* is the time at which the center of the unit pulse occurs, and Δt is the width of the unit pulse. The impulse response of the system to an impulse

$$\delta_D(t - t^*) = \lim_{\Delta t \to 0} p(t - t^*, \Delta t) \tag{A.1-2}$$

is then defined as

$$\bar{h}(t, t^*) = \lim_{\Delta t \to 0} \hbar(t, t^*, \Delta t) \tag{A.1-3}$$

The system is linear, and so the response to a unit pulse $r(t_k)$ units high is $\hbar(t, t_k, \Delta t) r(t_k)$. The system is linear and so the response to a sum of inputs is the sum of responses to the various inputs.[1] Thus we have [approximately, since the actual input is $r(t)$ and not Eq. (A.1-1)] the response to $\mathfrak{r}\,(t, t_k, \Delta t)$:

$$c(t, t_k, \Delta t) = \hbar(t, t_k, \Delta t) r(t_k)\, \Delta t \tag{A.1-4}$$

The total response is then

$$c(t, \Delta t) = \sum_{k=-\infty}^{\infty} c(t, t_k, \Delta t) = \sum_{=-\infty k}^{\infty} \hbar(t, t_k, \Delta t) r(t_k)\, \Delta t \tag{A.1-5}$$

The exact response is obtained when $\Delta t \longrightarrow 0$. We then have

$$c(t) = \lim_{\Delta t \to 0} c(t, \Delta t) = \int_{-\infty}^{\infty} \bar{h}(t, \tau) r(\tau)\, d\tau \tag{A.1-6}$$

It is necessary to place a restriction on the system that it be causal (or realizable), such that the impulse response is zero before the time at which the impulse is applied. This requires

$$\bar{h}(t, \tau) = 0 \qquad t < \tau \tag{A.1-7}$$

and thus Eq. (A.1-6) may be replaced by

$$c(t) = \int_{-\infty}^{t} \bar{h}(t, \tau) r(\tau)\, d\tau \tag{A.1-8}$$

[1] This is one of the definitions of linearity; that is, a linear system obeys the superposition principle.

which is a completely general expression for the impulse response (superposition integral) of a linear system.

We now restrict ourselves to a linear time-invariant system in which the impulse response is a function of the time difference between t and τ only, such that

$$\bar{h}(t, \tau) = h(t - \tau) \tag{A.1-9}$$

The impulse response of Eq. (A.1-8) may then be written in either of the equivalent forms

$$c(t) = \int_{-\infty}^{t} h(t - \tau)r(\tau)\, d\tau \tag{A.1-10}$$

or

$$c(t) = \int_{0}^{\infty} h(\tau)r(t - \tau)\, d\tau \tag{A.1-11}$$

A.2. The Laplace transform

Let $f(t)$ be a function of the real independent variable (generally time) t *defined only for* $t \geq 0$. *The Laplace transform of* $f(t)$, a complex function $F(s)$, where $s = \sigma + j\omega$, is defined by

$$F(s) = \mathfrak{L}\{f(t)\} = \int_{0}^{\infty} f(t)e^{-st}\, dt \tag{A.2-1}$$

There is generally a real number $\sigma_0(|\sigma_0| \leq \infty)$ such that

$$\int_{0}^{\infty} |f(t)|\, e^{-\sigma t}\, dt$$

exists for $\sigma > \sigma_0$ and does not exist for $\sigma < \sigma_0$. σ_0 is called the abscissa of absolute convergence of $F(s)$.

The Laplace transform possesses a unique inverse transform defined by

$$f(t) = \mathfrak{L}^{-1}\{F(s)\} = \frac{1}{2\pi j} \int_{\sigma - j\infty}^{\sigma + j\infty} F(s)e^{st}\, ds \tag{A.2-2}$$

where $\sigma > \sigma_0$. Equations (A.2-1) and (A.2-2) define what is often called a Laplace transform pair and may be derived by a variety of methods. Normally, Laplace transforms are obtained by integration of Eq. (A.2-1), and tables of Laplace transforms have been established. A few of the more important Laplace transforms are given in Table A.2-1. In practice, rather than use Eq. (A.2-2) to obtain the inverse transform, one normally uses partial fraction expansion of $F(s)$ and makes use of a table of Laplace transforms to obtain $f(t)$. Equation (A.2-2) is useful for many analytical purposes however.

TABLE A.2-1. Laplace Transform Table and Properties

$f(t)$	$F(s)$	(A.2-3)
$\delta_D(t-a)$	e^{-as}	(A.2-4)
1	$\dfrac{1}{s}$	(A.2-5)
$t^n(n \geq 0)$	$\dfrac{n!}{(s)^{n+1}}$	(A.2-6)
$t^n a^{-at}(n \geq 0)$	$\dfrac{n!}{(s+a)^{n+1}}$	(A.2-7)
$\sin at$	$\dfrac{a}{s^2+a^2}$	(A.2-8)
$\cos at$	$\dfrac{s}{s^2+a^2}$	(A.2-9)
$\sinh at$	$\dfrac{a}{s^2-a^2}$	(A.2-10)
$\cosh at$	$\dfrac{s}{s^2-a^2}$	(A.2-11)
$e^{-at}\sin bt$	$\dfrac{b}{(s+a)^2+b^2}$	(A.2-12)
$e^{-at}\cos b$	$\dfrac{s+a}{(s+a)^2+b^2}$	(A.2-13)
$a_1 f_1(t) + a_2 f_2(t)$	$a_1 F_1(s) + a_2 F_2(s)$	(A.2-14)
$\dfrac{df(t)}{dt}$	$sF(s) - f(0)$	(A.2-15)
$\dfrac{d^n f(t)}{dt^n}$	$s^n F(s) - \sum_{k=1}^{n} s^{k-1} \dfrac{d^{n-k} f(t)}{dt^{n-k}}\Big\|_{t=0}$	(A.2-16)
$\int_0^t f(\tau)\, d\tau$	$\dfrac{F(s)}{s}$	(A.2-17)
$f(t-\tau)\mathcal{S}(t-\tau)$	$e^{-s\tau}F(s)$	(A.2-18)
$e^{-\alpha t}f(t)$	$F(s+\alpha)$	(A.2-19)
$\int_0^\infty f_1(t-\tau) f_2(\tau)\, d\tau$	$F_1(s)F_2(s)$	(A.2-20)
$f_1(t)f_2(t)$	$\dfrac{1}{2\pi j}\int_{\sigma-j\infty}^{\sigma+j\infty} F_1(s-\eta)F_2(\eta)\, d\eta$	(A.2-21)
$f(at)$	$\dfrac{1}{a}F\left(\dfrac{s}{a}\right)$	(A.2-22)
$\lim_{t\to 0} f(t) = \lim_{s\to\infty} sF(s)$		(A.2-23)
$\lim_{t\to 0} f(t) = \lim_{s\to 0} sF(s)$		(A.2-24)

A.3. Partial-fraction expansion and solution of differential equations

In the determination of time functions from Laplace transforms, the method of partial-fraction expansions is quite useful. First we shall consider the case in which there are first-order poles only. A Laplace transform of the form

$$F(s) = \frac{C(s)}{D(s)} = \frac{\prod_{k=1}^{M} (s + c_i)}{\prod_{k=1}^{N} (s + d_i)} \tag{A.3-1}$$

where $M < N$ and where $d_i \neq d_j$ for $i \neq j$, is given. It turns out that Eq. (A.3-1) can be written in the partial-fraction form

$$F(s) = \sum_{k=1}^{N} \frac{a_k}{s + d_k} \tag{A.3-2}$$

where

$$a_k = (s + d_k)F(s)\,|_{s=-d_k} \qquad k = 1, 2, \ldots, N \tag{A.3-3}$$

If there are repeated roots, the partial-fraction-expansion technique becomes a bit more complicated. For

$$F(s) = \frac{C(s)}{\prod_{i=1}^{N} (s + d_i)^{M_i}} \tag{A.3-4}$$

where there is a pole of multiplicity M_k at $s = -d_k$, and where the numerator polynomial is of lower order than the denominator polynomial, the partial-fraction expansion is

$$F(s) = \sum_{i=1}^{N} \sum_{j=1}^{M_i} \frac{a_{ij}}{(s + d_i)^j} \tag{A.3-5}$$

where

$$a_{k\,M_k-i} = \left\{\frac{1}{i!}\frac{d^i}{ds^i}[(s + d_k)^{M_k}F(s)]\right\}\Bigg|_{s=-d_k} \tag{A.3-6}$$

This partial-fraction technique may be used to aid in the solution of differential equations. The differential equation

$$\sum_{i=0}^{N} \alpha_i \frac{d^i x(t)}{dt^i} = \sum_{i=0}^{M} \beta_i \frac{d^i y(t)}{dt^i} \tag{A.3-7}$$

where $y(t)$ is a known input, the system is initially unexcited, $M < N$, and where $d_N = c_M = 1$ without loss of generality, is considered. The Laplace transform of this equation is

$$\sum_{i=0}^{N} s^i \alpha_i X(s) = \sum_{i=0}^{M} s^i \beta_i Y(s) \tag{A.3-8}$$

The *transfer function* of the system is defined by

$$\frac{X(s)}{Y(s)} = \frac{\sum_{i=0}^{M} s^i \beta_i}{\sum_{i=0}^{N} s^i \alpha_i} = H(s) \tag{A.3-9}$$

Since the input $y(t)$ is known, $Y(s)$ is known or may be determined, and the Laplace transform of the output expressed as

$$X(s) = H(s)Y(s) = \frac{C(s)}{D(s)} \tag{A.3-10}$$

and the method of partial fractions used, if necessary, to decompose $C(s)/D(s)$ into simpler form such that the inverse transform may be readily determined and $y(t)$ thereby obtained.

A.4. The Fourier and bilateral Laplace transform

Let $f(t)$ be a function of the real independent variable (generally time) *t defined for all time*. The *Fourier transform of* $f(t)$, a complex function $F(\omega)$, is defined by

$$F(\omega) = \mathfrak{F}\{f(t)\} = \int_{-\infty}^{\infty} f(t)e^{-j\omega t}\,dt = \int_{-\infty}^{\infty} f(t)\cos\omega t\,dt - j\int_{-\infty}^{\infty} f(t)\sin\omega t\,dt \tag{A.4-1}$$

with the inverse Fourier transform defined by

$$f(t) = \mathfrak{F}^{-1}\{F(\omega)\} = \frac{1}{2\pi}\int_{-\infty}^{\infty} F(\omega)e^{j\omega t}\,d\omega \tag{A.4-2}$$

The Fourier transform obeys most of the Laplace transform properties given in Table A.2-1. The exceptions are the initial value theorem, Eq. (A.2-23), which does not exist for Fourier transforms since there is no particular significance to $t = 0$, and Eqs. (A.2-15) and (A.2-16), which should be replaced by

$$\mathfrak{F}\left\{\frac{d^n f(t)}{dt^n}\right\} = (j\omega)^n F(\omega) \tag{A.4-3}$$

Two facts are immediately apparent when comparing the Fourier transform pair, Eqs. (A.4.1) and (A.4-2), with the Laplace transform pair, Eqs. (A.2-1) and (A.2-2). The first is that the Laplace transform is a function of the variable $s = \sigma + j\omega$, whereas the Fourier transform is a function of $j\omega$ only. Also, the Laplace transform is defined for positive time only, whereas the Fourier transform is defined for all time. If a new variable s is substituted for $j\omega$ in Eqs. (A.4-1) and (A.4-2), there results

$$F(s) = \int_{-\infty}^{\infty} f(t)e^{-st}\,dt \tag{A.4-4}$$

$$f(t) = \frac{1}{2\pi j}\int_{-j\infty}^{j\infty} F(s)e^{st}\,ds \tag{A.4-5}$$

If the variable $s = j\omega$, the foregoing two relations are equivalent to those of Eqs. (A.4-1) and (A.4-2). Now if s is defined in terms of (the real numbers) σ and ω as

$$s = \sigma + j\omega \tag{A.4-6}$$

significant questions must be raised, since we have replaced the mathematics of real variables by those of complex variables. It turns out that there is a theorem of analytic continuation of complex-variable theory which states that if analytical continuation of $F(\omega)$ is identical to $F(s)$ in an arbitrarily small portion of the $j\omega$ axis of the s plane, then $F(s)$ is equal to analytical continuation of $F(j\omega)$ *for all* s. Since $\lim_{s \to j\omega} F(s) = F(\omega)$, we are justified in using the transform pair of Eqs. (A.4-4) and (A.4-5), which is commonly called the two-sided or bilateral Laplace transform pair. Since mathematical equivalence between the Fourier and two-sided Laplace transform exists, we are sure that the properties of the two transforms will be the same and can merely substitute the variable s for $j\omega$ in theorems and transforms for Fourier transforms to obtain bilateral Laplace transforms.

Since most transform tables are of the one-sided or unilateral Laplace transform, we shall now consider methods of obtaining Fourier or bilateral Laplace transforms from one-sided Laplace-transform tables. Consider the bilateral Laplace transform of a function $f(t)$ that is nonzero *only* for negative t. We have

$$F(s) = \int_{-\infty}^{0} f(t)e^{-st}\, dt$$

If we let $t = -\eta$, the foregoing becomes

$$F(s) = \int_{0}^{\infty} f(-\eta)e^{s\eta}\, d\eta$$

Therefore, we can obtain the bilateral Laplace transform from the unilateral Laplace transform in the following manner:

1. Break $f(t)$ into two parts:

$$f(t) = f^{+}(t) + f^{-}(t)$$

   where $f^{+}(t)$ is nonzero for $t \geq 0$ only, and $f^{-}(t)$ is nonzero for $t < 0$ only.
2. Find $F^{+}(s) = \mathcal{L}f^{+}(t)$.
3. Replace t by $-t$ in $f^{-}(t)$.
4. Find the unilateral Laplace transform of $f^{-}(-t)$.
5. Replace s by $-s$ to obtain the bilateral Laplace transform, $F^{-}(s) = \mathcal{L}f^{-}(t)$, of $f^{-}(t)$.
6. Write the total bilateral Laplace transform as

$$F(s) = F^{+}(s) + F^{-}(s)$$

Example ***A.4-1.*** We desire to obtain the bilateral Laplace transform of

$$C_x(\tau) = Ae^{-B|\tau|}$$

Use of the definition of the bilateral Laplace transform yields

$$C_x(s) = \int_{-\infty}^{\infty} Ae^{-B|\tau|}e^{-s\tau}\, d\tau = \int_{-\infty}^{0} Ae^{B\tau}e^{-s\tau}\, d\tau + \int_{0}^{\infty} Ae^{-B\tau}e^{-s\tau}\, d\tau$$

$$= \frac{A}{B-s} + \frac{A}{B+s} = \frac{2AB}{B^2 - s^2} = \frac{2AB}{B^2 + \omega^2}$$

This method would be quite complex if the time function were not so simple, and use of Laplace-transform tables would be preferable. Following the procedure just outlined, we have

1. $C_x(\tau) = C_x^{+}(\tau) + C_x^{-}(\tau)$
   $C_x^{+}(\tau) = Ae^{-B\tau} \qquad \tau \geq 0$
   $C_x^{-}(\tau) = Ae^{B\tau} \qquad \tau < 0$

2. $C_x^+(s) = \dfrac{A}{s + B}$
3. $C_x^-(-\tau) = Ae^{-B\tau} \qquad \tau > 0$
4. $\mathcal{L}\{C_x^-(-\tau)\} = \mathcal{L}\{Ae^{-B\tau}\} = \dfrac{A}{s + B}$
5. $\mathcal{L}\{C_x^-(\tau)\} = \dfrac{A}{-s + B}$
6. $C_x(s) = \dfrac{A}{B + s} + \dfrac{A}{B - s} = \dfrac{2AB}{B^2 - s^2}$

A.5. Parseval's theorem

A very useful tool in the work of this book is Parseval's theorem, which is

$$\int_{-\infty}^{\infty} f_1(t)f_2(t)\,dt = \frac{1}{2\pi j}\int_{-j\infty}^{j\infty} F_1(\pm s)F_2(\mp s)\,ds \tag{A.5-1}$$

where $F_1(s)$ and $F_2(s)$ are Laplace transforms of $f_1(t)$ and $f_2(t)$.

The proof of Parseval's theorem proceeds as follows. The left side of Eq. (A.5-1) may be written as

$$\int_{-\infty}^{\infty} [\mathcal{L}^{-1}F_1(s)]f_2(t)\,dt = \frac{1}{2\pi j}\int_{-\infty}^{\infty}\int_{-j\infty}^{j\infty} F_1(s)f_2(t)e^{st}\,ds\,dt$$

$$= \frac{1}{2\pi j}\int_{-j\infty}^{j\infty}\left[\int_{-\infty}^{\infty} f_2(t)e^{st}\,dt\right]F_1(s)\,ds = \frac{1}{2\pi j}\int_{-j\infty}^{j\infty} F_2(-s)F_1(s)\,ds \tag{A.5-2}$$

Clearly, the choice of $f_1(t)$ or $f_2(t)$ is arbitrary; so the theorem is proved.

Parseval's theorem allows us to replace integrals in the time domain with integrals in the frequency domain, and vice versa. A special but important case of the theorem is

$$\int_{-\infty}^{\infty} f(t)f(t)\,dt = \int_{-\infty}^{\infty} f^2(t)\,dt = \frac{1}{2\pi j}\int_{-j\infty}^{j\infty} F(s)F(-s)\,ds \tag{A.5-3}$$

Integrals of this form, where

$$F(s) = \frac{c(s)}{d(s)} = \frac{c_0 + c_1 s + c_2 s^2 + \cdots + c_{n-1}s^{n-1}}{d_0 + d_1 s + d_2 s^2 + \cdots + d_n s^n} \tag{A.5-4}$$

are tabulated for quotients of polynomials. The first three are

$$I_1 = \frac{c_0}{2d_0d_1} \qquad I_2 = \frac{c_0^2d_2 + c_2d_0}{2d_0d_1d_2}$$

$$I_3 = \frac{c_2^2d_0d_1 + (c_1^2 - 2c_0c_2)d_0d_3 + c_0^2d_2d_3}{2d_0d_3(-d_0d_3 + d_1d_2)} \tag{A.5-5}$$

A.6. Summary and references

In this Appendix we have presented some fundamental results that should be of value in a review of the essential features of operational mathematics and Laplace transforms, which are necessary for a study of correlation functions, power spectra, and the response of linear systems to random inputs. There are many references available on this important subject, and doubtlessly the reader of this book already is familiar with one or more of them. Particularly recommended are

(1) CHURCHILL, R. V., *Operational Mathematics*, McGraw-Hill, New York, 1958.

(2) COOPER, G. R., and C. D. MCGILLERN, *Methods of Signal and System Analysis*, Holt, Rinehart and Winston, New York, 1967.

(3) SCHWARTZ, R. J., and B. FRIEDLAND, *Linear Systems*, McGraw-Hill, New York, 1965.

Appendix B System Representation by State Variables

This appendix is intended to delineate the various state-variable and vector-matrix operations encountered in this text. The symbols for vectors are normally boldface lowercase Arabic or lowercase Greek letters, e.g., $\mathbf{x}$ or $\boldsymbol{\alpha}$. The symbols for matrices are boldface uppercase Arabic or uppercase Greek letters, e.g., $\mathbf{F}$ or $\boldsymbol{\Phi}$. Brackets [ ] are used around matrices only where clarity is improved. The transpose of a vector or matrix is indicated by a superscript following the symbol, e.g., $\mathbf{x}^T$ or $\mathbf{F}^T$. All vector symbols not followed by the transpose superscript are column vectors. The inverse of a nonsingular matrix is indicated by the superscript (-1) following the matrix symbol, e.g., $\mathbf{F}^{-1}$. The inverse transpose of a matrix is indicated by the superscript $(-T)$ following the matrix symbol, e.g., $\mathbf{F}^{-T}$. The basic symbols, definitions, and operations that we use are as follows.

B.1. Matrix algebra

Matrix We shall call a set of mn quantities, real or complex, arranged in a rectangular array of m columns and n rows a matrix of order (m, n), or $m \times n$. If $m = n$, we shall call the matrix a square matrix. A matrix is denoted in the following manner:

$$\mathbf{F} = \begin{bmatrix} f_{11} & f_{12} & \cdots & f_{1n} \\ f_{21} & f_{22} & \cdots & f_{2n} \\ \cdot & & & \\ \cdot & & & \\ \cdot & & & \\ f_{m1} & f_{m2} & \cdots & f_{mn} \end{bmatrix}$$

Row Vector A set of n quantities arranged in a row is a matrix of order $(1, n)$. We shall call it a row vector:

$$\mathbf{x}^T = [x_1 \; x_2 \; x_3 \ldots x_n]$$

Column Vector We shall call a set of n quantities arranged in a column a matrix of order $(n, 1)$ or a column vector:

$$\mathbf{x} = \begin{bmatrix} x_1 \\ x_2 \\ x_3 \\ \cdot \\ \cdot \\ \cdot \\ x_n \end{bmatrix}$$

Scalar A matrix of order (1, 1) will be called a scalar. A vector with only a single component is likewise a scalar.

Diagonal Matrix We shall define a diagonal matrix as a square matrix in which all elements off the main diagonal are zero:

$$\mathbf{F} = \begin{bmatrix} f_{11} & 0 & 0 \\ 0 & f_{22} & 0 \\ 0 & 0 & f_{33} \end{bmatrix}$$

Identity Matrix We shall define the identity matrix as a diagonal matrix having unit diagonal elements:

$$\mathbf{I} = \begin{bmatrix} 1 & 0 & 0 \\ 0 & 1 & 0 \\ 0 & 0 & 1 \end{bmatrix}$$

Null or Zero Matrix The matrix **[0]**, which has all zero elements, will be called a null or zero matrix.

Equality or Two Matrices We shall say that two matrices **F** and **G** are equal if, and only if, they are of the same order and all their corresponding elements are equal such that

$$[\mathbf{F}] - [\mathbf{G}] = [\mathbf{0}]$$

Singular and Nonsingular Matrices We shall call a square matrix [**F**], for which the determinant of its elements, det [**F**], is zero, a singular matrix. If det [**F**] $\neq 0$, we say that [**F**] is a nonsingular matrix.

Matrix Transpose We form the transpose of a matrix [**F**] by the interchange of rows and columns. Thus the transpose of the matrix [**F**] is

$$\mathbf{F}^T = \begin{bmatrix} f_{11} & f_{21} & f_{31} & \cdots & f_{n1} \\ f_{12} & & & & \\ \cdot & & & & \\ \cdot & & & & \\ \cdot & & & & \\ f_{1n} & & & & f_{nm} \end{bmatrix}$$

Orthogonal Matrix We shall call a matrix [**F**] such that $\mathbf{F}^T\mathbf{F} = \mathbf{F}\mathbf{F}^T = \mathbf{I}$ an orthogonal matrix.

Symmetric and Skew-Symmetric Matrices We shall define a symmetric matrix as one that is square and symmetrical about the main diagonal, $\mathbf{F}^T = \mathbf{F}$. $\mathbf{F}^T = -\mathbf{F}$ is the requirement for a skew-symmetric matrix.

Cofactor We shall define the cofactor of an element in the ith row and jth column of a matrix as the value of the determinant formed by writing the elements of the matrix as a determinant, then deleting the ith row and jth column, and giving to it the sign $(-1)^{i+j}$.

Adjoint Matrix We may obtain the adjoint matrix of a square matrix **F** by replacing each element by its cofactor and transposing. We shall denote the adjoint by [adj **F**].

Minor of a Matrix or Determinant The matrix or determinant obtained by suppressing m rows and m columns of the matrix **F** or the det[**F**] of the matrix **F** of order (n, n) will be called a minor of order $(n - m)$.

Characteristic Matrix, Characteristic Equations, and Eigenvalues The characteristic matrix of a square matrix **F** of order (n, n) and with constant elements is

$$[\mathbf{C}] = [\lambda\mathbf{I} - \mathbf{F}]$$

We shall call the equation $\det[\mathbf{C}] = \det[\lambda\mathbf{I} - \mathbf{F}] = 0$ the characteristic equation of the matrix $\mathbf{F}$. The n roots of the characteristic equation of $\det[\lambda\mathbf{I} - \mathbf{F}] = 0$, which will be of degree n in λ, will be called the eigenvalues of the matrix $\mathbf{F}$.

Addition of Matrices We shall perform the addition of two matrices $\mathbf{F}$ and $\mathbf{G}$ of the same order by adding their corresponding elements and by writing the sums as the corresponding elements of the resultant matrix $\mathbf{F} + (\mathbf{G} + \mathbf{H}) = (\mathbf{F} + \mathbf{G}) + \mathbf{H}$.

Multiplication of Matrices

Scalar Multiplication. A matrix $\mathbf{F}$ is multiplied by a scalar k if we multiply all the elements of $\mathbf{F}$ by k. We may easily show that scalar multiplication is commutative in that $k\mathbf{F} = \mathbf{F}k$. Thus we assume that the matrix is defined over a commutative field.

Multiplication of Two Vectors. We shall obtain multiplication of a row vector $\mathbf{x}^T$ by a column vector $\mathbf{y}$ by multiplying corresponding elements in each vector and then adding the products. Of course, the dimensions of $\mathbf{x}^T$ and $\mathbf{y}$ must be compatible. The product is a scalar and is denoted by

$$\mathbf{x}^T\mathbf{y} = [x_1, x_2, x_3, \ldots, x_n]\begin{bmatrix} y_1 \\ y_2 \\ y_3 \\ \cdot \\ \cdot \\ \cdot \\ y_n \end{bmatrix} = (x_1y_1 + x_2y_2 + x_3y_3 + \cdots + x_ny_n)$$

The scalar $\mathbf{x}^T\mathbf{y}$ is often called the inner product and given the symbol $\langle \mathbf{x}, \mathbf{y} \rangle$. This inner product can also be written with respect to a matrix as

$$\langle \mathbf{x}, \mathbf{P}\mathbf{y} \rangle = \mathbf{x}^T\mathbf{P}\mathbf{y}$$
$$\langle \mathbf{P}^T\mathbf{x}, \mathbf{y} \rangle = \mathbf{x}^T\mathbf{P}\mathbf{y}$$

When $\mathbf{x}$ and $\mathbf{y}$ are the same vector, the inner product is equivalent to the square of the euclidean norm:

$$\mathbf{x}^T\mathbf{x} = \|\mathbf{x}\|^2 = \langle \mathbf{x}, \mathbf{x} \rangle$$
$$\mathbf{x}^T\mathbf{P}\mathbf{x} = \|\mathbf{x}\|_{\mathbf{P}}^2 = \langle \mathbf{x}, \mathbf{P}\mathbf{x} \rangle$$

An interesting property of the norm is that $\|\mathbf{x} + \mathbf{y}\| \leq \|\mathbf{x}\| + \|\mathbf{y}\|$.

In a similar way, the outer product can be defined. In this case, it is not necessary that the **x** and **y** vectors have the same dimension. The outer product is defined by

$$\mathbf{x}{>}{<}\mathbf{y} = \mathbf{x}\mathbf{y}^T = \begin{bmatrix} x_1 \\ x_2 \\ \cdot \\ \cdot \\ \cdot \\ x_n \end{bmatrix} [y_1, y_2, \ldots, y_m]$$

$$= \begin{bmatrix} (x_1 y_1) & (x_1 y_2) & (x_1 y_3) & \cdots & (x_1 y_m) \\ (x_2 y_1) & (x_2 y_2) & \cdots & & \\ \cdot & & & & \\ \cdot & & & & \\ \cdot & & & & \\ (x_n y_1) & (x_n y_2) & \cdots & & (x_n y_m) \end{bmatrix}$$

Multiplication of Two Matrices. Our definition of multiplication will be such that two matrices can be multiplied together only when the number of columns of the first is equal to the number of rows of the second. The product of two matrices will be defined by the equation

$$H_{ij} = [\mathbf{FG}]_{ij} = \sum_{k=1}^{k=p} f_{ik} g_{kj}$$

where the order of the matrices **F**, **G**, and **H** are (m, p), (p, n), and (m, n), respectively. We may express this symbolically in the form $(m, p)\cdot(p, n) = (m, n)$.

Inverse of a Matrix We may show that $[\mathbf{F}]\,[\text{adj}\,\mathbf{F}] = [\text{adj}\,\mathbf{F}]\mathbf{F} = \mathbf{I}\det[\mathbf{F}]$, where **I** is the nth-order identity matrix and **F** is an nth-order square matrix. If **F** is a nonsingular matrix, $\det[\mathbf{F}] \neq 0$, and there exists a matrix **G** such that

$$\mathbf{FG} = \mathbf{GF} = \mathbf{I}, \qquad \mathbf{G} = \frac{[\text{adj}\,\mathbf{F}]}{\det[\mathbf{F}]} = \mathbf{F}^{-1}$$

The matrix **G** will be defined as the inverse matrix of **F**. We shall denote it by $\mathbf{F}^{-1}$. The inverse of a product matrix is the product of the inverses of the separate matrices in the reverse order. Thus we may write

$$(\mathbf{AB})^{-1} = \mathbf{B}^{-1}\mathbf{A}^{-1}$$

Transpose of a Product of Matrices If **F** and **G** are conformable matrices such that we may define the product **FG**, we may obtain the transpose of this product by

$$(\mathbf{FG})^T = \mathbf{G}^T\mathbf{F}^T$$

Cayley–Hamilton Theorem If **F** is a square matrix, we write

$$\Delta(\lambda) = \det[\lambda\mathbf{I} - \mathbf{F}] = 0$$

as its characteristic equation. We may show that

$$\Delta([\mathbf{F}]) = \mathbf{0}$$

This is a consequence of the Cayley–Hamilton theorem, which states that every square matrix satisfies its own characteristic equation in a matrix sense.

Transformation of a Square Matrix to Diagonal Form We are given the square matrix **F**. The equation $\mathbf{y} = \mathbf{Fx}$ expresses a linear transformation between the column vectors **x** and **y**. If we require that $\mathbf{y} = \lambda\mathbf{x}$, then $\lambda\mathbf{x} = \mathbf{Fx}$. The possible values of λ are $\lambda_1, \lambda_2, \ldots, \lambda_n$, which are the eigenvalues of **F**. In what is to follow, we shall assume that eigenvalues are distinct.

To each eigenvalue of **F** there corresponds an eigenvector $\mathbf{x}^i$ which satisfies the relation

$$\lambda_i \mathbf{x}^i = \mathbf{F}\mathbf{x}^i, \qquad i = 1, 2, 3, \ldots, n$$

If we construct the square matrix,

$$[\mathbf{X}] = [(\mathbf{x})^1 \,\vdots\, (\mathbf{x})^2 \,\vdots\, (\mathbf{x})^3 \,\vdots\, \cdots \,\vdots\, (\mathbf{x})^n]$$

we can write

$$\mathbf{XG} = \mathbf{FX}, \qquad \mathbf{G} = \begin{bmatrix} \lambda_1 & 0 & \cdots & 0 \\ 0 & \lambda_2 & \cdots & 0 \\ \cdot & & & \cdot \\ \cdot & & & \cdot \\ \cdot & & & \cdot \\ 0 & 0 & & \lambda_n \end{bmatrix} = \mathrm{diag}[\lambda]$$

and show that

$$\mathbf{F} = \mathbf{XGX}^{-1}$$

which is the required transformation which converts **F** to diagonal form. This tells us that if $\mathbf{y} = \mathbf{Fx}$, $\mathbf{x} = \mathbf{Xu}$, and $\mathbf{y} = \mathbf{Xv}$, then

$$\mathbf{v} = \mathbf{X}^{-1}\mathbf{y} = \mathbf{X}^{-1}\mathbf{Fx} = \mathbf{X}^{-1}\mathbf{FXu}$$

$$\mathbf{v} = \mathbf{Gu}$$

Quadratic Forms We shall define a quadratic form in $\mathbf{x}$ as

$$Q = \mathbf{x}^T\mathbf{F}\mathbf{x} = \sum_{i=1}^{n}\sum_{j=1}^{n} f_{ij}x_i x_j$$

where $\mathbf{F}$ is a square symmetric matrix. A quadratic form is said to be positive definite if $Q > 0$ for $\mathbf{x} \neq \mathbf{0}$. A square symmetric matrix is said to be positive definite if its corresponding quadratic form is positive definite. A necessary and sufficient test for a positive-definite matrix $\mathbf{F}$ is that the following inequalities be satisfied:

$$|f_{11}| > 0 \qquad \begin{vmatrix} f_{11} & f_{12} \\ f_{21} & f_{22} \end{vmatrix} > 0 \qquad \begin{vmatrix} f_{11} & f_{12} & f_{13} \\ f_{21} & f_{22} & f_{23} \\ f_{31} & f_{32} & f_{33} \end{vmatrix} > 0 \qquad \text{etc.}$$

If the matrix is positive semidefinite, the greater than signs in the foregoing are replaced by greater than or equal to signs. Negative-definite and negative-semidefinite matrices are defined by testing the matrix -$\mathbf{F}$ for positive definiteness.

B.2. Differentiation of matrices and vectors

Differentiation with Respect to a Scalar (Time) We define differentiation of a vector or matrix with respect to a scalar in the following way:

$$\mathbf{z} = \mathbf{z}(t) \qquad \frac{d\mathbf{z}}{dt} = \left[\frac{dz_1}{dt}\ \frac{dz_2}{dt} \cdots \frac{dz_m}{dt}\right]^T$$

$$\mathbf{F} = \mathbf{F}(t) \qquad \frac{d\mathbf{F}}{dt} = \begin{bmatrix} \frac{df_{11}}{dt} & \frac{df_{12}}{dt} & \cdots & \frac{df_{1n}}{dt} \\ \frac{df_{21}}{dt} & \frac{df_{22}}{dt} & \cdots & \frac{df_{2n}}{dt} \\ \vdots & \vdots & & \vdots \\ \frac{df_{m1}}{dt} & \frac{df_{m2}}{dt} & \cdots & \frac{df_{mn}}{dt} \end{bmatrix}$$

Differentiation with Respect to a Vector Differentiation of scalars or vectors with respect to vectors is defined as follows:

$$f = f(x) = \text{scalar} \qquad \frac{df}{d\mathbf{x}} = \left[\frac{df}{dx_1}\ \frac{df}{dx_2} \cdots \frac{df}{dx_n}\right]^T$$

This is often called a gradient and written $\nabla_{\mathbf{x}}\mathbf{f} = df/d\mathbf{x}$.

$$\mathbf{z} = \mathbf{z}(x) \qquad \frac{d\mathbf{z}^T}{d\mathbf{x}} = \begin{bmatrix} \frac{dz_1}{dx_1} & \frac{dz_2}{dx_1} & \cdots & \frac{dz_m}{dx_1} \\ \frac{dz_1}{dx_2} & \frac{dz_2}{dx_2} & \cdots & \frac{dz_m}{dx_2} \\ \vdots & \vdots & & \vdots \\ \frac{dz_1}{dx_n} & \frac{dz_2}{dx_n} & \cdots & \frac{dz_m}{dx_n} \end{bmatrix}$$

This is sometimes called the Jacobian matrix and written $J_{\mathbf{x}}\{\mathbf{z}(x)\}$, where

$$\left[\frac{dz}{dx}\right]_{ij} = [J_{\mathbf{x}}\{\mathbf{z}(\mathbf{x})\}]_{ij} = \left[\frac{\partial z_i}{\partial x_j}\right]$$

When we use the notation $\partial\mathbf{z}/\partial\mathbf{x}$, we mean by this the operation

$$\frac{\partial \mathbf{z}}{\partial \mathbf{x}} \triangleq \left[\frac{d\mathbf{z}^T}{d\mathbf{x}}\right]^T$$

It is also possible to obtain expressions for the derivatives of a matrix with respect to a vector. Since such operations are not needed in this text, we shall not develop these relations here.

Differentiation with Respect to Matrices It is convenient to define the ijth element of the derivative of the scalar $f(\mathbf{X})$ with respect to the matrix $\mathbf{X}$ as

$$\left[\frac{df(\mathbf{X})}{\partial \mathbf{X}}\right]_{ij} = \frac{\partial f(\mathbf{X})}{\partial [\mathbf{X}]_{ij}}$$

Operations Involving Partial Derivatives Consider the functions $f = f(y, x, t)$, $\mathbf{y} = \mathbf{y}(x, t)$, and $\mathbf{x} = \mathbf{x}(t)$. We define the following operations:

$$\frac{df}{d\mathbf{x}} = \left[\frac{\partial \mathbf{y}^T}{\partial \mathbf{x}}\right]\frac{\partial f}{\partial \mathbf{y}} + \frac{\partial f}{\partial \mathbf{x}}$$

$$\frac{df}{dt} = \left[\frac{\partial f}{\partial \mathbf{x}} + \left\{\frac{\partial \mathbf{y}^T}{\partial \mathbf{x}}\right\}\left(\frac{\partial f}{\partial \mathbf{y}}\right)\right]^T \frac{d\mathbf{x}}{dt} + \left[\frac{\partial f}{\partial \mathbf{y}}\right]^T \frac{\partial \mathbf{y}}{\partial t} + \frac{\partial f}{\partial t}$$

Similar operations on the vector function are defined by

$$\mathbf{z} = \mathbf{z}(\mathbf{y}, \mathbf{x}, t) \qquad \mathbf{y} = \mathbf{y}(\mathbf{x}, t) \qquad \mathbf{x} = \mathbf{x}(t)$$

$$\frac{d\mathbf{z}}{d\mathbf{x}} = \left[\frac{\partial \mathbf{z}}{\partial \mathbf{y}}\right]\left[\frac{\partial \mathbf{y}}{\partial \mathbf{x}}\right] + \frac{\partial \mathbf{z}}{\partial \mathbf{x}}$$

$$\frac{d\mathbf{z}}{dt} = \left[\frac{\partial \mathbf{z}}{\partial \mathbf{y}}\right]\left[\left\{\frac{\partial \mathbf{y}}{\partial \mathbf{x}}\right\}\frac{d\mathbf{x}}{dt} + \frac{\partial \mathbf{y}}{\partial t}\right] + \left[\frac{\partial \mathbf{z}}{\partial \mathbf{x}}\right]\frac{d\mathbf{x}}{dt} + \frac{\partial \mathbf{z}}{\partial t}$$

In much of our work we have need for series expansions of scalar functions of vectors:

$$f(\mathbf{x}) = f(\mathbf{x}_0) + \left[\frac{\partial f(\mathbf{x}_0)}{\partial \mathbf{x}_0}\right]^T (\mathbf{x} - \mathbf{x}_0) + \frac{1}{2}(\mathbf{x} - \mathbf{x}_0)^T \left[\frac{\partial}{\partial \mathbf{x}_0}\left(\frac{\partial f(\mathbf{x}_0)}{\partial \mathbf{x}_0}\right)\right]^T (\mathbf{x} - \mathbf{x}_0)$$
$$+ \text{ higher-order terms}$$

The Taylor series for the expansion of a vector function of a vector $\mathbf{x}$ about $\mathbf{x}_0$ is defined in exactly the same fashion. For the vector $\mathbf{x}$ we have

$$\mathbf{f}(\mathbf{x}) = f(\mathbf{x}_0) + \left[\frac{\partial \mathbf{f}(\mathbf{x}_0)}{\partial \mathbf{x}_0}\right]^T (\mathbf{x} - \mathbf{x}_0) + \text{ higher-order terms}$$

Trace of a Matrix We define the trace of $\mathbf{A}$, tr[$\mathbf{A}$], as being the sum of the principal diagonal elements. An alternative and equivalent definition is that the trace of matrix $\mathbf{A}$ is the sum of the eigenvalues of the matrix. It can easily be shown that

$$\text{tr}[\mathbf{A}^T\mathbf{B}] = \text{tr}[\mathbf{B}^T\mathbf{A}] = \text{tr}[\mathbf{A}\mathbf{B}^T] = \text{tr}[\mathbf{B}\mathbf{A}^T]$$

and

$$\text{tr}[\mathbf{A} + \mathbf{B}] = \text{tr}[\mathbf{A}] + \text{tr}[\mathbf{B}]$$

Also

$$\text{tr}[\mathbf{A}] = \text{tr}[\mathbf{A}^T]$$

If

$$\mathbf{A} = \mathbf{x}\mathbf{y}^T$$

then

$$\text{tr}[\mathbf{A}] = \mathbf{x}^T\mathbf{y}$$

where it has been assumed that all the matrix operations are admissible.

Derivatives of the Trace of a Matrix In some problems it is convenient to have notation that will enable us to differentiate scalar trace functions with respect to matrices. In previous work in this appendix, we defined differentiation of a scalar function with respect to a matrix.

The following list of gradient matrices of trace functions can easily be established:

$$\frac{d}{d\mathbf{X}}\operatorname{tr}[\mathbf{AX}] = \mathbf{A}^T$$

$$\frac{d}{d\mathbf{X}}\operatorname{tr}[\mathbf{AX}^T] = \mathbf{A}$$

$$\frac{d}{d\mathbf{X}}\operatorname{tr}[\mathbf{AXB}] = \mathbf{A}^T\mathbf{B}^T$$

$$\frac{d}{d\mathbf{X}}\operatorname{tr}[\mathbf{AX}^T\mathbf{B}] = \mathbf{BA}$$

$$\frac{d}{d\mathbf{X}}\operatorname{tr}[\mathbf{X}^2] = 2\mathbf{X}^T$$

$$\frac{d}{d\mathbf{X}}\operatorname{tr}[\mathbf{XX}^T] = 2\mathbf{X}$$

$$\frac{d}{d\mathbf{X}}\operatorname{tr}[\mathbf{AXBX}] = \mathbf{A}^T\mathbf{X}^T\mathbf{B}^T + \mathbf{B}^T\mathbf{X}^T\mathbf{A}^T$$

$$\frac{d}{d\mathbf{X}}\operatorname{tr}[\mathbf{AXBX}^T] = \mathbf{A}^T\mathbf{XB}^T + \mathbf{AXB}$$

$$\frac{d}{d\mathbf{X}}\operatorname{tr}[\mathbf{AX}^{-1}\mathbf{B}] = -(\mathbf{X}^{-1}\mathbf{BAX}^{-1})^T$$

$$\frac{d}{d\mathbf{X}}\operatorname{tr}[e^{\mathbf{X}}] = e^{\mathbf{X}}$$

Here it is assumed that the elements of **X** are independent.

Properties of Determinants We have defined the trace of a matrix as the sum of the eigenvalues of the matrix. It is convenient to give the product of the eigenvalues a name, and the name chosen is the determinant of the matrix. It is possible to show, using this property of the determinant, that the following are valid equations if the elements of **X** are independent:

$$\frac{d}{d\mathbf{X}}\det\mathbf{X} = [\det\mathbf{X}][\mathbf{X}]^{-T}$$

$$\frac{d}{d\mathbf{X}}\ln\det\mathbf{X} = \mathbf{X}^{-T}$$

$$\frac{d}{d\mathbf{X}}\det\mathbf{AXB} = [\det\mathbf{AXB}][\mathbf{X}]^{-T}$$

$$\frac{d}{d\mathbf{X}}\det\mathbf{X}^n = n[\det\mathbf{X}]^n[\mathbf{X}]^{-T}$$

B.3. Linear vector differential equations

In many cases, a sufficiently accurate model of a differential system is provided by the following state-variable equations, where **x** is an n vector, **u** is an m vector, **z** is an r vector, and **F**, **G**, and **H** are matrices of compatible orders.

$$\dot{\mathbf{x}} = \mathbf{F}(t)\mathbf{x}(t) + \mathbf{G}(t)\mathbf{u}(t)$$
$$\mathbf{z}(t) = \mathbf{H}(t)\mathbf{x}(t) + \mathbf{J}(t)\mathbf{u}(t)$$

If we assume that **F**, **G**, **H**, **J**, and **u** are piecewise continuous functions of time, the system of equations will have a unique solution of the form

$$\mathbf{x}(t) = \mathbf{\Phi}(t, t_0)\mathbf{x}(t_0) + \int_{t_0}^{t} \mathbf{\Phi}(t, \tau)\mathbf{G}(\tau)\mathbf{u}(\tau)\, d\tau$$
$$\mathbf{z}(t) = \mathbf{H}(t)\mathbf{x}(t) + \mathbf{J}(t)\mathbf{u}(t)$$

where the system transition matrix $\mathbf{\Phi}(t, \tau)$ satisfies the matrix differential equation [often called the fundamental matrix equation when $\mathbf{\Phi}(t, \tau)$ is called the fundamental matrix].

$$\frac{\partial \mathbf{\Phi}(t, \tau)}{\partial t} = \mathbf{F}(t)\mathbf{\Phi}(t, \tau)$$

for all t and τ. The initial condition matrix for the foregoing is

$$\mathbf{\Phi}(\tau, \tau) = \mathbf{I} \qquad \text{for all } \tau$$

From these properties, it follows immediately that

$$\mathbf{\Phi}^{-1}(t, \tau) = \mathbf{\Phi}(\tau, t) \qquad \text{for all } t, \tau$$
$$\mathbf{\Phi}(\tau_3, \tau_2)\mathbf{\Phi}(\tau_2, \tau_1) = \mathbf{\Phi}(\tau_3, \tau_1) \qquad \text{for all } \tau_1, \tau_2, \tau_3$$

In the particular case for which the system is stationary, $\mathbf{F}(t) = \mathbf{F}$ a constant, the solution for the transition matrix $\mathbf{\Phi}(t, \tau)$ can be immediately determined as

$$\mathbf{\Phi}(t, \tau) = e^{\mathbf{F}(t-\tau)}$$

where the matrix exponential is defined by

$$e^{\mathbf{A}t} = \mathbf{I} + \mathbf{A}t + \frac{\mathbf{A}^2 t^2}{2!} + \frac{\mathbf{A}^3 t^3}{3!} + \cdots$$

Also, where **F** and **G** are constant, that is, not functions of time, it is possible to write the vector Laplace transform of the system response as

$$\mathbf{X}(s) = (s\mathbf{I} - \mathbf{F})^{-1}[\mathbf{x}(t = 0) + \mathbf{G}\mathbf{U}(s)]$$

Thus, for the case of a stationary system, the state transition matrix $\mathbf{\Phi}(t, \tau) = \mathbf{\Phi}(t - \tau)$ is seen to be the inverse transform of $[s\mathbf{I} - \mathbf{F}]^{-1}$.

Example ***B.3-1.*** The transfer function of a linear system is

$$\frac{Z(s)}{U(s)} = \frac{3(s + 2)}{2(s + 1)(s + 3)}$$

and it is desired to obtain state-variable descriptions for the system. The first thing to note is that there is *no* unique state-variable description of the system. All the representations of Fig. B.3-1 are equivalent in that the input–output transfer function is the same. Yet the **F, G, H,** and **J** matrices are certainly different for each system. It is suggested that the reader verify that the given system matrices and

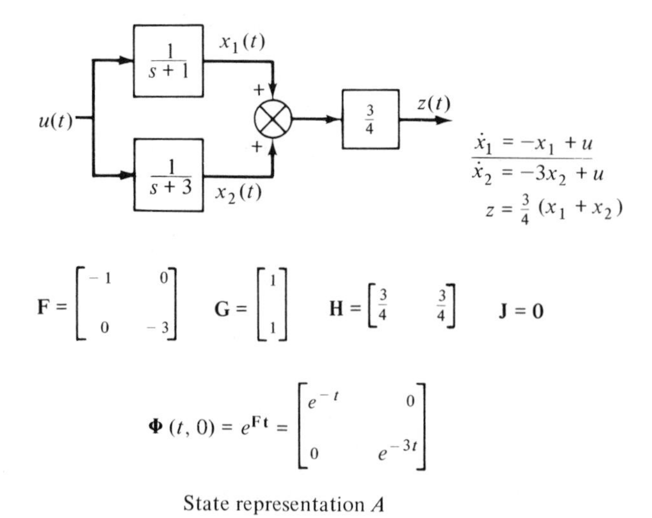

FIG. B.3-1 Single input-single output transfer functions and three of the many state-variable representations.

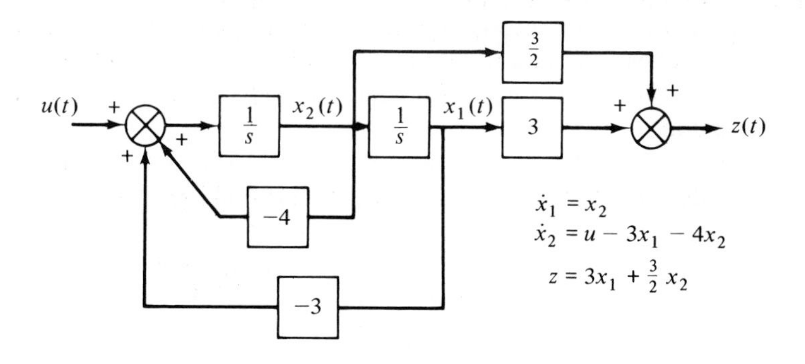

$$\dot{x}_1 = x_2$$
$$\dot{x}_2 = u - 3x_1 - 4x_2$$
$$z = 3x_1 + \tfrac{3}{2}x_2$$

$$\mathbf{F} = \begin{bmatrix} 0 & 1 \\ -3 & -4 \end{bmatrix} \quad \mathbf{G} = \begin{bmatrix} 0 \\ 1 \end{bmatrix} \quad \mathbf{H} = \begin{bmatrix} 3 & \tfrac{3}{2} \end{bmatrix} \quad \mathbf{J} = \mathbf{0}$$

$$\mathbf{\Phi}(t, 0) = e^{\mathbf{F}t} = \begin{bmatrix} \tfrac{3}{2}e^{-t} - \tfrac{1}{2}e^{-3t} & \tfrac{1}{2}e^{-t} - \tfrac{1}{2}e^{-3t} \\ \tfrac{3}{2}e^{-3t} - \tfrac{3}{2}e^{-t} & \tfrac{3}{2}e^{-3t} - \tfrac{1}{2}e^{-t} \end{bmatrix}$$

State representation *B*

$$u(t) \rightarrow \boxed{\frac{\tfrac{3}{2}}{s+3}} \rightarrow x_1(t) \rightarrow \boxed{\frac{1}{s+1}} \rightarrow x_2(t) \rightarrow \otimes \rightarrow z(t)$$

$$\dot{x}_1 = -3x_1 + \tfrac{3}{2}u$$
$$\dot{x}_2 = -x_2 + x_1$$
$$z = x_1 + x_2$$

$$\mathbf{F} = \begin{bmatrix} -3 & 0 \\ 1 & -1 \end{bmatrix} \quad \mathbf{G} = \begin{bmatrix} \tfrac{3}{2} \\ 0 \end{bmatrix} \quad \mathbf{H} = \begin{bmatrix} 1 & 1 \end{bmatrix} \quad \mathbf{J} = \mathbf{0}$$

$$\mathbf{\Phi}(t, 0) = e^{\mathbf{F}t} = \begin{bmatrix} e^{-3t} & 0 \\ \tfrac{1}{2}(e^{-t} - e^{-3t}) & e^{-t} \end{bmatrix}$$

State representation *C*

FIG. B.3-1 (*continued.*)

transition matrices are correct. It should also be noted that although each system has precisely the same response to a given input, the response of each system to an initial condition vector is different, since the state variables are chosen in a different manner for each representation.

Example ***B.3-2.*** We consider the homogeneous equations

$$\dot{\mathbf{x}}(t) = \mathbf{F}(t)\mathbf{x}(t) \qquad \mathbf{F}(t) = \begin{bmatrix} 0 & \dfrac{1}{(t+1)^2} \\ 0 & 0 \end{bmatrix}$$

and determine the state transition matrix.

The fundamental matrix equation

$$\frac{\partial \mathbf{\Phi}(t, \tau)}{\partial t} = \mathbf{F}(t)\mathbf{\Phi}(t) \qquad \mathbf{\Phi}(\tau, \tau) = \mathbf{I}$$

becomes in component form

$$\frac{\partial \Phi_{11}(t, \tau)}{\partial t} = \frac{1}{(t+1)^2} \Phi_{21}(t, \tau) \qquad \Phi_{11}(\tau, \tau) = 1$$

$$\frac{\partial \Phi_{12}(t, \tau)}{\partial t} = \frac{1}{(t+1)^2} \Phi_{22}(t, \tau) \qquad \Phi_{12}(\tau, \tau) = 0$$

$$\frac{\partial \Phi_{21}(t, \tau)}{\partial t} = 0 \qquad \Phi_{21}(t, t) = 0$$

$$\frac{\partial \Phi_{22}(t, \tau)}{\partial t} = 0 \qquad \Phi_{22}(t, t) = 1$$

We can immediately obtain the solution to these last two equations:

$$\Phi_{21}(t, \tau) = 0$$
$$\Phi_{22}(t, \tau) = 1$$

And thus with these solutions obtain

$$\Phi_{11}(t, \tau) = 1$$

as the solution to the first of the fundamental matrix-component equations. The remaining fundamental matrix equation becomes

$$\frac{\partial \Phi_{12}(t, \tau)}{\partial t} = \frac{1}{(t+1)^2} \qquad \Phi_{12}(\tau, \tau) = 0$$

This differential equation is easily solved to yield $\Phi_{12}(t, \tau) = -(t + 1)^{-1} + a$, where a is chosen such that $\Phi_{12}(\tau, \tau) = 0$. Thus $a = (\tau + 1)^{-1}$, and we have

$$\Phi_{12}(t, \tau) = \frac{1}{\tau + 1} - \frac{1}{t + 1} = \frac{t - \tau}{(\tau + 1)(t + 1)}$$

and the final state transition matrix is

$$\boldsymbol{\Phi}(t, \tau) = \begin{bmatrix} 1 & \dfrac{t - \tau}{(\tau + 1)(t + 1)} \\ 0 & 1 \end{bmatrix}$$

B.4. Linear vector difference equations

For discrete systems that are linear, an appropriate difference equation to characterize a system is

$$\begin{aligned} \mathbf{x}(k + 1) &= \mathbf{F}(k)\mathbf{x}(k) + \mathbf{G}(k)\mathbf{u}(k) \\ \mathbf{z}(k) &= \mathbf{H}(k)\mathbf{x}(k) + \mathbf{J}(k)\mathbf{u}(k) \qquad k = 0, 1, 2, 3, \ldots \end{aligned}$$

where $\mathbf{x}$, $\mathbf{u}$, and $\mathbf{z}$ are n, m, and r vectors, respectively, and where the orders of the matrices are compatible. k denotes the stage of the process. Thus $\mathbf{x}(\mathbf{k} + 1) = \mathbf{x}(t_{k+1})$ and $\mathbf{G}(k) = \mathbf{G}(t_k)$. The system of equations has a solution of the form

$$\begin{aligned} \mathbf{x}(k) &= \boldsymbol{\Phi}(k, l)\mathbf{x}(l) + \sum_{j=l}^{k-1} \boldsymbol{\Phi}(k, j + 1)\mathbf{G}(j)\mathbf{u}(j) \\ \mathbf{z}(k) &= \mathbf{H}(k)\mathbf{x}(k) + \mathbf{J}(k)\mathbf{u}(k) \end{aligned}$$

where the discrete-system transition matrix satisfies the matrix difference equation

$$\begin{aligned} \boldsymbol{\Phi}(k + 1, l) &= \mathbf{F}(k)\boldsymbol{\Phi}(k, l) \qquad \text{for all } k, l \\ \boldsymbol{\Phi}(k, k) &= \mathbf{I} \\ \boldsymbol{\Phi}(k, j) &= \prod_{i=j}^{k-1} \mathbf{F}(i) \end{aligned}$$

From these properties, it follows immediately that

$$\begin{aligned} \boldsymbol{\Phi}^{-1}(k, l) &= \boldsymbol{\Phi}(l, k) \qquad \text{for all } k \\ \boldsymbol{\Phi}(\alpha, \beta)\boldsymbol{\Phi}(\beta, \gamma) &= \boldsymbol{\Phi}(\alpha, \gamma) \qquad \text{for all } \alpha, \beta, \gamma \end{aligned}$$

In the particular case for which $\mathbf{F}$ is not a function of the stage, the solution to the foregoing equation is

$$\mathbf{\Phi}(k, l) = \mathbf{F}^{k-l}$$

B.5. References

(1) BELLMAN, R., *Introduction to Matrix Analysis*, McGraw-Hill, New York, 1960.

(2) GANTMACHER, F. R., *Theory of Matrices*, Vols. I and II, Chelsea, New York, 1959.

(3) KALMAN, R. E., "Mathematical Description of Linear Dynamical Systems," *J. S. I. A. M. Control*, Ser. A, Vol. I, No. 2, 159–192, 1963.

(4) OGATA, K., *State Space Analysis of Control Systems*, Prentice-Hall, Inc., Englewood Cliffs, N.J., 1967.

(5) PEASE, M. C., *Methods of Matrix Algebra*, Academic Press, New York, 1965.

(6) PIPES, L. A., "Matrices in Engineering," Chapter 13 of *Modern Mathematics for the Engineer*, E. F. Beckenback, ed., McGraw-Hill, New York, 1956.

(7) SCHULTZ, D. G., and J. L. MELSA, *State Functions and Linear Control Systems*, McGraw-Hill, New York, 1967.

(8) WIBERG, D. M., *State Space and Linear Systems*, Shaum's Outline Series, McGraw-Hill, New York, 1971.

Index